D0092972

Telecommunication System Engineering

Telecommunication System Engineering

Analog and Digital Network Design

Roger L. Freeman

A Wiley-Interscience Publication

JOHN WILEY & SONS

New York • Chichester • Brisbane • Toronto

Library of Congress Cataloging in Publication Data:

Freeman, Roger L
Telecommunication system engineering.

"A Wiley-Interscience publication."
Includes index.
1. Telecommunication systems—Design and construction. 2. Telephone systems—Design and construction. I. Title.

TK5103.F68 1980 621.38 79-26661
ISBN 0-471-02955-6

Printed in the United States of America

10 9 8 7 6 5 4 3 2 1

To my father, Andrew A. Freeman

To my father, Andrew A. Freeman

PREFACE

The purpose of this text is to present the general engineering considerations necessary for the design of practical telecommunication networks. The majority of today's networks are built primarily to serve the telephone subscriber. These same networks are being more and more extensively used to carry other types of information such as data, facsimile, and video. The first seven chapters of this text deal with conventional telephony. The remainder of the book covers digital communication, in particular, data systems and digital telephony.

I define telecommunication as a service that permits people or machines to communicate at a distance. It involves many disciplines that work together to form a system.

Traditionally, telecommunication is broken down into two major categories of engineering: transmission and switching. Each major category in itself is broken down into well-definable specialties or disciplines as shown below:

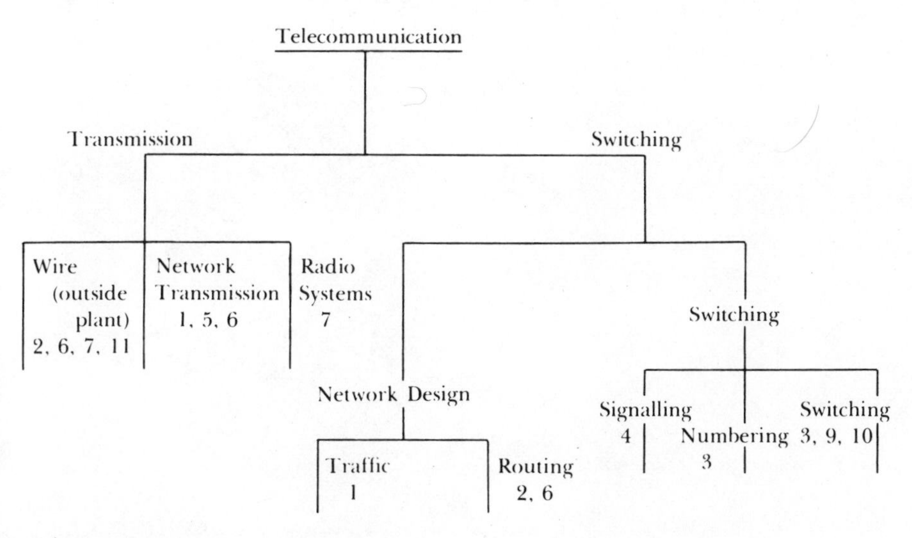

(Numbers are chapter references)

The advent of data communications on the one hand, and digital telephony, on the other, caused the distinct separation of disciplines to become rather hazy and ill defined. In fact, with integrated digital telephone networks and packet-switched data networks, the dividing lines disappeared almost entirely. The change is revolutionary.

I have organized the text to reflect both tradition and revolution. Conventional analog telecommunication has been with us since the late nineteenth century and will be with us for the remainder of the twentieth century at least. Digital systems are in their infancy and will mature over the next 20 to 30 years. My aim has been to give the reader a practical appreciation of both. Actually, I do not believe that a proper job of system engineering design can be done on a digital network without a solid background of conventional analog techniques.

ROGER L. FREEMAN

Sudbury, Massachusetts
February 1980

ACKNOWLEDGMENTS

I am deeply indebted to Peter Gerrand of Australia Telecommunications Research who encouraged me to pursue this effort and helped me to prepare the final outline. Subsequently, Peter was good enough to review several chapters. The work also benefited immensely from the review of many other friends and colleagues of mine in the telecommunications industry and academic community. Among these are Dr. Enric Vilar, Professor of Telecommunication Engineering at Portsmouth Polytechnic University (UK); Edmund Kiatiapov on assignment to ITT Laboratories Spain from LCT Velizy (France); Norman Doving, Fundación Chile, Santiago; Dr. Dan Varon, Larry Miller, and Hi Stevens from Raytheon's Equipment Division, Sudbury, Ma.; Dr. G. Shanholt and Donald J. Marsh from ITT's Telecommunication Technology Center, Stratford, Conn.; and Robert (Fritz) Gellerman from the Interamerican Development Bank. The assistance of Ray Fraser of Raytheon's Missile System Division is also appreciated. My wife, Paquita, is to be commended for her patience and fortitude during the long period of preparation and cycle of review.

R.L.F.

CONTENTS

Chapter 2 Local Networks, 43

Chapter 3 Conventional Switching Techniques in Telephony, 76

Chapter 9 Digital Transmission and Switching Systems, 343

Chapter 10 Data Networks and their Operation, 386

Telecommunication System Engineering

Chapter 1

SOME BASICS IN CONVENTIONAL TELEPHONY

1 BACKGROUND AND CONCEPT

Telecommunications deals with the service of providing electrical communication at a distance. The service is supported by an industry that depends on a large body of increasingly specialized scientists and engineers. The service may be private or open to public correspondence (i.e., access). The most cogent example of a service open to public correspondence is the telephone embodied in the telephone company, when based on private enterprise, or telephone administration, when government owned. Consider that by the early 1980s there will be more than 600 million telephones in the international network, with intercommunication between each and every telephone in that network.

A primary concern of this book is to describe the development of a telephone network and why it is built as it is. We also intend to show how it will expand and carry other than voice communication and how special services will evolve as based originally on the existing telephone network, where certain split-offs will occur at a later date. The bulk of the telecommunication industry is dedicated to the telephone network. Telecommunication engineering traditionally has been broken down into two basic segments, transmission and switching. This division is most apparent in telephony. Transmission concerns the carrying of an electrical signal from point X to point Y. Let us say that switching connects X to Y, rather than to Z. Until several years ago transmission and switching were two very separate and distinct disciplines. Today that distinction is disappearing. As we proceed, we deal with both disciplines and show in later chapters how the two are starting to meld together.

2 THE SIMPLE TELEPHONE CONNECTION

The common telephone as we know it today is a device connected to the outside world by a pair of wires. It consists of a handset and its cradle with a signaling device, consisting of either a dial or push buttons. The handset is made up of two electroacoustic transducers, the earpiece or receiver and the mouthpiece or transmitter. There is also a sidetone circuit that allows some of the transmitted energy to be fed back to the receiver.

The transmitter or mouthpiece converts acoustic energy into electric energy by means of a carbon granule transmitter. The transmitter requires a direct-current (dc) potential, usually on the order of 3 to 5 V, across its electrodes. We call this the *talk battery,* and in modern telephone systems it is supplied over the line (central battery) from the switching center. Current from the battery flows through the carbon granules or grains when the telephone is lifted from its cradle or goes "off hook." When sound impinges on the diaphragm of the transmitter, variations of air pressure are transferred to the carbon, and the resistance of the electrical path through the carbon changes in proportion to the pressure. A pulsating direct current results.

The typical receiver consists of a diaphragm of magnetic material, often soft iron alloy, placed in a steady magnetic field supplied by a permanent magnet, and a varying magnetic field caused by voice currents flowing through the voice coils. Such voice currents are alternating (ac) in nature and originate at the far-end telephone transmitter. These currents cause the magnetic field of the receiver to alternately increase and decrease, making the diaphragm move and respond to the variations. Thus an acoustic pressure wave is set up, more or less exactly reproducing the original sound wave from the distant telephone transmitter. The telephone receiver, as a converter of electrical energy to acoustic energy, has a comparatively low efficiency, on the order of 2 to 3%.

Sidetone is the sound of the talker's voice heard in his (or her) own receiver. Sidetone level must be controlled. When the level is high, the natural human reaction is for the talker to lower his voice. Thus by regulating sidetone, talker levels can be regulated. If too much sidetone is fed back to the receiver, the output level of the transmitter is reduced as a result of the talker lowering his voice, thereby reducing the level (voice volume) at the distant receiver and deteriorating performance.

To develop our discussion, let us connect two telephone handsets by a pair of wires, and at middistance between the handsets a battery is connected to provide that all-important talk battery. Such a connection is shown diagrammatically in Figure 1.1. Distance D is the overall separation of the two handsets and is the sum of distances d_1 and d_2; d_1 and d_2 are the distances from each handset to the central battery supply. The exercise is to extend the distance D to determine limiting factors given a fixed battery voltage, say, 48 V dc. We find that there are two limiting factors to the extension of the wire pair between the handsets. These are the IR drop, limiting the voltage across the handset transmitter, and the attenuation. For 19-gauge wire, the limiting distance is about 30 km, depending on the efficiency of the handsets. If the limiting characteristic is attenuation and

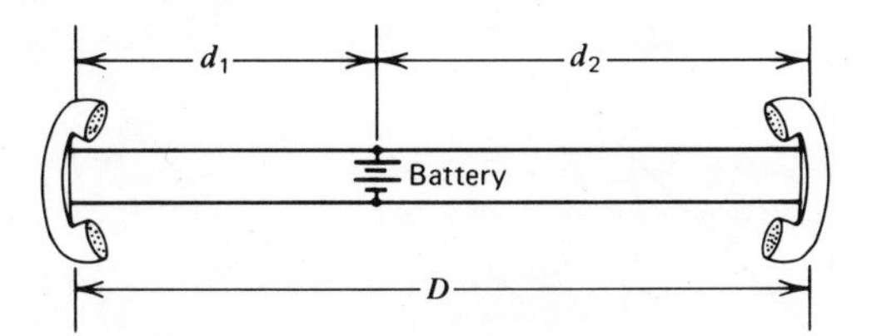

Figure 1.1 A simple telephone connection.

we desire to extend the pair farther, amplifiers could be used in the line. If the battery voltage was limiting, then the battery voltage could be increased. With the telephone system depicted in Figure 1.1, only two people can communicate. As soon as we add a third person, some difficulties begin to arise. The simplest approach would be to provide each person with two handsets. Thus party A would have one set to talk to B, another to talk to C, and so forth. Or the sets could be hooked up in parallel. Now suppose A wants to talk to C and doesn't wish to bother B. He then must have some method of selectively alerting the party to whom he wants to talk. As stations are added to the system, the alerting problem becomes quite complex. Of course, the proper name for this selection and alerting is *signaling.* If we allow that the pair of wires through which current flows is a loop, we are dealing with loops. Let us also call the holder of a telephone station a *subscriber.* The loops connecting them are subscriber loops.

Let us now look at an eight-subscriber system, each subscriber connected directly to every other subscriber. This is shown in Figure 1.2. When we connect each and every station with every other one in the system, this is called a *mesh* connection, or sometimes full mesh. Without the use of amplifiers and with 10-gauge copper wire size, the limiting distance is 30 km. Thus any connecting segment of the octagon may be no greater than 30 km. The only way we can justify a mesh connection of subscribers economically is when each and every subscriber wishes to communicate

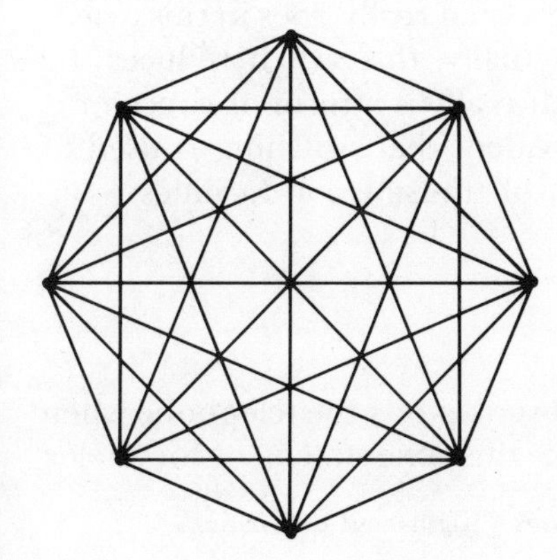

Figure 1.2 An 8-point mesh connection.

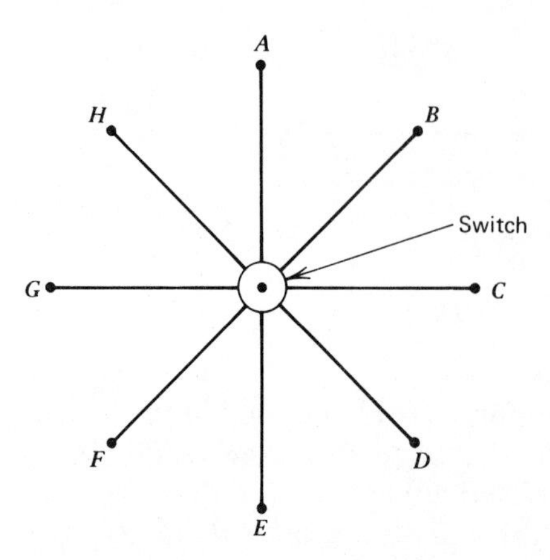

Figure 1.3 Subscribers connected in a star arrangement.

with every other subscriber in the network for virtually all the day (full period). As we know, however, most telephone subscribers do not use their telephones on a full-time basis. The telephone is used at what appears to be random intervals throughout the day. Further, the ordinary subscriber or telephone user will normally talk to only one other subscriber at a time. He will not need to talk to all other subscribers simultaneously.

If more subscribers are added and the network is extended beyond about 30 km, it is obvious that transmission costs will spiral, for that is what we are dealing with exclusively here—transmission. We are connecting each and every subscriber together with wire transmission means, requiring many amplifiers and talk batteries. Thus it would seem wiser to share these facilities in some way and cut down on the transmission costs. We now discuss this when switch and switching enter the picture. Let us define a *switch* as a device that connects inlets to outlets. The inlet may be a calling subscriber line and the outlet, the line of a called subscriber. The techniques of switching and the switch as a concept are widely discussed later in this text. Switching devices and how they work are covered in Chapters 3 and 9. Consider Figure 1.3, which shows our subscribers connected in a *star* network with a switch at the center. All the switch really does in this case is to reduce the transmission cost outlay. Actually, this switch reduces the number of links between subscribers, which really is a form of concentration. Later in our discussion it becomes evident that switching is used to concentrate traffic, thus reducing the cost of transmission facilities.

3 SOURCES AND SINKS*

Traffic is a term that quantifies usage. A subscriber *uses* the telephone when he wishes to talk to somebody. We can make the same statement for a telex

*The traffic engineer may wish to use the terminology "origins and destinations."

(teleprinter service) subscriber or a data-service subscriber. But let us stay with the telephone.

A network is a means of connecting subscribers. We have seen two simple network configurations, the mesh and star connections, in Figures 1.2 and 1.3. When talking about networks, we often talk of sources and sinks. A call is initiated at a traffic source and received at a traffic sink. Nodal points or nodes in a network are the switches.

4 TELEPHONE NETWORKS—INTRODUCTORY TERMINOLOGY

From our discussion we can say that a telephone network can be regarded as a systematic development of interconnecting transmission media arranged so that one telephone user can talk to any other within that network. The evolving layout of the network is primarily a function of economics. For example, subscribers share common transmission facilities; switches permit this sharing by concentration.

Consider a very simplified example. Two towns are separated by, say, 20 mi, and each town has 100 telephone subscribers. Logically, most of the telephone activity (the traffic) will be among the subscribers of the first town and among those of the second town. There will be some traffic, but considerably less, from one town to the other. In this example let each town have its own switch. With the fairly low traffic volume from one town to the other, perhaps only six lines would be required to interconnect the switch of the first town to that of the second. If no more than six people want to talk simultaneously between the two towns, a number as low as 6 can be selected. Economics has mandated that we install the minimum number of connecting telephone lines from the first town to the second to serve the calling needs between the two towns. The telephone lines connecting one telephone switch or exchange with another are called *trunks* in North America and *junctions* in Europe. The telephone lines connecting a subscriber to the switch or exchange that serves him are called *lines* or *subscriber lines.* Concentration is a line-to-trunk ratio. In the simple case above it was 100 lines to six trunks (or junctions), or about at 16:1 ratio.

A telephone subscriber looking into the network is served by a *local exchange.* This means that his telephone line is connected to the network via the local exchange or central office, in North American parlance. A local exchange has a serving area, which is the geographical area in which the exchange is located; all subscribers in that area are served by that exchange.

The term *local area,* as opposed to *toll area,* is that geographical area containing a number of local exchanges and inside which any subscriber can call any other subscriber without incurring tolls (extra charges for a call). Toll calls and long-distance calls are synonymous. For instance, a local call in North America, where telephones have detailed billing, shows up on the bill as a time-metered call or is covered by a flat monthly rate. Toll calls in North America appear as separate, detailed entries on the telephone bill. This is not so in most European countries and in those countries following European practice. In these countries there is no detailed billing on direct-distance dialed (subscriber-trunk-dialed) calls. All such subscriber-dialed calls, even international ones, are just metered, and the subscriber pays for the meter steps used per billing period, which is often 1 month or 2 months. In European practice a long-distance call, a toll call if you will, is one involving the dialing of additional digits (e.g., more than six or seven digits).

Let us call a network a *grouping of interworking telephone exchanges.* As the discussion proceeds, the differences between local networks and national networks are shown. Two other types of network are also discussed. These are specialized versions of a local network and are the rural network (rural area) and metropolitan network (metropolitan area).

5 ESSENTIALS OF TRAFFIC ENGINEERING

5.1 Introduction and Terminology

As we have already mentioned, telephone exchanges are connected by trunks or junctions. The number of trunks connecting exchange X with exchange Y are the number of voice pairs or their equivalent used in the connection. One of the most important steps in telecommunication engineering practice is to determine the number of trunks required on a route or connection between exchanges. We could say we are *dimensioning* the route. To correctly dimension a route, we must have some idea of its usage, that is, how many people will wish to talk at once over the route. The usage of a transmission route or a switch brings us into the realm of traffic engineering, and the usage may be defined by two parameters: (1) *calling rate,* or the number of times a route or traffic path is used per unit period; more properly defined, "the call intensity per traffic path during the busy hour"* and (2) *holding time,* or "the duration of occupancy of a traffic path by a call,"* or sometimes, "the average duration of occupancy of one or more paths by calls."* A *traffic path* is "a channel, time slot, frequency band, line, trunk, switch, or circuit over which individual communications pass in sequence."* *Carried traffic* is the volume of traffic actually carried by a switch, and *offered traffic* is the volume of traffic offered to a switch.

To dimension a traffic path or size a telephone exchange, we must know the traffic intensity representative of the normal busy season. There are weekly and daily variations in traffic within the busy season. Traffic is very random in nature. However, there is a certain consistency we can look for. For one thing, there usually is more traffic on Mondays and Fridays and a

**Reference Data for Radio Engineers* [1], p. 31-8.

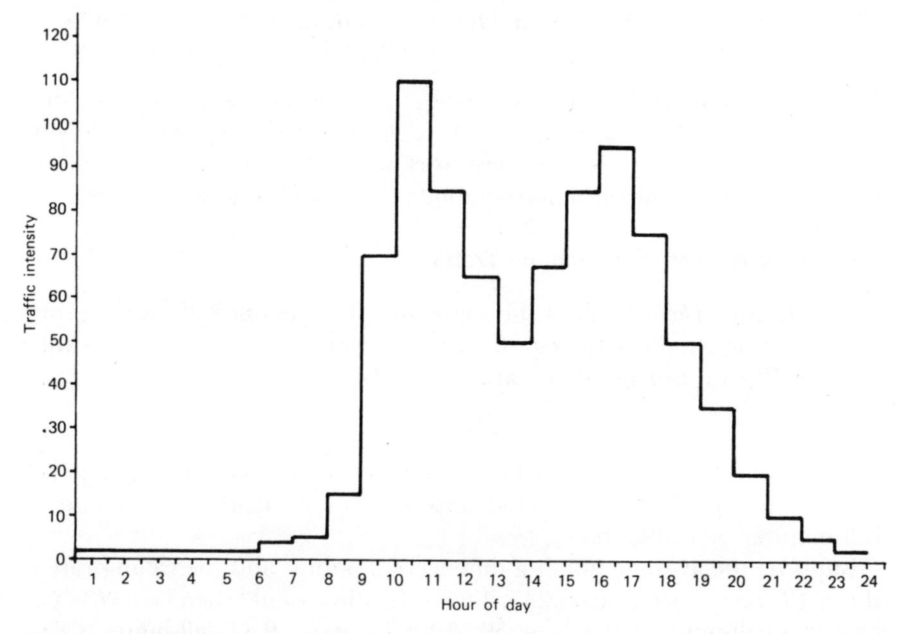

Figure 1.4 Bar chart of traffic intensity over a typical working day (US - mixed business and residential).

lower volume on Wednesdays. A certain consistency can also be found in the normal work-day hourly variation. Across the typical day the variation is such that a 1-h period shows greater usage than any other. From the hour with least traffic to the hour of greatest traffic, the variation can exceed 100 : 1. Figure 1.4 shows a typical hour-by-hour traffic variation for a serving switch in the United States. It can be seen that the busiest period, the *busy hour* (BH), is between 10 A.M. and 11 A.M. From one work day to the next, originating BH calls can vary as much as 20 or 25%. To these fairly "regular" variations, there are also unpredictable peaks caused by stock market or money market activity, weather, natural disaster, international events, sporting events, and so on. Normal system growth must also be taken into account. Nevertheless, suitable forecasts of BH traffic can be made. However, before proceeding, consider the four most common definitions of BH:

1. The average weekday reading over one or two weeks in the known busy season (normal practice for manual (operator) switched traffic).
2. The average of the BH traffic on the 30 busiest days of the year (defined as "mean busy hour traffic," CCITT Rec. Q.80).
3. The average of the BH traffic on the 10 busiest days of the year (North American standard).

4. The average BH traffic on the five busiest days of the year (referring to traffic on "exceptionally busy" days, CCITT Recs. Q.80 and Q.87).

When dimensioning telephone exchanges and transmission routes, we shall be working with BH traffic levels. The definition we accept would depend on what part of the world we were working in. For example, definition 4 would end up requiring more equipment than definitions 2 or 3.

5.2 Measurement of Telephone Traffic

If we define *telephone traffic* as the aggregate of telephone calls over a group of circuits or trunks with regard to the duration of calls as well as their number [2], we can say that traffic flow (A)

$$A = C \times T$$

where C is the calling rate per hour and T is the average holding time per call. From this formula it would appear that the traffic unit would be call-minutes or call-hours.

Suppose the average holding time were 2.5 mins and the calling rate in the BH for a particular day, 237. The traffic flow would then be 237×2.5, or 592.5 call-minutes (Cm), or 592.5/60, or about 9.87 call-hours (Ch).

The preferred unit of traffic is the erlang, named after the Danish mathematician, A. K. Erlang.[5] The erlang is a dimensionless unit. One erlang of traffic intensity on one traffic circuit means a continuous occupancy of that circuit. Considering a group of circuits, traffic intensity in erlangs is the number of call seconds per second or the number of call hours per hour. If we knew that a group of 10 circuits had a call intensity of 5 erlangs, we would expect half of the circuits to be busy at the time of measurement.

Other traffic units are not dimensionless. For instance: *call-hour* (Ch)—1 Ch is the quantity represented by one or more calls having an aggregate duration of 1 h; *call-second* (Cs)—1 Cs is the quantity represented by one or more calls having an aggregate duration of 1 s; *"cent" call-second* (ccs)—1 ccs is the quantity represented by one 100-s call or by an aggregate of 100 Cs of traffic; (the term ccs derives from "cent" call seconds; with cent representing 100 from the French); and *equated busy hour call* (EBHC)—a European unit of traffic intensity; 1 EBHC is the average intensity in one or more traffic paths occupied in the BH by one 2-min call or for an aggregate duration of 2 min. Thus we can relate our terms as follows:

$$1 \text{ erlang} = 30 \text{ EBHC} = 36 \text{ ccs} = 60 \text{ Cmin}$$

assuming a 1-h time-unit interval.

5.3 Congestion, Lost Calls, and Grade of Service

Assume that an isolated telephone exchange serves 5000 subscribers and that no more than 10% of the subscribers wish service simultaneously.

Therefore, the exchange is dimensioned with sufficient equipment to complete 500 simultaneous connections. Each connection would be, of course, between any two of the 5000 subscribers. Now let subscriber No. 501 attempt to originate a call. He cannot because all the connecting equipment is busy, even though the line he wishes to reach may be idle. This call from subscriber 501 is termed a *lost call* or *blocked call.* He has met congestion. The probability of meeting congestion is an important parameter in traffic engineering of telecommunication systems. If congestion conditions are to be met in a telephone system, we can expect that those conditions will usually be met during the BH. A switch is engineered (dimensioned) to handle the BH load. But how well? We could, indeed, far overdimension the switch such that it could handle any sort of traffic peaks. However, that is uneconomical. So with a well-designed switch, during the busiest of BHs we may expect some moments of congestion such that additional call attempts will meet blockage. *Grade of service* expresses the probability of meeting congestion during the BH and is expressed by the letter p. A typical grade of service is $p = .01$. This means that an average of one call in 100 will be blocked or "lost" during the BH. Grade of service, a term in the Erlang formula, is more accurately defined as the *probability of congestion.* It is important to remember that lost calls (blocked calls) refer to calls that fail at *first* trial. We discuss reattempts (at dialing) later, that is, the way blocked calls are handled.

We exemplify grade of service by the following problem. If we knew that there were 354 seizures (lines connected for service) and six blocked calls (lost calls) during the BH, what was the grade of service?

$$\text{Call congestion} = \frac{\text{Number of lost calls}}{\text{Total number of offered calls}}$$

$$= \frac{6}{354 + 6} = \frac{6}{360}$$

or

$$p = 0.017$$

The average grade of service for a network may be obtained by adding the grade of service contributed by each constituent switch, switching network, or trunk group. The *Reference Data For Radio Engineers* [Ref. 1, Section 31] states that the grade of service provided by a particular group of trunks or circuits of specified size and carrying a specified traffic intensity, is the probability that a call offered to the group will find available trunks already occupied on first attempt. That probability depends on a number of factors, the most important of which are being: (1) the distribution in time and duration of offered traffic (e.g., random or periodic arrival and constant or exponentially distributed holding time), (2) the number of traffic sources—limited or high (infinite), (3) the availability of trunks in group to traffic sources—full or restricted availability, and (4) the manner in which lost calls are "handled."

Several new concepts are suggested in these four factors. These must be explained before continuing.

5.4 Availability

Switches were previously discussed as devices with lines and trunks, but better terms for describing a switch are "inlets" and "outlets". When a switch has full availability, each inlet has access to any outlet. When not all the free outlets in a switching system can be reached by inlets, the switching system is referred to as one with "limited availability." Examples of switches with limited and full availability are shown in Figures 1.5A and 1.5B. Of course, full availability switching is more desirable than limited availability but is more expensive for larger switches. Thus full availability switching is found only in small switching configurations. *Grading* is one method of improving the traffic-handling capacities of switching configurations with limited availability. Grading is a scheme for interconnecting switching subgroups to make the switching load more uniform.

5.5 "Handling" of Lost Calls

In conventional telephone traffic theory three methods are considered for the handling or dispensing of lost calls: (1) lost calls held (LCH), (2) lost calls cleared (LCC), and (3) lost calls delayed (LCD). The LCH concept assumes that the telephone user will immediately reattempt his call on receipt of a congestion signal and will continue to redial. He hopes to seize connection equipment or a trunk as soon as switching equipment becomes

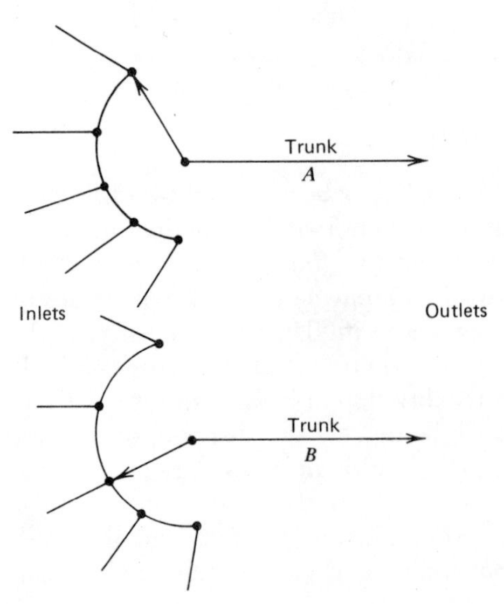

Figure 1.5A An example of a switch with limited availability.

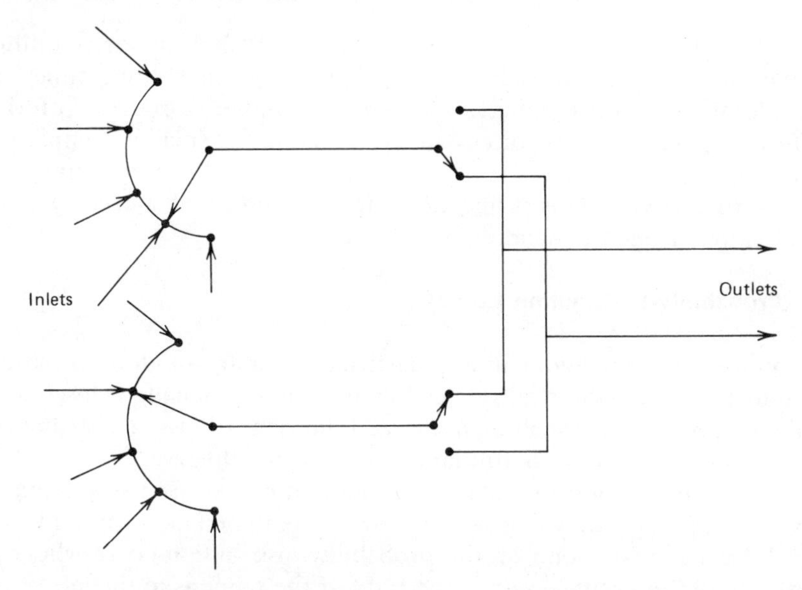

Figure 1.5B An example of a switch with full availability.

available for his call to be handled. It is the assumption in the LCH concept that lost calls are held or waiting at the user's telephone. This concept further assumes that such lost calls extend the average holding time theoretically, and in this case the average holding time is zero, and all the time is waiting time. The principal traffic formula used in North America is based on the LCH concept.

The LCC concept, which is used primarily in Europe or those countries accepting European practice, assumes that the user will hang up and wait some time interval before reattempting if he hears the congestion signal on his first attempt. Such calls, it is assumed, disappear from the system. A reattempt (after the delay) is considered as initiating a new call. The Erlang formula is based on this criterion.

The LCD concept assumes that the user is automatically put in queue (a waiting line or pool). For example, this is done when the operator is dialed. It is also done on modern, computer-controlled switching systems, generally referred to under the blanket term *stored program control* (SPC). The LCD category may be broken down into three subcategories, depending on how the queue or pool of waiting calls is handled. The waiting calls may be handled last in first out, first in line first served, or at random.

5.6 Infinite and Finite Traffic Sources

We can assume that traffic sources are infinite or finite. For the infinite-traffic-sources case, the probability of call arrival is constant and does not depend on the state of occupancy of the system. It also implies an infinite number of call arrivals, each with an infinitely small holding time. An example of finite traffic sources is when the number of sources offering traffic to a group of trunks or circuits is comparatively small in comparison to the number of circuits. We can also say that with a finite number of sources, the arrival rate is proportional to the number of sources that are not already engaged in sending a call.

5.7 Probability-Distribution Curves

Telephone-call originations in any particular area are random in nature. We find that originating calls or call arrivals at an exchange closely fit a family of probability-distribution curves following a Poisson distribution. The Poisson distribution is fundamental to traffic theory.

Most of the common probability distribution curves are two-parameter curves; that is, they may be described by two parameters, mean and variance. The mean is a point on the probability-distribution curve where an equal number of events occur to the right of the point as to the left of the point. "Mean" is synonymous with "average." We define mean as the x-coordinate of the center of the area under the probability-density curve for the population. The small Greek letter mu (μ) is the traditional indication of the mean; $\bar{x}$ is also used.

The second parameter used to describe a distribution curve is the dispersion, which tells us how the values or population are dispersed about the center or mean of the curve. There are several measures of dispersion. One is the familiar *standard deviation,* where "The standard deviation s of a sample of n observations $x_1, x_2, \ldots, x_n$ is

$$s = \sqrt{\frac{1}{n-1}\sum_{i=1}^{n}(x_i - \bar{x})^2}$$

The *variance* V of the sample values is the square of s. The parameters for dispersion s and s^2, the standard deviation and variance, respectively, are usually denoted σ and σ^2 and give us an idea of the squatness of a distribution curve. Mean and standard deviation of a normal distribution curve are shown in Figure 1.6, where we can see that σ^2 is another measure of dispersion, the variance, or essentially the average of the squares of the distances from mean aside from the factor $n/(n-1)$.

We have introduced two distribution functions describing the probability of distribution, often called the *distribution* of or just $f(x)$. Both functions are used in traffic engineering. But before proceeding, the variance-to-mean ratio (VMR) must also be introduced. Sometimes VMR is called the *coefficient of overdispersion.* The formula for VMR is

$$\alpha = \frac{\sigma^2}{\mu}$$

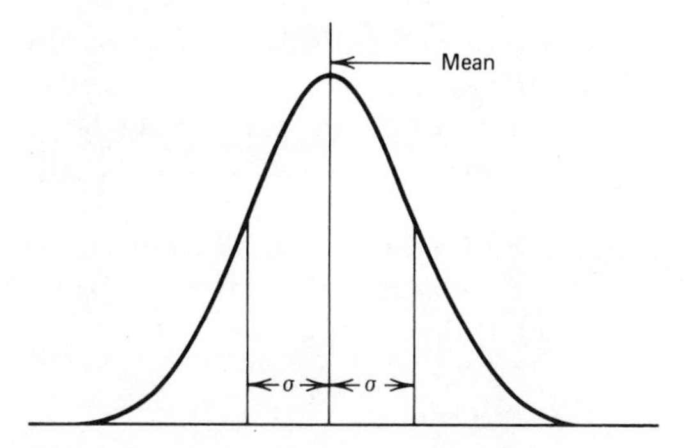

Figure 1.6 A normal distribution curve showing the mean and the standard deviation.

5.8 Smooth, Rough, and Random Traffic

Traffic probability distributions can be divided into three distinct categories: (1) smooth, (2) rough, and (3) random. Each may be defined by α, the VMR. For smooth traffic, α is less than 1. For rough traffic, α is greater than 1. When α is equal to 1, the traffic distribution is called *random*. The Poisson distribution function is an example of random traffic where the VMR = 1. Rough traffic tends to be peakier than random or smooth traffic. For a given grade of service more circuits are required for rough traffic because of the greater spread of the distribution curve (greater dispersion).

Smooth traffic behaves like random traffic that has been filtered. The filter is the local exchange. The local exchange looking out at its subscribers sees call arrivals as random traffic, assuming that the exchange has not been overdimensioned. The smooth traffic is the traffic on the local exchange outlets. The filtering or limiting of the peakiness is done by call blockage during the BH. Of course, the blocked traffic may actually overflow to alternative routes. Smooth traffic is characterized by a positive binomial distribution function, perhaps better known to traffic people as the *Bernoulli distribution.* An example of the Bernoulli distribution is as follows [6]. If we assume subscribers make calls independently of each other and that each has a probability p of being engaged in conversation, then if n subscribers are examined, the probability that x of them will be engaged is

$$B(x) = C_x^n p^x(1-p)^{n-x}; \qquad 0 < x < n$$

$$\text{Its mean} = np$$

$$\text{Its variance} = np(1-p)$$

Where the symbol C_x^n means the number of ways that x entities can be taken n at a time. Smooth traffic is assumed in dealing with small groups of subscribers, the number 200 is often used as the breakpoint [6]. That is, groups of subscribers are considered small when the subscribers number less than 200. And as mentioned, smooth traffic is also used with carried traffic. In this case the rough or random traffic would be the offered traffic.

Let's consider the binomial distribution for rough traffic. This is characterized by a negative index. Therefore, if the distribution parameters are k and q, where k is a positive number representing a hypothetical number of traffic sources and q represents the occupancy per source and may vary between 0 and 1, then

$$R'(x,k,q) = \binom{x+k-1}{k-1} q^x (1-q)^k$$

where R' is the probability of finding x calls in progress for the parameters k and q [2]. Rough traffic is used in dimensioning toll trunks with alternative routing. The symbol B (Bernoulli) is used by traffic engineers for smooth traffic and R for rough traffic. Although P may designate probability, in traffic engineering it designates Poissonian, and hence we have "P" tables such as those in Table 1.2.

The Bernoulli formula is

$$B'(x,s.h) = C_s^x \, h^x (1-h)^{s-x}$$

where C_s^x indicates the number of combinations of s things taken x at a time, h is the probability of finding the first line busy of an exchange, $1-h$ is the probability of finding the first line idle, and s is the number of subscribers. The probability of finding two lines busy is h^2, the probability of finding s lines busy is h^s, and so on. We are interested in finding the probability of x of the s subscribers with busy lines.

The Poisson probability function can be derived from the binomial distribution assuming the number of subscribers s very large and the calling rate per line h low* such that the product $sh = m$ remains constant and letting s increase to infinity in the limit

$$P(x) = \frac{m^x}{x!} e^{-m}$$

where

$$x = 0, 1, 2, \ldots$$

For most of our future discussion, we consider call-holding times to have a negative exponential distribution in the form

$$P = e^{-t/h}$$

where t/h is the average holding time and in this case P is the probability of a call lasting longer than t, some arbitrary time interval.

*For example, less than 50 millierlangs (mE).

6 ERLANG AND POISSON TRAFFIC FORMULAS

When dimensioning a route, we want to find the number of circuits that serve the route. There are several formulas at our disposal to determine that number of circuits based on the BH traffic load. In Section 5.3 four factors were discussed that will help us to determine which traffic formula to use given a particular set of circumstances. These factors primarily dealt with (1) call arrivals and holding-time distribution, (2) number of traffic sources, (3) availability, and (4) handling of lost calls.

The Erlang B loss formula is probably the most common one used today outside the United States. Loss here means the probability of blockage at the switch due to congestion or to "all trunks busy" (ATB). This is expressed as grade of service E_B or the probability of finding x channels busy. The other two factors in the Erlang B formula are the mean of the *offered* traffic and the number of trunks or servicing channels available. Thus

$$E_B = \frac{A^n/n!}{1 + A - A^2/2! + \cdots + A^n/n!}$$

where n = number of trunks or servicing channels
A = mean of the offered traffic
E_B = grade of service using the Erlang B formula

This formula assumes that

- Traffic originates from an infinite number of sources.
- Lost calls are cleared assuming a zero holding time.
- The number of trunks or servicing channels is limited.
- Full availability exists.

At this point in our discussion of traffic we suggest that the reader learn to differentiate between time congestion and call congestion when dealing with grade of service. *Time congestion,* of course, refers to the decimal fraction of an hour during which all trunks are busy simultaneously. *Call congestion,* on the other hand, refers to the number of calls that fail at first attempt, which we term *lost calls.* Keep in mind that the Erlang B formula deals with offered traffic, which differs from carried traffic by the number of lost calls.

Table 1.1 is based on the Erlang B formula and gives trunk-dimensioning information for some specific grades of service, from .001 to .05 and from 1 to 150 trunks. Table 1.1 uses traffic-intensity units UC and TU, where TU is in erlangs assuming BH and UC is in ccs (100 call seconds); 1 erlang = 36 ccs (based on a 1-h time interval). To exemplify the use of Table 1.1, suppose that a route carried 16.68 erlangs of traffic with a desired grade of service of .001; then 30 trunks would be required. If the

Table 1.1 Trunk-loading Capacity, Based on Erlang B Formula, Full Availability

	Grade of Service 1 in 1000		Grade of Service 1 in 500		Grade of Service 1 in 200		Grade of Service 1 in 100		Grade of Service 1 in 50		Grade of Service 1 in 20	
Trunks	UC	TU	UC	TU	UC	TU	UC	TU	UC	TU	UC	TU
1	0.04	0.001	0.07	0.002	0.2	0.005	0.4	0.01	0.7	0.02	1.8	0.05
2	1.8	0.05	2.5	0.07	4	0.11	5.4	0.15	7.9	0.22	14	0.38
3	6.8	0.19	9	0.25	13	0.35	17	0.46	22	0.60	32	0.90
4	16	0.44	19	0.53	25	0.70	31	0.87	39	1.09	55	1.52
5	27	0.76	32	0.90	41	1.13	49	1.36	60	1.66	80	2.22
6	41	1.15	48	1.33	58	1.62	69	1.91	82	2.28	107	2.96
7	57	1.58	65	1.80	78	2.16	90	2.50	106	2.94	135	3.74
8	74	2.05	83	2.31	98	2.73	113	3.13	131	3.63	163	4.54
9	92	2.56	103	2.85	120	3.33	136	3.78	156	4.34	193	5.37
10	111	3.09	123	3.43	143	3.96	161	4.46	183	5.08	224	6.22
11	131	3.65	145	4.02	166	4.61	186	5.16	210	5.84	255	7.08
12	152	4.23	167	4.64	190	5.28	212	5.88	238	6.62	286	7.95
13	174	4.83	190	5.27	215	5.96	238	6.61	267	7.41	318	8.83
14	196	5.45	213	5.92	240	6.66	265	7.35	295	8.20	350	9.73
15	219	6.08	237	6.58	266	7.38	292	8.11	324	9.01	383	10.63
16	242	6.72	261	7.26	292	8.10	319	8.87	354	9.83	415	11.54
17	266	7.38	286	7.95	318	8.83	347	9.65	384	10.66	449	12.46
18	290	8.05	311	8.64	345	9.58	376	10.44	414	11.49	482	13.38
19	314	8.72	337	9.35	372	10.33	404	11.23	444	12.33	515	14.31
20	339	9.41	363	10.07	399	11.09	433	12.03	474	13.18	549	15.25
21	364	10.11	388	10.79	427	11.86	462	12.84	505	14.04	583	16.19
22	389	10.81	415	11.53	455	12.63	491	13.65	536	14.90	617	17.13

23	415	11.52	442	12.27	483	13.42	521	14.47	567	15.76	651	18.08
24	441	12.24	468	13.01	511	14.20	550	15.29	599	16.63	685	19.03
25	467	12.97	495	13.76	540	15.00	580	16.12	630	17.50	720	19.99
26	493	13.70	523	14.52	569	15.80	611	16.96	662	18.38	754	20.94
27	520	14.44	550	15.28	598	16.60	641	17.80	693	19.26	788	21.90
28	546	15.18	578	16.05	627	17.41	671	18.64	725	20.15	823	22.87
29	573	15.93	606	16.83	656	18.22	702	19.49	757	21.04	858	23.83
30	600	16.68	634	17.61	685	19.03	732	20.34	789	21.93	893	24.80
31	628	17.44	662	18.39	715	19.85	763	21.19	822	22.83	928	25.77
32	655	18.20	690	19.18	744	20.68	794	22.05	854	23.73	963	26.75
33	683	18.97	719	19.97	774	21.51	825	22.91	887	24.63	998	27.72
34	711	19.74	747	20.76	804	22.34	856	23.77	919	25.53	1033	28.70
35	739	20.52	776	21.56	834	23.17	887	24.64	951	26.43	1068	29.68
36	767	21.30	805	22.36	864	24.01	918	25.51	984	27.34	1104	30.66
37	795	22.03	834	23.17	895	24.85	950	26.38	1017	28.25	1139	31.64
38	823	22.86	863	23.97	925	25.69	981	27.25	1050	29.17	1175	32.63
39	851	23.65	892	24.78	955	26.53	1013	28.13	1083	30.08	1210	33.61
40	880	24.44	922	25.60	986	27.38	1044	29.01	1116	31.00	1246	34.60
41	909	25.24	951	26.42	1016	28.23	1076	29.89	1149	31.92	1281	35.59
42	937	26.04	981	27.24	1047	29.08	1108	30.77	1182	32.84	1317	36.58
43	966	26.84	1010	28.06	1078	29.94	1140	31.66	1215	33.76	1353	37.57
44	995	27.64	1040	28.88	1109	30.80	1171	32.54	1248	34.68	1388	38.56
45	1024	28.45	1070	29.71	1140	31.66	1203	33.43	1282	35.61	1424	39.55
46	1053	29.26	1099	30.54	1171	32.52	1236	34.32	1315	36.53	1459	40.54
47	1083	30.07	1129	31.37	1202	33.38	1268	35.21	1349	37.46	1495	41.54
48	1111	30.88	1159	32.20	1233	34.25	1300	36.11	1382	38.39	1531	42.54
49	1141	31.69	1189	33.04	1264	35.11	1332	37.00	1415	39.32	1567	43.54
50	1170	32.51	1220	33.88	1295	35.98	1364	37.90	1449	40.25	1603	44.53

Table 1.1 Continued

Trunks	Grade of Service 1 in 1000		Grade of Service 1 in 500		Grade of Service 1 in 200		Grade of Service 1 in 100	
	UC	TU	UC	TU	UC	TU	UC	TU
51	1200	33.33	1250	34.72	1327	36.85	1397	38.80
52	1229	34.15	1280	35.56	1358	37.72	1429	39.70
53	1259	34.98	1310	36.40	1390	38.60	1462	40.60
54	1289	35.80	1341	37.25	1421	39.47	1494	41.50
55	1319	36.63	1371	38.09	1453	40.35	1527	42.41
56	1349	37.46	1402	38.94	1484	41.23	1559	43.31
57	1378	38.29	1432	39.79	1516	42.11	1592	44.22
58	1408	39.12	1463	40.64	1548	42.99	1625	45.13
59	1439	39.96	1494	41.50	1579	43.87	1657	46.04
60	1468	40.79	1525	42.35	1611	44.76	1690	46.95
61	1499	41.63	1556	43.21	1643	45.64	1723	47.86
62	1529	42.47	1587	44.07	1675	46.53	1756	48.77
63	1559	43.31	1617	44.93	1707	47.42	1789	49.69
64	1590	44.16	1648	45.79	1739	48.31	1822	50.60
65	1620	45.00	1679	46.65	1771	49.20	1855	51.52
66	1650	45.84	1710	47.51	1803	50.09	1888	52.44
67	1681	46.69	1742	48.38	1835	50.98	1921	53.35
68	1711	47.54	1773	49.24	1867	51.87	1954	54.27
69	1742	48.39	1804	50.11	1900	52.77	1987	55.19
70	1773	49.24	1835	50.98	1932	53.66	2020	56.11
71	1803	50.09	1867	51.85	1964	54.56	2053	57.03
72	1834	50.94	1898	52.72	1996	55.45	2087	57.96

73	1865	51.80	1929	53.59	2029	56.35	2120	58.88
74	1895	52.65	1960	54.46	2061	57.25	2153	59.80
75	1926	53.51	1992	55.34	2093	58.15	2186	60.73
76	1957	54.37	2024	56.21	2126	59.05	2219	61.65
77	1988	55.23	2055	57.09	2159	59.96	2253	62.58
78	2019	56.09	2087	57.96	2191	60.86	2286	63.51
79	2050	56.95	2118	58.84	2223	61.76	2319	64.43
80	2081	57.81	2150	59.72	2256	62.67	2353	65.36
81	2112	58.67	2182	60.60	2289	63.57	2386	66.29
82	2143	59.54	2213	61.48	2321	64.48	2420	67.22
83	2174	60.40	2245	62.36	2354	65.38	2453	68.15
84	2206	61.27	2277	63.24	2386	66.29	2487	69.08
85	2237	62.14	2308	64.13	2419	67.20	2521	70.02
86	2268	63.00	2340	65.01	2452	68.11	2554	70.95
87	2299	63.87	2372	65.90	2485	69.02	2588	71.88
88	2330	64.74	2404	66.78	2517	69.93	2621	72.81
89	2362	65.61	2436	67.67	2550	70.84	2655	73.75
90	2393	66.48	2468	68.56	2583	71.76	2688	74.68
91	2425	67.36	2500	69.44	2616	72.67	2722	75.62
92	2456	68.23	2532	70.33	2650	73.58	2756	76.56
93	2488	69.10	2564	71.22	2682	74.49	2790	77.49
94	2519	69.98	2596	72.11	2715	75.41	2823	78.43
95	2551	70.85	2628	73.00	2748	76.32	2857	79.37
96	2582	71.73	2660	73.90	2781	77.24	2891	80.31
97	2614	72.61	2692	74.79	2814	78.16	2925	81.24
98	2645	73.48	2724	75.68	2847	79.07	2958	82.18
99	2677	74.36	2757	76.57	2880	79.99	2992	83.12
100	2709	75.24	2789	77.47	2913	80.91	3026	84.06

Table 1.1 Continued

Trunks	Grade of Service 1 in 1000		Grade of Service 1 in 500		Grade of Service 1 in 200		Grade of Service 1 in 100	
	UC	TU	UC	TU	UC	TU	UC	TU
101	2740	76.12	2821	78.36	2946	81.83	3060	85.00
102	2772	77.00	2853	79.26	2979	82.75	3094	85.95
103	2804	77.88	2886	80.16	3012	83.67	3128	86.89
104	2836	78.77	2918	81.05	3045	84.59	3162	87.83
105	2867	79.65	2950	81.95	3078	85.51	3196	88.77
106	2899	80.53	2983	82.85	3111	86.43	3230	89.72
107	2931	81.42	3015	83.75	3145	87.35	3264	90.66
108	2963	82.30	3047	84.65	3178	88.27	3298	91.60
109	2995	83.19	3080	85.55	3211	89.20	3332	92.55
110	3027	84.07	3112	86.45	3244	90.12	3366	93.49
111	3059	84.96	3145	87.35	3277	91.04	3400	94.44
112	3091	85.85	3177	88.25	3311	91.97	3434	95.38
113	3122	86.73	3209	89.15	3344	92.89	3468	96.33
114	3154	87.62	3242	90.06	3378	93.82	3502	97.28
115	3186	88.51	3275	90.96	3411	94.74	3536	98.22
116	3218	89.40	3307	91.86	3444	95.67	3570	99.17
117	3250	90.29	3340	92.77	3478	96.60	3604	100.12
118	3282	91.18	3372	93.67	3511	97.53	3639	101.07
119	3315	92.07	3405	94.58	3544	98.45	3673	102.02
120	3347	92.96	3437	95.48	3578	99.38	3707	102.96
121	3379	93.86	3470	96.39	3611	100.31	3741	103.91
122	3411	94.75	3503	97.30	3645	101.24	3775	104.86
123	3443	95.64	3535	98.20	3678	102.17	3809	105.81

124	3475	96.54	3568	99.11	3712	103.10	3843	106.76
125	3507	97.43	3601	100.02	3745	104.03	3878	107.71
126	3540	98.33	3633	100.93	3779	104.96	3912	108.66
127	3572	99.22	3666	101.84	3812	105.89	3946	109.62
128	3604	100.12	3699	102.75	3846	106.82	3981	110.57
129	3636	101.01	3732	103.66	3879	107.75	4015	111.52
130	3669	101.91	3765	104.57	3912	108.68	4049	112.47
131	3701	102.81	3797	105.48	3946	109.62	4083	113.42
132	3733	103.70	3830	106.39	3980	110.55	4118	114.38
133	3766	104.60	3863	107.30	4013	111.48	4152	115.33
134	3798	105.50	3896	108.22	4047	112.42	4186	116.28
135	3830	106.40	3929	109.13	4081	113.35	4221	117.24
136	3863	107.30	3961	110.04	4114	114.28	4255	118.19
137	3895	108.20	3994	110.95	4148	115.22	4289	119.14
138	3928	109.10	4027	111.87	4181	116.15	4324	120.10
139	3960	110.00	4060	112.78	4215	117.09	4358	121.05
140	3992	110.90	4093	113.70	4249	118.02	4392	122.01
141	4025	111.81	4126	114.61	4283	118.96	4427	122.96
142	4058	112.71	4159	115.53	4316	119.90	4461	123.92
143	4090	113.61	4192	116.44	4350	120.83	4496	124.88
144	4122	114.51	4225	117.36	4384	121.77	4530	125.83
145	4155	115.42	4258	118.28	4418	122.71	4564	126.79
146	4188	116.32	4291	119.19	4451	123.64	4599	127.74
147	4220	117.22	4324	120.11	4485	124.58	4633	128.70
148	4253	118.13	4357	121.03	4519	125.52	4668	129.66
149	4285	119.03	4390	121.95	4552	126.46	4702	130.62
150	4318	119.94	4423	122.86	4586	127.40	4737	131.58

Source: Courtesy of GTE Automatic Electric Company (Bulletin No. 485).

grade of service were reduced to .05, the 30 trunks could carry 24.80 erlangs of traffic. When sizing a route for trunks or an exchange, we often come up with a fractional number of servicing channels or trunks. In this case we would opt for the next highest integer because we cannot install a fraction of a trunk. For instance, if calculations show that a trunk route should have 31.4 trunks, it would be designed for 32 trunks.

The Erlang B formula, based on lost calls cleared, has been standardized by the CCITT (CCITT Rec. Q.87) and has been generally accepted outside the United States. In the United States the Poisson formula [2] is favored. This formula is often called the *Molina formula*. It is based on the LCH concept. Table 1.2 provides trunking sizes for various grades of service deriving from the P formula; such tables are sometimes called "P" tables, (Poisson) and assume full availability. We must remember that the Poisson equation also assumes that traffic originates from a large (infinite) number of independent subscribers or sources (random traffic input), with a limited number of trunks or servicing channels and LCH.

It is not as straightforward as it may seem when comparing grades of service between Poisson and Erlang B formulas (or tables). The grade of service $p = .01$ for the Erlang B formula is equivalent to a grade of service of .005 when applying the Poisson (Molina) formula. Given these grades of service, assuming LCC with the Erlang B formula permits up to several tenths of erlangs of less traffic when dimensioning up to 22 trunks, where the two approaches equate (e.g., where each formula allows 12.6 erlangs over the 22 trunks). Above 22 trunks Erlang B permits the trunks to carry somewhat more traffic and at 100 trunks, 2.7 erlangs more than for the Poisson formula under the LCH assumption.

7 WAITING SYSTEMS (QUEUEING)

A short discussion follows regarding traffic in queueing systems. Queueing or waiting systems, when dealing with traffic, are based on the third assumption, namely, lost calls delayed (LCD). Of course, a queue in this case is a pool of callers waiting to be served by a switch. The term *serving time* is the time a call takes to be served from the moment of arrival in the queue to the moment of being served by the switch. For traffic calculations in most telecommunication queueing systems, the mathematics is based on the assumption that call arrivals are random and Poissonian. The traffic engineer is given the parameters of offered traffic, the size of the queue, and a specified grade of service and will determine the number of serving circuits or trunks required.

The method by which a waiting call is selected to be served from the pool of waiting calls is called *queue discipline*. The most common discipline is the first-come–first served discipline, where the call waiting longest in the queue is served first. This can turn out to be costly because of the equipment required to keep order in the queue. Another type is random selection, where the time a call has waited is disregarded and those waiting are selected in random order. There is also the last-come–first-served discipline and bulk service discipline, where batches of waiting calls are admitted, and there are also priority-service disciplines, which can be preemptive and nonpreemptive. In queueing systems the grade of service may be defined as the probability of delay. This is expressed as $P(t)$, the probability that a call is not being immediately served and has to wait a period of time greater than t. The average delay on all calls is another parameter that can be used to express grade of service, and the length of queue is another.

The probability of delay, the most common index of grade of service for waiting systems when dealing with full availability and a Poisonnian call arrival process, is calculated by using the Erlang C formula, which assumes an infinitely long queue length. Syski [3] provides a good guide to Erlang C and other, more general, waiting systems.

8 DIMENSIONING AND EFFICIENCY

By definition, if we were to dimension a route or estimate the required number of servicing channels, where the number of trunks (or servicing channels) just equaled the Erlang load, we would attain 100% efficiency. All trunks would be busy with calls all the time or at least for the entire BH. This would not even allow several moments for a trunk to be idle while the switch decided the next call to service. In practice, if we engineered our trunks, trunk routes, or switches this way, there would be many unhappy subscribers.

On the other hand, we do, indeed, want to size our routes (and switches) to have a high efficiency and still keep our customers relatively happy. The goal of our previous exercises in traffic engineering was just that. The grade of service is one measure of subscriber satisfaction. As an example, let us assume that between cities X and Y there are 100 trunks on the interconnecting telephone route. The tariffs, from which the telephone company derives revenue, are a function of the erlangs of carried traffic. Suppose we allow a dollar per erlang-hour. The very upper limit of service on the route is 100 erlangs. If the route carried 100 erlangs of traffic per day, the maximum return on investment would be 2400 dollars a day for that trunk route and the portion of the switches and local plant involved with these calls. As we well know, many of the telephone company's subscribers would be unhappy because they would have to wait excessively to get calls through from X to Y. How, then, do we optimize a trunk route (or serving circuits) and keep the customers as happy as possible?

Remember from Table 1.1, with an excellent grade of service of .001, that we relate grade of service to subscriber satisfaction and that the 100 circuits could carry up to 75.24 erlangs during the BH. Assuming the route

Table 1.2 Trunk Loading Capacity Based on Poisson Formula, Full Availability

	Grade of Service 1 in 1000		Grade of Service 1 in 100		Grade of Service 1 in 50		Grade of Service 1 in 20		Grade of Service 1 in 10	
Trunks	UC	TU	UC	TU	UC	TU	UC	TU	UC	TU
1	0.1	0.003	0.4	0.01	0.7	0.02	1.9	0.05	3.8	0.10
2	1.6	0.05	5.4	0.15	7.9	0.20	12.9	0.35	19.1	0.55
3	6.9	0.20	16	0.45	20	0.55	29.4	0.80	39.6	1.10
4	15	0.40	30	0.85	37	1.05	49	1.35	63	1.75
5	27	0.75	46	1.30	56	1.55	71	1.95	88	2.45
6	40	1.10	64	1.80	76	2.10	94	2.60	113	3.15
7	55	1.55	84	2.35	97	2.70	118	3.25	140	3.90
8	71	1.95	105	2.90	119	3.30	143	3.95	168	4.65
9	88	2.45	126	3.50	142	3.95	169	4.70	195	5.40
10	107	2.95	149	4.15	166	4.60	195	5.40	224	6.20
11	126	3.50	172	4.80	191	5.30	222	6.15	253	7.05
12	145	4.05	195	5.40	216	6.00	249	6.90	282	7.85
13	166	4.60	220	6.10	241	6.70	277	7.70	311	8.65
14	187	5.20	244	6.80	267	7.40	305	8.45	341	9.45
15	208	5.80	269	7.45	293	8.15	333	9.25	370	10.30
16	231	6.40	294	8.15	320	8.90	362	10.05	401	11.15
17	253	7.05	320	8.90	347	9.65	390	10.85	431	11.95
18	276	7.65	346	9.60	374	10.40	419	11.65	462	12.85
19	299	8.30	373	10.35	401	11.15	448	12.45	492	13.65
20	323	8.95	399	11.10	429	11.90	477	13.25	523	14.55
21	346	9.60	426	11.85	458	12.70	507	14.10	554	15.40
22	370	10.30	453	12.60	486	13.50	536	14.90	585	16.25

23	395	10.95	480	13.35	514	14.30	566	15.70	616	17.10
24	419	11.65	507	14.10	542	15.05	596	16.55	647	17.95
25	444	12.35	535	14.85	572	15.90	626	17.40	678	18.85
26	469	13.05	562	15.60	599	16.65	656	18.20	710	19.70
27	495	13.75	590	16.40	627	17.40	686	19.05	741	20.60
28	520	14.45	618	17.15	656	18.20	717	19.90	773	21.45
29	545	15.15	647	17.95	685	19.05	747	20.75	805	22.35
30	571	15.85	675	18.75	715	19.85	778	21.60	836	23.20
31	597	16.60	703	19.55	744	20.65	809	22.45	868	24.10
32	624	17.35	732	20.35	773	21.45	840	23.35	900	25.00
33	650	18.05	760	21.10	803	22.30	871	24.20	932	25.90
34	676	18.80	789	21.90	832	23.10	902	25.05	964	26.80
35	703	19.55	818	22.70	862	23.95	933	25.90	996	27.65
36	729	20.25	847	23.55	892	24.80	964	26.80	1028	28.55
37	756	21.00	876	24.35	922	25.60	995	27.65	1060	29.45
38	783	21.75	905	25.15	951	26.40	1026	28.50	1092	30.35
39	810	22.50	935	25.95	982	27.30	1057	29.35	1125	31.25
40	837	23.25	964	26.80	1012	28.10	1088	30.20	1157	32.14
41	865	24.05	993	27.60	1042	28.95	1120	31.10	1190	33.05
42	892	24.80	1023	28.40	1072	29.80	1151	31.95	1222	33.95
43	919	25.55	1052	29.20	1103	30.65	1183	32.85	1255	34.85
44	947	26.30	1082	30.05	1133	31.45	1214	33.70	1287	35.75
45	975	27.10	1112	30.90	1164	32.35	1246	34.60	1320	36.65
46	1003	27.85	1142	31.70	1194	33.15	1277	35.45	1352	37.55
47	1030	28.60	1171	32.55	1225	34.05	1309	36.35	1385	38.45
48	1058	29.40	1201	33.35	1255	34.85	1340	37.20	1417	39.35
49	1086	30.15	1231	34.20	1286	35.70	1372	38.10	1450	40.30
50	1115	30.95	1261	35.05	1317	36.60	1403	38.95	1482	41.15

Table 1.2 Continued

Trunks	Grade of Service 1 in 1000		Grade of Service 1 in 100		Grade of Service 1 in 50	
	UC	TU	UC	TU	UC	TU
51	1143	31.75	1291	35.85	1349	37.45
52	1171	32.55	1322	36.70	1380	38.35
53	1200	33.35	1352	37.55	1410	39.15
54	1228	34.10	1382	38.40	1441	40.05
55	1256	34.90	1412	39.20	1472	40.90
56	1285	35.70	1443	40.10	1503	41.75
57	1313	36.45	1473	40.90	1534	42.60
58	1342	37.30	1504	41.80	1565	43.45
59	1371	38.10	1534	42.60	1596	44.35
60	1400	38.90	1565	43.45	1627	45.20
61	1428	39.65	1595	44.30	1659	46.10
62	1457	40.45	1626	45.15	1690	46.95
63	1486	41.30	1657	46.05	1722	47.85
64	1516	42.10	1687	46.85	1752	48.65
65	1544	42.90	1718	47.70	1784	49.55
66	1574	43.70	1749	48.60	1816	50.45
67	1603	44.55	1780	49.45	1847	51.30
68	1632	45.35	1811	50.30	1878	52.15
69	1661	46.15	1842	51.15	1910	53.05
70	1691	46.95	1873	52.05	1941	53.90
71	1720	47.80	1904	52.90	1973	54.80
72	1750	48.60	1935	53.75	2004	55.65
73	1779	49.40	1966	54.60	2036	56.55

74	1809	50.25	1997	55.45	2067	57.40
75	1838	51.05	2028	56.35	2099	58.30
76	1868	51.90	2059	57.20	2130	59.15
77	1898	52.70	2091	58.10	2162	60.05
78	1927	53.55	2122	58.95	2194	60.95
79	1957	54.35	2153	59.80	2226	61.85
80	1986	55.15	2184	60.65	2258	62.70
81	2016	56.00	2215	61.55	2290	63.60
82	2046	56.85	2247	62.40	2321	64.45
83	2076	57.65	2278	63.30	2354	65.40
84	2106	58.50	2310	64.15	2386	66.30
85	2136	59.35	2341	65.05	2418	67.15
86	2166	60.15	2373	65.90	2451	68.10
87	2196	61.00	2404	66.80	2483	68.95
88	2226	61.85	2436	67.65	2515	69.85
89	2256	62.65	2467	68.55	2547	70.75
90	2286	63.50	2499	69.40	2579	71.65
91	2317	64.35	2530	70.30	2611	72.55
92	2346	65.15	2562	71.15	2643	73.40
93	2377	66.05	2594	72.05	2674	74.30
94	2407	66.85	2625	72.90	2706	75.15
95	2437	67.70	2657	73.80	2739	76.10
96	2468	68.55	2689	74.70	2771	76.95
97	2498	69.40	2721	75.60	2803	77.85
98	2528	70.20	2752	76.45	2836	78.80
99	2559	71.10	2784	77.35	2868	79.65
100	2589	71.90	2816	78.20	2900	80.55

Source: Courtesy of GTE Automatic Electric Company (Bulletin No. 485).

did carry 75.24 erlangs for the BH, it would earn $75.24 for that hour and something far less than $2400 per day. If the grade of service were reduced to .01, 100 trunks would bring in $84.06 for the busy hour. Note the improvement in revenue at the cost of reducing grade of service. Another approach to saving money is to hold the Erlang load constant and decrease the number of trunks and switch facilities accordingly as the grade of service is reduced. For instance, 70 Erlangs of traffic at $p = .001$ requires 96 trunks and at $p = 0.01$, only 86 trunks.

8.1 Alternative Routing

One method of improving efficiency is to use alternative routing (called *alternate routing* in North America). Suppose that we have three serving areas, X, Y, and Z, served by three switches, X, Y, and Z as shown in Figure 1.7.

Let the grade of service be .005 (1 in 200 in Table 1.1). We found that it would require 67 trunks to carry 50 erlangs of traffic during the BH to meet that grade of service between X and Y. Suppose that we reduced the number of trunks between X and Y, still keeping the BH traffic intensity at 50 erlangs. We would thereby increase the efficiency on the X–Y route at the cost of reducing the grade of service. With a modification of the switch at X, we could route the traffic bound for Y that met congestion on the X–Y route via Z. Then Z would route this traffic on the Z–Y link. Essentially, this is alternative routing in its simplest form. Congestion probably would only occur during very short peaky periods in the BH, and chances are that these peaks would not occur simultaneously with peaks in traffic intensity on the Z–Y route. Further, the added load on the X–Z–Y route would be very small. Some idea of traffic peakiness that would overflow onto the secondary route ($X + Z + Y$) is shown in Figure 1.8.

One of the most accepted methods of dimensioning switches and trunks using alternative routing is the equivalent random group (ERG) method developed by Wilkinson [11]. The Wilkinson method uses the mean M and

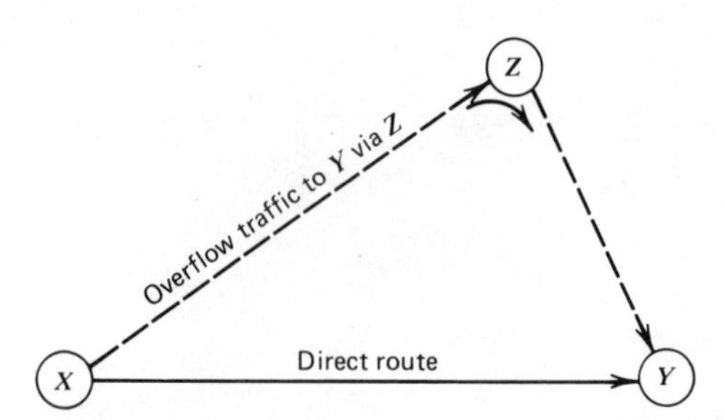

Figure 1.7 Simplified diagram of the alternative routing concept (solid line direct route, dashed line alternative route carrying the overflow from X to Y).

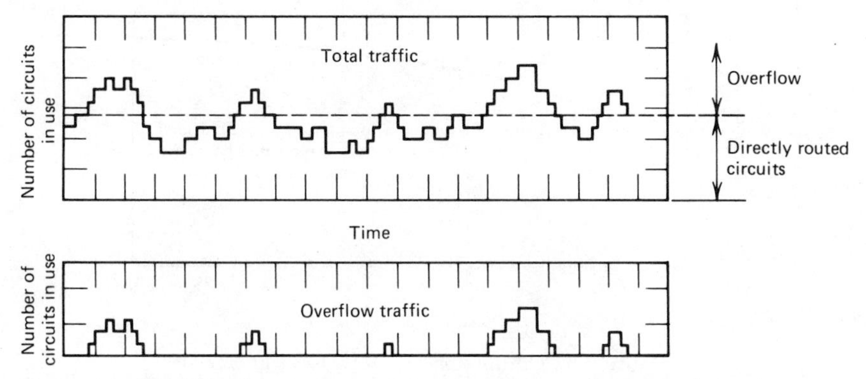

Figure 1.8 Traffic peakiness, the peaks representing overflow onto alternative routes.

the variance V. Here the *overflow traffic* is the "lost" traffic in the Erlang B calculations, which were discussed earlier. Let M be the mean value of that overflow, and A be the random traffic offered to a group of n circuits (trunks). Then

$$V = M\left(1 - M + \frac{A}{1 + n + M - A}\right)$$

When the overflow traffic from several sources is combined and offered to a single second (or third, fourth, etc) choice of a group of circuits, both the mean and the variance of the combined traffic are the arithmetical sums of the means and variances of the contributors.

The basic problem in alternative routing is to optimize circuit group efficiency (e.g., to dimension a route with an optimum number of trunks). Thus we are to find what circuit quantities result in minimum cost for a given grade of service, or to find the optimum number of circuits (trunks) to assign to a direct route allowing the remainder to overflow on alternative choices. There are two approaches to the optimization. The first method is to solve the problem by successive approximations, and this lends itself well to the application of the computer [12]. Then there are the manual approaches, two of which are suggested in the annex to CCITT Rec. Q.88.

8.2 Efficiency versus Circuit Group Size

In the present context a *circuit group* refers to a group of circuits performing a specific function. For instance, all the trunks (circuits) routed from X to Y in Figure 1.7 make up a circuit group, irrespective of size. This circuit group should not be confused with the "group" used in transmission-engineering carrier systems.

If we assume full loading, it can be stated that efficiency improves with circuit group size. From Table 1.1, given $p = .01$, 5 erlangs of traffic

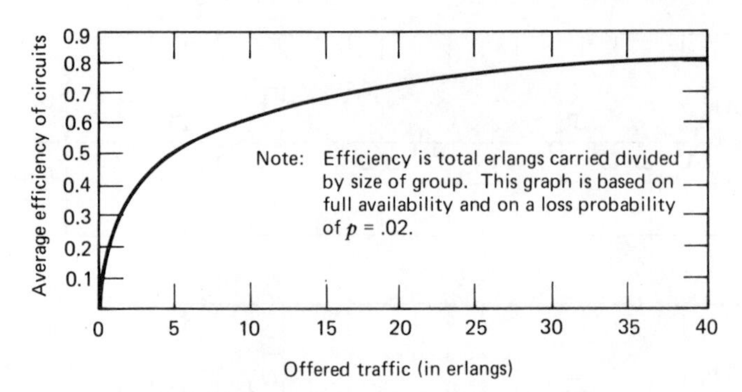

Figure 1.9 Group efficiency increases with size.

require a group with 11 trunks, more than a 2:1 ratio of trunks to erlangs, and 20 erlangs require 30 trunks, a 3:2 ratio. Note how the efficiency has improved. One hundred and twenty trunks will carry 100 erlangs, or six trunks for every 5 erlangs for a group of this size. Figure 1.9 shows how efficiency improves with group size.

9 BASES OF NETWORK CONFIGURATIONS

9.1 Introductory Concepts

A network in telecommunications may be defined as a method of connecting exchanges so that any one subscriber in the network can communicate with any other subscriber. For this introductory discussion, let us assume that subscribers access the network by a nearby local exchange. Thus the problem is essentially how to connect exchanges efficiently. There are three basic methods of connection in conventional telephony: (1) mesh, (2) star, and (3) double and higher-order star (see Section 2 of this chapter). The mesh connection is one in which each and every exchange is connected by trunks (or junctions) to each and every other exchange as shown in Figure 1.10A. A star connection utilizes an intervening exchange, called a *tandem exchange,* such that each and every exchange is interconnected via a *single* tandem exchange. An example of a star connection is shown in Figure 1.10B. A double star configuration is one where sets of pure star subnetworks are connected via higher-order tandem exchanges as shown in Figure 1.10C. This trend can be carried still further, as we see later on, when hierarchical networks are discussed.

As a general rule we can say that mesh connections are used when there are comparatively high traffic levels between exchanges such as in metropolitan networks. On the other hand, a star network may be applied when traffic levels are comparatively low.

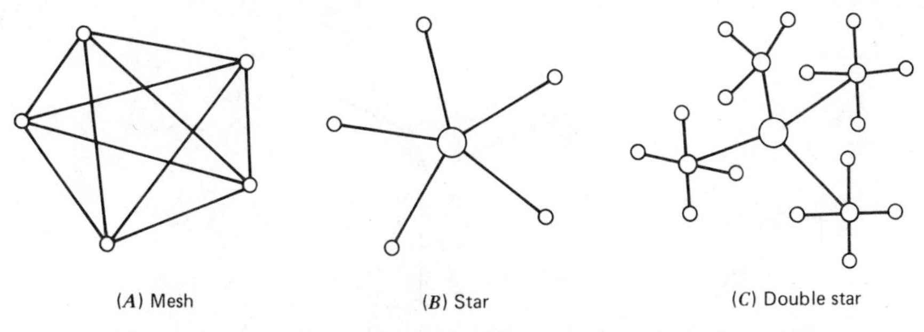

Figure 1.10 Examples of star, double star and mesh configurations.

Another factor that leads to star and multiple-star network configurations is network complexity in the trunking outlets (and inlets) of a switch in a full mesh. For instance, an area with 20 exchanges would require 380 traffic groups (or links) and 100 exchanges, 9900 traffic groups. This assumes what are called *one-way groups.* A one-way group is best defined considering the connection between two exchanges, A and B. Traffic originating at A to B is carried in one group and the traffic originating at B bound for A, in another group, as shown in the following diagram.

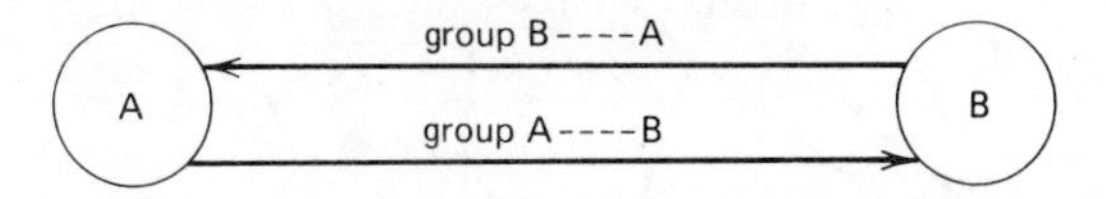

Thus, in practice, most networks are compromises between mesh and star configurations. For instance, outlying suburban exchanges may be connected to a nearby major exchange in the central metropolitan area. This exchange may serve nearby subscribers and may be connected in mesh to other large exchanges in the city proper. Another example is the city's long-distance exchange, which is a tandem exchange looking into the national long-distance network, whereas the major exchanges in the city are connected to it in mesh. An example of a real-life compromise between mesh, star, and multiple-star configurations is shown in Figure 1.11.

9.2 Hierarchical Networks

To bring order out of this confusion, hierarchical networks evolved. That is, a systematic network was developed that reduces the trunk group outlets (and inlets) of a switch to some reasonable amount, permits the handling of high traffic intensities on certain routes where necessary, and allows for overflow and a certain means of restoral in certain circumstances. Consider Figure 1.12, which is a simplified example of a higher-order star network. The term "order" here is significant and leads to the discussion of hierarchical networks.

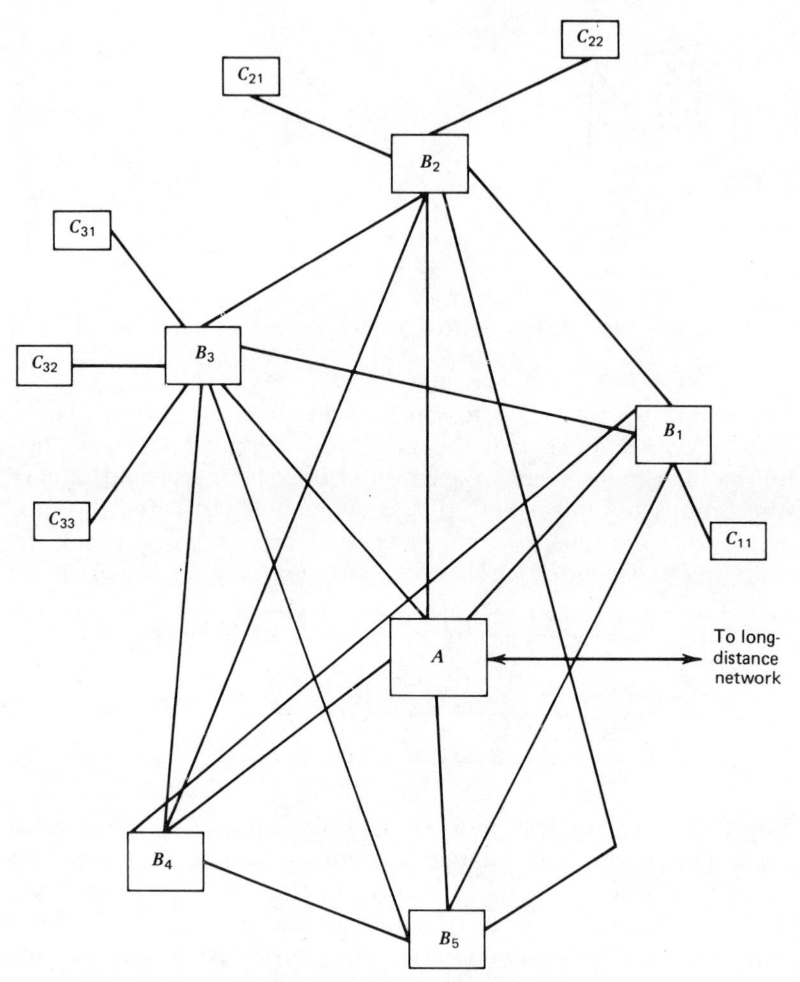

Figure 1.11 A typical telephone network serving a small city as an example of a compromise between mesh and star configuration.
A is a class 4(ATT), primary center (CCITT)
B is a class 5 exchange, a local exchange
C may be a satellite exchange or a concentrator.

A hierarchical network has levels giving orders of importance of the exchanges making up the network, and certain restrictions are placed on traffic flow. For instance, in Figure 1.12 there are three levels or ranks of exchange. The smallest boxes in the diagram are the lowest-ranked exchanges, which have been marked with a "3" to indicate the third level or rank. Note the restrictions (or rules) of traffic flow. As the figure is drawn, traffic from $3A_1$ bound for $3A_2$ would have to flow through exchange $2A_1$. Likewise, traffic from exchange $2A_2$ to $2A_3$ would have to flow through

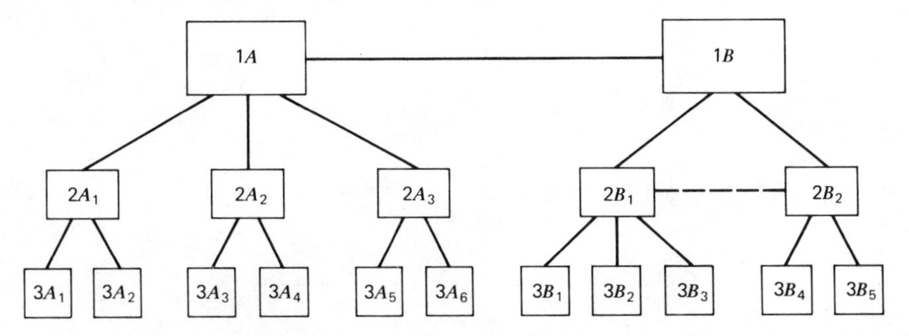

Figure 1.12 A higher order star network.

exchange 1*A*. Carrying the concept somewhat further, traffic from any *A* exchange to any *B* exchange would necessarily have to be routed through exchange 1*A*.

The next consideration is the high-usage route. For instance, if we found that there were high traffic intensities between $2B_1$ and $2B_2$, trunks and switch gear might well be saved by establishing a high-usage route between the two (shown in dashed line). Thus we might call the high-usage route a *highly traveled shortcut.* Of course, high-usage routes could be established between any pair of exchanges in the network if traffic intensities and distances involved proved this strategy economical. When high-usage routes are established, traffic between the exchanges involved will first be offered to the high-usage route and overflow would take place through hierarchical structure. Or as shown in our Figure 1.12, up to the next level and down. If routing is through the highest level in the hierarchy, we call this route the *final route.* Figure 1.12 shows traffic routed between exchanges $2B_1$ and $2B_2$ via exchange 1*B* being routed on the final route.

9.3 The ATT and CCITT Hierarchical Networks

Two types of hierarchical network exist today, each serving about 50% of the world's telephones. These are the ATT network, generally used in North America, and the CCITT network, typically used in Europe or areas of the world under European influence. Frankly, there is really little difference from the routing viewpoint. Each has five levels or ranks in the hierarchy, although CCITT allows for a sixth level. The basic difference is in the nomenclature used. Figure 1.13 illustrates the ATT hierarchy and Figure 1.14, the CCITT hierarchy.

Particularly in Europe, the terminology distinguishes between tandem exchanges and transit exchanges. Although both perform the same function, the switching of trunks, a tandem exchange serves the local area, as shown at the bottom of Figure 1.14 and figures in the lowest levels of hierarchy. A transit exchange switches trunks in the toll or long-distance

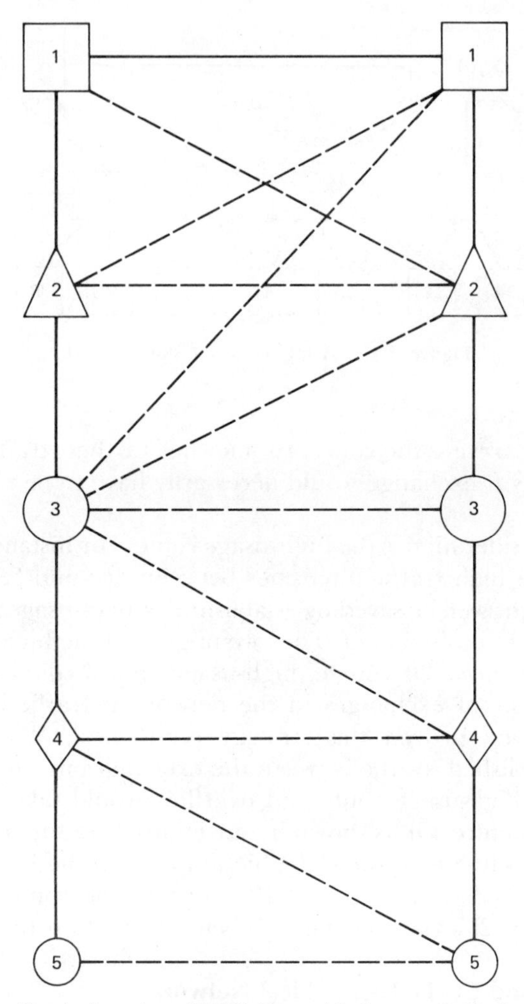

Figure 1.13 The North American (ATT) hierarchical network (dashed lines show high-usage trunks). Note how the two highest ranks are connected in mesh.

area. Also, in CCITT international routing schemes, we should expect to see the term "CT," meaning "central transit" in French or central tránsito in Spanish. In English the term is simply *transit exchange.* The CCITT usually places a number after CT, as follows: CT1, the highest-order transit exchange in CCITT routing; CT2, the next-to-highest order; and CT3, the third order from the top. You will sometimes see the term *junction,* which is a British (UK) term for a trunk serving the local area. In CCITT terminology, trunks serve as higher-order connections. Primary centers are collecting centers (exchanges) for traffic to interconnect the toll or long-distance network. The term *center* may be related to "central," meaning a switching node or exchange, usually of higher order.

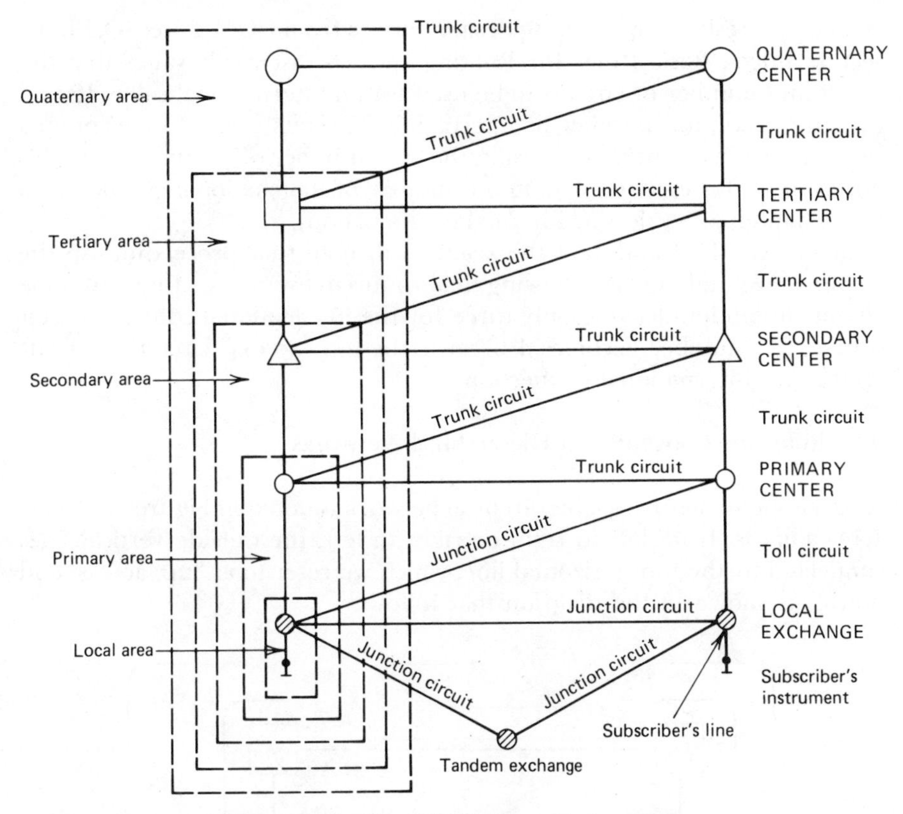

Figure 1.14 The CCITT hierarchical structure.

Figure 1.13 presents the ATT "routing pattern." The highest order or rank in the hierarchy is the class 1 center and the lowest, a class 5 office ("office" is taken from the North American term "central office"). It should be noted that a high-usage (HU) trunk group may be established between any two switching centers regardless of location or rank, whenever the traffic volume justifies it. The table that follows clarifies the comparative nomenclature of the two types of hierarchy, with the highest rank at the top.

North American		CCITT
Class 1.	Regional center	Quaternary center
Class 2.	Sectional center	Tertiary center
Class 3.	Primary center	Secondary center
Class 4.	Toll center (toll point)	Primary center
Class 5.	End office	Local exchange

The first restraint on routing design derives from CCITT Rec. Q.13, the section titled "Basic Rules for Routing," which essentially states that the maximum number of circuits to be used for an international call is 12 and that the maximum number of international circuits is six. In exceptional cases and for a low number of calls, the total number of circuits may be 14, but even in this case the maximum number of international circuits is six (see Chapter 6, Section 3, for further discussion).

In Figures 1.13 and 1.14 the reader will note that proceeding up the chain, across and down* following final routes in every case, there are nine circuits in tandem leaving only three for the international connection. Of course, this number becomes 4 because the top "across" circuit is considered as an international connection.

9.4 Rules for Conventional Hierarchical Networks

A backbone structure to a hierarchical network is noted in Figures 1.13 and 1.14. That is, from left to right or right to left, the outside vertical lines connected by the top horizontal line, which we refer to as "up, across, and down" as shown in the diagram that follows.

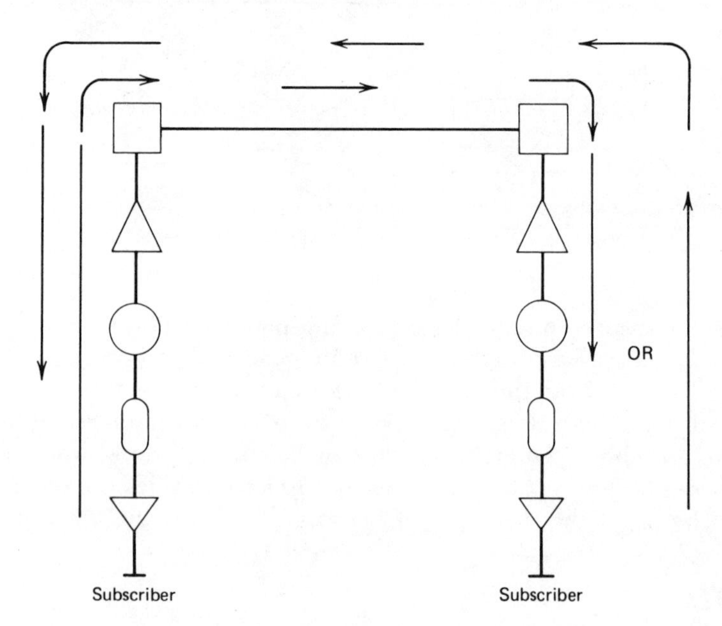

The CCITT terms these routes "theoretical final routes." For our argument, a final route is a route from which no overflow is permitted. A hierarchical network is characterized by a full set of final routes from

*See also Section 9.4 for further discussion of hierarchical network rules and "up, across, and down."

source to sink. Any other routes are supplementary to the pure hierarchy, regardless of whether overflow is permitted on them.

A hierarchical system of routing leads to simplified switch design. A common expression used when discussing hierarchical routing and star connections is that lower-rank exchanges "home" on higher-rank exchanges. If a call is destined for an exchange of lower rank in its chain, the call proceeds down the chain. Likewise, if a call is destined for another exchange outside the chain, it proceeds up the chain. Or when such high usage routes exist, a call may be routed on a route additional or supplementary to the pure hierarchy, proceeding to the distant transit center and then descending to the destination. Of course, at the highest level in a pure hierarchy the call crosses from one chain over to the other. In hierarchical networks, only the order of each switch in the hierarchy and those additional links (high-usage routes) that provide access need be known. In such networks administration is simplified, and storage or routing information is reduced when compared to the full-mesh type of network, for example.

The CCITT (Rec. Q.13) suggests the far-to-near criterion whereby the first choice route in advancing a call is to advance the call as far as possible from its origin using the backbone route to measure distances. The second choice is the next best and so forth. The additional routes (diagonal dashed lines in Figures 1.13 and 1.14) result in decreasing the number of links traversed, thus minimizing the total links traversed in a call connection and improving transmission and signaling characteristics and, on long international connections, keeping the total number of links required to 12 or below, as we mentioned previously.

One weakness in conventional hierarchical structured networks is circuit security. When we say "security" in this context, we mean that the loss of one (or several) links due to fire, explosion, natural disaster, cutting, or sabotage will not cause the full breakdown of communication in the network. Rather, we mean that communication can be maintained, perhaps with reduced capability and increased blockage, but nevertheless maintained. Now consider a hierarchical network. Higher in the network, the nodes of each rank become fewer and fewer. For instance take the United States and Canada. There are only 10 nodes of class 1 in the United States and two in Canada. Even with these 12 nodes completely interconnected in mesh, as they actually are in practice, the loss of one or more nodes or links at this level in the hierarchy might seriously jeopardize communications. Thus the current tendency is to reduce the number of levels or ranks in the hierarchy, thus increasing the number of higher-level nodal points. Improved circuit security is the result when they are fully interconnected (mesh), offering more combinations of alternative route configurations. The desirability of this trend is obvious. It becomes highly feasible as the network becomes modernized, replacing electromechanical switching with computer-controlled switching. We see future large national networks with only three levels of hierarchy.

9.5 Homing Arrangements and Interconnecting Network in North America

We spoke of "homing" of a dependent switching node to the next-higher rank nodal point in purely hierarchical networks. The use of HU routes so widely used in practice makes the rule really valid only on the backbone final routes. For example, consider the North American network where class 5, the lowest ranking switch *may* be served directly from any higher ranking location. Possible homing arrangements for each class (rank) of switching node are shown in Table 1.3.

Final trunk groups in this network are engineered for a lost-call probability of .01. As we have seen, direct high-usage trunk groups are provided between switching nodal points of any rank (class) when the volume of traffic and economics warrant these groups and where automatic alternative routing features are available. The rule is that high-usage trunk groups carry most of, but not all, the offered traffic in the BH. The proportion of the offered traffic carried on a direct high-usage trunk group ordinarily is determined, in part, by the relative costs of the direct route and the alternative route. The economic factors to be considered are the additional switching costs on the alternative routes. In some instances high-usage trunk groups may be designed on a "no-overflow" basis with the .01 lost-call objective. Homing arrangements are not changed in this case, and these trunks are called "full groups." Such full groups effectively truncate or limit the hierarchical chain of final routes for the traffic offered them.

In the North American network the number of trunks connected in the final route chains from a class 4 switch to another class 4 switch is not allowed to exceed 7. By adding these to the trunk at each end connecting to the end office (class 5), the maximum number of trunks that can be connected in tandem is 9. It is estimated that the probability of a call traversing all the final route links in two complete routing chains be only several calls out of millions. Of course, calls between high-intensity traffic locations (nodes) rarely encounter multiple switches. However, calls between low-intensity traffic points use multiple switching, or what we call "tandem switching."

Table 1.3 Homing Arrangements in North America

Rank and Class	May Home at Switches of Classes
End office, 5	Class 4, 3, 2, or 1
Toll center, 4	Class 3, 2, or 1
Primary center, 3	Class 2 or 1
Sectional center, 2	Class 1
Regional center, 1	All regional centers mutually interconnected

10 ROUTING METHODS

There are generally three methods of routing calls from source to sink through one, several, or many intermediate switching nodes. As we have seen, there may be many possible patterns through which a given call can traverse. The problem is to *decide* how the call should proceed through the many possible path combinations in the network. The three methods are: (1) right-through routing, (2) own-exchange routing, and (3) computer-controlled routing (with common-channel signaling). In right-through routing the originating exchange determines the route from source to sink. Alternative routing is not allowed at intermediate switching points. However, the initial outgoing circuit group may be arranged so that one or more alternative routes are presented. Because of its inherent limitations in alternative routing and the requirement that a change in network configuration or the addition of new exchanges entail alteration in each existing complex switch (i.e., switches with translators), right-through routing is limited almost exclusively to the local area.

Own-exchange routing allows for changes in routing as the call proceeds to its destination. This routing system is particularly suited to networks with alternative routing and changes in routing patterns in response to changes in load configuration. Another advantage in own-exchange routing is that when new exchanges are added or the network is modified, minimal switch modifications are required in the network. One disadvantage is the possibility of establishing a closed routing loop where a call may be routed such that it is eventually routed back to its originating exchange or other exchange through which it has already been routed in attempting to reach its destination. However, a hierarchical routing system ensures that such loops cannot be generated. If routing loops are established in an operating network, there can be disastrous consequences, as the reader can appreciate.

Conventional telephone networks have signaling information for a particular call carried on the same path (pair of wires or their equivalent) that carries the speech, often called the conversation path. Signaling, as we discuss later in considerable detail, is the generation and transmission of information that sets up a desired call and routes it through the network to its destination. New and more modern computer controlled networks often use a separate path to carry the required signaling information. In this case the computer in the originating exchange or originating long-distance exchange can "optimally" route the call through the network on a separate signaling path. The originating computer would have a "map in memory" of the network with updated details of network conditions such as traffic load at the various nodes and trunks, outages, and so forth. The necessary adaptive information is broadcast on the separate path that connects the various computers in the network. This is computer-controlled routing.

Such routing is termed "routing with common-channel signaling" and with adaptive network management signals.

11 VARIATIONS IN TRAFFIC FLOW

In networks covering large geographic expanses and even in cases of certain local networks, there may be a variation in time of day of the BH or in the direction of traffic flow. In the United States business traffic peaks during several hours before and after the noon lunch period on weekdays, and social calls peak in early evening. Traffic flow tends to be from suburban living areas to urban centers in the morning, and the reverse in the evening.

In national networks covering several time zones where the differences in local time may be appreciable, long-distance traffic tends to be concentrated in a few hours common to BH peaks at both ends. In such cases it is possible to direct traffic so that peaks of traffic in one area fall into valleys of traffic in another. The network design can be made more economical if configured to take advantage of these phenomena, particularly in the design and configuration of direct routes versus overflow.

12 BOTH-WAY CIRCUITS

We defined one-way circuits in Section 9.1. Here traffic from *A* to *B* is assigned to one group of circuits and traffic from *B* to *A*, on another separate group. In both way (or two-way) operation a circuit group may be engineered to carry traffic in both directions. The individual circuits in the group may be used in either direction, depending on which exchange seizes the circuit first.

In engineering networks it is most economical to have a combination of one-way and both-way circuits on longer routes. Signaling and control arrangements on both-way circuits are substantially more expensive. However, when dimensioning a system for a given traffic intensity, less circuits are needed in both-way operation, with notable savings on low-intensity routes (i.e., below about 10 erlangs in each direction). For long circuits, both-way operation has obvious advantages when dealing with a noncoincident BH. During overload conditions both-way operation is also advantageous because the direction of traffic flow in these conditions is usually unequal.

The major detriment to two-way operation, besides its increased signaling cost, is the possibility of double seizure. This occurs when both ends seize a circuit at the same time. There is a period of time when double seizure can occur in a two-way circuit; this extends from the moment the circuit is seized to send a call and the moment when it becomes blocked at the other end. Signaling arrangements can help to circumvent this problem. Likewise, switching arrangements can be made such that double seizure can occur only on the last free circuit of a group. This can be done by arranging in turn the sequence of scanning circuits so that the sequence on one end of a two-way circuit is reversed from that of the other end. Of course, great care must be taken on circuits having long propagation times such as satellite and long undersea cable circuits. By extending the time between initial seizure and blockage at the other end, these circuits are the most susceptible, just because a blocking signal takes that much longer to reach the other end.

13 QUALITY OF SERVICE

Quality of service appears at the outset to be an intangible concept. However, it is very tangible for a telephone subscriber unhappy with his or her service. The concept of service quality must be mentioned early in any all-encompassing text on telecommunications systems. The system engineer should never once lose sight of the concept, no matter what segment of the system he or she may be responsible for. Quality of service also means *how happy* the telephone company (or other common carrier) is keeping the customer. For instance, we might find that about half the time the customer dials, his call goes awry or he cannot get a dial tone or he cannot hear what is being said by the party at the other end. All these impact on quality of service. So we begin to find that quality of service is an important factor in many areas of the telecommunications business and means different things to different people. In the old days of telegraphy, a rough measure of how well the system was working was the number of service messages received at a switching center. In modern telephony we now talk about service observing (see Chapter 3, Section 18).

The transmission engineer calls quality of service "customer satisfaction," which is commonly measured by how well the customer can hear the calling party. It is called *reference equivalent,* which is measured in decibels (dB). In our discussion of traffic, lost calls certainly constitute one measure of service quality and if measured in decimal quantity, one target figure would be $p = .01$. Other items to be listed under service quality are as follows:

- Delay before receiving dial tone ("dial-tone delay").
- Post dial(ing) delay (time from completion of dialing a number to first ring of telephone called).
- Availability of service tones (busy tone, telephone out of order, ATB, etc.).
- Correctness of billing.
- Reasonable cost to customer of service.

- Responsiveness to servicing requests.
- Responsiveness and courtesy of operators.
- Time to installation of new telephone, and

by some, the additional services offered by the telephone company. One way or another each item, depending on service quality goal, will impact the design of the system.

Furthermore, each item on the list can be quantified—usually statistically, such as reference equivalent, or in time, such as time taken to install a telephone. In some countries this can be measured in years. Good reading can be found in CCITT Recs. Q.60, Q.60 bis, and Q.61.

REFERENCES

1. International Telephone and Telegraph Corporation, *Reference Data for Radio Engineers,* 5th ed., Howard W. Sams, Indianapolis, 1968.
2. Ramses R. Mina, "The Theory and Reality of Teletraffic Engineering," *Telephony,* a series of articles (April 1971).
3. R. Syski, *Introduction to Congestion Theory in Telephone Systems,* Oliver and Boyd, Edinburgh, 1960.
4. G. Dietrich et al., *Teletraffic Engineering Manual,* Standard Electric Lorenz, Stuttgart, Germany.
5. E. Brockmeyer et al., "The Life and Works of A. K. Erlang," *Acta Polytechnica Scandinavia,* The Danish Academy of Technical Sciences, Copenhagen, 1960.
6. *A Course in Telephone Traffic Engineering,* Australian Post Office, Planning Branch, 1967.
7. Arne Jensen, *Moe's Principle,* The Copenhagen Telephone Company, Copenhagen, Denmark, 1950.
8. *Networks,* Laboratorios ITT de Standard Eléctrica SA, Madrid, 1973 (limited circulation).
9. *Local Telephone Networks,* The Internationa Telecommunications Union, Geneva, 1968.
10. *Electrical Communication System Engineering Traffic,* U.S. Department of the Army, TM-11-486-2, August 1956.
11. R. I. Wilkinson. "Theories for Toll Traffic Engineering in the USA," *BSTJ,* **35,** (March 1956).
12. *Optimization of Telephone Trunking Networks with Alternate Routing,"* ITT Laboratories of Standard Eléctrica (Spain), Madrid, 1974 (limited circulation).
13. John Riordan, *Stochastic Service Systems,* Wiley, New York, 1962.
14. Leonard Kleinrock, *Queueing Systems,* Vols. 1 and 2, Wiley, New York, 1975.
15. Thomas L. Saaty, *Elements of Queueing Theory with Applications,* McGraw Hill, New York, 1961.
16. J. E. Flood, *Telecommunications Networks,* IEE Telecommunications Series 1, Peter Peregrinus, L. London, 1975.
17. *Notes on Distance Dialing,* American Telephone and Telegraph Company, New York, 1975.
18. *National Telephone Networks for the Automatic Service,* International Telecommunications Union–CCITT, Geneva, 1964.
19. D. Bear, *Principles of Telecommunication Traffic Engineering,* IEE Telecommunications Series 2, Peter Peregrinus, London, 1976.

Chapter 2

LOCAL NETWORKS

1 INTRODUCTION

The importance of local network design, whether standing on its own merit or part of an overall national network, cannot be overstressed. In comparison to the long-distance sector, the local sector is not the big income producer per capita invested, but there would be no national network without it. Telephone companies or administrations invest, on the average, more than 50% in their local areas. In the larger, more developed countries the investment in local plant may reach 65 to 70% of total plant investment.

The local area as distinguished from the long distance or national network was discussed in Section 4 of Chapter 1. In this chapter we are more precise in defining the local area itself. Let us concede that the local area includes the following: the subscriber plant, local exchanges, the trunk plant interconnecting these exchanges as well as those trunks connecting a local area to the next level of network hierarchy, and the class 4 exchange (USA) or primary center (CCITT).

To further emphasize the importance of the local area, consider Table 2.1, which was taken from Ref. 1 (CCITT). Figure 2.1 is a simplified diagram of a local network with five local exchanges and illustrates the makeup of a typical small local area.

The design of such a network (Figure 2.1) involves a number of limiting factors, the most important of which is economic. Investment and its return are not treated in this text. However, our goal is to build the most economical network assuming an established quality of service. Considering both quality of service and economy, certain restraints will have to be placed on the design. For example, we will want to know:

- Geographic extension of the local area of interest.
- Number of inhabitants and existing telephone density.
- Calling habits.
- Percentage of business telephones.
- Location of existing telephone exchanges and extension of their serving areas.

Table 2.1 Average Percentage of Investments in Public Telephone Equipment

Item	Average for 16 Countries
Subscriber plant	13%
Outside plant for local networks	27%
Exchanges	27%
Long-distance trunks	23%
Buildings and land	10%

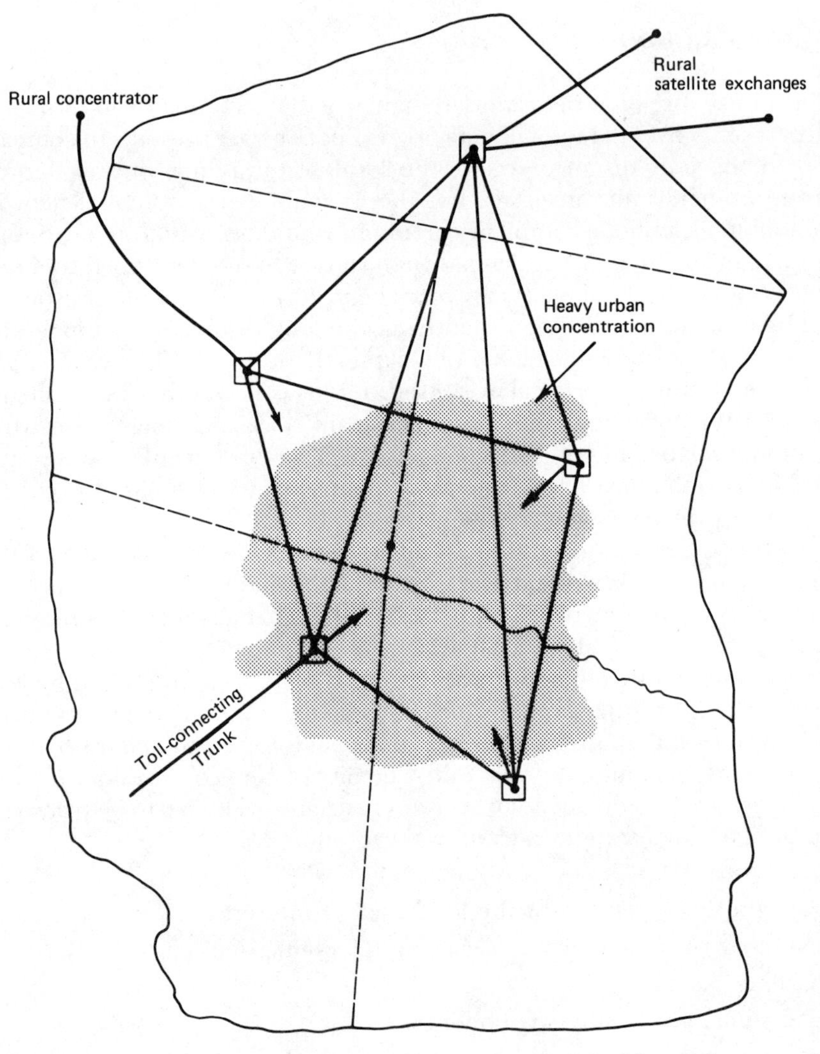

Figure 2.1 A sample local area (arrows represent trunk pull; dashed lines delineate serving areas).

- Trunking scheme.
- Present signaling and transmission characteristics.

Each of these restraints or limiting factors are treated separately, and interexchange signaling and switching per se are dealt with later in separate chapters. Let us also assume that each exchange in the sample will be capable of serving up to 10,000 subscribers. Also assume that all telephones in the area have seven-digit numbers, the last four of which are the subscriber number of the respective serving area of each exchange. The reasoning behind these assumptions becomes apparent in later chapters.

A further assumption is that all subscribers are connected to their respective serving exchanges by wire pairs, resulting in some limiting subscriber loop length. This leads to the first constraining factor dealing with transmission and signaling characteristics. In general terms, the subscriber should be able to hear the distant calling party reasonably well (transmission) and to "signal" his serving switch. These items are treated at length in the following section.

2 SUBSCRIBER LOOP DESIGN

2.1 General

The pair of wires connecting the subscriber to his local serving switch has been defined as the *subscriber loop*. It is a dc loop in that it is a wire pair supplying a metallic path for the following:

- Talk battery for the telephone transmitter (Chapter 1, Section 2).
- An ac ringing voltage for the bell on the telephone instrument supplied from a special ringing source voltage.
- Current to flow through the loop when the telephone instrument is taken out of its cradle ("off hook"), telling the serving switch that it requires "access," thus causing a line seizure at that switch.
- The telephone dial that, when operated, makes and breaks the dc current on the closed loop, which indicates to the switching equipment the number of the distant telephone with which communication is desired.

The typical subscriber loop is powered by means of a battery feed circuit at the switch. Such a circuit is shown in Figure 2.2. Telephone battery source voltage has been fairly well standardized at −48 V dc.

2.2 Subscriber Loop Length Limits

The two basic criteria that are considered when designing subscriber loops, and which limit their length, are attenuation limits (covered under

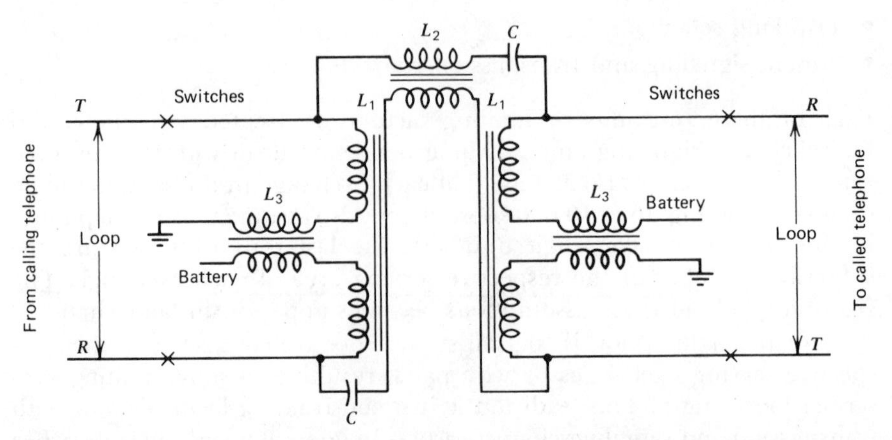

Figure 2.2 Battery feed circuit [7]. *Note:* Battery and ground are fed through inductors L_2 and L_1 through switch to loops. Copyright © 1961 by Bell Telephone Laboratories, Inc.

what we call *transmission design*) and signaling limits (covered under what we call *resistance design*). Attenuation in this case refers to loop ac loss at reference frequency measured in decibels (or nepers). Reference frequency is 1000 Hz in North America and 800 Hz in Europe and many other parts of the world. As a telephone loop is extended in length, its loss at reference frequency increases. It follows that at some point as the loop is extended, level will be attenuated such that the subscriber cannot hear *sufficiently well.*

Likewise, as a loop is extended in length while the battery (supply) voltage is kept constant, the effectiveness of signaling is ultimately lost. This limit is a function of the IR drop of the line. We know that R, the resistance, increases as length increases. With today's modern telephone sets, the first feature to suffer is usually "signaling," particularly that area of loop signaling called "supervision." In this case it is a signal sent to the switching equipment requesting "seizure" of a switch circuit and at the same time indicating to callers that the line is busy. "Off hook" is the term more commonly used to describe this signal condition. When a telephone is taken "off hook" (i.e., out of its cradle), the telephone loop is closed and current flows, closing a relay at the switch. If current flow is insufficient, the relay will not close or will close and open intermittently ("chatter") such that line seizure cannot be effected.

Signaling limits are a function of the conductivity of the loop conductor and its diameter or gauge. For this introductory discussion, we can consider that the transmission limits are controlled by the same parameters. Consider a copper conductor. The larger the conductor, the greater the ability to conduct current, and thus the longer the loop conductors may be for signaling purposes. As copper is expensive, we cannot make the conductor as large as we would wish and extend subscriber cable loops over long distances. This is an economic constraint. Before we go further, we describe a method of measuring what a subscriber considers as "hearing sufficiently well."

2.3 The Reference Equivalent

2.3.1 Definition

"Hearing sufficiently well" on a telephone connection is a subjective matter under the blanket heading of "customer satisfaction." Various methods have been derived over the years to rate telephone connections regarding customer (subscriber) satisfaction. Regarding the received telephone signal, subscriber satisfaction is affected by the level (signal power), the signal-to-noise ratio, and the response or attenuation frequency characteristic. A common rating system internationally in use today for grading of customer satisfaction is the *reference equivalent* system. This system considers only the first criterion (viz, level). It must be emphasized that subscriber satisfaction is subjective. To measure satisfaction, the world regulative body for telecommunications, the International Telecommunications Union, devised a system of rating the level sufficient to "satisfy" using the familiar decibel as the unit of measurement. The reference equivalent is broken down into two basic parts. The first is a subjective value in decibel rating of a particular type of subset. The second part is simply the losses (measured at 800 Hz) end-to-end of the intervening network. To determine the reference equivalent of a particular circuit, we add algebraically the decibel value assigned to the subset to the losses of the connecting circuit. Let us look at how the reference-equivalent system was developed, keeping in mind again that it is a subjective measurement dealing with the likes and dislikes of the "average" human being. A standard for reference equivalent was determined in Europe by a team of qualified personnel in a laboratory. A telephone connection, intended to be the most efficient telephone system known, was established in the laboratory. The original reference system or unique master reference consisted of the following:

- A solid-back telephone transmitter.
- Bell telephone receiver.
- Interconnecting these, a "zero-decibel-loss" subscriber loop.
- Connecting the loop, a manual central battery, 22-V dc telephone exchange (switch).

To avoid ambiguity of language, the test team used a test language which consisted of logatoms. A "logatom" is a one-syllable word consisting of a consonant, a vowel, and another consonant.

More accurate measurement methods have been developed since. A more modern reference system is now available at the ITU (International

Telecommunications Union) laboratory in Geneva, Switzerland, called the NOSFER. From this master reference, field test standards are available to telephone companies, administrations, and industry to establish the reference equivalents of telephone subsets in use. These field test sets are calibrated for equivalence with the NOSFER. The NOSFER is made up of a standard telephone transmitter, a receiver, and a network. The reference equivalent of a subscriber's subset, together with the associated subscriber line and feeding bridge, is a quantity obtained by balancing the loudness of receiver speech signals and is expressed relative to the whole or to a corresponding part of the NOSFER (or field) reference system.

2.3.2 Application

Most telephone companies or administrations consider that a standard telephone subset is used. The objective is to measure the capabilities of these subsets regarding loudness. Thus type tests are run on the subsets against calibrated field standards. As mentioned earlier, these may be done on the set alone or on the set plus a fixed length of subscriber loop and feed bridge of known characteristics. The tests are subjective and are carried out in a laboratory. The microphone or transmitter and the earpiece or receiver are each rated separately and are called the *transmit reference equivalent* (TRE) and the *receive reference equivalent* (RRE), respectively. The unit of measurement is the decibel, and negative values indicate that the reference equivalent is better than the laboratory standard.

In telephone systems the *overall reference equivalent* (ORE) is the more common measurement. Simply, this is the sum of the TRE, the RRE, and the losses of the intervening network. Now consider the simplified telephone network shown in the following diagram.

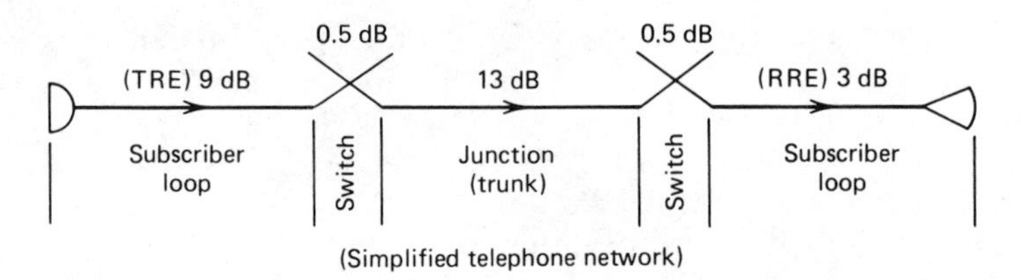

(Simplified telephone network)

The reference equivalent for this circuit is 26 dB, including a 0.5-dB loss for each switch. As defined previously, a junction or trunk is a circuit connecting two switches (exchanges). These may or may not be adjacent. The circuit shown in the preceding diagram may be called a *small transmission plan.* For this discussion, we can define a transmission plan as a method of assigning losses end-to-end on a telephone circuit. Further on in this text we discuss why all telephone circuits, at least all conventional analog circuits, must be lossy. The reference equivalent is a handy device for rating such a plan regarding subscriber satisfaction (see CCITT Recs. P.42 and P.72).

For more in-depth studies of transmission quality (subscriber satisfaction), the reader should review the AEN (articulation reference equivalent) method described in CCITT Rec. P.12, P.12A, and P.13.

When studying transmission plans or developing them, we usually consider that all sections of a circuit are symmetrical. Let us examine the one shown in the preceding diagram. On each end of a circuit we have a subscriber loop. Thus the same loss is assigned to each loop in the plan, which may not be the case at all in real life. From the local exchange to the first long-distance exchange, variously termed *junctions* or *toll-connecting trunks,* a loss is assigned that is identical at each end, and so forth.

To maintain the symmetry regarding reference equivalent of telephone subsets, we use the term $(T + R)/2$. As we see from the preceding diagram, the TRE and the RRE of the subset have different values. We get the $(T + R)/2$ by summing the TRE and RRE and dividing by 2. This is done to arrive at the desired symmetry. Table 2.2 gives reference-equivalent data

Table 2.2 Reference Equivalents for Subscriber Sets in Various Countries

Country	Sending (dB)	Receiving (dB)
With limiting subscriber lines and exchange feeding bridges		
Australia	14.0[a]	6.0[a]
Austria	11.0	2.6
France	11.0	7.0
Norway	12.0	7.0
Germany	11.0	2.0
Hungary	12.0	3.0
Netherlands	17.0	4.0
United Kingdom	12.0	1.0
South Africa	9.0	1.0
Sweden	13.0	5.0
Japan	7.0	1.0
New Zealand	11.0	0.0
Spain	12.0	2.0
Finland	9.5	0.9
With no subscriber lines		
Italy	2.0	−5
Norway	3.0	−3
Sweden	3.0	−3
Japan	2.0	−1
United States (loop length 1000 ft, 83 Ω)	5.0	−1[b]

[a]Minimum acceptable performance.
[b]Freeman [8].
Source: CCITT, *Local Telephone Networks,* ITU, Geneva, July 1968, and *National Telephone Networks for the Automatic Service.*

on a number of standard subscriber subsets used in various parts of the world.

It is stated in CCITT Rec. G.121 that the reference equivalent from the subscriber set to an international connection should not exceed 20.8 dB (TRE) and to the subscriber subset at the other end from the same point of reference (RRE), 12.2 dB (the intervening losses are already included in these figures). By adding 12.2 dB and 20.8 dB, we find 33 dB to be the ORE recommended as a maximum* for an international connection. Table 2.3 should be of interest in this regard. Supplementing the table, it should also be noted that when the overall reference equivalent drops to about 6 dB, subscribers begin to complain that calls are too loud.

The reader may ask why international connections are being discussed in a chapter on local networks. The answer is simple and the concept very important. All calls originate and terminate in a local area. If we are to follow CCITT Rec. G.121, no calls (or very few) should exceed an ORE of 33 dB. In fact, a national transmission plan should reflect the obvious results of Table 2.3 in that we could improve subscriber satisfaction by reducing the ORE from the 33-dB level. And more than half of the 33 dB can be attributed to the local area(s), source, and sink. The limitation, the constraint on local area design, is obvious; the total transmission loss should be under 9 dB (approximately half of 18 dB), with no more than 6 or 7 dB (depending on the transmission plan) for subscriber loop loss. A transmission plan assigns losses to the various segments of a telephone network to meet an ORE goal among other factors.

Consider the following drawing of a simplified subscriber loop.

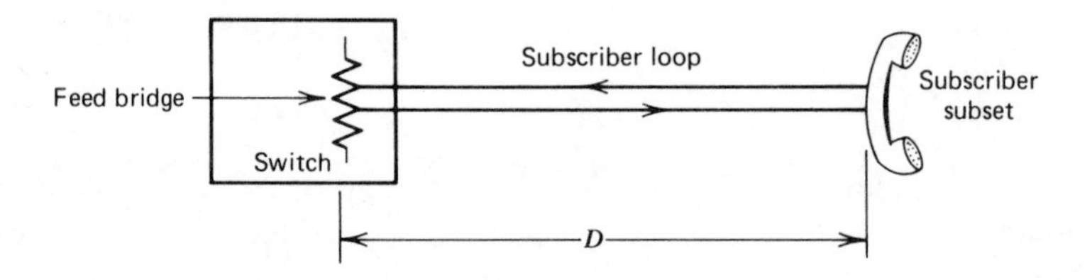

Distance D, the loop length, is most important. We know from this diagram that D must be limited because of attenuation of the voice signal. Likewise, there is a limit to D due to dc resistance, so signaling the local switch can be effected.

The attenuation limit would be taken from the national transmission plan, and for our discussion, 6 dB is assigned as the limit (referenced to 800 Hz). For the loop resistance limit, we must look to the switch. For instance, many conventional crossbar switches will accept up to 1300 Ω.† From this figure we subtract 50 Ω, the nominal resistance for the telephone

*This is the recommendation for 97% of the connections made in a country of average size.

†Many semielectronic switches will accept 1800-Ω loops and, with special line equipment, 2400 Ω.

Table 2.3 British Post Office Survey of Subscribers for Percentage of Unsatisfactory Calls

Overall Reference Equivalent (dB)	Unsatisfactory Calls (%)
40	33.6
36	18.9
32	9.7
28	4.2
24	1.7
20	0.67
16	0.228

subset in series with the loop, leaving a 1250-Ω limit for the wire pair if we disregard the feed bridge resistance. Therefore, in the paragraphs that follow, the figures 6 dB (attenuation limit for loop)* and 1250 Ω (resistance limit) are used.

2.4 Basic Resistance Design

To calculate the dc loop resistance for copper conductors, the following formula is applicable:

$$R_{dc} = \frac{0.1095}{d^2}$$

where R_{dc} is the loop resistance in Ω per mile (statute) and d is the diameter of the conductor (in inches).

If we wish a 10-mi loop and allow 125 Ω per mile of loop (for the stated 1250-Ω limit), what diameter of copper wire would be needed?

$$125 = \frac{0.1095}{d^2}$$

$$d^2 = \frac{0.1095}{125}$$

$$d = 0.03 \text{ in. or } 0.76 \text{ mm (round off to } 0.80 \text{ mm)}$$

Using Table 2.4, we can compute maximum loop lengths for 1250-Ω signaling resistance. As an example, for a 26-gauge loop, we have

$$\frac{1250}{83.5} = 14.97 \text{ or } 14{,}970 \text{ feet}$$

This, then, is the signaling limit, and not the loss (attenuation) limit, or what some call the "transmission limit," referred to in Section 2.5. As the reader

*In the United States this value may be as high as 9 dB.

Table 2.4 Loss and Resistance per 1000 ft of Subscriber Cable[a]

Cable Gauge	Loss/1000 ft (dB)	Ω/1000 ft
26	0.51	83.5
24	0.41	51.9
22	0.32	32.4
19	0.21	16.1

[a]Cable is low-capacitance type (i.e., $\leq$ 0.075 nF/mi).

has certainly inferred by now, resistance design is a method of designing subscriber loops using resistance limits as a basis or limiting parameter.

2.5 Basic Transmission Design

Attenuation or loop loss is the basis of transmission design of subscriber loops. The attenuation of a wire pair varies with frequency, resistance, inductance, capacitance, and leakage conductance. Also, resistance of the line will depend on temperature. For open-wire lines, attenuation may vary by ±12% between winter and summer conditions. For buried cable, which we are more concerned with in this context, variations due to temperature are much less.

Table 2.4 gives losses of some common subscriber cable per 1000 ft. If we are limited to 6-dB (loss) on a subscriber loop, then by simple division we can derive the maximum loop length permissible for transmission design considerations for the wire gauges shown.

$$26 \qquad \frac{6}{0.51} = 11.7 \text{ kft}$$

$$24 \qquad \frac{6}{0.41} = 14.6 \text{ kft}$$

$$22 \qquad \frac{6}{0.32} = 19.0 \text{ kft}$$

$$19 \qquad \frac{6}{0.21} = 28.5 \text{ kft}$$

2.6 Loading

In many situations it is desirable to extend subscriber loop lengths beyond the limits described in Sections 2.2 through 2.5. Common methods to attain longer loops without exceeding loss limits are to increase conductor diameter, use amplifiers and/or loop extenders,* and use inductive loading.

*A loop extender is a device that increases battery voltage on a loop that extends its signaling range. It may also contain an amplifier, thereby extending transmission loss limits as well.

Table 2.5 Code for Load-coil Spacing

Code Letter	Spacing (ft)	Spacing (m)
A	700	213.5
B	3000	915.0
C	929	283.3
D	4500	1372.5
E	5575	1700.4
F	2787	850.0
H	6000	1830.0
X	680	207.4
Y	2130	649.6

Inductive loading tends to reduce transmission loss on subscriber loops and other types of voice pair at the expense of good attenuation-frequency response beyond 3000 Hz. Loading a particular voice-pair loop consists of inserting series inductances (loading coils) into the loop at fixed intervals. Adding load coils tends to decrease the velocity of propagation and increase impedance. Loaded cables are coded according to the spacing of the load coils. The standard code for load coils regarding spacing is shown in Table 2.5.

Loaded cables typically are designated 19-H-44, 24-B-88, and so forth. The first number indicates the wire gauge, with the letter taken from Table 2.5 and indicative of the spacing, and the third item is the inductance of the coil in millihenries (mH). For instance, 19-H-66 is a cable commonly used for long-distance operation in Europe. Thus the cable has 19-gauge voice pairs loaded at 1830-m intervals with coils of 66-mH inductance. The most commonly used spacings are B, D, and H.

Table 2.6 will be useful for calculation of attenuation of loaded loops for a given length. For example, for 19-H-88 (last entry in the table) cable, the attenuation per kilometer is 0.26 dB (0.42 dB/statute mile). Thus for our 6-dB loop loss limit, we have 6/0.26, limiting the loop to 23 km in length (14.3 statute miles). When determining signaling limits in loop design, about 15 Ω per load coil should be added as if the coils were series resistors.

2.7 Summary of Limiting Conditions—Transmission–Signaling

We have been made aware that the size of an exchange serving area is limited by factors of economy involving signaling and transmission. Signaling limitations are a function of the type of exchange and the diameter of the subscriber pairs and their conductivity, whereas transmission is influenced by pair characteristics. Both limiting factors can be extended, but that extension costs money, particularly when there may be many

Table 2.6 Some Properties of Cable Conductors

Diameter (mm)	AWG No.	Mutual Capacitance (nF/km)	Type of Loading	Loop Resistance (Ω/km)	Attenuation at 1000 Hz (dB/km)
0.32	28	40	None	433	2.03
		50	None		2.27
0.40		40	None	277	1.62
		50	H-66		1.42
		50	H-88		1.24
0.405	26	40	None	270	1.61
		50	None		1.79
		40	H-66	273	1.25
		50	H-66		1.39
		40	H-88	274	1.09
		50	H-88		1.21
0.50		40	None	177	1.30
		50	H-66	180	0.92
		50	H-88	181	0.80
0.511	24	40	None	170	1.27
		50	None		1.42
		40	H-66	173	0.79
		50	H-66		0.88
		40	H-88	174	0.69
		50	H-88		0.77
0.60		40	None	123	1.08
		50	None		1.21
		40	H-66	126	0.58
		50	H-88	127	0.56
0.644	22	40	None	107	1.01
		50	None		1.12
		40	H-66	110	0.50
		50	H-66		0.56
		40	H-88	111	0.44
0.70		40	None	90	0.92
		50	H-66		0.48
		40	H-88	94	0.37
0.80		40	None	69	0.81
		50	H-66	72	0.38
		40	H-88	73	0.29
0.90		40	None	55	0.72
0.91	19	40	None	53	0.71
		50	None		0.79
		40	H-44	55	0.31
		50	H-66	56	0.29
		50	H-88	57	0.26

Source: ITT, *Telecommunication Planning Documents—Outside Plant* [7].

thousands of pairs involved. The decision boils down to:

1. If the pairs to be extended are few, they should be extended.
2. If the pairs to be extended are many, it probably is worthwhile to set up a new exchange area.

These economies are linked to the cost of copper. The current tendency is to reduce the wire gauge wherever possible or even resort to the use of aluminum as the pair conductor.

3 OTHER LIMITING CONDITIONS

The size of an exchange area obviously will depend greatly on subscriber (or potential subscriber) density and distribution. The subscriber traffic is another factor to be considered. The CCITT offers the values in Table 2.7 for subscriber line traffic intensity.

Exchange sizes are often in units of 10,000 lines. Although the number of subscribers initially connected should be considerably smaller than when an exchange is installed, 10,000 is the number of subscribers that may be connected when an exchange reaches "exhaust," where it is filled and no more subscribers can be connected.

Ten thousand is not a magic number, but it is a convenient one. It lends itself to crossbar unit size and is a mean unit for subscriber densities in suburban areas and midsized towns in fairly well developed countries. More important, though, is its significance in telephone numbering (the assignment of telephone numbers). Consider a seven-digit number. Now break that down into a three-number group and a four-number group. The first three ciphers, that is the first three dialed, identify the local exchange. The last four identify the individual subscriber and is called the subscriber number. Note the breakdown in the following sample:

For the subscriber number there are 10,000 number combination possibilities, from 0000 to 9999. Of course there are up to 1000 possibilities for the exchange identifier. Numbering is discussed at length in Chapter 3 as a consideration under switching.

The foregoing discussion does not preclude exchanges larger than 10,000 lines. But we still deal in units of 10,000 lines, at least in conventional telephony. The term "wire center" is often used to denote a single location housing one or more 10,000 line exchanges. Some wire centers house up to 100,000 lines with a specific local serving area. Wire centers

Table 2.7 Average Occupation Time during the Busy Hour per Subscriber Line

Subscriber Type	BH Traffic Intensity (erlangs)
Residence	0.01–0.04
Business	0.03–0.06
PABX	0.1 –0.6
Coin box	0.07

with an ultimate capacity of up to 140,000 lines can be economically justified under certain circumstances. Subscriber density, of course, is the key. Nevertheless, many exchanges will have extended loops requiring some sort of special conditioning such as larger-gauge wire pairs, loading, loop extenders, amplifiers, and the application of carrier techniques (Chapter 5). Leaving aside rural areas, 5 to 25% of an exchange's subscriber loops may well require such conditioning or may be called "long loops."

4 SHAPE OF A SERVING AREA

The shape of a serving area has considerable effect on optimum exchange size. If a serving area has sharply angular contours, the exchange size may have to be reduced to avoid excessively long loops (e.g., revert to the use of more exchanges in a given local geographical area of coverage).

There is an optimum trade-off between exchange size, and we mean here the economies of large exchanges (centralization) and the high cost of long subscriber loops. An equation that can assist in determining the trade-off is as follows, which is based on uniform subscriber density, a circular serving area A of radius r such that

$$C = \frac{A}{\pi r^2}(a + b\,d\pi\, r^2) + Ad(f + gL)$$

where C is the total cost of exchanges, which decreases when r increases and to which is added the cost of subscriber loops, which increases with r, and L is the average loop length, which may be related as $L = (2r/3)$, the straight-line distance. To determine the minimum cost of C with respect to r, the equation is differentiated and the result is set equal to zero. Thus

$$\frac{2Aa}{\pi r^3} = \frac{2A\,dg}{3}$$

and this may be fairly well approximated by

$$r = \left(\frac{a}{dg}\right)^{1/3}$$

The cost of exchange equipment is $a + bn$, where n is the number of lines, and the cost of subscriber loops is equal to $(f + gL)$ where, as we stated, L is the average loop length given a uniform density of subscribers, d; and a, b, f, and g are constants.

Since r varies as the cube root, its value does not change greatly for wide ranges of values of d. One flaw is that loops are seldom straight-line distances, and this can be compensated for by increasing g in the ratio (average loop length). This theory is simplified by making exchange areas into circles.

If an entire local area is to be covered, fully circular exchange serving areas are impractical. Either the circles will overlap or uncovered spaces will result, neither of which is desirable. There are then two possibilities: square serving areas or hexagonal serving areas. Of the two, the hexagonal more nearly approaches a circle. The size of the hexagon can vary with density with a goal of 10,000 lines per exchange as the ultimate capacity. Again, a serving area could have a wire center of 100,000 lines or more, particularly in heavily populated metropolitan areas.

Besides the hexagon, full coverage of local areas may only be accomplished using serving areas of equal triangles or squares. This assumes, of course, that the local area was *ideally* divided into identical geometric figures and would only apply under the hypothetical situation of nearly equal telephone density throughout. A typical hexagon subdivision is shown in the diagram that follows.

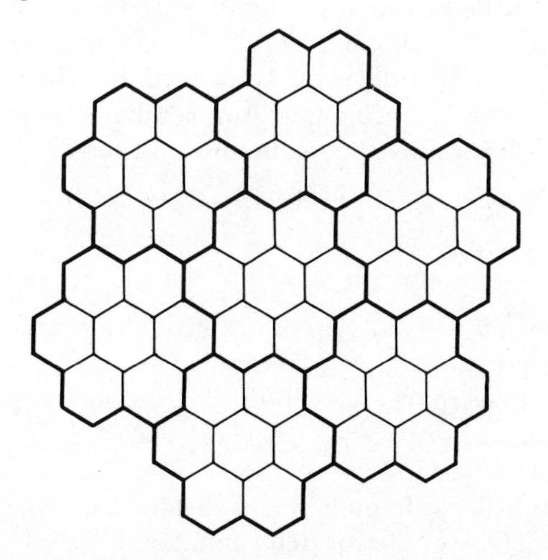

The routing problem then arises. How should the serving areas with their respective local exchanges be interconnected? From our previous discussion we know that two extremes are offered, mesh and star. We are also probably aware that as the number of exchanges involved increases, full mesh becomes very complicated and is not cost-effective. Certainly it is not as cost-effective as a simple hierarchical network of two or three levels permitting high-usage (HU) connections between selected nodes. For instance, given the hexagon formation in the preceding diagram, a full mesh or two-level star network can be derived, as shown in the diagram that follows.

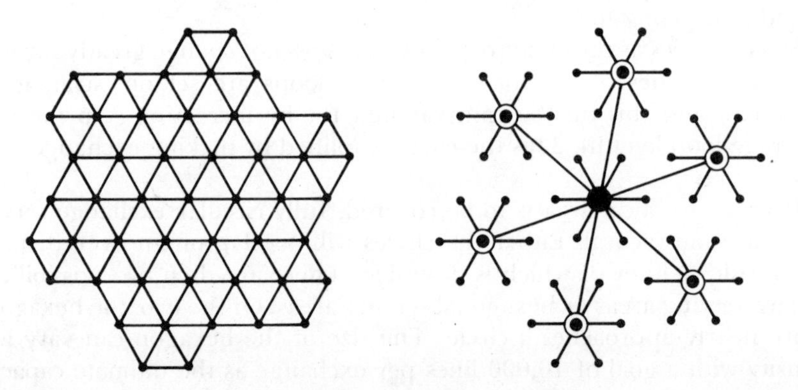

From the routing pattern in the preceding diagram it should be noted that fan outs of 6 and 8 allow symmetry, whereas fan outs of 5 and 7 lead to inequalities. A fan out of 4 is usually too small for economic routing.

It will be appreciated that some of the foregoing assumptions are rarely found in any real telecommunications environment. Uniform telephone density was one unreal assumption, and the implication of uniform traffic flow was another. Serving areas are not uniform geometric figures, exchanges seldom may be placed at serving-area centers, and routing will end up as a mix of star and mesh. Small local areas serving five to 15 exchanges or even more may well be fully mesh connected. Some cities are connected in full mesh with over 50 exchanges. But as telephone growth continues, tandem routing will become the economic alternative.

5 EXCHANGE LOCATION

A fairly simple, straightforward method for determination of the theoretical optimum exchange location is described in Ref. 1, Chapter 6. Basically, the method determines the center of subscriber density much in the same way the center of gravity would be calculated. In fact, other publications call it the center-of-gravity method.

Using a map to scale, a defined area is divided into small squares of 100 to 500 m on a side. One guide for determination of side length would be to use a standard length of the side of a standard city block in the serving area of interest. The next step is to write in the total number of subscribers in each of the blocks. This total is the sum of three figures: (1) existing subscribers, (2) waiting list, and (3) forecast of subscribers for 15 or 20 years into the future. It follows that the squares used for this calculation should coincide with the squares used in the local forecast. The third step is to

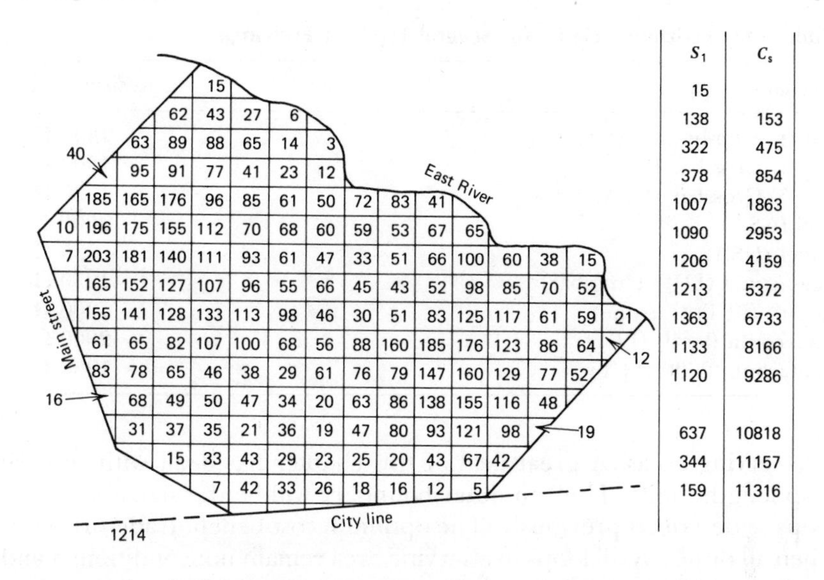

S_2	17	1104	1264	1211	1060	885	615	489	556	722	927	1072	770	399	242	33
C_s		1121	3195	3546	4686	5491	6106	6595	7151	7873	8800	9872	10642	11041	11283	11316

Figure 2.3 Sample wire centering exercise using "center-of-gravity" method.

trace two lines over the subscriber area. One is a horizontal line that has approximately the same number of total subscribers above the line as below. The second is a vertical line where the number of subscribers to the left of the line is the same as that to the right. The point of intersection of these two lines is the theoretical optimum center or exchange location. A sample of this method is shown in Figure 2.3, where S_1 is the sum of subscribers across a single line and C_s is the cumulative sum.

Now that the ideal location is known, where will the real optimum location be? This will depend considerably on secondary parameters such as availability of buildings and land; existing and potential cable or feeder runs; the so-called trunk pull; and layout of streets, roads, and highways. "Trunk pull" refers to the tendency to place a new exchange near the one or several other exchanges with which it will be interconnected by trunks (junctions). Of course, this situation occurs on the fringes of urban areas where a new exchange location will tend to be placed nearer to the more populated area, thereby tending to shorten trunk routes. This is illustrated in Figure 2.1 and discussed in more detail below.

The preceding discussion assumed a bounded exchange area; in other words, the exchange area boundaries were known. Assume now that exchange locations are known and that the boundaries are to be determined. What follows is also valid for redistributing an entire area and cutting it up

Table 2.8 Resistance Limits for Several Types of Exchange

Exchange Type	Resistance limit
No. 1. Step-by-step (USA)	1300 Ω
No. 1. Crossbar (USA)	1300 Ω
No. 5. Crossbar (USA)	1520 Ω
ESS (USA)	2000 Ω
Panel (USA)	785 Ω
Pentaconta (ITT) (crossbar)	1250 Ω
Rotary (ITT) (Europe)	1200 Ω
Metaconta (local) (ITT)	2000 Ω
Pentaconta 2000 (ITT)	1250 Ω

into serving areas. A great deal of this chapter has dealt with subscriber loop length limits. Thus an outer boundary will be the signaling limits of loops as described previously. The optimum cost-benefit trade-off is found when all or nearly all loops in a serving area remain nonconditioned and of small diameter, say, 26 gauge. We note that with H-66 loading the outer boundary would be just under 5 km in this case. It also would be desirable to have a hexagonal area if possible. In practice, however, natural boundaries may well be the most likely real boundaries of a serving area. "Main street," "East river," and "City line" in Figure 2.3 illustrate this point. In fact, these boundaries may set the limits such that they may be considerably greater or less than the maximum signaling (supervisory) limits suggested earlier. Of course, there are two types of serving areas where the argument does not hold. These are rural areas and densely populated urban areas. For the rural areas, we can imagine very large serving areas and for urban areas of dense population, considerably smaller serving areas than those set out with maximum supervisory signaling limits.

To determine boundaries of serving areas when dealing with an exchange that is already installed and a new exchange, we could use the so-called ratio technique. Again, we use signaling (supervisory) limits as the basis. As we are aware, these limits are basically determined by the type of exchange and copper wire gauge utilized for subscriber loops. Table 2.8 gives resistance limits for several of the more common telephone exchanges found in practice.

The ratio method is as follows. Given an existing exchange A and a new exchange B that will be established on a cable route from A, assume that A is a step-by-step exchange, and B a Pentaconta exchange. The distance from A to B along the cable route can be computed by equating distance to the sum of the resistances of exchange A and B. Use 26-gauge wire in this case and take the resistances from Table 2.8. From Table 2.6 using H-66 loading, resistance can be equated to 273 Ω/km. Sum the resistance limits of exchanges $(A + B)$:

$$A + B = 1250 + 1300 = 2550\ \Omega$$

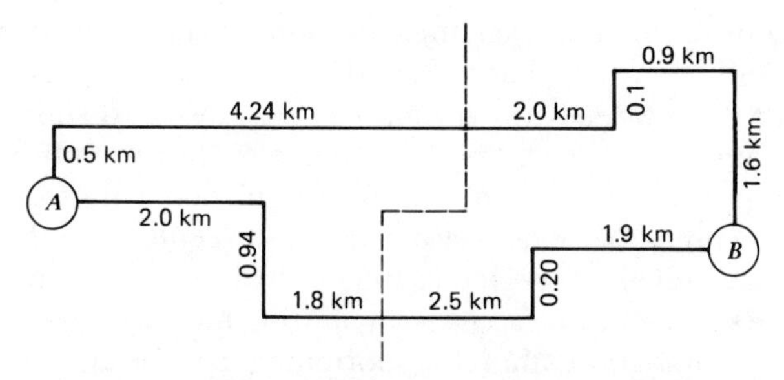

Figure 2.4 Determining serving area boundaries with ratio method.

Then divide by 273:

$$2550/273 \simeq 9.34 \text{ km (distance } A \text{ to } B)$$

With no other factors influencing the decision, the boundary would be established along the feeder (cable) route (distance from A):

$$D_A = 1300 \times 9.34/(1250 + 1300)$$

or $\simeq$ 4.74 km from A. This exercise is shown diagrammatically in Figure 2.4.

Continue the exercise and examine the exchange serving area at A. Assume the area to be a square with A at its center. Allow for *non*-crow-fly feeder routes, we can assume the square has 8 km on a side as shown in the following diagram.

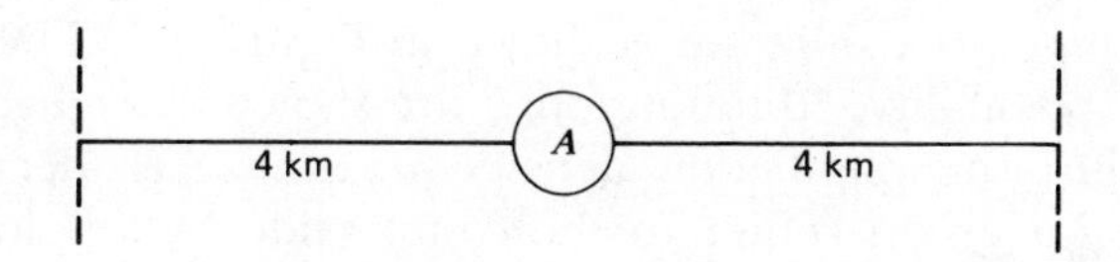

or 64 km². If the largest exchange wanted were to have 100,000 subscribers and the smallest, 10,000 at the end of the forecast period (15 to 20 years), then subscriber density would be 10,000/64 to 100,000/64 or 156 subscribers per km² and 1560 subscribers per km².

When serving areas exceed 100,000 lines at the end of a forecast period in densely populated urban areas, breaking up of the areas into smaller ones with exchanges of smaller capacity should be considered. One problem with large conventional exchanges is size; another is cable entry and mainframe size. The latter is more evident with new SPC [electronic switching system (ESS)] exchanges in that cable entry and mainframe is much larger than the exchange itself. However, the really major problem is reliability or survivability. If a very large exchange is knocked out as a result of fire, explosion, sabotage, or other disaster, it is much more catastrophic than the loss of a smaller single exchange.

Another consideration in exchange location is what is called "trunk pull" discussed briefly above. This is a secondary factor in exchange placement and refers to the tendency, in certain circumstances, to shift a proposed exchange location to shorten trunks (junctions). Trunk pull (see Fig. 2.1) becomes a significant factor only where population fringe areas around urban centers and trunks extend toward the city center and the exchanges in question may well be connected in full mesh in their respective local area. Some possible saving may be accrued by shifting the exchange more toward the center of population, thereby shortening trunks at the expense of lengthening subscriber loops in the direction of more sparse population, even with the implication of conditioning on well over 10% of the loops. The best way to determine if a shift is worthwhile is to carry out an economic study comparing costs in PWAC (present worth of annual charges) [18, 19], comparing the proposed exchange as located by the center-of-gravity method to the cost of the shifted exchange. Basic factors in the comparison are in the cost of subscriber plant, trunk plant, plant construction, and cost of land (or buildings). The various methods of handling the sparsely populated sections of the fringe serving area in question should also be considered. For instance, we might find that in those areas, much like rural areas, there may be a tendency for people to bunch up in little villages or other small centers of population. This situation is particularly true in Europe. In this case the system designer may find considerable savings in telephone-plant costs by resorting to the use of satellite exchanges or concentrators. Trunk connections between satellite or concentrator and main exchange may be made by using carrier techniques. Concentrator and satellite exchanges are discussed in Chapter 3 and carrier techniques, in Chapter 5.

A typical fringe-area situation is shown in Figure 2.5. The "bunching" and the other possibility, "thinning out," are shown in Figure 2.5. "Thinning out" is just the population density per unit area decreases as we proceed from an urban center to the countryside. A topological line of population density of 10 inhabitants per square kilometer (26 inhabitants per square mile) is a fair guideline for separation of the rural part of the fringe area from the urban–suburban part. Of course, in the latter part we would have to resort to widespread conditioning, to the use of concentrator–satellites, or to subscriber carrier or subscriber pulse-code modulation (PCM) (Chapter 9).

6 UNIGAUGE DESIGN

The administration of a serving area is often quite costly. For one thing, administration costs can be reduced by ensuring that all subscribers have the same type of telephone set. Cable-pair assignment is another factor. Following resistance design or transmission design approaches to sub-

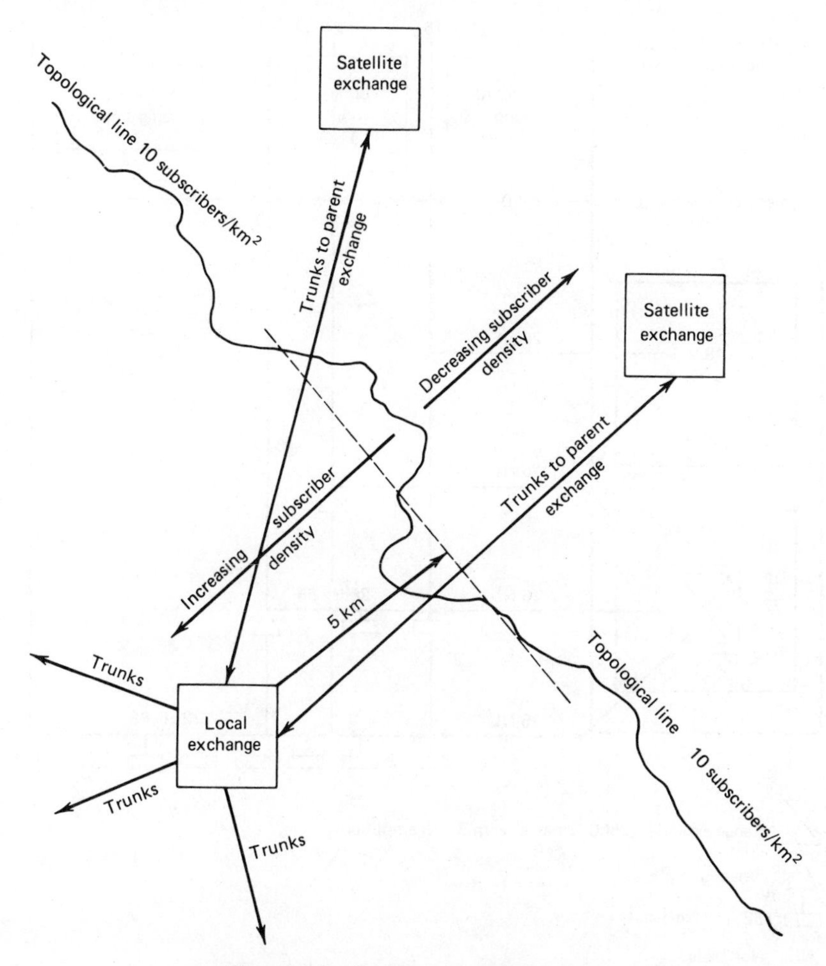

Figure 2.5 Fringe-area considerations.

scriber plant engineering will bring about a "mix" of various gauges of wire in loops. We also want to keep wire gauge as small as possible to minimize the investment in copper. If loading and other conditioning are to be used, "systemization" of their use will also help plant administration. "Unigauge design" is one approach to this systemization, and it can be applied in the urban, suburban and fringe areas discussed earlier, as well as in many rural applications. Loops up to 52 kft (17.3 km) in length can be handled.

A typical layout of a subscriber plant based on unigauge design is shown in Figure 2.6. The example is taken from the Bell System (USA) [3]. It can be seen in Figure 2.6 that subscribers within 15,000 ft (5000 m) of the switch are connected over loops made up of 26-gauge nonloaded cable fed with the standard −48-V battery. Loop connection at the switch is conventional. Unigauge was developed in the United States, where 80% of all

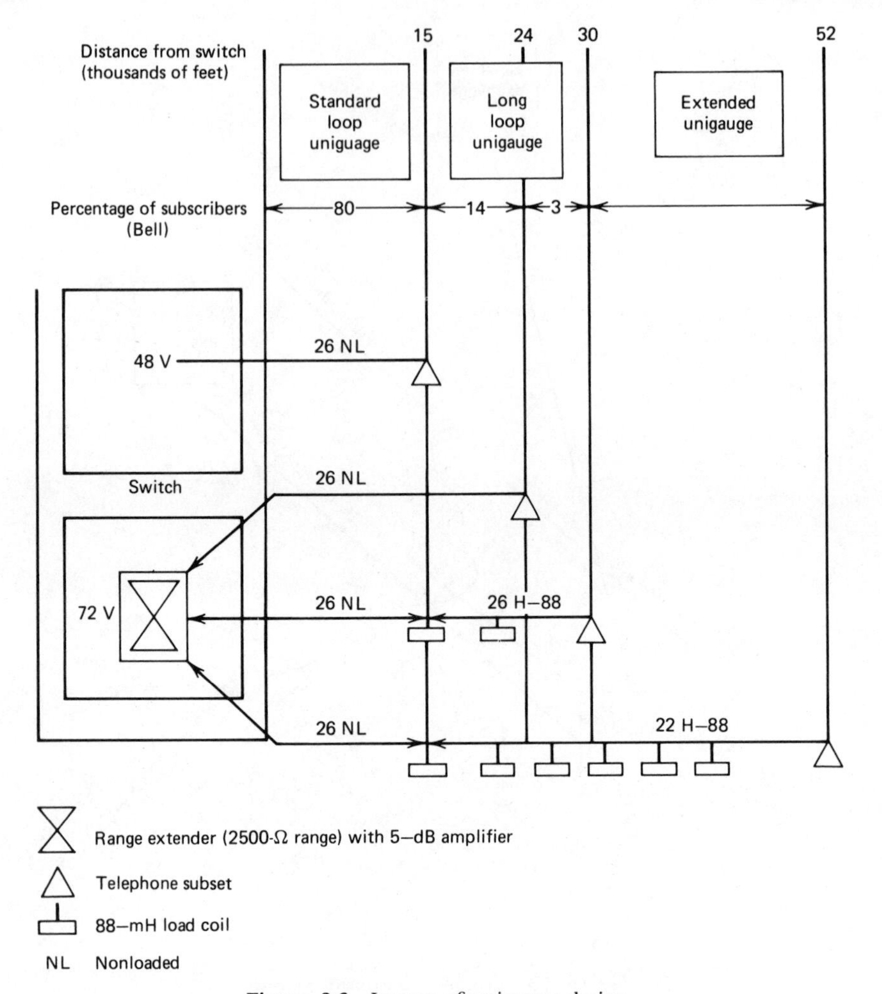

Figure 2.6 Layout of unigauge design.

subscribers are within this distance of their serving exchange. Loops 15,000 to 30,000 ft long (5000 to 10,000 m) are called *unigauge loops.* Subscribers in the range of 15,000 to 24,000 ft (5000 to 8000 m) from the switch are connected by 26-gauge, nonloaded cable as well but require a range extender to provide sufficient voltage for supervision and signaling. Figure 2.6 shows a 72-V range extender equipped with an amplifier that gives a midband gain of 5 dB. The output of the amplifier "emphasizes" the higher frequencies. This offsets that additional loss suffered at the higher frequencies of the voice channel on the long, nonloaded loops. To extend the loops to 30,000 ft (10,000 m), 88 mH loading coils are added at the 15,000- and 21,000-ft points (5000- and 7000-m points).

For long loops, more than 15,000 ft (5000 m) the range extender–

amplifier combinations are not connected on a line-for-line basis. It is standard practice to equip four or five subscriber loops with only one range extender–amplifier, which is used on a shared basis. When the subscriber goes "off hook" on a long, unigauge loop, a line is seized and the range extender is switched in. This concentration is another point in favor of unigauge because of the economics involved. It should also be noted that the switch is facing into long (15,000-ft) nonloaded sections providing a fairly uniform impedance for all conditions when an amplifier is switched in. This is an important factor with regard to stability, as we discuss later.

Loops measuring more than 30,000 ft (10,000 m) long may also use the unigauge principle and are often referred to as "extended unigauge." Such a loop is equipped with 26-gauge nonloaded cable from the switch out to the 15,000-foot point. Beyond 15,000 ft (5000 m), 22-gauge cable is used with H-88 loading. As with all loops measuring more than 15,000 ft in length following the unigauge principle, a range extender–amplifier is switched in when the loop is in use. The loop length for this combination is up to 52,000 ft. Loops longer than 52,000 ft (17.3 km) may also be installed by using a gauge with a diameter larger than 22.

Another possibility is to replace switch line relays with ones that are sensitive to up to 2500 Ω of loop resistance. Such a modification is done on long loops only. The 72 V supplied by the range extender is for pulsing. Ringing voltage (to ring the distant telephone) is superimposed on the line only when the subscriber's subset is in the "on-hook" condition. Besides the notable savings in the expenditure on copper, unigauge displays some small improvements in transmission characteristics over older design methods of subscriber loops:

- Unigauge has a slightly lower average loss, when we look at a statistical distribution of subscribers.
- There is 15-dB average return loss* on the switch side of an amplifier, compared with an average of 11 dB for older design methods.

7 OTHER LOOP DESIGN TECHNIQUES

Fine-gauge and "minigauge" techniques essentially are refinements of the unigauge concept. In each case the principal object is to reduce the amount of copper in the subscriber plant. Obviously, one method is to use still smaller gauge pairs on shorter loops. For instance, the use of gauges as small as 32 is being considered. Another approach is to use aluminum as the conductor. When aluminum is used, a handy rule of thumb to follow is that the ohmic and attenuation losses of aluminum may be equated to copper in that aluminum wire should always be the next "standard gauge"

*Return loss is discussed in Chapter 5.

larger than its copper counterpart if copper were to be used. Some of the more common gauges are compared in the table that follows.

Copper	Aluminum
19	17
22	20
24	22
26	24

Aluminum has some drawbacks as well. The major ones are summed up as follows:

- It should not be used on the first 500 yd (m) of cable where the cable has a larger diameter (and here we mean more loops before branching).
- It is more difficult to splice than copper.
- It is more brittle.
- Because the equivalent conductor is larger than its copper counterpart, an equivalent aluminum cable with the same conductivity–loss characteristics will have a smaller pair count in the same sheath.

8 DESIGN OF LOCAL AREA TRUNKS (JUNCTIONS)

Exchanges in a common local area are interconnected by trunks, called junctions in the United Kingdom. Depending on length and certain other economic factors, these trunks use voice frequency (VF) transmission over wire pairs formed up in cables. In view of the relatively small number of such trunk circuits in comparison to the total number of subscriber lines* in the area, it is generally economical to minimize attenuation in this portion of the network.

One approach used by some telephone companies or administrations is to allot $\frac{1}{3}$ of the total end-to-end reference equivalent to each subscriber's loop and $\frac{1}{3}$ to the trunk network. Figure 2.7 illustrates this concept. For instance, if the transmission plan called for a 24-dB ORE, then $\frac{1}{3}$ of 24 dB, or 8 dB, would be assigned to the trunk plant. Of this we may assign 4 dB to the four-wire portion of the long distance (toll) network, leaving 4 dB for local VF trunks or 2 dB at each end. The example has been highly simplified, of course. For the toll-connecting trunks (e.g., those trunks connecting the local network to the toll network), if a good return loss cannot be maintained on all or nearly all connections, the losses on two-wire

*Because of the inherent concentration in local switches, approximately one trunk is allotted from eight to 25 subscribers, depending on design.

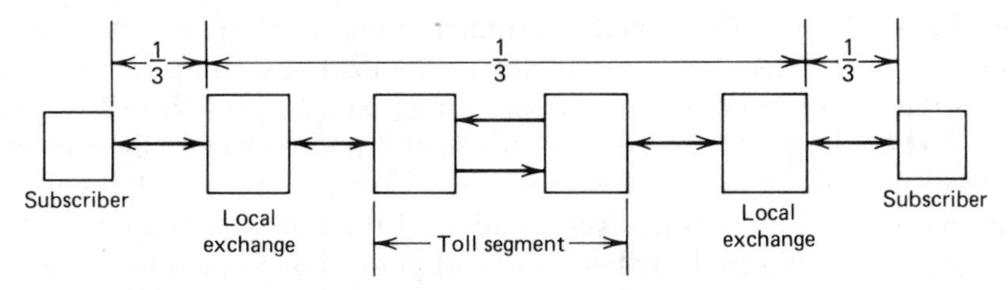

Figure 2.7 One-thumb rule for network loss assignment.

toll-connecting trunks may have to be increased to reduce possibilities of echo and singing. Sometimes the range of loss for these two-wire circuits must be extended to 5 or 6 dB. It is just these circuits into which the four-wire toll network looks directly. Two-wire and four-wire circuits, echo and singing, are discussed in Chapter 5. Thus it can be seen that the approach to the design of VF trunks varies considerably from that used for subscriber loop design. Although we must ensure that signaling limits are not exceeded, the transmission limits will almost always be exceeded well before the signaling limit. The tendency to use larger-diameter cable on long routes is also evident.

One major difference from the subscriber-loop approach is in the loading. If loading is to be used, the first load coil is installed at distance $D/2$, where D is the normal separation distance between load points. Take the case of H loading, for instance. The distance between load points is 1830 m (Table 2.5), but the first load coil from the exchange is placed at $D/2$, or 915 m from the exchange. Then, if an exchange is bypassed, a full load section exists. This concept is illustrated in Figure 2.8.

Now consider this example. A loaded 500-pair VF trunk cable extends across town. A new switching center is to be installed along the route where 50 pairs are to be dropped and 50 pairs inserted. It would be desirable to establish the new switch midway between load points. At the switch 450 circuits will bypass the new switching center. Using this $D/2$ technique, these circuits will need no conditioning; they will be fully loaded sections (i.e., $D/2 + D/2 = 1D$, a fully loaded section). Meanwhile, the 50 circuits

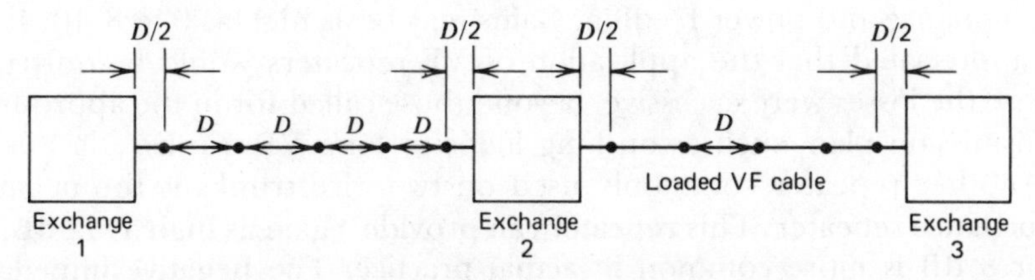

Figure 2.8 Loading of VF trunks (junctions).

entering from each direction are terminated for switching and need conditioning, so each electrically resembles a fully loaded section. However, the physical distance from the switch out to the first load point is $D/2$ or, in the case of H loading, 915 m. To make the load distance electrically equivalent to 1830 m, line build-out (LBO) is used. This is done simply by adding capacity to the line. Suppose the location of the new switching center was such that it was not midway between load points but some other fractional distance. For the section consisting the shorter distance, LBO is used. For the other, longer run, often a half load coil is installed at the switching center, and LBO is added to trim up the remaining electrical distance.

By this time the reader has gathered what conditioning is in this context. For loaded trunks and loops, conditioning is the action taken to change the electrical characteristics of the line by adding capacitance or inductance. It may be done at line midpoint where there is access such as at a load-coil location. However, it is more commonly done at the switching center because of accessibility of cable pairs at the mainframe associated with the center.

9 VOICE-FREQUENCY

In telephone terminology, *voice-frequency repeaters* imply the use of *unidirectional* amplifiers at voice frequency on VF trunks. On a two-wire trunk we must resort to four-wire transmission techniques at the repeater. Two-wire and four-wire transmission are discussed in Chapter 5. Thus on a two-wire trunk two amplifiers must be used on each pair with a hybrid in and a hybrid out. The hybrid converts the two-wire circuit to a four-wire circuit, and vice-versa. A simplified block diagram of a VF repeater is shown in Figure 2.9. The gain of a VF repeater can be run up as high as 20 or 25 dB, and originally they were used on 50-mi, 19-gauge loaded cable in the long-distance (toll) plant. Today they are seldom found on long-distance circuits but do have application on local trunk circuits, where the gain requirements are considerably less. Trunks using VF repeaters have the repeater's gain adjusted to the equivalent loss of the circuit minus the 4-dB loss to provide the necessary singing margin (see Chapter 5). In practice, a repeater is installed at each end of the trunk circuit to simplify maintenance and power feeding. Gains may be as high as 6 to 8 dB. It can be appreciated that the application of VF repeaters would be on trunks where the losses were excessive, beyond those called for in the appropriate transmission plan, such as on long loops or long VF trunks.

Another repeater commonly used on two-wire trunks is the negative-impedance repeater. This repeater can provide a gain as high as 12 dB, but 7 or 8 dB is more common in actual practice. The negative impedance repeater requires an LBO at each port and is a true two-way, two-wire repeater. The repeater action is based on regenerative feedback of two amplifiers. The advantage of negative impedance repeaters is that they are

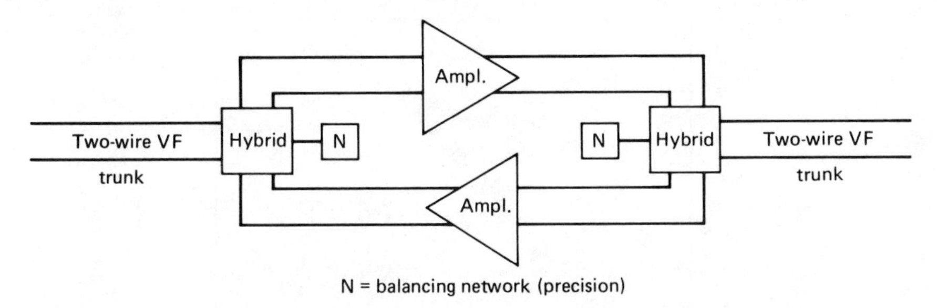

Figure 2.9 Simplified block diagram of VF repeater.

transparent to dc signaling. On the other hand, VF repeaters require a composite arrangement to pass the dc signaling. This consists of a transformer bypass.

10 TANDEM ROUTING

The local-area trunking scheme evolutionally has been mesh connection of exchanges, and in many areas of the world it remains full mesh. We said initially that mesh connection is desirable and viable for heavy traffic flows. As traffic flows reduce, going from one situation to another, the use of tandem routing in the local area becomes an interesting, economical alternative.

Further, it can be shown that a local trunk network can be optimized, under certain circumstances, with a mix of tandem, high-usage (overflow to alternative routes), and direct connection (mesh). We often refer to these three possibilities as THD. The system designer wishes to determine, on a particular trunk circuit, if it should be "Tandem," "High-usage," or "Direct" (T, H, or D). This determination is based on incremental cost of the trunk (junction), making the total network costs as low as possible. Such incremental cost can be stated:

$$B = c + (bl)$$

where c is the cost of switching equipment per circuit, b is the incremental costs for trunks per mile or kilometer, and l is the length of the trunk (or junction) circuit.

To carry out a THD decision for a particular trunk route, the input data required are the *offered* traffic between the local exchanges in question and the grade of service. We can now say that T, H, or D decision is to a greater extent determined by the offered traffic A between the exchanges and the cost ratio ϵ where

$$\epsilon = \frac{B}{B_1 + B_2}$$

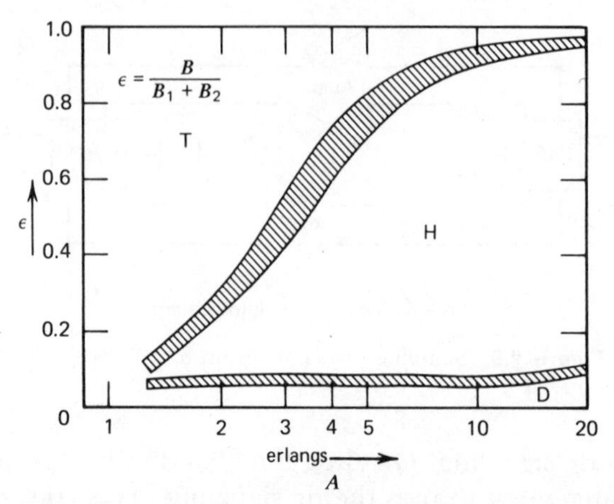

Figure 2.10 THD diagram (Courtesy of the International Telecommunication Union–CCITT).

where B is the cost for the direct route, with B_1 from exchange 1 to the proposed tandem and B_2 from the proposed tandem exchange to exchange 2. for incremental costs between direct and tandem routing. Of course, before starting such an exercise, the provisional tandem points must be known. Figure 2.10, which was taken from Ref. 4, may be helpful as a decision guide.

To approximate the number of high-usage (HU) trunk circuits required, the following formula may be used:

$$F(n,A) = AE(n,A) - E(n+1, A) = \epsilon\psi$$

where E = grade of service
n = number of high-usage circuits
A = offered traffic between exchanges
$F(n,A)$ = "improvement" function (i.e., increase in traffic carried on high-usage trunk group on increase of number of these trunk circuits from n to $n+1$)
ϵ = cost ratio (as previously)
ψ = efficiency of incremental trunks (marginal utilization (of magnitude 0.6 to 0.8).

However, a still better approximation will result if the following formula is used:

$$F(n,A) = \epsilon[1 - 0.3(1 - \epsilon^2)]$$

The following exercise emphasizes several practical points on the application of tandem routing in the local area. Assume that there are 120,000 subscribers in a certain local area. If we allow a calling rate of 0.05 erlangs

Table 2.9 Traffic Table—Full-Mesh Connection (120,000 Subscribers)

Number of Exchanges	Average Size of Exchange	Originated Traffic per Exchange (erlangs)	Average Traffic to Each Distant Exchange (erlangs)
10	12,000	600	60
12	10,000	500	42
15	8000	400	27
20	6000	300	15
30	4000	200	6.7
40	3000	150	3.8

per subscriber (see Table 2.7) and assume full mesh connection, we can assemble Table 2.9. On the basis of Table 2.9 it would not be reasonable to expect any appreciable local area trunk economy through tandem working with less than 30 to 40 exchanges in the area. Of course, Table 2.9 assumes unity of community of interest with the more distant exchanges so that a few attractive tandem routings might exist in the 20-to-30-exchange range. Areas smaller than those in our sample will naturally reduce the number of exchanges at which tandem working is viable, and a larger area and higher calling rate will increase this number. The results of any study to determine feasibility of using tandem exchanges is very sensitive to the number of exchanges in the local area and considerably less sensitive to the size of the area and calling rate. Hence we can see that the economy of tandem routing is least in areas of dense subscriber population, where exchanges can be placed without exceeding subscriber loop length limits and where the trunks will be short. Relatively sparsely populated areas are more favorable and are likely to show relatively low community-of-interest factors between distant points in the area.

11 DIMENSIONING OF TRUNKS

A primary effort for the systems engineer in the design of the local trunk network is the dimensioning the trunks of that network. Here we simply wish to establish the economic optimum number of trunk circuits between exchanges X and Y. If we are given the traffic (in erlangs) between the two exchanges and the grade of service, we can assign the number of trunk circuits between X and Y. As discussed in Chapter 1, it is assumed that all traffic values are BH values. Once given this input information, the erlang

Table 2.10 Sample Traffic Matrix[a]

	Traffic (ERLANGS)								
From/to	*A*	*B*	*C*	*D*	*E*	Toll	Total Orig.	Lines Working	Traffic per line
A	21	20	65	2.5	2.5	1.5	54	9200	0.059
B	22	80	13	6.0	5.0	4.0	130	26000	0.050
C	5	11	7	2.5	2.5	1.0	29	7500	0.039
D	2	7	1	0.3	0.2	0.5	11	3000	0.035
E	2	5	2	0.2	0.3	0.5	10	2800	0.035
Toll	2	3	1	0.5	0.5	—	7	—	—
Total	54	126	30.5	12	11	7.5	241	48500	0.050
Total–line	0.059	0.048	0.041	0.040	0.037	0.0015	—	—	—

[a]For 8-year forecast period.

value, the number of trunk circuits can be derived from Table 1.1 (no overflow assumed).

It can be appreciated by the reader that the trunk intensities used in the design of trunk routes have allowed for growth, that is, the increase of traffic from the present with the passage of time. This increase is attributable to several factors: (1) the increase in the number of telephones in the area that generate traffic, (2) the probable increase in telephone usage, and (3) the possibility of a change in character of the area in question, such as from rural to suburban, or from residential to commercial. Thus the designer must use properly forecast future traffic values. These values in practice are for the forecast period 5 to 8 years in the future. The art of arriving at these figures is called *forecasting*. Present traffic values should always be available as a base or point of departure.

Suppose we use a sample local area with five exchanges: A, B, C, D, and E. For an 8-year forecast period we could possibly come up with a traffic matrix as that shown in Table 2.10.

Thus from the traffic matrix in the direction exchange C to exchange B there is a traffic intensity of 11 erlangs. Applying the 11 erlangs to Table 1.1, we see that 23 circuits would be required. For the distance B to C, the traffic intensity is 13 erlangs, and again from Table 1.1, 26 circuits would be required. These circuit figures suppose a grade of service of $p = .001$. For a grade of service of $p = .01$, 19 and 22 circuits, respectively, would be required.

The preceding discussion of routing implied an on-paper routing that would probably vary in the practice of actual cable lays and facility drawings to something quite different. In very simplified terms, these differences are shown in Figure 2.11.

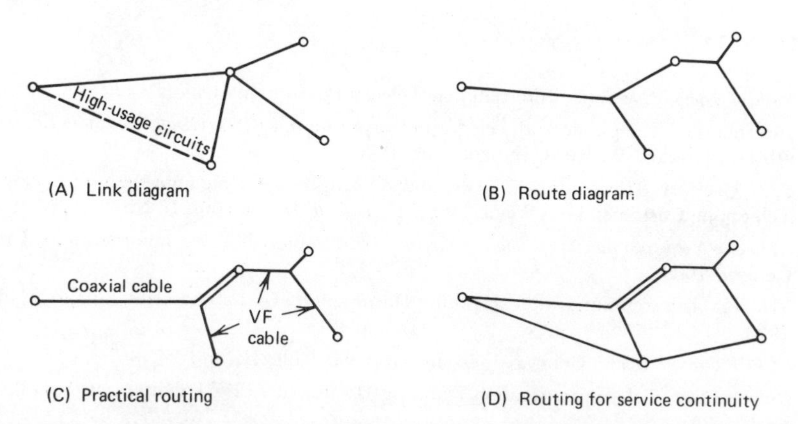

Figure 2.11 Routing diagrams and practical routing (HU = high usage): (A) link diagram; (B) route diagram; (C) practical routing; (D) routing for service continuity.

12 COMMUNITY OF INTEREST

We have referred to the community-of-interest concept in passing. This is a method used as an aid to estimate calling rate and traffic distribution for a new exchange and its connecting trunks. The community-of-interest factor K can be defined as follows:

$$\text{Traffic } A\text{–}B = \text{Kx} \frac{\text{Traffic originating at } A \times \text{traffic originating at } B}{\text{Total originating traffic in area}}$$

If all subscribers are equally likely to call all others, the proportion results:

$$\frac{\text{Traffic originating at } B}{\text{Traffic originating in area}} = \frac{\text{Traffic terminating at } B}{\text{Traffic terminating in area}}$$

This is the expected proportion of all A traffic that is directed to B. In this case $K = 1$, corresponding to equal community of interest between all subscribers. The condition of $K > 1$ or $K < 1$ indicates a greater or lesser interest than average between exchanges A and B. The factor K is affected by the type of area, whether residential or business, as well as the distances between exchanges. For instance, in metropolitan areas K may range from 4 for calls originating and terminating on the same exchange to 0.25 for calls on opposite sides of the city or metropolitan area.

The K factor is a useful reference for the installation of a new exchange where the serving area of the new exchange will be cut out from serving areas of other exchanges. The community-of-interest factors may then be taken from traffic data from the old exchanges, and the values averaged and then applied to the new exchange. The same principle may be followed for exchange extensions in a multiexchange area.

REFERENCES

1. *Local Telephone Networks,* International Telecommunications Union, Geneva, 1968.
2. International Telephone and Telegraph Corporation, *Reference Data for Radio Engineers,* 6th ed., Howard W. Sams, Indianapolis, 1976.
3. P. A. Gresh, L. Howson, A. F. Lowe, and A. Zarouni, "A Unigauge Design Concept for Telephone Customer Loop Plant," *IEEE Com. Tech. J.,* vol. com. 16 No. 2 (April 1968).
4. *National Networks for the Automatic Service,* International Telecommunications Union, Geneva, 1968.
5. *Networks,* Telecommunications Planning Documents, ITT Laboratories, (Spain), Madrid, 1973.
6. *CCITT Orange Books,* Geneva, 1976, in particular Vols. III and VI.
7. *Outside Plant,* Telecommunication Planning Documents, ITT Laboratories (Spain), Madrid, 1973.
8. Roger L. Freeman, *Telecommunication Transmission Handbook,* Wiley, New York, 1975.
9. Y. Rapp, "Algunos Puntos de Vista Económicos para el Planeamiento a Largo Plazo de la Red Telefónica," L. M. Ericsson Stockholm, 1964.

10. *Placement of Exchanges in Urban Areas—Computer Program,* ITT Laboratories (Spain), Madrid, 1974.

11. J. C. Emerson, *Local Area Planning,* Telecommunications Planning Symposium [ITT Laboratories (Spain)], Boksburg, South Africa, 1972.

12. L. Alvarez Mazo and P. H. Williams, *Influence of Different Factors on the Optimum Size of Local Exchanges,* [ITT Laboratories (Spain)], Boksburg, South Africa, 1972.

13. J. C. Emerson, *Factors Affecting the Use of Tandem Exchanges in the Local Area,* [ITT Laboratories (Spain)], Boksburg, South Africa, 1972.

14. IEEE ComSoc, *The International Symposium on Subscriber Loops and Services,* Atlanta, Ga., 1977.

15. IEEE ComSoc, *Second International Symposium of Subscriber Loops and Services,* London, 1976.

16. J. E. Flood, *Telecommunications Networks,* IEE Telecommunications Series 1, Peter Peregrinus, London, 1975.

17. Y. Rapp, "Planning of Exchange Locations and Boundaries in Multi-exchange Networks," Ericsson Tech., 1962, Vol. 18, p. 94.

18. "Telecommunications Planning," ITT Laboratories (Spain), Madrid, 1974.

19. O. Smidt, "Engineering Economics," Telephony Publishing Co., Chicago, 1970.

Chapter 3

CONVENTIONAL SWITCHING TECHNIQUES IN TELEPHONY

1 SWITCHING IN THE TELEPHONE NETWORK

A network of telephones consists of pathways connecting switching nodes so that each telephone in the network may connect with any other telephone for which the network provides service. Today there are hundreds of millions of telephones in the world, and nearly each and everyone can communicate with any other one. Chapter 1 discussed the two basic technologies in the engineering of a telephone network, transmission and switching. Transmission allows any two subscribers in the network to be heard satisfactorily. Switching permits the network to be built economically by concentration of transmission facilities. These facilities are the pathways connecting the switching nodes.

Switching is a complex subject. To do justice to it, several volumes could be written. The intention of this chapter is to give an overview and appreciation of telephone switching. Such important factors as functional description, desirable features, trends in technology, and operational requirements are discussed.

Switching establishes a path between two specified terminals, which we call *subscribers* in *telephony*. The term *subscriber* implies a public telephone network. There is, however, no reason why these same system criteria cannot be used on private or quasipublic networks. Likewise, there is no reason why that network cannot be used to carry information other than speech telephony. In fact, in later chapters these "other" applications are discussed, as well as modifications in design and features specific for special needs.

A switch sets up a communication path on demand and takes it down when the path is no longer needed. It performs logical operations to establish the path and automatically charges the subscriber for usage. A commercial switching system satisfies, in broad terms, the following user requirements:

1. Each user has need for the capability of communicating with any other user.

2. The speed of connection is not critical, but the connection time should be relatively small compared to holding time or conversation time.
3. The grade of service, or the probability of completion of a call, is also not critical but should be high. Minimum acceptable percentage of completed calls during the BH may average as low as 95%, although the general grade of service goal for the system should be 99%* (equivalent to $p = .01$).
4. The user expects and assumes conversation privacy but usually does not specifically request it, nor, except in special cases, can it be guaranteed.
5. The primary mode of communication for most users will be voice (or the voice channel).
6. The system must be available to the user at any time he may wish to use it.

2 NUMBERING—ONE BASIS OF SWITCHING

A telephone subscriber looking into a telephone network sees before him a repeatedly branching tree of links. At each branch point there are multiple choices. Assume that our calling subscriber wishes to contact one particular distant subscriber. To reach that subscriber, a connection is built up utilizing one choice at each branch point. Of course, some choices lead toward the desired end point, and others lead away from it. Alternative paths are also presented. A call is directed through this maze, which we call a *telephone network*, by a telephone *number*. It is this *number* that activates the switch or switches at the "maze" branch points.

Actually, a telephone number performs two important operations: (1) it routes the call; and (2) it activates the necessary apparatuses for proper call charging. Each telephone subscriber is assigned a distinct number, which is cross-referenced in the telephone directory with his (or her) name and address and in his local serving exchange with a distinct subscriber line.

If a subscriber wishes to make a telephone call, he lifts his receiver "off hook" and awaits a dial tone that indicates readiness of his serving switch to receive instructions. These "instructions" are the number that the subscriber dials (or the buttons that he punches) giving the switch certain information necessary to route and charge this subscriber's call to the distant subscriber with whom he wishes to communicate.

A subscriber number is the number to be dialed or called to reach a subscriber in the same local (serving) area. Remember that our definition of a local "serving area" is the area served by a single switch (exchange).

*See CCITT Rec. Q.95; $p = .01$ per link on an international connection.

If we had a switch with a capacity of

100 lines, it could serve up to
100 subscribers and we could
assign telephone number
00 through 99.

If we had a switch with a capacity of

1000 lines, it could serve up to
1000 subscribers and we could
assign telephone numbers
000 through 999.

If we had a switch with a capacity of

10,000 lines, it could serve up to
10,000 subscribers and we could
assign telephone numbers
0000 through 9999.

Thus the critical points occur where the number of subscribers reaches numbers such as 100, 1000, and 10,000.

In most present switching systems there is a top limit to the number of subscribers that may be served by one switching unit. Increase beyond this number is either impossible or uneconomical. A given switch unit is usually most economical when operating with the number of subscribers near the maximum of its design. However, it is necessary for practical purposes to hold some spare capacity in reserve. As we proceed in the discussion of switching, we consider exchanges with seven-digit subscriber numbers such as:

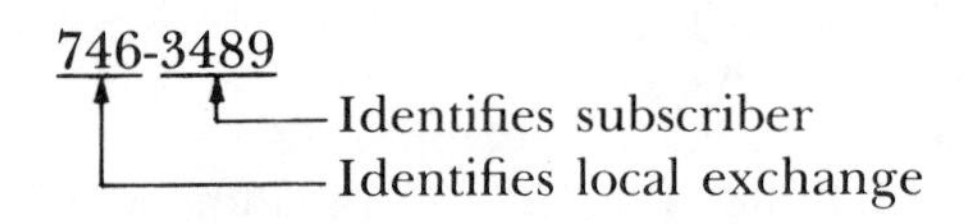

The subscriber is identified by the last four digits permitting up to 10,000 subscribers, 0000 through 9999, allowing for no blocked numbers such as:

746-0000

The calling area has a capacity of 999 exchanges, again allowing for no blocked numbers such as:

000
911 (emergency number—USA)

Section 16 of this chapter presents a more detailed discussion of numbering.

3 CONCENTRATION

One key to switching and network design is concentration. A local switching exchange concentrates traffic. This concept is often depicted as shown in the following diagram.

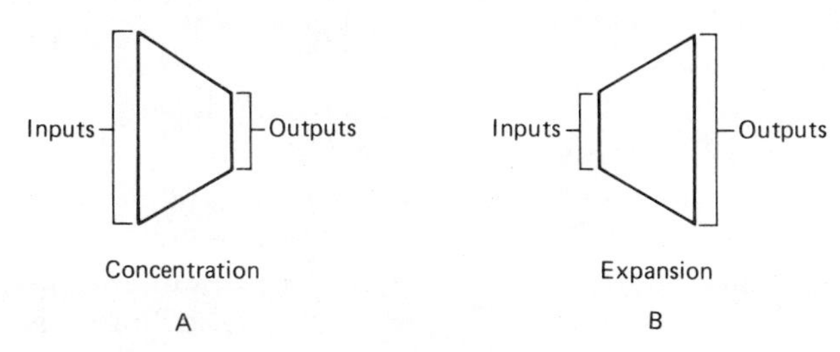

Let us dwell on the term *concentration* a bit more. Concentration reduces the number of switching paths or links *within* the exchange and the number of trunks connecting the local exchange to other exchanges. A switch also performs the function of expansion to provide all subscribers served by the exchange with access to incoming trunks and local switching paths.

Consider trunks in a long-distance network. A small number of trunks is inefficient not only in terms of loading (see Chapter 1, Section 8.2), but also in terms of economy. Cost is amortized on a per circuit basis; 100 trunks in a link or traffic relation are much more economical than 10 on the same link.* Tandem exchanges concentrate trunks in the local area for traffic relations (links) from sources of low traffic intensity, particularly below 20 erlangs, improving trunk efficiency.

4 BASIC SWITCHING FUNCTIONS

In a local exchange means are provided to connect each subscriber line to any other in the same exchange. In addition, any incoming trunk can connect to any subscriber line and any subscriber, to any outgoing trunk. The switching functions are remotely controlled by the calling subscriber, whether he is a local or long-distance subscriber. These remote instructions are transmitted to the exchange by "off-hook," "on-hook," and dial information. There are eight basic functions of a conventional switch or exchange:

- Interconnection.
- Control.

*Assuming an efficient traffic loading in each case.

- Alerting.
- Attending.
- Information receiving.
- Information transmitting.
- Busy-testing.
- Supervisory.

Consider a typical manual switching center (Figure 3.1) where the eight basic functions are carried on for each call. The important interconnecting function is illustrated by the jacks appearing in front of the operator, subscriber-line jacks and jacks for incoming and outgoing trunks. The interconnection is made by double-ended connecting cords, connecting subscriber-to-subscriber or subscriber-to-trunk. The cords available are always less than half the number of jacks appearing on the board, because one interconnecting cord occupies two jacks (by definition). Concentration takes place at this point on a manual exchange. Distribution is also carried out because any cord may be used to complete a connection to any of the terminating jacks. The operator is alerted by a lamp when there is an incoming call requiring connection. This is the attending–alerting function. The operator then assumes the control function determining an idle connecting cord and plugging into the incoming jacks. She then determines call destination, continuing her control function by plugging the cord into the terminating jack of the called subscriber or proper trunk to terminate her portion of control of the incoming call. Of course, before plugging into the terminating jack, she carries out a busy-test function to determine that the called line–trunk is not busy. To alert the called subscriber that there is a call, she uses the manual ring-down by connecting the called line to a ringing current source as shown in Figure 3.1. Other signaling means are usually used for trunk signaling if the incoming call is destined for another exchange. On such a call the operator performs the information function orally or by dialing the call information to the next exchange.

The supervision function is performed by lamps to show when a call is completed and the cord taken down. The operator performs numerous control functions to set up a call, such as selecting a cord, plugging it into the originating jack of the calling line, connecting her headset to determine calling information, selecting (and busy-testing) the called subscriber jack, and then plugging the other end of the cord into the proper terminating jack and alerting the called subscriber by ring-down. Concentration is the ratio of the field of incoming jacks to cord positions. Expansion is the number of cord positions to outgoing (terminating) jacks. The terminating jacks and originating jacks can be interchangeable. The called subscriber at another moment in time may become a calling subscriber. On the other hand, incoming and outgoing trunks may be separated. In this case they would be one-way circuits. If not separated, they would be both-way circuits, accepting both incoming and outgoing traffic.

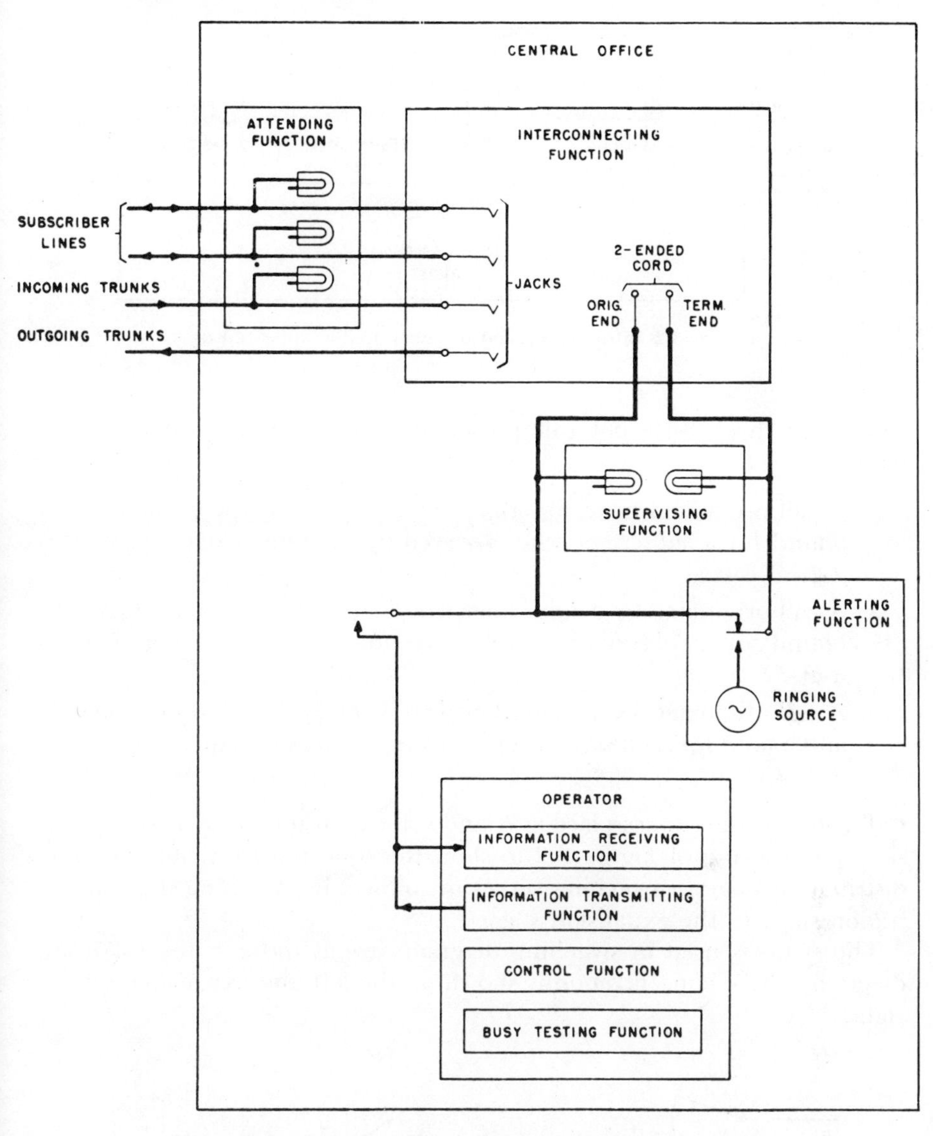

Figure 3.1 Manual exchange illustrating switching functions.

5 SOME INTRODUCTORY SWITCHING CONCEPTS

All telephone switches have, as a minimum, three functional elements: concentration, distribution, and expansion. Concentration (and expansion) was briefly introduced in Chapter 1, Section 2 to explain the basic rationale of switching. Viewing a switch another way, we can say that it has originating-line appearances and terminating-line appearances. These are shown in the simplified conceptual drawing in Figure 3.2. The Figure 3.2

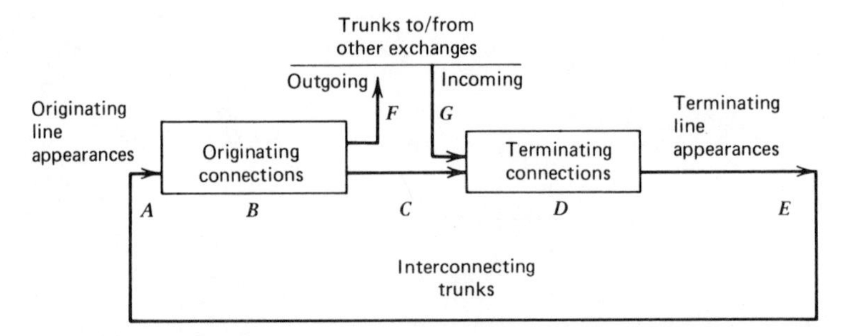

Figure 3.2 Originating and terminating line appearances.

shows the three different call possibilities of a typical local exchange (switch):

1. A call originated by a subscriber who is served by the exchange and bound for a subscriber who is served by the same exchange (routes *A–B–C–D–E*).
2. A call originated by a subscriber who is served by the exchange and bound for a subscriber who is served by another exchange (routes *A–B–F*).
3. A call originated by a subscriber who is served by another exchange and bound for a subscriber served by the exchange in question (routes *G–D–E*).

Call concentration takes place in *B* and call expansion, at *D*. Figure 3.3 is simply redrawing of Figure 3.2 to show the concept of distribution. The distribution stage in switching serves to connect by switching the concentration stage to the expansion stage.

The symbols used in switching diagrams are as those in the following diagram, where concentration is shown on the left and expansion, on the right.

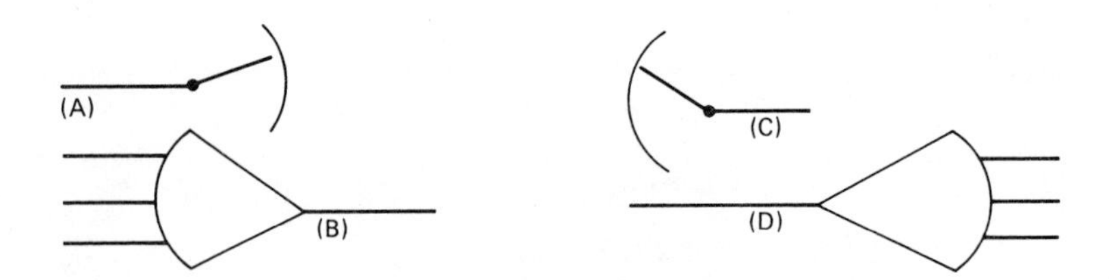

The number of inputs to a concentration stage is determined by the number of subscribers connected to the exchange. Likewise, the number of outputs of the expansion stage is equal to the number of connected subscribers whom the exchange serves. The outputs of the concentration stage are less than the inputs. These outputs are called *trunks* and are formed in

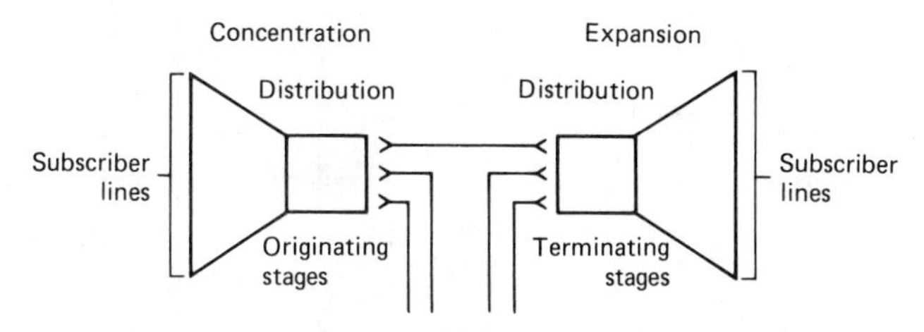

Figure 3.3 The concept of distribution.

groups; thus we refer to *trunk groups*. The sizing or dimensioning of the numbers of the trunks per group is a major task of the systems engineer. The number is determined by the erlangs of traffic originated by the subscribers and the calling rate (see Chapter 1, Sections 6 and 8).

A group selector is used in distribution switching to switch one trunk (between concentration and expansion stages) to another and is often found in switches that not only switch subscriber lines, but also where trunk switching is required. Such a requirement may be found in small cities where a switch may carry out a dual function, both subscriber switching and trunk switching from concentrators or other local switches. A group selector alone is a tandem switch. Figure 3.4 illustrates the group-selector principle.

6 TYPES OF ELECTROMECHANICAL SWITCHES

There are basically three types of electromechanical switches. Sometimes these types are broken down into gross-motion categories and fine-motion

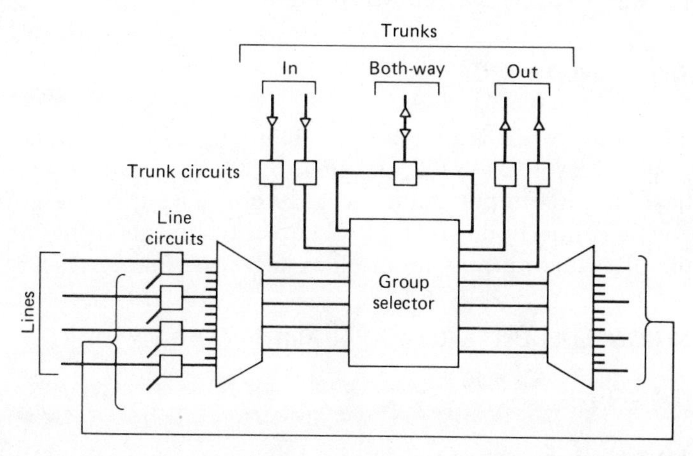

Figure 3.4 The group selector concept where both lines and trunks can be switched by the same matrix.

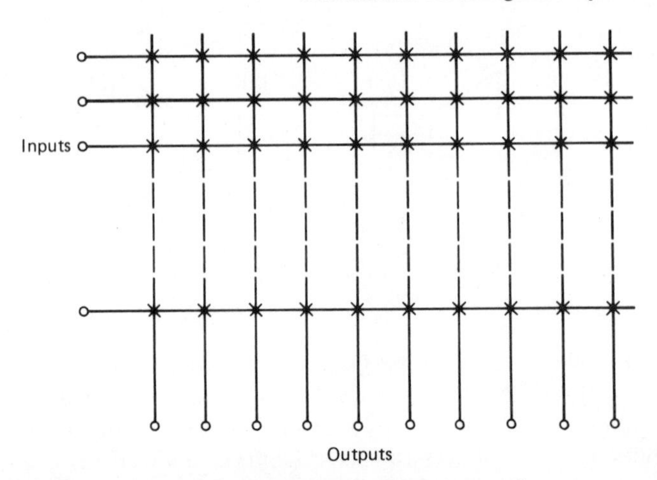

Figure 3.5 Typical diagram of a cross-point matrix.

categories. All are relay (or equivalent) operated. In the gross-motion category the oldest (patented in 1891), is the Strowger switch or step-by-step switch (S×S). The second type of gross-motion switch is the rotary switch such as the ITT 7D. Both switches are still used widely. The switches are operated electromechanically by ratchets such as a stepping relay. Conveniently, the relays (S×S) were in steps of 10 to fit our decimal-number system. They were termed *gross-motion switches* because of the space traversed between terminals.

Fine-motion switches are typified by the crossbar switch. This is a coordinate switch or matrix. A speech-path connection proceeding through a switch is made by cross points. Similar matrices can be constructed of reed relays or solid-state cross points. Such matrices may be represented by a block diagram, such as that shown in Figure 3.5.

7 MULTIPLES AND LINKS

A multiple multiplies. It is a method of obtaining several outputs from one input. Thus access is extended; (see Figure 3.6A), or the reverse may be true, where multiple inputs gain access to one output (see Figure 3.6B). Links provide connection for a multiple of switch inputs from one stage to a multiple of switch outputs in another stage (see Figure 3.7).

8 DEFINITIONS: DEGENERATION, AVAILABILITY, AND GRADING

8.1 Degeneration

Degeneration can be expressed by the following ratio:

$$\text{Degeneration on a link} = \frac{\text{Variance of offered traffic}}{\text{Mean of offered traffic}}$$

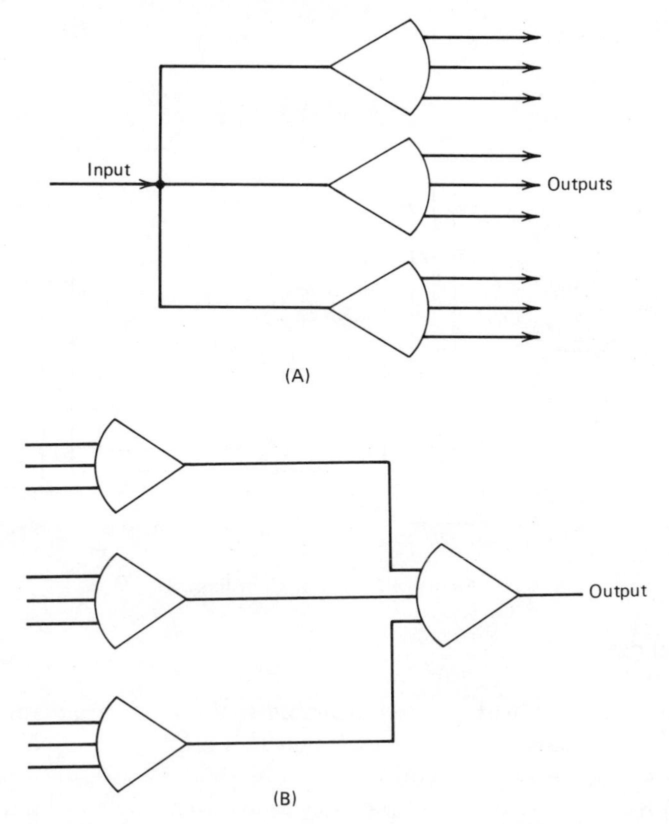

Figure 3.6 Examples of "multiple": (A) single inputs to multiple outputs; (B) multiple inputs to a single output.

Degeneration is a measure of the extent to which the traffic on a given link varies from pure random traffic. For pure random traffic, degeneration (the preceding ratio), equals 1. For overflow traffic, the variance is equal to, or in the majority of cases, greater than the mean. The more degenerate the traffic, the heavier will be the demand during peak periods, and the greater will be the number of transmission facilities that will be required.

8.2 Availability

At a switching array, availability describes the number of outlets a free inlet is able to reach and test for free or busy condition:

1. *Full availability:* Every free inlet is at all times able to test every outlet.
2. *Limited availability.* The absence of full availability. The availability at a switching array can be assigned a value, namely, the number of outlets available to each inlet (see Section 8.3).

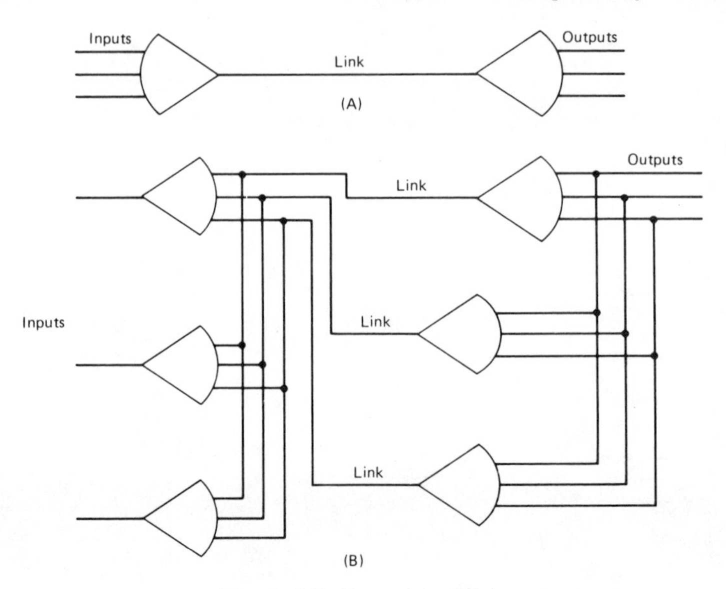

Figure 3.7 Examples of links.

8.3 Grading

At a switching array with limited availability, the inlets are arranged into groups, called *grading groups*. All the inlets in a grading group always have access to the same outlets. Grading is a method of assigning outlets to grading groups in such a way that they assist each other in handling the traffic (see Figure 3.8).

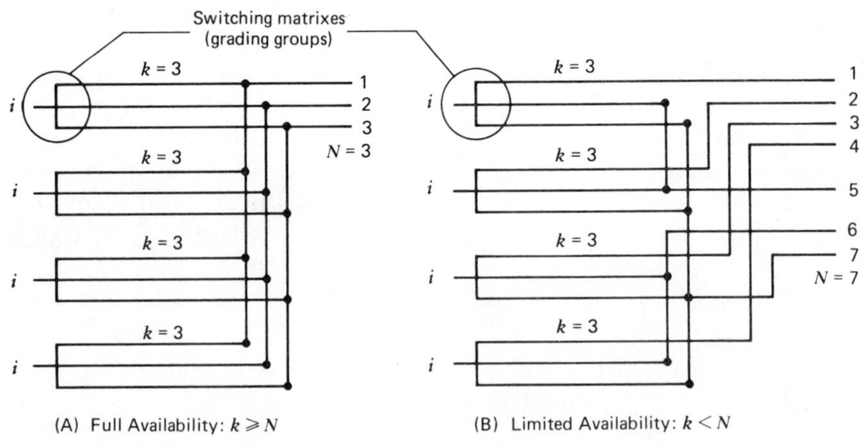

Figure 3.8 Full and limited availability. i = number of inlets per switching matrix (grading group), m = number of switching matrices (grading groups) (4, in A and B), N = number of outlets (three in A; seven in B), k = availability (number of outlets per switching matrix (grading group)). (*Note:* This is a simplified illustration. In a typical switching array, k would equal 10 or even 20.)

9 CONVENTIONAL ELECTROMECHANICAL SWITCHES

9.1 Step-by-Step or Strowger Switch

A step-by-step switch is conveniently based on a stepping relay with 10 levels. In its simplest form, which uses direct progressive control, dial pulses from the subscribers telephone activate the switch, with each pulse stepping the switch one level. If the subscriber dialled a 3, three pulses are generated by the subscriber subset and transmitted to the switch. The switch then steps to level 3. We can then imagine the second digit dialed passing to the second stepping relay bank, and the third, to the third bank. A dialed number, say 375, may be stepped through three sets of banks of 10 as shown in the following diagram.

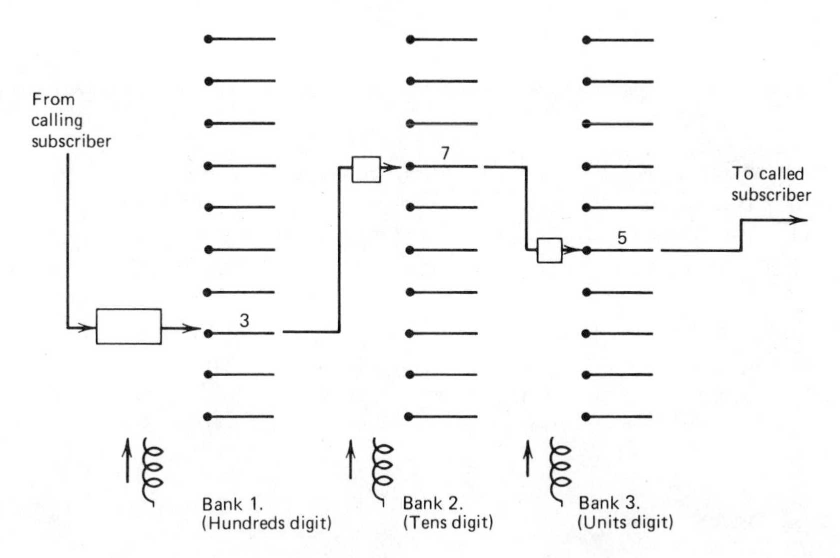

To save space and reduce post dial delay, the step-by-step switch evolved into a two-motion switch, and two banks became one. For the first digit dialed, the switch steps vertically as shown in the preceding diagram. For the second digit, the same bank steps horizontally. Thus a bank covers 10 × 10 or 100 digits, and two banks in series can cover 10,000 (0 through 9999).

The line-finder technique is used in more modern S×S (post-1928) switches. Line finders are more simple switches, with several available per group of incoming subscriber lines. On an incoming call, when a subscriber goes "off hook," a line finder automatically seeks the line desiring service and extends the connection to a line selector. The line finder provides the first stage of concentration. The line finder, once connected, supplies dial tone to the calling subscriber. A 1000-line S×S switch is illustrated in Figure 3.9. The connector in Figure 3.9 is called the *final selector* in Europe. To extend the switch in Figure 3.9 to 10,000 lines, a second selector stage is added between the selector and connector stages. The connector stage

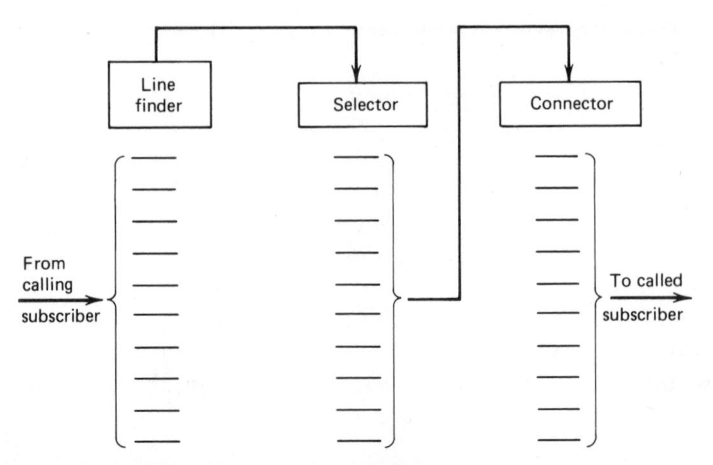

Figure 3.9 A step-by-step switch showing line finders, selectors, and connectors serving 1000 lines.

shown in Figure 3.9 may also be called the *final connecting stage.* Normally, the last two digits dialed by the calling subscriber control the connector, which typically has access to 100 lines. The connector is more complex than are the preceding stages. It must busy-test the called line and if busy, return a "busy-back" (the audible "line-busy" signal) to the calling subscriber. If the called line is idle, it must apply ringing current to the called line. It will supply talk battery to both calling and called subscriber once the called subscriber goes "off hook." It provides supervision, holding the talk path in operation until one or both conversing parties go "on-hook."

9.2 Crossbar Switch

Crossbar switching dates from 1938 and will reach a peak of installed lines sometime in the 1980s. Its life will be extended by using stored-program control (SPC) rather than hard-wired control in the more conventional crossbar configuration. The crossbar is actually a matrix switch used to establish the speech path. An electrical contact is made by actuating a horizontal and a vertical relay. Consider the switch in Figure 3.10. To make contact at point B_4 on the matrix, horizontal relay B and vertical relay 4 must close to establish connection. Such a closing is usually momentary but sufficient to cause "latching." Two forms of latching are found in conventional crossbar practice, mechanical and electrical. The latch keeps the speech path connection until an "on-hook" condition results, freeing the horizontal and vertical relays to establish other connections, whereas connection B_4 in Figure 3.10 has been "busied out."

Private industry has fielded numerous types of crossbar configuration.

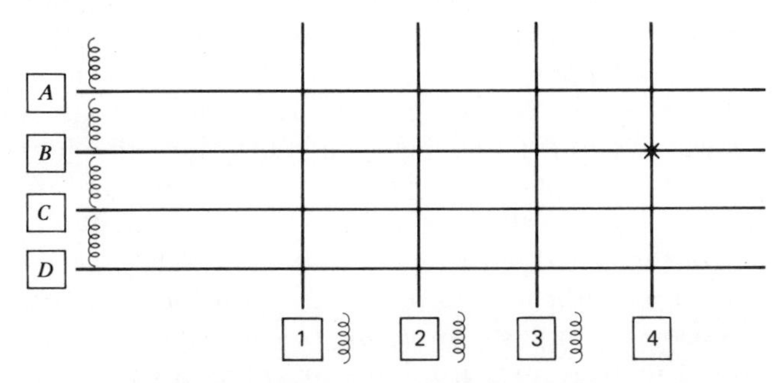

Figure 3.10 The crossbar concept.

Early crossbar switches were made up of basic 10 × 10 switching matrices or blocks. Northern Electric (Canada) has a 10 × 20 (20 vertical) matrix in a miniswitch configuration. The ITT Pentaconta 1000 has 22 vertical and 14 horizontal bars, one of which is used as a changeover bar to provide for switching 52 outlets. The output of the basic switch block matrix for Pentaconta has 50 lines. Modern local crossbar switches, such as Pentaconta 1000 or the ATT No. 5 crossbar, handle up to 10,000 subscribers in a basic switch. In the case of Pentaconta 1000, this can be extended to 20,000. A basic 10,000-line Pentaconta has 22 primary selection switch blocks.

10 SYSTEM CONTROL

10.1 Introduction

The basic function of the control system in a switch is to establish a path through the switch matrix. Thus the control system must know the calling and called ports on the matrix and be able to find a free path between them. There are two methods of establishing a path, progressive control and common control.

10.2 Progressive Control

As the term indicates, "progressive control" implies that a speech path is set up progressively through the switch, stage by stage. At each stage the selecting action chooses a group of identical paths that lead toward the ultimate call destination in the switch. Hunting action selects one of a group of competing circuits as the call progresses. The control system has no foreknowledge of conditions ahead in the next stage or step in the call setup. The call could run into blocking or a busy line and must wait through this extensive setup process before returning a "busy-back" or congestion signal.

If the control system functions directly as a result of the subscriber dial pulses, it is called *direct progressive control*. This was the case in our previous discussion of step-by-step switching. In more modern progressive control systems a register is interposed between (buffering) the subscriber dial digits. The register accepts dial digits, interprets them, and then controls switch functions. This is called *register progressive control*. With direct progressive control the called telephone number is directly associated with a specific path through the selection tree. Because the number is decimal based, progression through the switching tree is carried out in branches of 10 as shown in the diagram that follows. In direct progressive control the switching timing is in direct sequence and in synchronysm with the subscriber-dialed digits. The principal advantage of direct progressive control is economy. Control circuitry is minimized and such switches may still be found in small-community dial offices (exchanges). A functional block diagram of a direct-control switch is shown in Figure 3.11.

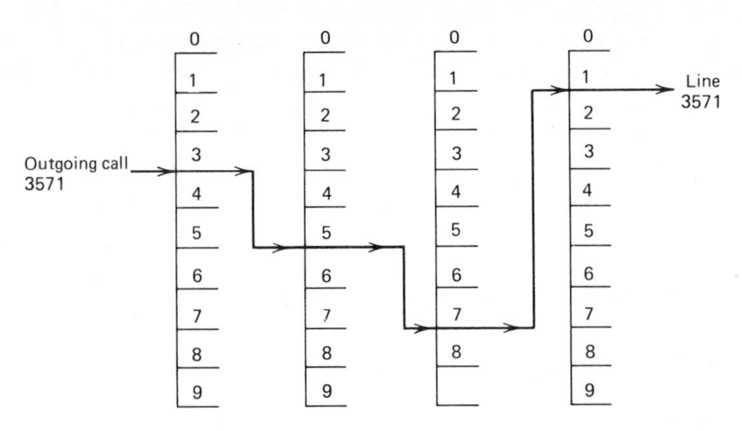

Disadvantages are that direct progressive control switches require a certain amount of overbuild to provide a satisfactory grade of service because of the stage-by-stage hunt and choose requirement without foreknowledge that a certain path is busy. The result is a lower efficiency than that in other systems on interconnecting paths. It can, however, be more tolerant of overloads concomitant with a poorer grade of service. Of course, direct progressive control requires subscriber lines to terminate directly on switch connector stages in strict correspondence with the subscribers' numbers. There is also an inherent inflexibility of trunk assignments that must conform directly with the exchange identifying digits.

Register progressive control eliminates many of the disadvantages of direct progressive control. It buffers incoming dial information, thus providing number translation. On step-by-step installations register progressive control consists of an A-unit hunter and a director. The A-unit hunter connects the incoming line to a free A-digit selector. This unit steps the selector to the group level corresponding to the first dialed digit and then

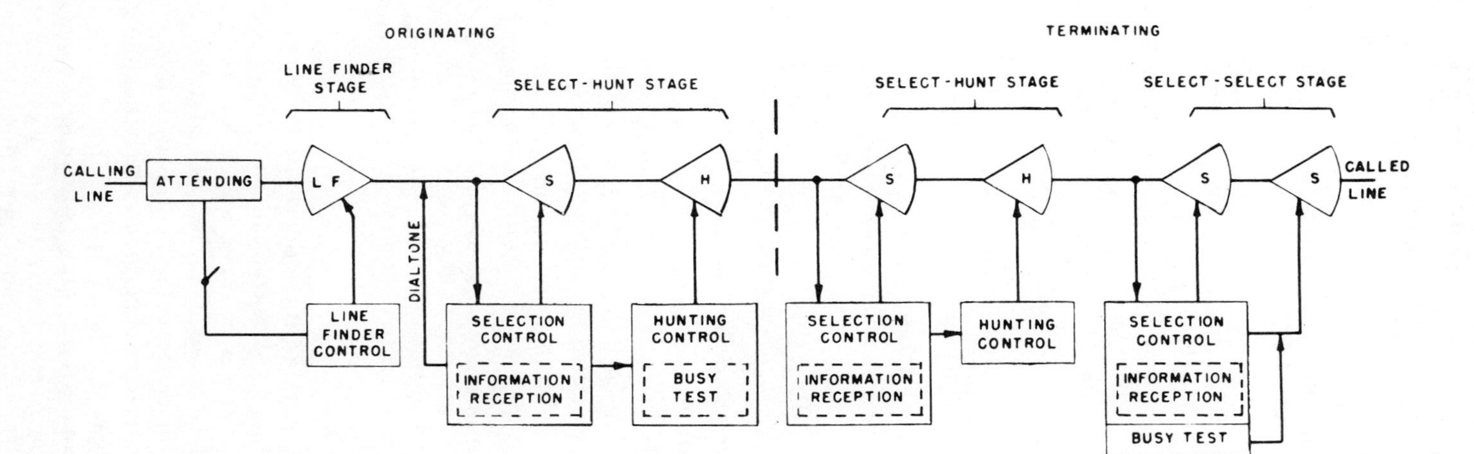

Figure 3.11 Simplified diagram of control circuits for direct progressive control. Copyright © American Telephone and Telegraph Company, 1961.

rotates to a free director. The remaining two digits of the exchange code (see Section 16 of this chapter) and the four digits of the subscriber number are registered in the director. The first two remaining digits route the call to (own) exchange or proper distant exchange. The director translates digits two and three to a code that is more versatile for switch operation and stores the last four digits of the number dialed. The translated code is then transmitted (second and third digits of called numbers) to the first code selector, and the call is extended to the proper exchange. The remainder of the digits (i.e., the subscriber number) are then transmitted as is for the indicated exchange to act on. Once these operations are completed, the A-digit hunter, selector and the director are freed to act on another incoming call. The setup of the call proceeds as in direct progressive control.

It will be noted that a seven-digit telephone number was used in the preceding example. The call setup proceeds as follows:

1. Seven incoming dialed digits.
2. A unit selector hunts/finds free director.
3. The director then:
   a. Acts on the first digit.
   b. Translates second and third digits.
   c. Transmits second and third digits.
   d. Registers fourth through seventh digits.
   e. Transmits digits four through seven (untranslated).
4. Second and third translated digits are acted on by the first selector for own exchange or indicated distant exchange.
5. Fourth through seventh digits are fed to the second selector at own exchange or the call is routed to a distant exchange.
6. A-Digit switch and director is released for the next incoming call.

Register progressive control was introduced in 1923 in the United Kingdom with director system. This first attempt at number translation lacked the capability of alternative routing when congestion occurred. Call routing remained inflexible.

As switching evolved, some of these deficiencies were eliminated or, at least, alleviated by adding further translation or by associating register-translation with the later stages of switches as well as the first stages.

To reduce post dial delay a switch may be made to start a call set up once the first three dialed digits are received. Initial translations and trunk selection can take place during the period in which the calling subscriber is dialing the last four digits.

As automatic telephone service grew, switching engineers were confronted with the problem of numerous types of switch in a common area. Some would be older progressive control and others, register progressive

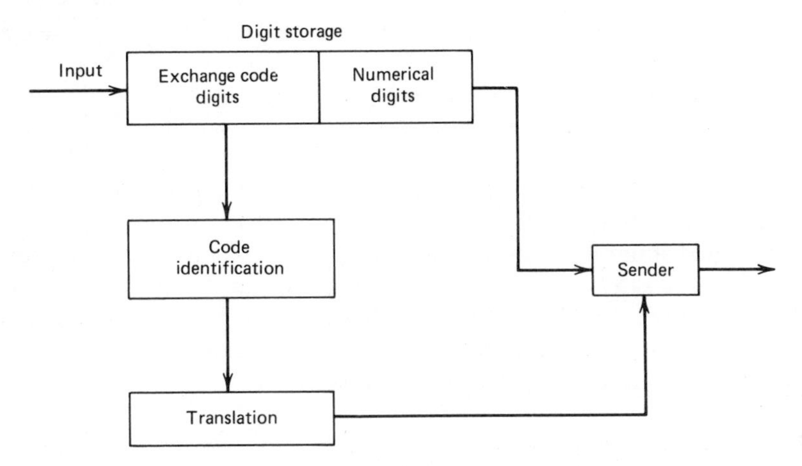

Figure 3.12 Interexchange control register.

control and still others, common control. Even greater variance occurred in internal switch codes, the digital codes used in a switch to internally set up and route a call.

Switching design engineers have since resorted to the technique of the control register with information transmitting function added. The register translator (control register) carries out three consecutive steps in this case: (1) information reception, (2) internal control signaling, and (3) information transmission. The translator in this case determines from the exchange code (the first three dialed digits), the outgoing trunk group, *and* the type of signals required by the group and the amount of information to be transmitted. This flexibility in design and operation of multiexchange areas was vastly improved. The interexchange control register concept is shown in Figure 3.12. Outgoing trunk-control registers facilitate the use of tandem exchange operation. Also, incoming trunk calls to an exchange so equipped can be handled directly from a distant exchange if they are similarly equipped.

Interexchange control registers enabled alternative routing. When first-attempt routing is blocked, this same technique can be used for second-attempt routing through an exchange by setting up a speech path with switching components different from the first setup that failed. The sender in Figure 3.12 generates and transmits signaling information to distant exchanges.

10.3 Common Control (Hard-Wired)

10.3.1 General

"Common control" is an ambiguous term. Any control circuitry in a switch that is used for more than one switching device may be termed "common control." We try to distinguish common control from the progressive control referred to in the previous section, yet common control is used in the register translators of progressive control.

For purposes of this discussion, common control is defined as providing a means of control of the interconnecting switch network, first identifying the input and output terminals of the network that are free and then establishing a path between them. This implies a busy-test of a speech path before establishment of the path. Common control may cover the entire switch or separate control for the originating and terminating halves. Common-control systems employ markers, which are discussed subsequently. Such marker systems are most applicable to grid switching networks (e.g., networks internal to the switch).

10.3.2 Grid Networks

A grid in switching networks may be defined as a combination of two or more stages of switch blocks connected together to accommodate numbers of input and output circuits as needed to meet the speech-path switching requirements of a given exchange design. The grid network differs from others previously discussed because it can be considered bidirectional.

This means that the input could also be considered an output and an output, an input. Thus parts of the network may be used for both originating and terminating purposes. Concentration and expansion can be combined, and thus a subscriber line has only one appearance on the network. A basic two-stage primary secondary grid network is shown in Figure 3.13 and an elemental three-stage grid, in Figure 3.14. Within each block in a grid network any input can connect to any output. If the blocks in Figure 3.13 had 10 inputs each, with 10 blocks in each stage, and fulfilling the basic requirement for at least one access from each primary group block to each secondary group, then at least 100 links would be required.

The fundamental difference between progressive and grid networks is in the number of paths a specific call looks ahead to inside the network on call setup. The progressive network is a hunt–select network, where the "tree" principle holds (see Section 10.2). As a call progresses through the network, the path is built up stage by stage. At each stage it looks at a choice of 10 branches ahead of it. Only when looking at the grid network as a whole is there a choice of 10 paths. On a stage-by-stage basis in the grid network, however, there is only a choice of one path for a given call at any intermediate stage. As we see later on, this path has been selected and busy-tested before actually being set up.

Conventional crossbar switches for large switching centers are usually four-stage grid networks. Such a network is shown in Figure 3.15. It is actually made up of two stages of primary and secondary grids. The two stages are connected by what are called *junctors.* Note that only one junctor as a minimum is required per secondary switch of the input grid to connect

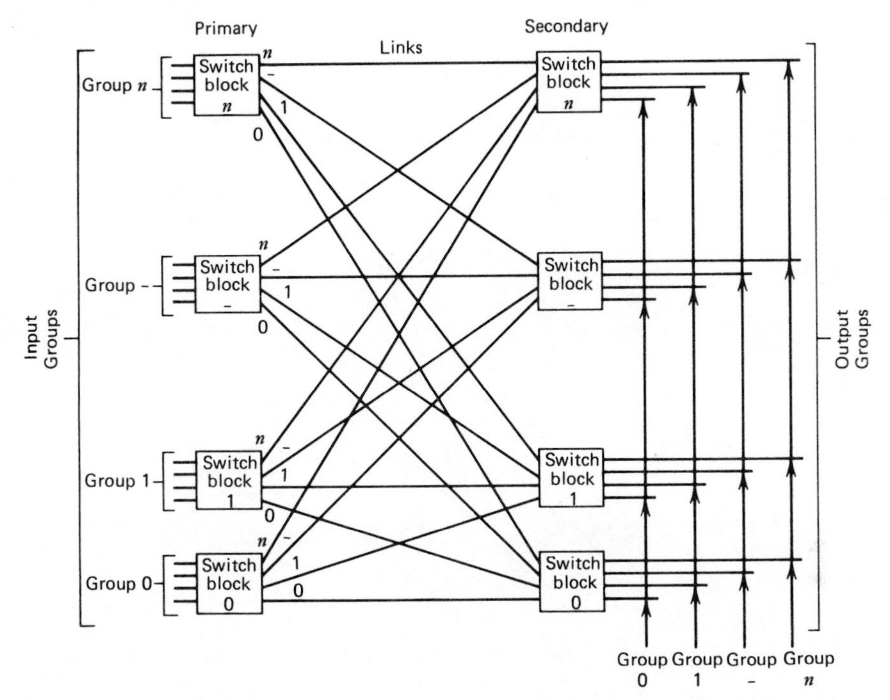

Figure 3.13 Basic two-stage grid network. Copyright © American Telephone and Telegraph Company, 1961.

to each input grid of the primary switch of the output grid. One junctor matches any pair of originating and terminating links. As can be seen in Figure 3.15, it is particularly important that traffic balance be carefully maintained on the grid inputs and outputs since the junctors from each input grid are divided equally among all grids. However, the busy-test and trunk-hunting functions of common-control systems provide for many different path setups to service a particular call.

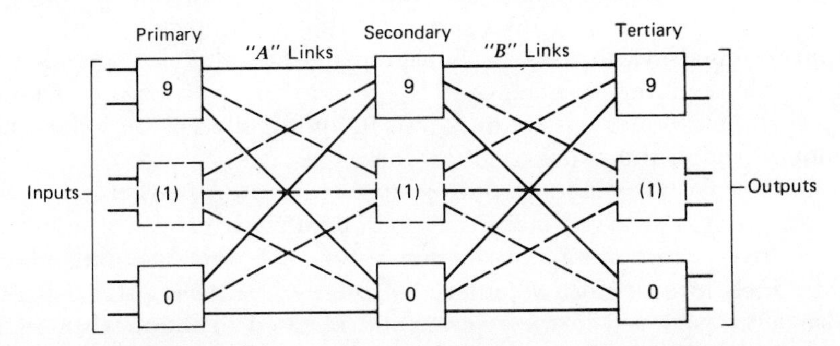

Figure 3.14 An elemental three-stage grid. Copyright © American Telephone and Telegraph Company, 1961.

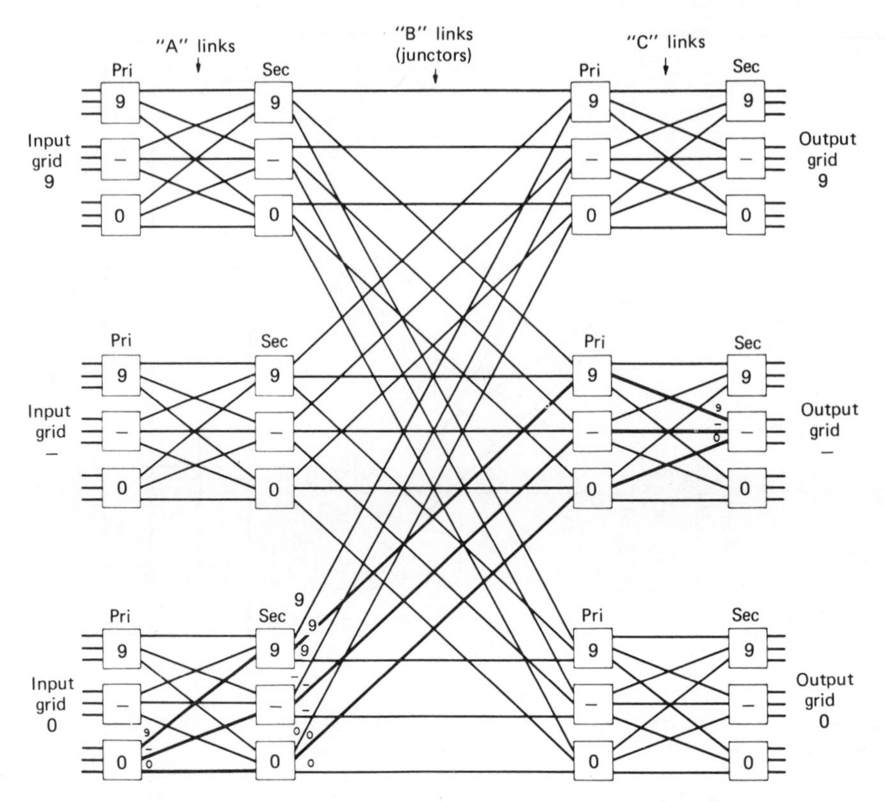

Figure 3.15 Typical four-stage grid network. Copyright © American Telephone and Telegraph Company, 1961.

10.3.3 Common-Control Principles

The "marker" sets aside common-control systems as we define them from other more generalized common-control systems discussed previously. On grid circuits the characteristics are such that with specified input and output terminals, various sets of linkage paths exist that can provide connection between the terminals. It is the marker, with terminal points identified, that locates a path, busy-tests it, and finally sets up a particular channel through the switch grid network.

A marker always works with one or more registers. A marker is a rapidly operating device serving many calls per minute. It cannot wait on the comparatively slow input information supplied by an incoming line or trunk. Such information, whether dial pulses or interregister signaling information, is stored in the register and released to the marker on demand. The register may receive the entire dialed number and store it before dumping to the marker or may take only the exchange code or area

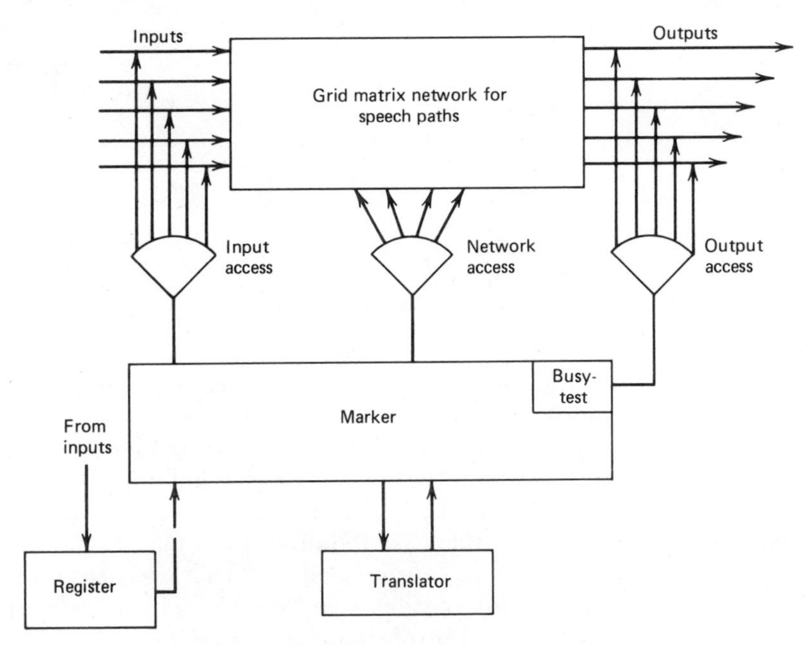

Figure 3.16 Functional diagram of a marker.

code plus exchange code (see Section 16, this chapter) which is sufficient to identify a trunk group. The register will also identify the location of the call input or set up a control path to the input for the marker. Figure 3.16 illustrates the basic functions of a marker.

Whereas the register provides the dialed exchange number to the marker, the translator provides the marker information on access to the proper trunk group. Of course, the translator may also provide other information such as the type of signaling required on that trunk group. The reader must bear in mind that modern common-control switches, particularly grid-type switches, use a control code that differs from the numerical dialed code. The dialed code is the directory number (DN) of the called subscriber. The equipment number (EN) is the number the equipment uses and is often a series of five one- or two-digit numbers to indicate location on the grid matrix of the speech-path network. The equipment codes are arbitrary and require changes from time to time to accommodate changes in number assignments. The changes are done on a patch field associated with the translator. One equipment number code is the "two out of five" code, which has 10 possibilities to correspond to our decimal base number system, that used on the subscriber dial (or push buttons). As the number implies, combinations of five elements are taken two at a time. This code is shown in Table 3.1.

The number combinations are additive, corresponding to the decimal equivalent except the 4–7 combination that adds to 11 (not to zero). For

Table 3.1 Two Out of Five Code

Digit	Two out of Five 0–1–2–4–7	Digit	Two out of Five 0–1–2–4–7
1	0–1	6	2–4
2	0–2	7	0–7
3	1–2	8	1–7
4	0–4	9	2–7
5	1–4	0	4–7

transmission through the exchange, the numbers are represented by two out of five possible audio frequencies.

On conventional exchanges one marker cannot serve all incoming calls, particularly during the busy hour (BH). Good marker holding time per call is on the order of half a second. Call attempts may be much greater than two per second, so several markers are required. Each marker must have access from any input grid to any output grid, and thus we can see where markers might compete. This possibility of "double seizure" can cause blockage. Therefore, only one control circuit is allowed into a specific grid at any one time.

To reduce marker holding time, fast action switches are required. This is one reason why common (marker) control is more applicable to fine-motion switches such as the crossbar switch. The control circuits themselves must also be fast acting. Older switches thus used all-relay devices and some vacuum tubes. Newer switches use solid-state control circuitry, which is more reliable and even faster acting.

Figure 3.17 is a simplified functional diagram of the ATT (North American) No. 5 crossbar system. The originating and terminating networks are combined. Two different types of marker are used, the dial-tone marker

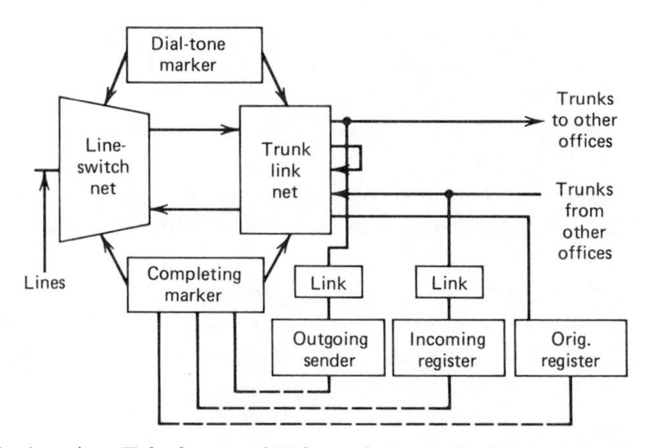

Figure 3.17 American Telephone and Telegraph No. 5 crossbar; a typical marker application. Copyright © American Telephone and Telegraph Co., 1961.

and the completing marker. A typical No. 5 crossbar has a maximum of four dial-tone markers and eight call-completing markers. The dial-tone marker sets up connections between the calling subscriber line and an originating register. Call-completing markers carry out the remainder of the control functions such as trunk selection, identification of calling and called line terminals, channel busy-test and selection, junctor group and pattern control, route advance, sender link control, trunk charge control when applicable, overall timing, automatic message accounting (AMA) information, and trouble recorder control.

The marker is also used for exchange troubleshooting and fault location. It is the most convenient check of the switching network because it samples the entire network at frequent intevals. Many markers provide trouble-recording circuits. Trouble on one call setup attempt may have no effect on another. The marker provides a trouble output, returns to service, and reattempts the call. Reattempts by a marker are usually limited to two so that the marker can return to service other calls.

If a marker finds an "all-trunks-busy" (ATB) condition on a tandem or transit call, it can ask for instructions to reroute the call. It then handles the call as a first attempt. If again an ATB condition is found and a second alternative route is available, it can make a call attempt on the second rerouting. When all possible routes are busy, the marker returns an ATB signal to the calling subscriber.

11 ROTARY SWITCHING SYSTEMS

Rotary switches are very similar in operation to step-by-step systems. One difference is that they do not have both vertical and rotary motion, a feature found in the Strowger S×S switch. Rotary systems are progressive switches with line finders and selectors. The layout is similar to S×S, but the switch is power driven with electric motors, generally continuous running. All rotary switches have register storage of dialed digits and incorporate features of common control similar to register progressive control described in Section 10.2. Some of the later versions are nearly as versatile as the marker common control described in Section 10.3.3.

Today rotary switching is used essentially in Europe or in countries using European switches and is principally manufactured by ITT, with its 7A-E; L.M. Ericcson, with its AGF; and Siemens, with its EMD system.

12 STORED-PROGRAM CONTROL (SPC)

12.1 Introduction

Stored-program control is a broad term designating switches where common control is carried out to a greater extent or entirely by computerware.

Computerware can be a full-scale computer, minimicrocomputer, microprocessors, or other electronic logic circuits. Control functions may be entirely carried out by a central computer in one extreme for centralized processing or partially or wholly by distributed processing utilizing microprocessors. Software may be hardwired or programmable. Telephone switches are logical candidates for digital computers. A switch is digital in nature,* as it works with discrete values. Most of the control circuitry such as the marker work in a binary mode.

The conventional crossbar marker requires about half a second to service a call. Up to 40 expensive markers are required on a large exchange. Strapping points on the marker are available to laborously reconfigure the exchange for subscriber change, new subscribers, changes in traffic patterns, reconfiguration of existing trunks or their interface, and so on.

Replacing register markers with programable logic—a computer, if you will—permits one device to carry out the work of 40. A simple input sequence on the keyboard of the computer teleprinter replaces strapping procedures. System faults are printed out as they occur and circuit status, may be printed out periodically. Due to the speed of the computer, post dial delay is reduced and so on. Computer-controlled exchanges permit numerous new service offerings such as conference calls, abbreviated dialing, "camp on," call forwarding, and incoming-call signal to a busy line.

12.2 Basic Functions of Stored Program Control

There are four basic functional elements of an SPC switching system: (1) switching matrix, (2) call store (memory), (3) program store (memory), and (4) central processor.

The switching matrix may be made up of electromechanical crosspoints such as in the crossbar switch, reed, correed, or ferreed cross points or switching semiconductor diodes, often SCR (silicon controlled rectifier). An SCR matrix is shown in Figure 3.18.

The call store is often referred to as the "scratch pad" memory. This is a temporary storage of incoming call information ready for use, on command from the central processor. It also contains availability and status information of lines, trunks and service circuits, and internal switch-circuit conditions. Circuit status information is brought to the memory by a method of scanning. All speech circuits are scanned for a busy/idle condition.

The program store provides the basic instructions to the controller (central processor). In many installations translation information is held in this store such as DN to EN translation and trunk signaling information.

A simplified functional diagram of a basic (full) SPC system is shown in Figure 3.19.

*This important concept is the basis for the argument set forth in Chapter 9 regarding the rationale for digital switching.

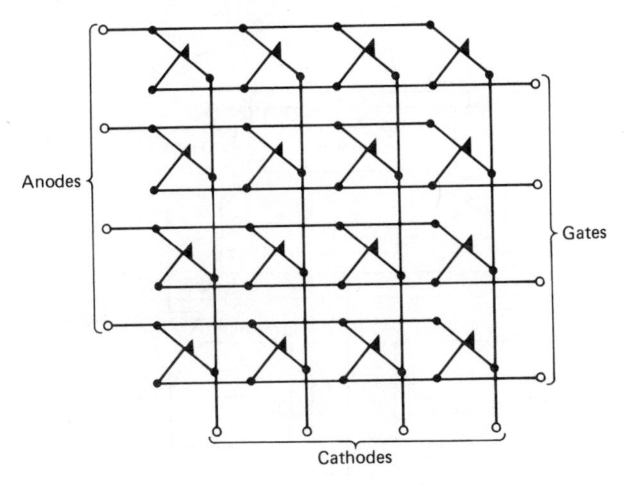

Figure 3.18 An SCR cross-point matrix.

12.3 Additional Functions in a Stored-Program-Control Exchange

Figure 3.20 is a conceptual block diagram of a typical North American SPC exchange. As we can see, an SPC exchange can be broken down into three functional levels: (1) line and trunk switching network, (2) input–output equipment, and (3) common-control equipment. Figure 3.20 is an expansion of Figure 3.19, showing, in addition, scanner circuitry and signal distribution.

The control network executes the orders given by the central control processor (computer). These orders usually consist of instructions such as "connect" or "release," along with location information on where to carry out the action in the switching network. In the ATT ESS system the network control circuits are classified in three major functional categories:

1. Selectors, which set up and release a connection on receipt of location information.

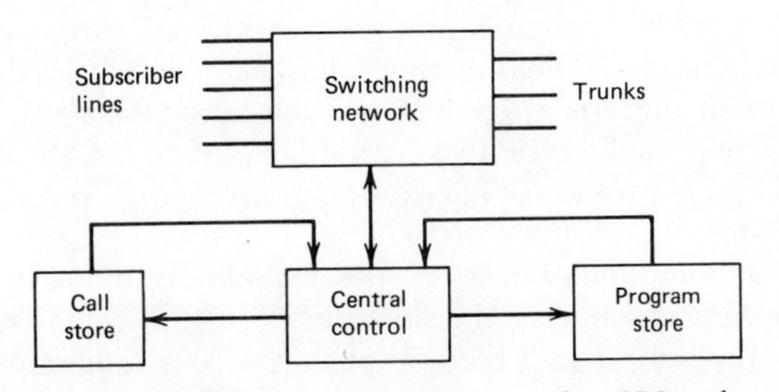

Figure 3.19 Simplified functional diagram of an SPC exchange.

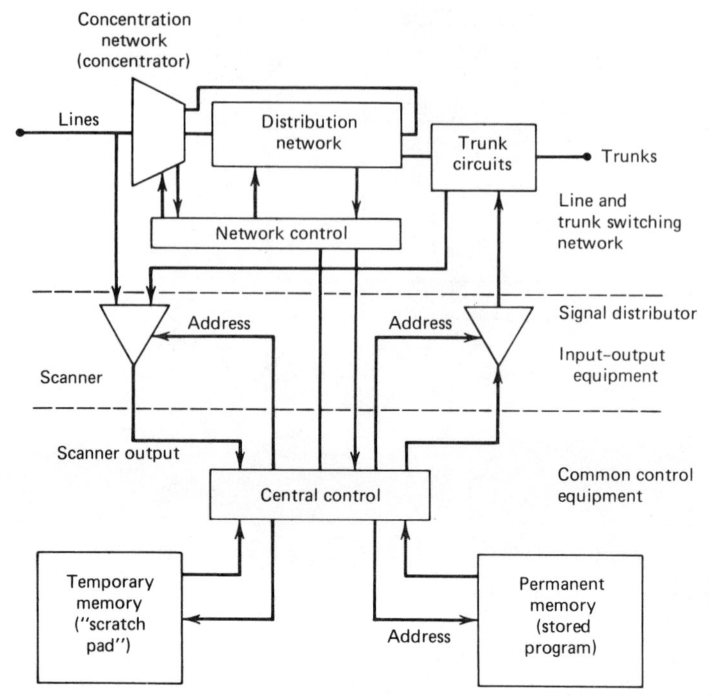

Figure 3.20 Conceptual block diagram of a typical North American semielectronic, stored-program-control exchange.

2. Identifiers, which determine the location (called an "address") of a network terminal on one side that is to be connected to a known terminal (address) on the other side of the network.
3. Enablers, which are circuits sequentially enabling junctors to be connected together.

The input–output equipment consists of line scanner and signal distributor. Both circuits operate under control of the central control processor. The scanner and distributor carry out the "time-sharing" concept of SPC. The term "time sharing" is the basis where one (or several) computers can control literally thousands of circuits, with each circuit being served serially. The concept of "holding time," important in SPC, is the time taken to serve a circuit, and "delay" is the time each circuit must wait to be served. A computer read–write cycle in a typical SPC is 2 to 5 μs, with a scanning rate of 2 to 5 μs per terminal, line scanning during digit reception every 10 μs, and 100-ms supervisory scan. The scanner is an input circuit used for sampling the states (idle or busy) of subscriber lines, trunks, and switch test points to permit monitoring the operation of the system. The signal distributor, on the other hand, is an output circuit directing output signals to various points in the system. In the ESS the signal distributor is primarily used on trunk circuits for supervisory and signaling actions.

Common-control equipment, as mentioned previously, is made up of the central processor, call store, and program store. The three units can be considered as making up the control computer, which is capable of transmitting orders to the system as well as detecting signals from the system. Stored-program-control systems with centralized control that we have discussed above have a human interface (I/O = input–output) with the central controller. This, in many instances, is a teleprinter (i.e., keyboard send, printer receive) or keyboard with a visual display unit. The installation adds many advantages and conveniences not found in more conventional switching installations. Several of these are as follows:

- Rerouting and reallocation of trunks.
- Traffic statistics.
- Renumbering of lines, subscriber move.
- Changes in subscriber class.
- Exchange status.
- Fault finding.
- Charge records.

All these functions may be carried out via the I/O equipment connected to the central processor. In the basic ATT ESS exchanges there two teleprinter accesses.

12.4 A Typical Stored-Program-Control Exchange—European Design

The ITT Metaconta is a joint French–Spanish development being installed worldwide. Its switching network can handle up to 32,000 lines. The control section can handle up to about 60,000 BH call attempts depending on the facilities offered, as well as the type and number of different signaling systems to be handled.

The central processor is made up of dual ITT 1600 16-bit binary parallel minicomputers under a load-sharing principle. The memory is of the random access type with a word length of 17 bits (16 bits + 1 parity bit) and a maximum capacity of $2^{16} = 65{,}536$ words, consisting of eight blocks of 8192 words. The memory is a ferrite core type with an access time of 0.35 μs and a cycle time of 0.85 μs.

A Metaconta L exchange uses two classes of operational programs, consisting of a resident package and an on-demand package. The *resident package* has four types of computer program.

- Call processing programs that control all the call treatment as well as charging and statistics recording.
- Man–machine communication programs that handle communications between the system and operating personnel and also permit the loading of on-demand programs.

- On-line test and defensive programs that observe abnormal behavior of the system and either restore normal conditions or make appropriate decisions to disconnect faulty devices.
- Start-up and recovery programs that handle the system status evolution. In this group are programs that permit taking over call processing when one machine fails.

The on-demand package contains programs that have to be called up by the operating personnel through teleprinter messages. The on-demand programs carry out such functions as traffic observation, charging information dump, call tracing, and routine tests. Service features and facilities offered by Metaconta L, listed in Tables 3.2 and 3.3, are broken down into

Table 3.2 Service Features and Facilities—Subscriber Related

Line classes	Subscriber facilities
One-party lines	Abbreviated numbering
With dial or push-button-type subscriber sets	One or two digits
With or without special charging categories	For individual lines, or groups of lines, or for all lines
Two-party lines	Transfer of terminating calls
With or without secrecy, separate ringing, separate charging, revertive call facility	Conversation hold and transfer
Multiparty lines (up to 10 main stations)	Calling party's ring back
Without secrecy	Toll-call offering
With selective or semiselective ringing	Hot line
PABXs	Automatic wake-up
Unlimited number of trunks	Doctor-on-duty service
Uni- or bidirectional trunks	Do-not-disturb service
With or without in-dialing	Absentee service
Coin-box lines	Immediate time and charge information
Local traffic	Conference calls
Toll traffic	Centrex facilities (optional)
Special applications	
Restricted lines	
To own exchange	
To urban, regional, national, or toll areas	
To some specified routes	
Priority lines	
Toll essential	
Essential	
Priority during emergency, overload situations, etc.	

Table 3.3 Service Features and Facilities—Administration Related

Administration Facilities
Interoffice signaling
Direct-current signaling codes (step-by-step, rotary 7A and 7D, R6 with register or direct control, North America dc codes)
Alternating-current pulse signaling codes
MF signaling codes for register-controlled exchanges (MFCR 2 code, MF Socotel, North America MF codes)
Direct data transmission over common signaling link between processor-controlled exchanges of the time or space division types
Charging
Control of charge indicators at subscriber premises
Metering on a single fee or a multifee basis
Free number service
Numbering
Full flexibility for equipment number—directory number translation
For local calls, national toll calls, international toll calls
Private automatic branch exchanges with direct inward dialing
Routing
Prefix translation for outgoing or transit calls
Alternate or overflow routing on route busy or congestion condition
Resignaling on route busy or congestion condition
Called side release control
Maintenance
Plug-in boards
Automatic fault detection and identification means by diagnostic programs
Operation
Generalized use of teleprinters
Possibility of a remote centralized maintenance and operation center

two areas, subscriber related and administration related. Figure 3.21 is a functional block diagram of the Metaconta L system.

12.5 Evolutionary Stored Program Control and Distributed Processing

12.5.1 Introduction

Presently installed switching plant represents an extremely large investment yet to be fully amortized. This plant is essentially made up of step-by-step, rotary, and crossbar equipment. The peak in crossbar installations measured in lines served will not be reached until well into the 1980s. In 1973 SPC served less than 10% of ATT subscribers. One current approach is to modify existing common-control exchanges with some limited SPC techniques and another is to install conventional crossbar with SPC control, either partial or full. The ITT Pentaconta 2000 is an example. Still another

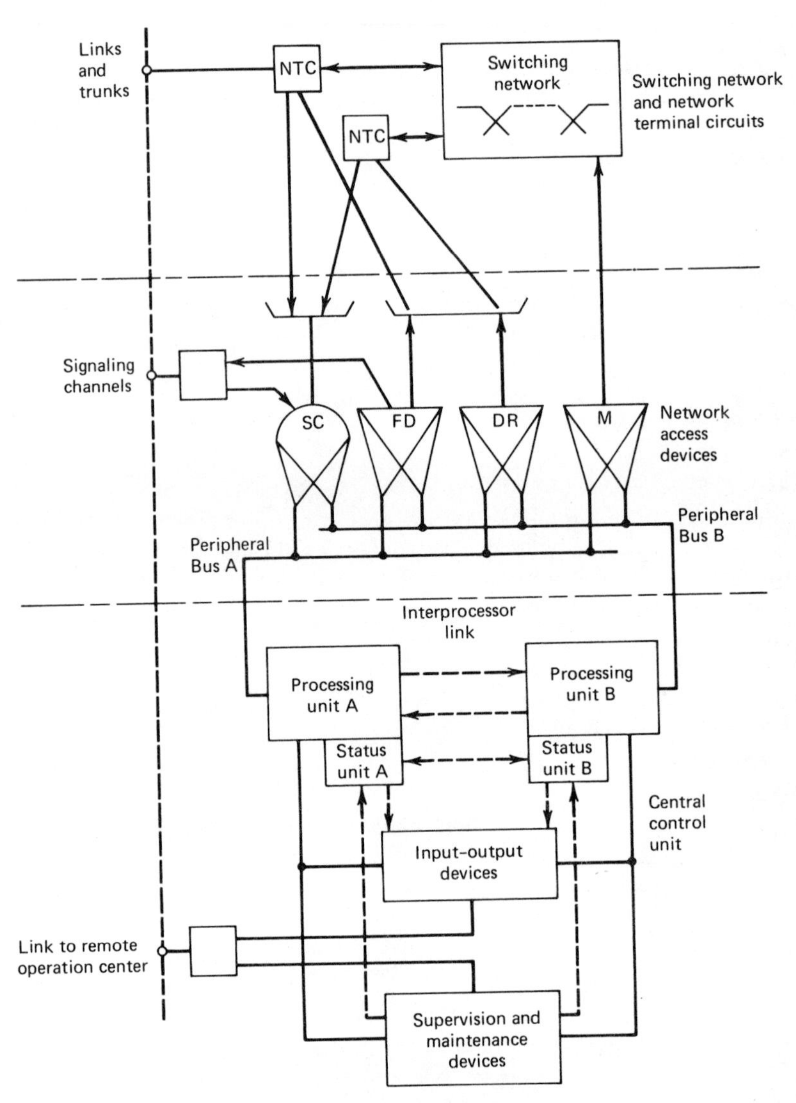

Figure 3.21 Block schematic of Metaconta L showing switching network and network terminal circuits, network access devices, and the central control unit (SC = scanner, FD = fast driver, DR = slow driver, M = marker, NTC = network terminal circuit). Courtesy of the International Telephone and Telegraph Corporation.

approach is to resort to distributed processing rather than use a comparatively large computer as described in Sections 12.3 and 12.4.

12.5.2 Modification of Existing Plant

Consider conventional electromechanical switching with common control (in its widest meaning). Processing by relays is distributed in registers, markers, translators, directors, and finders. Whereas implementation of new services and features available in full SPC systems is costly and difficult, or even impossible because of the performance limitations of conventional electromechanical design, it is feasible or even desirable to substitute, on a limited basis, microprocessor control to replace hard-wired relay-based control elements such as registers and translators. In essence, software will replace much of the hardware in the control systems in electromechanical switches. Many administrations and telephone companies have found this to be a cost-effective alternative to full SPC implementation. The resulting modified equipment will provide many of the new services of SPC as well as cost savings in maintenance, administration, and operation of existing installations. The basic design of crossbar and S×S features distributed control and goes against the main concept of full SPC, which features completely centralized control.

There are other incompatibilities as well, one of which is environmental. Relay circuitry can tolerate wide ranges of temperature and humidity; solid state cannot. Relay circuits use comparatively high voltage (i.e., 48 V) and current (mA) ranges. Solid state, particularly integrated circuits (ICs), use low voltage, low current. "Low" in this case means < 5 V and several microamperes per circuit. Electromechanical systems are essentially immune to electromagnetic interference (EMI) and in many cases are good generators of EMI. Solid-state circuitry, on the other hand, is particularly susceptible to EMI. Replacement of electromechanical control circuits with solid-state programmable units just cannot be done at random. Interfacing costs is a very major consideration.

Survivability is another argument against full SPC with centralized control. There is a certain amount of redundancy in the control circuitry of a conventional crossbar switch. Calls can be processed several at a time, each with its own marker, registers, and translator. There is "graceful degradation" in service when one control unit fails. However, let the singular central control fail in a full SPC switch, and catastrophic failure results. This is one important reason why the Metaconta L has two central processors.

To render the system cost-effective, the design engineer must select the point of interface between the conventional electromechanical switch and the SPC modules to be installed. One consideration for interface point would be the concentration links of the register–receiver–sender–translator group.

A major area that is a prime candidate for replacement of hardware by a software-oriented device [processor(s)] is in the routing and control sector of an electromechanical switch. It is here where most changes are required in a switch during its lifetime. These are the administrative and functional changes of subscriber data, routing, and signaling. By implementing SPC, many of the advantages are achieved by carrying out changes in the stored programs rather than the laborious strapping required on the electromechanical units. There is also the advantage of substitution of one for many. One control processor can replace 64 or more conventional registers, although it is desirable to use two processors for the additional reliability provided. One processor in this case is used for backup with automatic switchover on failure of the operational processor. As a switch grows (i.e., more and more subscribers are added) during its lifetime, additional processors may be added, as needed, to share the load.

One approach to centralization is to have the several control processors be served by a common memory for the switch as a whole. This simplifies the problem of programming new information or modifying existing programs. It also simplifies the I/O such that one keyboard teleprinter can serve the entire switch.

Administration and maintenance procedures are often carried out by a separate processor in a modified electromechanical switch. Connection is made to the routing and control processor bus as well as electromechanical elements. A separate I/O is also provided.

12.5.3 Distributed Processing—Another Approach

The concept of distributed processing has evolved with the availability of low-cost microprocessors. The approach is economical because it overcomes some of the limitations of the central processor–common-control technique described in Section 12.4. One limitation is that of speed in time sharing. As more calls enter an exchange, the faster the central controller must process them. Processing time is a function of cost, considering even that a computer can operate up to several thousand times faster than electromechanical control. Distributed processing removes much of the burden of ultrarapid serial processing. Of course, reliability is increased notably and chances of catastrophic failure are decreased with distributed microprocessor control.

There are two principles basic to distributed control: (1) free communication between control processors via data buses and (2) one or several processor(s), termed *central control*, acting as system coordinator (not master). Control is divided on a functional basis between different types of control units, and a number of similar control units of each type are provided to share the work load. Such a modular approach permits the addition of common modules as an exchange expands service. The TXE4 [17] is an example of a local telephone exchange using distributed control. This exchange is a British development and is manufactured by Standard Telephones and Cables Ltd. Figure 3.22 is a simplified block diagram of the TXE4, showing the switching network, which comprises reed relay matrices, and is the control network, which is microprocessor based.

The basic control system is made up of cyclic stores, main control units, markers, and supervisory processing units. The cyclic stores are "read-only" to the system. They contain the data dealing with subscriber and trunk terminations, routing codes, and numbering information relating to

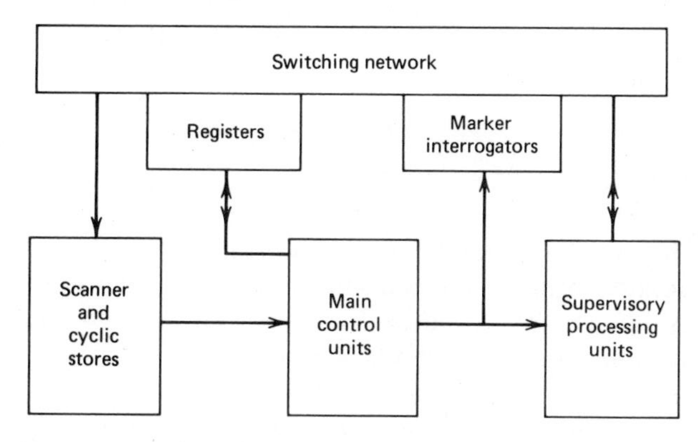

Figure 3.22 Simplified block diagram of the TXE4 exchange.

a particular exchange. A cyclic store uses a semiconductor memory and large shift registers up to 13*k* bits in length, and up to 60*k* bits are accommodated on one printed-circuit board. Data from the cyclic stores are broadcast to all main control units on a cyclic basis, providing the necessary multiple access. Up to 20 main control units are provided, depending on exchange size. These units supervise all call-connection processes that receive data from the cyclic stores, provide communications with subscriber lines and trunks, and instruct markers and supervising control units. Main control units have some self-checking capability and can be rendered out of service on failure. Note in Figure 3.22 that registers are independent and can handle complete digits and perform their own time-out functions, relieving the main control unit of this burden.

The markers are also microprocessor based, and up to 80 are provided on the TXE4. The main controller identifies the two ends of a required connection and the type of marking sequence to be used. The marker then finds the possible free paths in associated interrogation equipment and sets up a busy-test of the best path in accordance with its internal program. The marker also has self-checking capability and will remove itself from service when a fault condition is discovered.

Supervisory processing units provide call supervision. A unit monitors the state of each call under its jurisdiction as well as its service requirements, reserving a specific area in its storage for each junctor under supervision. Junctors are scanned in rotation, as is the memory area, to determine change in call state and then process this information and issue an instruction sequence to the junctor when necessary.

Data "highways" provide communication between processors and can become complex and expensive, particularly with the call-charging functions and the additional service facilities usually offered by the SPC. The TXE4 uses overlay processors connected to the necessary data-highway ports that can communicate with the main control units. The overlay processors carry out dedicated functions, relieving the main control units of these responsibilities.

13 CONCENTRATORS AND SATELLITES

Concentrators or "line concentrators" consolidate subscriber loops, are remotely operated, and are the concentration (and expansion) portion of a switch placed at a remote location. This is the conventional meaning of a concentrator. The concentrator may use S×S or crossbar facilities for the concentration matrix. Control may be by conventional relay or by a hard-wired processor. A typical crossbar type is shown in Figure 3.23, where 100 subscriber loops are consolidated to 20 trunks plus two trunks for control from a nearby exchange. Concentrators are used in sparsely populated areas that require a long trunk connection to the nearest exchange. In effect, the concentrators extend the serving area of an exchange. However, all calls originating in the concentrator serving area must be serviced by the parent exchange. Conventional concentrators can serve up to several hundred subscribers.

Concentration can also be effectively carried out using carrier techniques, either PCM (Chapter 9) or FDM (Chapter 5).

A satellite switch originates and terminates calls from a parent exchange. It differs from a concentrator in that local calls (i.e., calls originating and terminating inside the same satellite serving area) are served by the satellite and do not have to traverse the parent exchange. A block of numbers is

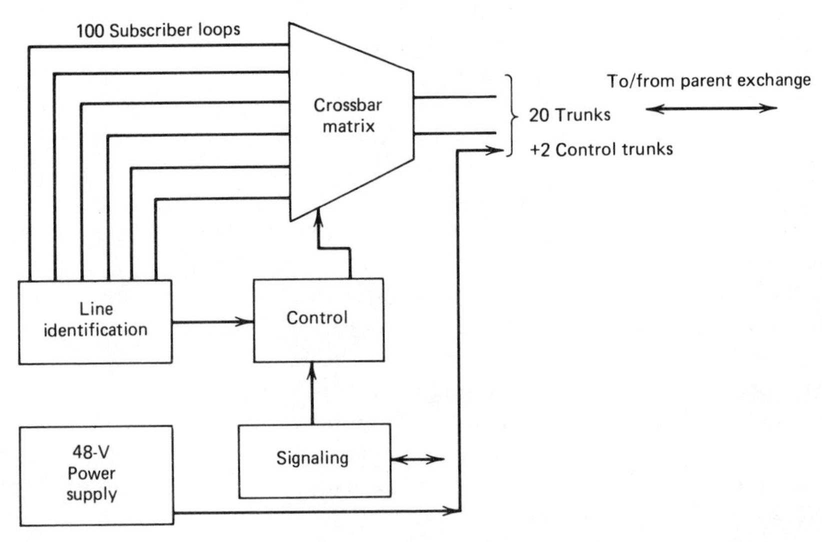

Figure 3.23 A crossbar concentrator.

assigned to the satellite serving area and is usually part of the basic number block assigned to the parent exchange. Because of its numbering arrangement, the satellite can discriminate between local calls and calls to be handled by the parent exchange. The satellite can be regarded as a component of the parent exchange that has been dislocated and moved to a distant site. This is still another method of extending the serving area of a main exchange to reduce the cost of serving small groups of subscribers, which, under ordinary circumstances, could only be served by excessively long loops to the parent exchange. Satellites range in size from 300 to 2000 lines. Concentrators are usually more cost-effective than satellites when serving 300 lines or fewer.

14 CALL CHARGING—EUROPEAN VERSUS NORTH AMERICAN APPROACHES

In Europe and in countries following European practice, telephone call charging is simple and straightforward. Each subscriber line is equipped with a meter with a stepping motor at the local exchange. Calls are metered on a time basis. The number of pulses per second actuating the meter are derived from the exchange code or the area code of the dialed number. A local call to a neighbor (same exchange) may be 1 pulse (one step) per minute, a call to a nearby city 3 pulses, or a call to a distant city 10 pulses per minute. International direct-dial calls require checking the digits of the country code that will then set the meter to pulse at an even higher rate. All completed calls are charged; thus the metering circuitry must also sense call supervision to respond to call completion, that is, when the called subscriber goes "off hook" (which starts the meter pulses) and also to respond to call termination (i.e., when either subscriber goes "on hook") to stop meter pulses. Many administrations and telephone companies refer to this as the *Karlson method* of pulse metering. Some form of number translation is required, in any event, to convert dialed number information to key the metering circuit for the proper number of pulses per unit period.

Pulse metering or "flat rate" billing requires no further record keeping other than periodic meter readings. Charging equipment cost is minimal, and administrative expenses are nearly inconsequential. Detailed billing on toll calls is used in North America (and often flat rate on local calls). Detailed billing is comparatively expensive to install and requires considerable administrative upkeep. Such billing information is recorded by determining first the calling subscriber number and then the called number. Also, of course, there is the requirement of timing call duration.

Automatic message accounting (AMA) is a term used in North American practice. When the accounting is done locally, it is called *local automatic message accounting* (LAMA), and when carried out at a central location, *centralized automatic message accounting* (CAMA). Sometimes operator intervention is required, especially when automatic number identification

(ANI) is not available or for special lines such as hotels. With automatic identified outward dialing (AIOD), independent data links are sometimes used, or trunk outpulsing may be another alternative.

Only some of the complexities have been described in North American billing practice. One overriding reason why detailed billing is provided on long-distance service is customer satisfaction. In European practice it is difficult to verify a telephone bill. A customer can complain that it is too high but can do little to prove it. Ordinarily subscriber-dialed long-distance service is lumped with local service, and a bulk amount for the number of meter steps incurred for the intervening billing period is shown on the bill. With detailed billing, individual subscriber-dialed calls can be verified and the bulk local charge is separate. In this case a subscriber has verification for a complaint if it is a valid one.

15 TRANSMISSION FACTORS IN ANALOG SWITCHING

15.1 Introduction

Transmission in telephony deals with the delivery of a "quality" signal to the far-end user. If the signal is speech, it should suffer minimum distortion as well as meet certain requirements regarding level and signal-to-noise ratio. If the signal is data, it should meet a certain minimal error rate. A switch placed in a transmission path is one more element that will affect signal quality as it traverses the network. On many calls, whether speech, data, or facsimile, there could be as many as 13 switches in tandem. Thus the extent to which a switch distorts transmission must be specified.

15.2 Basic Considerations

Normally a switch sets up a speech path. In space division switching, the type of switching discussed in this chapter, this is a metallic path made up of wire and relay or semiconductor cross points. On the path certain signals are added and withdrawn. This is the signaling that controls the switching. Thus we would expect an analog, space-division switch to act on the signal passing through it in the following ways:

1. It will attenuate the signal; this is called *insertion loss.*
2. It will add noise to the signal, thus deteriorating signal-to-noise ratio.
3. As longitudinal balance is improved, the tendency for added noise is decreased.
4. It will be a source of impulse noise, especially with gross motion switches.
5. It will tend to distort the attenuation response.
6. It will tend to deteriorate envelope delay characteristics.

7. It will have a return loss characteristic when looking into the switch from either direction. This may affect echo characteristics.
8. It will be a source of crosstalk.
9. It will be a source of absolute delay
10. and it will be a source of harmonic distortion.

15.3 Zero Test-Level Point

Signal level in telephony (see Chapter 5, Section 2.3) is a most important consideration. In a telephone network we often deal with relative levels measured in dBr, which can be related to the familiar dBm as follows:

$$\text{dBm} = \text{dBm } 0 + \text{dBr}$$

where dBm 0 is an absolute unit of power in dBm referred to the 0 TLP.

The 0 TLP can be located anywhere in the network. North American practice places it at the outgoing switch of the local switching network, whereas CCITT Rec. G.122 sets levels outgoing from the first international switch in a country (the CT; see Chapter 6) at −3.5 dBr and entering the same switch from the international network at −4.0 dBr. The CCITT then leaves it to the national telephone administration to set the 0 TLP, provided that it meets the relative levels just specified.

System gain is usually achieved and levels adjusted at the inputs and outputs of carrier equipment (FDM; see Chapter 5) to meet the requirements of a national transmission plan.

15.4 Transmission Specifications

Table 3.4 lists transmission characteristics for four-wire switches. Loss dispersion deals with the variation of loss through a switch. The loss of a four-wire switch (Table 3.4) is 0.5 dB. On same paths through the switch the loss may only be 0.4 dB and on others, 0.6 dB. Therefore, the loss dispersion is 0.2 dB. This factor is of particular importance when dealing with network stability. If loss dispersion is too great, singing may occur particularly when circuits operate near their singing margin. Return loss at local exchanges (two-wire) should be at least 11 dB (North American practice), with a standard deviation of 3 dB in the band 500 to 2500 Hz and from 250 to 3200 Hz, 6 dB with a standard deviation of 2 dB. Return loss measurements on a local exchange are made against a 900-Ω resistor in series with a 2.14 microfarad capacitor [12].

Longitudinal balance refers to balance to ground for both sides of the circuit. Open telephone circuit measurements from one of the two wire leads to ground can display a 100-V peak, and it is not unusual to find several milliamperes of current through a 1000-Ω terminating resistor to ground. The objective is to balance the two sides to ground, producing no net voltage between the two sides. When there is imbalance, the noise

Table 3.4 Transmission Characteristics for Four-wire Switches

Item	1 ((CCITT)Q.45)[c]	2 (Ref.18)[c]
Loss	0.5 dB	0.5 dB
Loss, dispersion[a]	<0.2 dB	<0.2 dB
Attenuation/ frequency response	300–400 Hz: −0.2/+0.5 dB 400–2400 Hz: −0.2/+0.3 dB 2400–3400 Hz: −0.2/+0.5 dB	−0.1/+0.2 dB −0.1/+0.2 dB −0.1/+0.3 dB
Impulse noise	5 in 5 min above −35 dBm0	5 in 5 min, 12 dB above floor of random noise[d]
Noise		
Weighted	200 pWp	25 pWp
Unweighted	1000,000 pW	3000 pW
Unbalance against ground	300–600 Hz: 40 dB 600–3400 Hz: 46 dB	300–3000 Hz: 55 dB 3000–3400 Hz: 53 dB
Crosstalk		
Between go and return paths	60 dB	65 dB } in the band 200–3200 Hz
Between any two paths	70 dB	80 dB } in the band 200–3200 Hz
Harmonic distortion	—	50 dB down with a −10 dBm signal for second harmonic, 60 dB down for third harmonic at 200 Hz: 15 dB
Impedance variation with frequency[b]	300–600 Hz: 15 dB 600–3400 Hz: 20 dB	300 Hz: 18 dB 500–2500 Hz: 20 dB 3000 Hz: 18 dB 3400 Hz: 15 dB

[a]Dispersion loss is the variation in loss from calls with the highest loss to those with the lowest loss. This important parameter affects circuit stability.
[b]Expressed as return loss.
[c]Reference frequency, where required, is 800 Hz for column 1 and 1000 Hz for column 2.
[d]Taken from standard measurement techniques for impulse noise.

produced is often referred to as *metallic noise*. The method of measurement of longitudinal balance must be stated.

16 NUMBERING CONCEPTS FOR TELEPHONY

16.1 Introduction

Numbering was introduced in Section 2 as a basic element of switching in telephony. This section discusses, in greater detail numbering as it impacts the design of a telephone network.

16.2 Definitions

There are four elements to an international telephone number. CCITT Rec. E.161 recommends that not more than 12 digits make up an international number. These 12 digits exclude the international prefix, which is that combination of digits used by a calling subscriber to a subscriber in another country to obtain access to the automatic outgoing international equipment; thus we have 12 digits maximum made up of four elements. For example, dialing from Madrid to a specific subscriber in Brussels requires only 10 digits (inside the 12-digit maximum).

07	32	2	4561234
International prefix	Country code	National significant number Trunk Code (Area Code)	Subscriber Number

Thus the international number is

32 2 456 1234

According to CCITT international usage (Rec. E.160, E.161, and E.162), we define the following:

Numbering Area (Local Numbering Area). This is the area in which any two subscribers use the same dialing procedure to reach another subscriber in the telephone network. Subscribers belonging to the same numbering area may call one another by simply dialing the subscriber number. If they belong to different numbering areas, they must dial the trunk prefix plus the trunk code in front of the subscriber number.

Subscriber number. This is the number to be dialed or called to reach a subscriber in the same local network or numbering area.

Trunk prefix (toll-access code). This is a digit or combination of digits to be dialed by a calling subscriber making a call to a subscriber in his own country but outside his own numbering area. The trunk prefix provides access to the automatic outgoing trunk equipment.

Trunk Code (Area Code). This is a digit or combination of digits (not including the trunk prefix) characterizing the called numbering area within a country.

Country Code. This is the combination of one, two, or three digits characterizing the called country.

Local Code. This is a digit or combination of digits for obtaining access to an adjacent numbering area or to an individual exchange (or exchanges) in that area. The national significant number is not used in this situation.

From Madrid, dialing a subscriber in Copenhagen requires nine digits, Brussels 10, near London (Croydon) 10, Harlow (England) outskirts 11, another in Harlow 10, and New York City 11 (not including the international prefix). This raises the concepts of uniform and nonuniform numbering as well as some ambiguity.

The CCITT defines *uniform numbering* as a numbering scheme in which the length of the subscriber numbers is uniform inside a given numbering area. It defines nonuniform numbering as a scheme in which the subscriber numbers vary within a given numbering area. With uniform numbering, we feel that each subscriber, by using the same number of digits, may be reached inside a numbering area and from one numbering area to another inside national boundaries. Theoretically, this is true for North America (north of the Rio Grande River), where a subscriber may dial seven digits and always seven digits for that subscriber to reach any other subscriber inside the calling area. Ten digits is required to reach any subscriber outside the numbering area. Note that in North America the trunk code is called the *area code*. This arrangement is shown in the following formula as it was used at the end of 1973, prior to the introduction of "interchangeable codes" discussed briefly further in this section.

Area code	Telephone number (subscriber number)
$N\frac{0}{1}X$*	*NNX-XXXX*

where X is any number from 0 through nine, N is any number from two through nine, and 0/1 is the number 0 (zero) or 1 (one).

Following a plan set up as in this formula, the ATT (the major North American telephone administration) provided for an initial arrangement of 152 area codes, each with a capacity of 540 exchange codes. Remember that there are 10 digits involved in dialing between areas including all the 50 United States, Canada, Puerto Rico, and parts of Mexico.

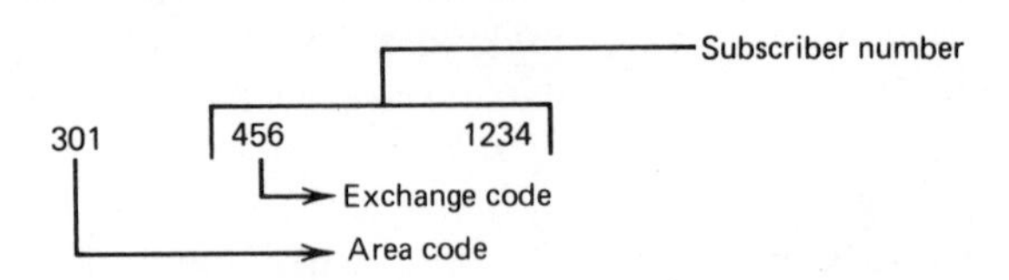

Subsequently, the 540 exchange codes were expanded to 640 when ANC (all number calling) was introduced. This simply means that the use of

*Excludes the combination $N11$.

letters was eliminated and digit combinations 55, 57, 95, and 97 could be utilized where names and resulting letters could not be structured from such combinations, and the code group NN0 was also added. This provided needed number relief. This relief was not sufficient to meet telephone growth on the continent, and further code relief had to be provided. This is carried out by realigning code areas, introduction of interchangeable codes with code areas where required, and splitting existing code areas and introduce new area codes [13].

It should be appreciated that a switch must be able to distinguish between receipt of calls bound both out of and within the calling area. The introduction of "interchangeable" codes precludes the ability of a switch to determine whether a seven-digit number (bound for a subscriber inside the same calling area) or a 10-digit number (for a call bound for a subscriber in another calling area) can be expected. This was previously based on the presence of a "0" or "1" in the second digit position. Now the switch must use either of two different methods to distinguish seven digit numbers: the "timing method," where the exchange is designed so that the equipment waits for 3 to 5 s after receiving seven digits to distinguish between seven- and 10-digit calls (if one or more digits are received in the waiting period, the switch then expects a 10-digit call) or the "prefix method," which utilizes the presence of either a "1" or "0" prefix that identifies the call as having a 10-digit format. This, of course, is the addition of an extra digit.

The prefix method described here for ATT is the "trunk prefix" defined previously. Such a trunk prefix is used in Spain, for example. In the automatic service Spain uses either six or seven digits for the subscriber number. Numbering areas with high population density and high telephone growth have seven digits such as Madrid, Barcelona, and Valencia. For numbering areas of low telephone density, six digits are used. However, when dialing between numbering areas, the subscriber always dials nine digits. This is also referred to as a *uniform system.* Inside numbering areas the subscriber number is made up of six or seven digits. Area codes (trunk codes) consist of one or two digits. The ninth digit is used for toll access. To dial Madrid, 91 + 7 digits are used, and to dial Huelva (a small province in southwestern Spain), 955 + 6 digits are used. The United Kingdom presents an example of nonuniform numbering where subscriber numbers in the same numbering area may be five, six, or seven digits. Dialing from one area to another often involves different procedures.

16.3 Factors Affecting Numbering

16.3.1 General

In telecommunications network design there are numerous trade-offs between economy and operability. Operability covers a large realm, one aspect of which is the human interface. The subscriber must use his telephone, and its use should be easy to understand and apply. Uniform numbering and number length notably improve operability.

Number assignment should leave as large a reserve of numbers as possible for growth. The simplest method to accomplish this tends toward longer numbers and nonuniform numbers. Another goal is to reduce switching costs. One way is to reduce number analysis, that is, the number of digits to be analyzed by a switch for proper routing and charging. We find that an economical analysis becomes increasingly necessary as the network becomes more complex with more and more direct routes.

16.3.2 Routing

Consider the scheme for number analysis in routing shown in Figure 3.24. Only one-digit analysis is required to route any call from exchange X to any station in the network. The first digit selects the required outlet. However, if a direct route was established between X and Z_1, two-digit analysis would be required. If a direct route was established to Z_2, three-digit analysis would be necessary. Note that some freedom of number assignment has

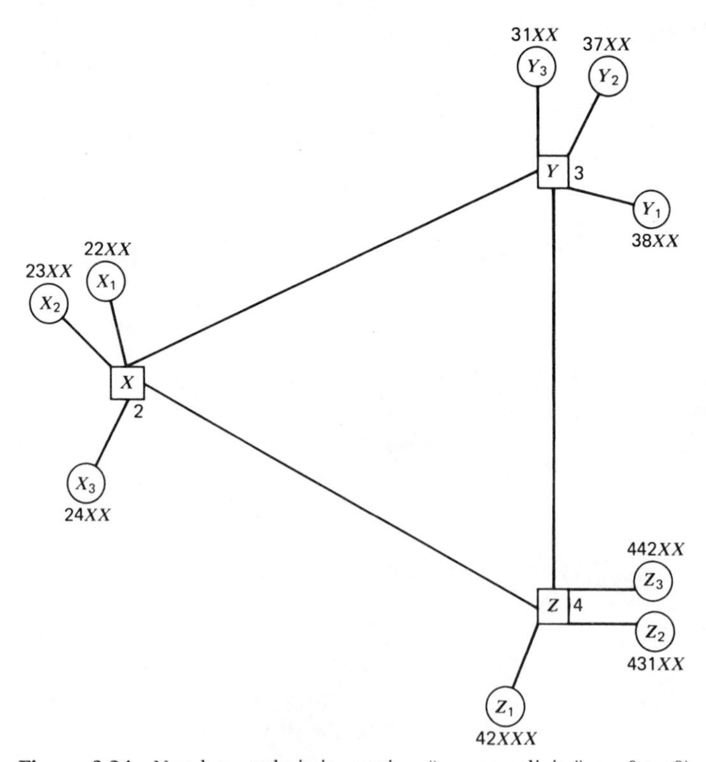

Figure 3.24 Number analysis in routing (λ = any digit from 0 to 9).

been lost because all numbers beginning with digit 2 home on X, 3 on Y, and 4 on Z. With the loss of such freedom, we have gained simplification of switches. We could add more digit analysis capability at each exchange and gain more freedom of digit assignment.

This brings up one additional point, geographical significance and the definition of numbering areas. When dealing with trunk codes, "geographical significance" means that neighboring call areas are assigned digits beginning with the same number. In Spain there is geographical significance; in the United States there is not. In the northwestern corner of Spain all trunk codes begin with 8. Aside from Barcelona all codes in northeastern Spain begin with 7. In Andalusia all begin with 5, and that groups around Valencia, which begins with 6; Alicante is 65 and Murcia is 68. In the United States, New York City is 212, just to the north is 914, across the Hudson River is 301, and Long Island is 516.

However, both countries try to use political administrative boundaries to coincide with call-area boundaries. In the United States in many cases we have a whole state or a grouping of counties, but never crossing state boundaries. In Spain we are dealing with provinces, with each province bearing its own call-area number (although some provinces share the same number). Nevertheless, boundaries do coincide with political demarcations. This eases subscriber understanding and simplifies tariffing procedures.

16.3.3 Tariffs (Charging)

The billing of telephone calls is automated in the automatic service. There are two basic methods for charging, bulk billing and detailed billing (see Section 14 of this chapter). Detailed billing is essentially a North American practice. Detailed billing includes in the subscriber's bill a listing of each toll call made, number called, date and time, charge time, and individual charge entry for each toll call. Such a form of billing requires extensive number analysis, which is usually carried out by data-processing equipment in centralized locations.

When charges are determined typically by accumulated meter steps, billing is defined as bulk billing. The subscriber is periodically presented with a bill indicating the total of steps incurred for the period at a certain money rate per step. Bulk billing is much more economical to install and administer. It does lead to much greater subscriber misunderstanding and in the long run, to dissatisfaction.

With bulk billing, stepping meters are part of the switching equipment. The switching equipment determines the tariff and call duration and actuates the meters accordingly. For bulk billing, the relationship between numbering and billing is much closer than in the case of detailed billing. Ideally, tariff zones should coincide with call areas (routing areas). Care should be taken in the compatibility of charging for tariff zones and the numerical series used in numbering. It follows that the larger is the tariff zone, the less numerical analysis will be required.

16.3.4 Size of Numbering Area and Number Length

Size relates not only to geographical dimension, but also to the number of telephones encompassed in the area. Areas of very large geographical size present problems in network and switch design. Principally, large areas show a low level of community of interest between towns. In this case switching requires more digit analysis, and tariff problems may be complicated. If the area has a large telephone population, a larger number of digits may be required, with implied longer dialing times and more postdial delay for analyses. However, large areas are more efficient for number assignment.

With smaller numbering areas, short-length subscriber numbers may be used for those with a higher community of interest. Smaller digit storage would be required for intraarea calls with shorter holding times for switch control units. Smaller areas do offer less flexibility, particularly in the future when uneven growth takes place and forecasts are in error. In the mid-1970s North America faced a situation requiring reconfiguration and the addition of new areas and also the use of an extra digit or access code.

One report [10] recommends, as a goal, that a numbering area be no greater than 70,000 km^2 nor have less than 100,000 subscribers at the end of a numbering-plan validity period. A subscriber should be able to dial a shorter number when calling other subscribers inside his own numbering area. All other subscribers would be reached by dialing the national number. This is done by adding a simple trunk code in front of the subscriber number. The use of a trunk prefix has advantages in switch design. This should be a single digit, preferably "0." Zero is recommended because few if any subscriber numbers start with zero. Thus when a switch receives the initial digit as zero, it is prepared to receive the longer toll number for interarea dialing, whereas if it receives any other digit, it is prepared for receipt of a subscriber number for intraarea dialling.

For dialed international calls, the CCITT recommends number length no greater than 12 digits. A little research will show that few international calls require 12 digits. Number length, of course, deals with number of subscribers, code blocks reserved, immediate and future spare capacity, call-area size, and trunk prefix assignment. Uniform and nonuniform numbering are also factors. Other factors are 40-year forecast accuracy, subscriber habits, routing and translation facility availability, switching system capabilities, and existing numbering scheme.

16.4 In-Dialing

Numbers, particularly in uniform numbering systems, must be set aside for PABXs with in-dialing capability. This is particularly true when the PABX numbering scheme is built into the national scheme for in-dialing. For example, suppose a PABX main number were 543-7000, with an extension to that PABX of 678. To dial the PABX extension directly from the outside, we would dial 543-7678, and the 543-7*XXX* block would be lost for use except for 999 extension possibilities for that PABX. If it were a small installation, the numbering block loss could be reduced by 543-7600 (PABX main number), extension 78—543-7678. The PABX would be limited to 99 (or 98) direct in-dialed extensions. If the PABX added digits onto the end of the main number, there would be no impact on national or area numbering. For example:

Main PABX number:	543-7000
PABX extension number:	678
Extension dialed directly from outside:	543-7000678

Here we face the CCITT limitation of 12 digits for international dialing; 10 of the 12 have been used. We must also consider digit storage in switches handling such calls. Of course, the realm of nonuniform numbering has been entered. A third method suggested is shown in the following example:

Main PABX number:	53-87654
PABX extension number:	789
Extension dialed directly from outside:	503-8789

Here the first and third digits locate the local exchange on which the PABX is a subscriber. That local exchange must then analyze the fourth, fifth, and possibly the sixth digits. Disadvantages are that translation is always required and there is some loss to the numbering system.

17 TELEPHONE TRAFFIC MEASUREMENT

Statistics on telephone traffic is mandatory both at the individual exchange level and for the network as a whole. Traffic measurements provide the required statistics when carried out in accordance with a well-organized plan. The statistics include traffic volume (erlangs or ccs) and its distribution by type of subscriber and service on each route and circuit group and its variation daily, weekly, and seasonally. In the process of traffic measurement, congestion is indicated, if present, and switch and network efficiency can be calculated. The most common measurement of "efficiency" is grade of service.

Congestion and its causes tell the systems engineer whether an exchange is overdimensioned, underdimensioned, or of improper traffic balance. Traffic statistics provide a concrete base or starting point from which to forecast growth. It gives past traffic evolution of subscriber-generated traffic growth by type of subscriber and class of service. It also provides a prediction of the evolution of local traffic between exchanges of national long-distance (toll) traffic and international traffic. Traffic measurements (plus forecasts) supply the data necessary to dimension new exchanges and for the extension of existing ones. It is especially important when a new exchange is to replace an existing exchange or when a new exchange will replace several existing exchanges.

Traffic measurement involves a number of parameters, including seizures (call attempts), completed calls, traffic intensity (involving holding times), and congestion. The term "seizures" indicates the number of times a switching unit or groups are seized without taking into account holding time. Seizures may be equated to call attempts and give an expression of how much exchange control equipment is being worked. The number of completed calls is of interest in the operation and administration of an exchange. Of real interest are statistics on uncompleted calls that are not attributed to busy lines or lack of answer or that can be attributed to the specified grade of service. However, at a particular exchange a completed call means only that the switch in question has carried out its function, which does not necessarily imply that a connection has been established between two subscribers. The intensity of traffic or traffic volume, a most important parameter, directly provides a measure of usage of a circuit. It is especially useful in those switching units involved directly in the speech network. On the other hand, it is not directly indicative of grade of service. Approximate grade of service should be taken from the traffic tables used to dimension the exchange by using measured traffic intensity as an input. The approximation is most accurate when traffic intensity approaches dimensioned intensity.

Congestion involves three characteristics: "all circuits busy," overflow, and dial-tone delay. "All circuits busy" is an indication of the number of times, and eventually the duration, where all units of a switching group are handling traffic simultaneously. Thus it represents an index of the real grade of service. Its use is particularly effective for those switching groups that are operating near maximum capacity. For an overdimensioned exchange, "all circuits busy" index is useless and tells us nothing. Overflow is an index of the number of call attempts that have not been able to proceed due to congestion. For networks with alternative routing, overflow tells us the number of offered calls not handled by specific switching equipment group. Dial-tone delay is directly indicative of overall grade of service that an exchange provides its subscribers, particularly in the preselection switching stages. Dial-tone delay is normally expressed in the time required in getting dial tone compared to a fixed time, usually 3 s—as a percentage of total calls.

Traffic measurements should be made through the busy hour (BH), which can be determined by reading amperage of exchange battery over the estimated period of occurrence, say, 9:30 to 12:30 P.M., every 10 min (or by peg count usage devices, microprocessors on lines, equipment or trunk circuits). These measurements should be done daily for at least 3 weeks. Traffic measurements should be carried out for the work week, one week per month for an entire year. The means of measurement depends on available equipment and exchange type. This may range from simple observation to fully automatic traffic-measuring equipment with recorders on magnetic or paper tape. The engineer may have to resort to the use of electromechanical counters and/or electromechanical or electronic hunting devices with which many exchanges, particularly those with common control, are equipped.

18 DIAL-SERVICE OBSERVATION

Dial-service observation provides an index or measurement of the telephone service provided to the customer (see Chapter 1, Section 13). It gives an administration or telephone company a sample measure of maintenance required (or how well it is being carried out), load balance, and adequacy of installed equipment. The number of customer-originated calls sampled per day is in the order of 1 to 1.5% of the total calls originated; 200 observations per exchange per day is a minimum [20].

Traditionally, service observation has been done manually, requiring the presence of an observer. Service observation positions have some forms of automation such as automatic recording of a calling number and/or the mark sensing or keypunching to be employed for computer-processing input. Tape recorders may be placed at exchanges to automatically record calls on selected lines. The tapes are then periodically sent to the service observation desk. Tapes usually have a "time hack" to record the time and the duration of calls. A service observation desk usually serves many exchanges or an entire local area. Tandem exchanges are good candidates because they concentrate the service function. The results of service observation are intended to represent average service; thus care should be taken to ensure that subscribers selected represent the average customer. The following is one list of data to be collected per line observed:

- Total call attempts (local and long distance).
- Completed calls.
- Ineffective calls due to calling subscriber.
  Incomplete call.
  Late dialing.
  Unavailable number dialed.
  Call abandoned prematurely.

- Ineffective calls due to called subscriber.
  Busy.
  No answer.
- Ineffective calls due to equipment.
  Wrong number.
  Congestion.
  No ringing, busy tone, no answer.
  Interruptions.
- Transmission quality (level, distortion, noise, etc.).
- Wrong number dialed by subscriber.
- Dial-tone delay.
- Postdial delay.

Results of service observation are computerized and summary tables published monthly or quarterly. Individual tables are often made for each exchange, relating it to the remainder of the network. Quality of route can be determined since observations identify the called local exchange. A similar set of observations may be made for the long-distance (toll) service exclusively.

There is a tendency to fully automate service observation without any human intervention. The major weakness to this approach is the lack of subjective observation of factors such as types of noise and its effect on a call, subscriber behavior to certain stimuli, and so forth.

REFERENCES AND BIBLIOGRAPHY

1. *Switching Systems*, American Telephone and Telegraph Company, New York, 1961.
2. Marvin Hobbs, *Modern Communication Switching Systems,* Tab Books, Blue Ridge Summit, Pa., 1974.
3. T. H. Flowers, *Introduction to Exchange Systems,* Wiley, New York, 1976.
4. Amos F. Joel, "What is Telecommunication Circuit Switching?," *Proc. IEEE,* **65** (9) (September 1977).
5. *Fundamental Principles of Switching Circuits and Systems,* American Telephone and Telegraph Company, New York, 1963.
6. International Telephone and Telegraph Corporation, *Reference Data for Radio Engineers*, 6th ed., Howard W. Sams, Inc., Indianapolis, 1976.
7. J. G. Pearce, *Electronic Switching,* Telephony Publishing Company, Chicago, 1968.
8. J. P. Dartois, "Metaconta L Medium Size Local Exchanges," *Electr. Commun.,* **48,** (3) (1973).
9. J. G. Pearce, "The New Possibilities of Telephone Switching." *Proc. IEEE,* **65** (9) (September 1977).
10. "Numbering," Telecommunication Planning, ITT Laboratories (Spain), Madrid, 1973.
11. CCITT, Q Recommendations (Rec. Q), *Orange Books,* ITU, Geneva, 1977.

12. L. F. Goeller, *Design Background for Telephone Switching,* Lees ABC of the Telephone, Geneva, Ill., 1977.

13. *Notes on Distance Dialing,* American Telephone and Telegraph Company, New York 1975.

14. Bruce E. Briley and Wing N. Toy, "Telecommunication Processors," *Proc. IEEE,* **65** (9) (September 1977).

15. T. H. Flowers, "Processors and Processing in Telephone Exchanges," *Proc. IEEE,* **119, (3) (March 1972).**

16. Enn Aro, "Stored Program Control-assisted Electromechanical Switching—An Overview," *Proc. IEEE,* **65** (9) (September 1977).

17. P. J. Hiner, "TXE4 and the New Technology." *Telecommunications* (January 1977).

18. USITA Symposium, April 1970, Open Questions 18–37.

19. Roger L. Freeman, *Telecommunication Transmission Handbook,* Wiley, New York, 1975.

20. L. Alvarez Mazo and M. Poza Martinez, "Dial Service Observation," Telecommunication Planning Symposium, STC (SA) Boksburg South Africa, June 1972.

21. L. Alvarez Mazo, "Network Traffic Measurements," Telecommunication Planning Symposium, STC (SA) Boksburg, South Africa, June 1972.

Chapter 4

SIGNALING FOR TELEPHONE NETWORKS

1 INTRODUCTION

In a switched telephone network signaling conveys the intelligence needed for one subscriber to interconnect with any other in that network. Signaling tells the switch that a subscriber desires service and then gives the local switch the data necessary to identify the required distant subscriber and thence properly route the call. It also provides supervision of the call along its path. Signaling also gives the subscriber certain status information such as dial tone, busy tone (busy back), and ringing. Metering pulses for call charging may also be considered a form of signaling.

There are several classifications of signaling:

1. General.
    a. Subscriber signaling.
    b. Interswitch signaling.
2. Functional.
    a. Audible–visual.
    b. Supervisory.
    c. Address signaling.

Figure 4.1 shows a more detailed breakdown of these functions.

It should be appreciated that on many or even on most telephone calls more than one switch is involved in call routing. Therefore, switches must interchange information among switches in fully automatic service. Address information is provided between modern switching machines by interregister signaling and the supervisory function by line signaling. The audible/visual category of signaling functions inform the calling subscriber regarding call *progress,* as shown in Figure 4.1. The *alerting* function informs the called subscriber of a call waiting or an extended "off-hook" condition of his (or her) handset. Signaling information can be conveyed by

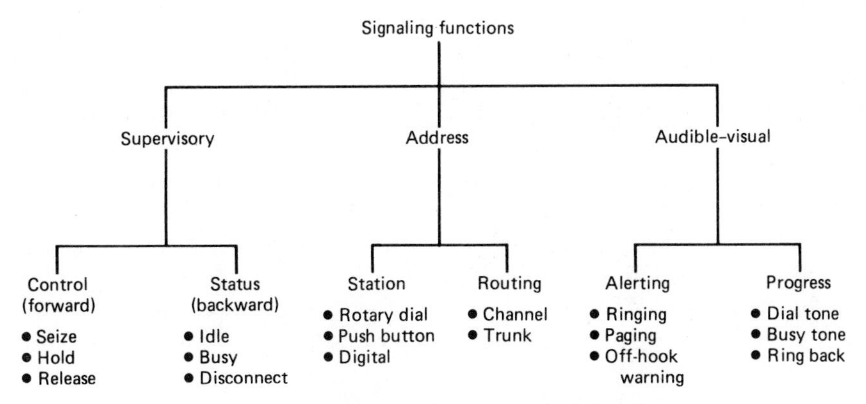

Figure 4.1 A functional breakdown of signaling.

a number of means from subscriber to switch or between (and among) switches. Signaling information can be transmitted by means such as:

- Duration of pulses (pulse duration bears a specific meaning).
- Combination of pulses.
- Frequency of signal.
- Combination of frequencies.
- Presence or absence of a signal.
- Binary code.
- For dc systems, the direction or level of transmitted current.

2 SUPERVISORY SIGNALING

Supervisory signaling provides information on line or circuit condition and indicates whether a circuit is in use. It informs the switch and interconnecting trunk circuits information on line condition such as calling party "off hook," calling party "on hook," called party "off hook," and called party "on hook." The terms "on hook" and "off hook" were derived from the position of the receiver of an old-fashioned telephone set in relation to the mounting, in this case a hook, provided for it. If a subscriber subset is on hook, the conductor (subscriber loop) between the subscriber and his local exchange is open and no current is flowing. For the reverse or off-hook condition, there is a dc shunt across the line, and current is flowing. These terms have also been found convenient for designating the two signaling conditions of a trunk (junction). Usually, if a trunk is not in use, the on-hook condition is indicated toward both ends. Seizure of the trunk at the calling end initiates an off-hook signal transmitted toward the called end.

The reader must appreciate that supervisory information–status must be maintained end to end on every telephone call. It is necessary to know when a calling subscriber lifts his telephone off hook, thereby requesting service. It is equally important that we know when the called subscriber answers (i.e., lifts his telephone off hook), for that is when we may start metering the call to establish charges. It is also important to know when the called and calling subscribers return their telephones to the on-hook condition. Charges stop, and the intervening trunks comprising the talk path as well as the switching points are then rendered idle for use by another pair of subscribers. During the period of occupancy of a talk path end to end, we must know that this particular path is busy, is occupied so that no other call attempt can seize it.

Dialing of a subscriber line is merely interruption of the subscriber loop's off-hook condition, often called "make and break." The "make" is a current flow condition (or off hook), and the "break" is the no-current condition (or on hook). How do we know the difference between supervisory and dialing? Primarily by duration—the on-hook interval of a dial pulse is relatively short and is distinguishable from an on-hook disconnect signal (subscriber hangs up), which is transmitted in the same direction for a longer duration. Thus the switch is sensitized to duration to distinguish between supervisory and dialing of a subscriber loop. Figure 4.2 is a simplified diagram of a subscriber loop showing its functional signaling elements.

2.1 E and M Signaling

Probably the most common form of trunk supervision is E and M signaling. Yet it only becomes true E and M signaling where the trunk interfaces with the switch (see the following diagram).

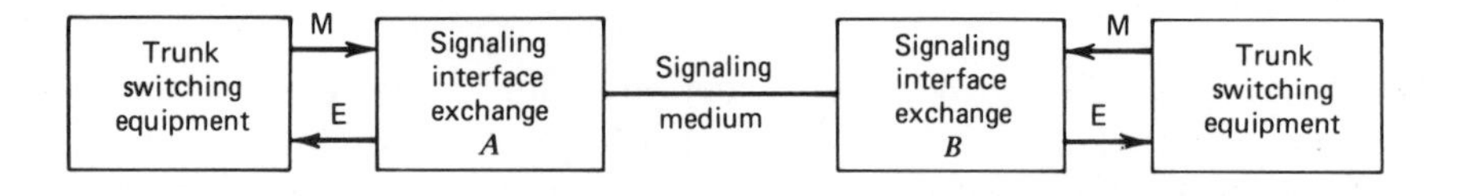

E-lead and M-lead signaling systems are semantically derived from historical designation of signaling leads on circuit drawings covering these systems. Historically, the E and M signaling interface provided two leads between the switch and what we may call *trunk-signaling equipment* (signaling interface). One lead is called the "E-lead," which carries signals *to* the switching equipment. Such signal directions are shown in the preceding diagram, where we see that signals from switch *A* and switch *B* leave *A* on the M-lead and are delivered to *B* on the E-lead. Likewise, from *B* to *A*, supervisory information leaves *B* on the M-lead and is delivered to *A* on the E-lead.

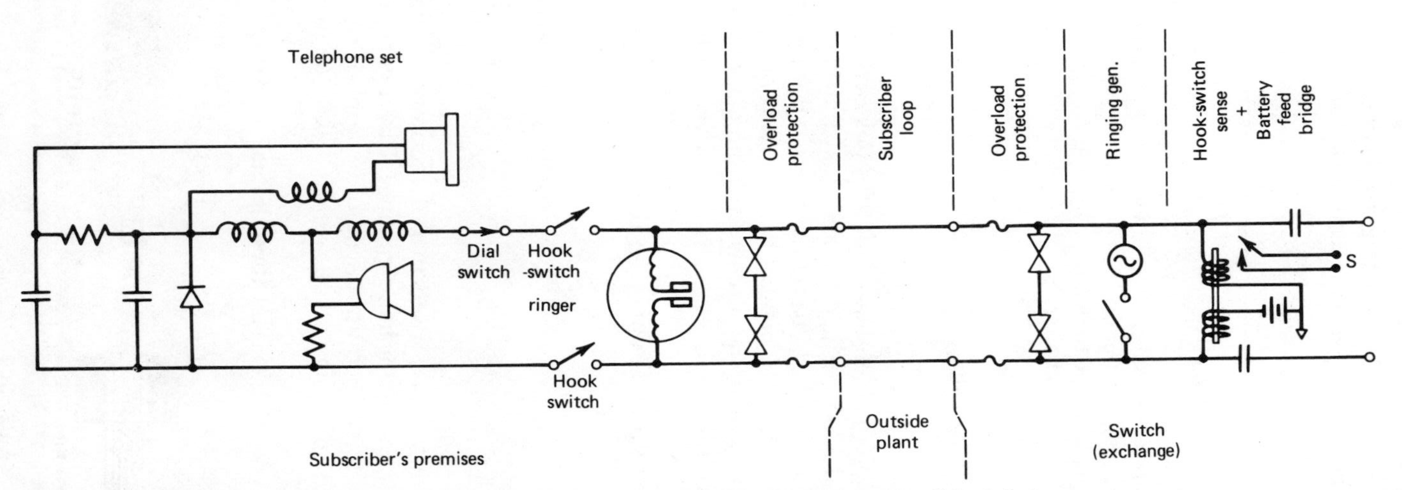

Figure 4.2 Signaling with a conventional telephone subset. Note functions of hook switch, dial, and ringer.

For conventional E and M signaling (referring to electromechanical exchanges), the following supervisory conditions are valid:

Direction		Condition at *A*		Condition at *B*	
Signal *A* to *B*	Signal *B* to *A*	M-Lead	E-Lead	M-Lead	E-Lead
On hook	On hook	Ground	Open	Ground	Open
Off hook	On hook	Battery	Open	Ground	Ground
On hook	Off hook	Ground	Ground	Battery	Open
Off hook	Off hook	Battery	Ground	Battery	Ground

2.2 Reverse Battery Signaling

Another method of supervisory signaling used on metallic pair trunks is *reverse battery signaling.* The on-hook and off-hook conditions are given by the polarity across the loop (e.g., direction of current flow). Polarity in the case of dc trunk loops refers to battery and ground state. Terminology in signaling often refers back to manual switchboards or, specifically, to the plug used with these boards and its corresponding jack as shown in Figure 4.3. The off-hook condition places the battery on the tip and ground on the ring of the plug. The on-hook condition has reverse current conditions with ground on the tip and battery on the ring.

3 AC SIGNALING

3.1 General

Up to this point we have reviewed two of the most employed means of supervisory trunk signaling (or line signaling). Direct-current signaling has notable limits on distance because it cannot be applied directly to carrier systems (Chapter 5) and is limited on metallic pairs due to the IR drop of the lines involved.

There are many ways to extend these limits, but from a cost effective-

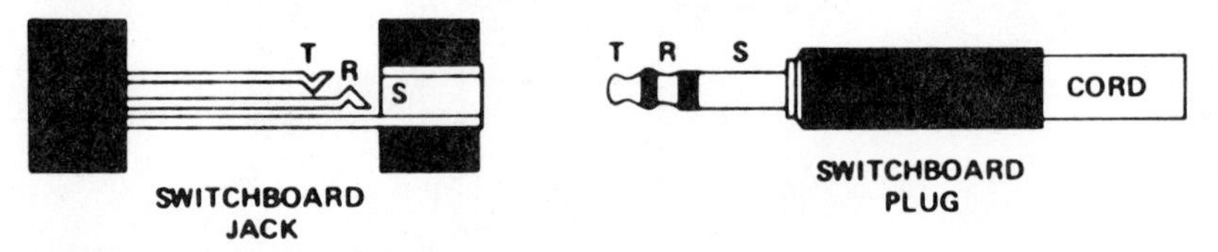

Figure 4.3 Switchboard plug with corresponding jack (R, S, and T are ring, sleeve, and tip, respectively).

ness standpoint there is a limit that we cannot afford to exceed. On trunks exceeding dc capabilities some form of ac signaling will be used. Traditionally, ac signaling systems are divided into three categories: low frequency, in band, and out-band (out-of-band) systems.

3.2 Low-Frequency AC Signaling Systems

An ac signaling system operating below the limits of the conventional voice channel (i.e., < 300 Hz) are termed *low frequency.* Low-frequency signaling systems are one-frequency systems, typically 50 Hz, 80 Hz, 135 Hz, or 200 Hz. It is impossible to operate such systems over carrier-derived channels (see Chapter 5) because of the excessive distortion and band limitation introduced. Thus low-frequency signaling is limited to metallic pair transmission systems. Even on these systems, cumulative distortion limits circuit length. A maximum of two repeaters may be used, and depending on the type of circuit (open wire, aerial cable, or buried cable) and wire gauge, a rough rule of thumb is a distance limit of 80 to 100 km.

3.3 In-Band Signaling

In-band signaling refers to signaling systems using audiotone, or tones inside the conventional voice channel, to convey signaling information. In-band signaling is broken down into three categories: (1) one-frequency (SF or single frequency), (2) two-frequency (2VF), and (3) multifrequency (MF). As the term implies, in-band signaling is where signaling is carried out directly in the voice channel. As the reader is aware, the conventional voice channel as defined by the CCITT occupies the band of frequencies from 300 to 3400 Hz. Single-frequency (SF) and two-frequency (2VF) signaling systems utilize the 2000-to-3000 Hz portion, where less speech energy is concentrated.

3.3.1 Single-Frequency Signaling

Single-frequency signaling is used almost exclusively for supervision. In some locations still it is used for interregister signaling, but the practice is diminishing in favor of more versatile methods such as MF. The most commonly used SF frequency is 2600 Hz, particularly in North America. On two-wire trunks 2600 Hz is used in one direction and 2400 Hz in the other. A diagram showing application of SF signaling on a four-wire trunk is shown in Figure 4.4.

3.3.2 Two-Frequency Signaling

Two-frequency (2VF) signaling is used for both supervision (line signaling) and address signaling. We often associate SF and 2VF supervisory signaling

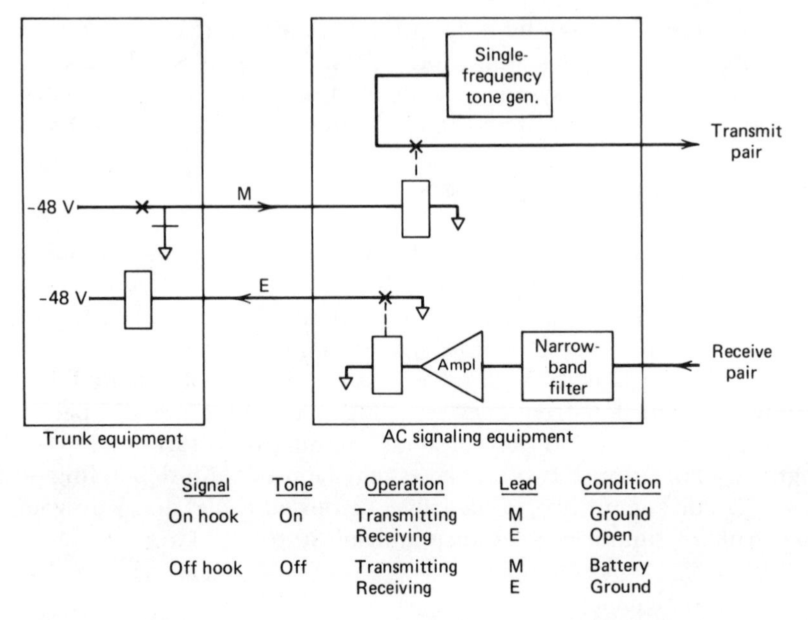

Signal	Tone	Operation	Lead	Condition
On hook	On	Transmitting	M	Ground
		Receiving	E	Open
Off hook	Off	Transmitting	M	Battery
		Receiving	E	Ground

Figure 4.4 Functional block diagram of a single-frequency signaling circuit (*Note:* Wire pairs, "receive" and "transmit," derive from carrier-equipment "receive" and "transmit" channels.)

systems with carrier (FDM multiplex) operation. Of course, when we discuss such types of line signaling (supervision), we know that the term "idle" refers to the on-hook condition and "busy," to the off-hook condition. Thus for such types of line signaling that are governed by audio tones of which SF and 2VF are typical, we have the conditions of "tone on when idle" and "tone on when busy." The discussion holds equally well for in-band and for out-of-band signaling methods.

However, for in-band signaling, supervision is by necessity carried out only when the call is being set up and when the call is being taken down or terminated ("gone on hook"). A major problem with in-band signaling is the possibility of "talk down," which refers to the premature activation or deactivation of supervisory equipment by an inadvertent sequence of voice tones through normal usage of the channel. Such tones could simulate the SF tone. Chances of simulating a 2VF tone set are much less likely. To avoid the possibility of talk-down on SF circuits, a time-delay circuit or slot filters to bypass signaling tones may be used. Such filters do offer some degradation to speech unless they are switched out during conversation.

Thus it becomes apparent why some administrations and telephone companies have turned to the use of 2VF for supervision. For example, a typical 2VF line signaling arrangement is the CCITT No. 5 code, where $f1$ (one of the two frequencies) is 2400 Hz and $f2$ is 2600 Hz. 2VF signaling is also used widely for address signaling (see Section 4.1 of this chapter).

3.4 Out-of-Band Signaling

With out-of-band signaling, supervisory information is transmitted out of band (e.g., > 3400 Hz). In all cases it is a single-frequency system. Some out-of-band systems use "tone on when idle," indicating the on-hook condition, whereas others use "tone off." The advantage of out-of-band signaling is that either system may be used, tone on or tone off when idle. Talk-down cannot occur because all supervisory information is passed out of band, away from the speech-information portion of the channel.

The preferred CCITT out-of-band frequency is 3825 Hz, whereas 3700 Hz is commonly used in the United States. It also must be kept in mind that out-of-band signaling is used exclusively on carrier systems, not on wire trunks. On the wire side, inside an exchange, its application is E and M signaling. In other words, out-of-band signaling is one method of extending E and M signaling over a carrier system.

In the short run, out-of-band signaling is attractive in terms of both economy and design. One drawback is that when channel patching is required, signaling leads have to be patched as well. In the long run the signaling equipment required may indeed make out-of-band signaling even more costly because of the extra supervisory signaling equipment and signaling lead extensions required at each end and at each time that the carrier (FDM) equipment demodulates to voice. The major advantage of out-of-band signaling is that continuous supervision is provided, whether tone on or tone off during the entire telephone conversation. In-band SF signaling and out-of-band signaling are illustrated in Figure 4.5A,B. An example of out-of-band signaling is the CCITT R-2 System (CCITT Rec. Q.351) (see Table 4.1).

4 ADDRESS SIGNALING—INTRODUCTION

Address signaling originates as dialed digits from a calling subscriber, whose local switch accepts these digits and, using that information, directs

Table 4.1 R-2 Line Signaling (3825 Hz)

Circuit State	Direction: Forward (Go)	Direction: Backward (Return)
Idle	Tone on	Tone on
Seized	Tone off	Tone on
Answered	Tone off	Tone off
Clear back	Tone off	Tone on
Release	Tone on	Tone on or off
Blocked	Tone on	Tone off

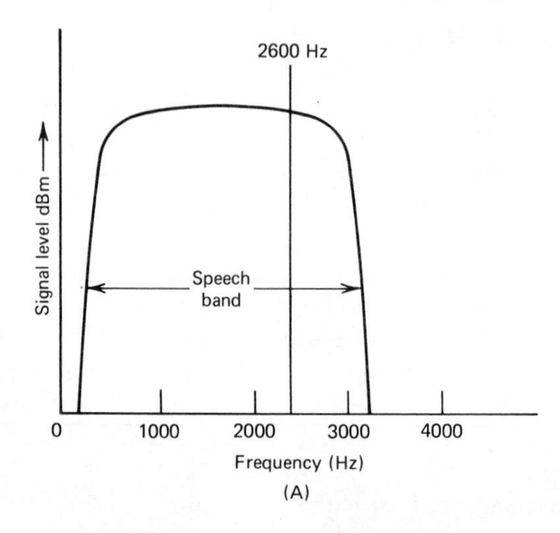

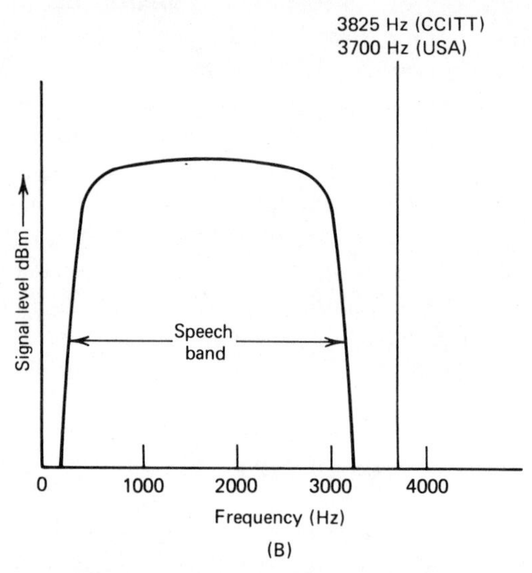

Figure 4.5 Single-frequency signaling: (A) in band; (B) out of band.

the telephone call to the desired distant subscriber. If more than one switch is involved in the call setup, signaling is required between switches (both address and supervisory). Address signaling between switches in conventional systems is called *interregister signaling.*

The paragraphs that follow discuss various more popular standard ac signaling techniques such as 2VF, MF (multifrequency) pulse, MF tone, and the more advanced common-channel signaling. Although interregister signaling is stressed where appropriate, some supervisory techniques are also reviewed.

Table 4.2 CCITT Signaling Code System No. 4

Digit	Number	Combination Elements			
		1	2	3	4
1	1	*y*	*y*	*y*	*x*
2	2	*y*	*y*	*x*	*y*
3	3	*y*	*y*	*x*	*x*
4	4	*y*	*x*	*y*	*y*
5	5	*y*	*x*	*y*	*x*
6	6	*y*	*x*	*x*	*y*
7	7	*y*	*x*	*x*	*x*
8	8	*x*	*y*	*y*	*y*
9	9	*x*	*y*	*y*	*x*
0	10	*x*	*y*	*x*	*y*
Call operator code 11	11	*x*	*y*	*x*	*x*
Call operator code 12	12	*x*	*x*	*y*	*y*
Spare code (see CCITT Rec. Q.104)	13	*x*	*x*	*y*	*x*
Incoming half-echo suppressor required	14	*x*	*x*	*x*	*y*
End of pulsing	15	*x*	*x*	*x*	*x*
Spare code	16	*y*	*y*	*y*	*y*

Sending duration of binary elements 35 ± 7 ms. Sending duration of blank elements between binary elements 35 ± 7 ms. Element *x* is 2040 Hz; element *y* is 2400 Hz.

4.1 Two-Frequency Pulse Signaling

Two-frequency (2VF) signaling is commonly used as an interregister mode of signaling employing the speech band for the transmission of information. It may also be used for line signaling.* There are various methods of using two-voice frequencies to transmit signaling information. For example, CCITT No. 4 uses 2040 Hz and 2400 Hz to represent binary 0 and 1, respectively. It uses a four-element code, permitting 16 different coded characters as shown in Table 4.2. With the CCITT No. 4 code both interregister and line signaling utilize the 2VF technique. Interregister signaling in this case is a pulse-type signaling, and line signaling utilizes the combination of the two frequencies and the duration of signal to convey the necessary supervisory information as shown in Table 4.3.

As we mentioned, line signaling with the CCITT No. 4 code is based on signal duration as well as frequency. The line-signaling format uses both

*Supervision (interswitch).

Table 4.3 CCITT No. 4 Line Signaling

Forward Signals	
Terminal seizing	*Px*
Transit seizing	*Py*
Numerical signals	As in Table 4.2
Clear forward	*Pxx*
Forward transfer	*Pyy*
Backward Signals	
Proceed to send	
Terminal	*x*
International transit	*y*
Number received	*p*
Busy flash	*pX*
Answer	*pY*
Clear back	*Px*
Release guard	*Pyy*
Blocking	*Px* (congestion)
Unblocking	*Pyy*

tone frequencies, 2040 Hz and 2400 Hz. Each line signal consists of an initial *prefix* (P) signal followed by a control signal element, called a *suffix*. The P signal consists of both frequencies (2VF), and the suffix signal consists of one frequency, where x is 2040 Hz and y is 2400 Hz (see Table 4.3). Now consider the durations of the following signal elements used for line signaling:

$$P = 150 \pm 30 \text{ ms}$$
$$x, y = 100 \pm 20 \text{ ms (each)}$$
$$xx\ yy = 350 \pm 70 \text{ ms (each)}$$

This set of values refers to transmitted signal duration, that is, as transmitted by the signaling sender.

Let us see how these signals are used in the line (interswitch supervision) signaling (see Table 4.3).

The supervisory functions "clear forward" and "release guard" do the reverse of "call setup." They take the call down or disconnect, readying the circuit for the next user. "Clear back" is another example.

4.2 Multifrequency Signaling

Multifrequency (MF) signaling is in wide use today for interregister signaling. It is an in-band method utilizing five or six tone frequencies, two at a time. Multifrequency signaling works equally well over metallic and carrier (FDM) systems. Four commonly used MF signaling systems follow with a short discussion for each. Tables 4.10 and 4.11 show the North American MF push-button codes and subscriber audible tones, respectively.

Table 4.4 Basic SOCOTEL MF Signaling Code

Tone Frequencies (Hz)[a]	Digit
700 + 900	1
700 + 1100	2
900 + 1100	3
700 + 1300	4
900 + 1300	5
1100 + 1300	6
700 + 1500	7
900 + 1500	8
1100 + 1500	9
1300 + 1500	0

[a]The 1700-Hz frequency is also used for signaling system check and when more code groups are required by a telephone company or national administration. When 1700 Hz is used for coding, 1900 Hz is used for checking the system; 1700 Hz and/or 1900 Hz may also be used for control purposes.

4.2.1 SOCOTEL

SOCOTEL is an interregister signaling system used principally in France, areas of French influence, and Spain with some modifications. The frequency pairs and their digit equivalents are shown in Table 4.4. Line signaling used with SOCOTEL may be dc, 50 Hz, or 2000 Hz. The same frequencies are used in both directions.

4.2.2 Multifrequency Signaling in North America—The R-1 Code

The MF signaling system principally used in the United States and Canada is recognized by the CCITT as the R-1 code. It is a two-out-of-five frequency-pulse system. Additional signals for control functions are provided by combinations using a sixth frequency. Table 4.5 shows digits or other applications and their corresponding frequency combinations as well as a brief explanation of "other applications."

4.2.3 CCITT No. 5 Signaling Code

Interregister signaling with the CCITT No. 5 code is very similar in makeup to the North American R-1 code. Variations with R-1 are shown in Table 4.6. The CCITT No. 5 Line Signaling Code is shown in Table 4.7.

4.2.4 The R-2 Code

The R-2 code is listed by CCITT (Rec. Q.361) as a European regional signaling code. Taking full advantage of combinations of two-out-of-six

Table 4.5 The R-1 Code (North American MF)

Digit	Frequency Pair (Hz)	
1	700 + 900	
2	700 + 1100	
3	900 + 1100	
4	700 + 1300	
5	900 + 1300	
6	1100 + 1300	
7	700 + 1500	
8	900 + 1500	
9	1100 + 1500	
10 (0)	1300 + 1500	
Use	**Frequency Pair**	**Explanation**
KP	1100 + 1700	Preparatory for digits
ST	1500 + 1700	End of pulsing sequence
STP	900 + 1700	Used with TSPS (traffic service position system)
ST2P	1300 + 1100	Used with TSPS (traffic service position system)
ST3P	700 + 1700	Used with TSPS (traffic service position system)
Coin collect	700 + 1100	Coin control
Coin return	1100 + 1700	Coin control
Ringback	700 + 1700	Coin control
Code 11	700 + 1700	Inward operator (CCITT No. 5)
Code 12	900 + 1700	Delay operator
KP1	1100 + 1700	Terminal call
KP2	1300 + 1700	Transit call

[a]Pulsing of digits is at the rate of about seven digits per second with an interdigital period of 68 ± 7 ms. For intercontinental dialing for CCITT No. 5 code compatibility, the R-1 rate is increased to 10 digits per second. The KP pulse duration is 100 ms.

Table 4.6 CCITT No. 5 Code[a] Showing Variations with R-1 Code

Signal	Frequencies (Hz)	Remarks
KP1	1100 + 1700	Terminal traffic
KP2	1300 + 1700	Transit traffic
1	700 + 900	
2	700 + 1100	
3–0	Same as Table 4.5	
ST	1500 + 1700	
Code 11	700 + 1700	Code 11 operator
Code 12	900 + 1700	Code 12 operator

[a]Line signalling for CCITT No. 5 code is 2 VF, with f_1 2400 Hz and f_2, 2600 Hz. Line-signaling conditions are shown in Table 4.7.

Table 4.7 Line-signal Code CCITT No. 5

Signal	Direction	Frequency	Sending Duration	Recognition Time (ms)
Seizing	⟶	$f1$	Continuous	40 ± 10
Proceed to send	⟵	$f2$	Continuous	40 ± 10
Busy flash	⟵	$f2$	Continuous	125 ± 25
Acknowledgment	⟶	$f1$	Continuous	125 ± 25
Answer	⟵	$f1$	Continuous	125 ± 25
Acknowledgment	⟶	$f1$	Continuous	125 ± 25
Clear back	⟵	$f2$	Continuous	125 ± 25
Acknowledgment	⟶	$f1$	Continuous	125 ± 25
Forward Transfer	⟶	$f2$	850 ± 200 ms	125 ± 25
Clear forward	⟶	$f1 + f2$	Continuous	125 ± 25
Release guard	⟵	$f1 + f2$	Continuous	125 ± 25

tone frequencies, 15 frequency-pair possibilities are available. This number is doubled in each direction by having meaning groups I and II in the forward direction and groups A and B in the backward direction (see Table 4.8).

Groups I and A are said to be of primary meaning and groups II and B, secondary. The change from primary to secondary meaning is commanded by the backward signal A-3 or A-5. Secondary meanings can be changed back to primary meanings only when the original change from primary to secondary was made by the use of the A-5 signal. Referring to Table 4.8, the 10 digits to be sent in the forward direction in the R-2 system are in group I and are index numbers 1 through 10 in the table. The index 15 signal (group A) indicates "congestion in an international exchange or at its output." This is a typical backward information signal giving circuit status information. Group B consists of nearly all "backward information" and in particular deals with subscriber status.

The R-2 line-signaling system has two versions: the one used on analog networks is discussed here; the other, on digital (PCM) networks, is briefly covered in Chapter 9. The analog version is an out-of-band, tone-on-when-idle system. Table 4.9 shows the line conditions in each direction, forward and backward. Note that the code takes advantage of a signal sequence that has six characteristic operating conditions.

Let us consider several of these conditions.

Seized. The outgoing exchange (call-originating exchange) removes the tone in the forward direction. If seizure is immediately followed by release,

Table 4.8 European R-2 System

Index No. for Groups I/II and A/B	Frequencies (Hz)						
	1380	1500	1620	1740	1860	1980	Forward Direction I/II
	1140	1020	900	780	660	540	Backward Direction A/B
1	*x*	*x*					
2	*x*		*x*				
3		*x*	*x*				
4	*x*			*x*			
5		*x*		*x*			
6			*x*	*x*			
7	*x*				*x*		
8		*x*			*x*		
9			*x*		*x*		
10				*x*	*x*		
11	*x*					*x*	
12		*x*				*x*	
13			*x*			*x*	
				x		*x*	
15					*x*	*x*	

removal of the tone must be maintained for at least 100 ms to assure that it is recognized at the incoming end.

Answered. The incoming end removes the tone in the backward direction. When another link of the connection using tone-on-when-idle continuous signaling precedes the outgoing exchange, the "tone-off" condition must be established on the link as soon as it is recognized in this exchange.

Table 4.9 Line Conditions for R-2 Code

Operating Condition of the Circuit	Signaling Conditions	
	Forward	Backward
1. Idle	Tone on	Tone on
2. Seized	Tone off	Tone on
3. Answered	Tone off	Tone off
4. Clear back	Tone off	Tone on
5. Release	Tone on	Tone on or off
6. Blocked	Tone on	Tone off

Table 4.10 North American Push-Button Codes

Digit	Dial Pulses (Breaks)	Multifrequency Push-button Tones
0	10	941, 1336 Hz
1	1	697, 1209 Hz
2	2	697, 1336 Hz
3	3	697, 1474 Hz
4	4	770, 1209 Hz
5	5	770, 1336 Hz
6	6	770, 1477 Hz
7	7	852, 1209 Hz
8	8	852, 1336 Hz
9	9	852, 1477 Hz

Clear back. The incoming end restores the tone in the backward direction. When another link of the connection using tone-on-when-idle continuous signaling precedes the outgoing exchange, the "tone-off" condition must be established on this link as soon as it is recognized in this exchange.

Clear-forward. The outgoing end restores the tone in the forward direction.

Blocked. At the outgoing exchange the circuit stays blocked as long as the tone remains off in the backward direction.

4.2.5 Subscriber Tones and Push-Button Codes (North America)

Subscriber subsets in many places in the world are either dial or push button. The push-button type is more versatile, and more rapid dialing may be accomplished by a subscriber. Table 4.10 compares digit dialed, dial pulses (breaks), and MF push-button tones. Table 4.11 shows the audible

Table 4.11 Audible Tones Commonly Used in North America

Tone	Frequencies (Hz)	Cadence
Dial	350 + 440	Continuous
Busy (station)	480 + 620	0.5 s on, 0.5 s off
Busy (network congestion)	480 + 620	0.2 s on, 0.3 s off
Ring return	440 + 480	2 s on, 4 s off
Off-hook alert	Multifreq. howl	1 s on, 1 s off
Recording warning	1400	0.5 s on, 15 s off
Call waiting	440	0.3 s on, 9.7 s off

tones commonly used in North America. Functionally, these are the call-progress tones presented to the subscriber.

5 COMPELLED SIGNALING

In many of the signaling systems discussed thus far, signal element duration is an important parameter. For instance, in a call setup an initiating exchange sends a 100-ms seizure signal. Once this signal is received at the distant end, the distant exchange sends a "proceed-to-send" signal back to the originating exchange; in the case of the R-1 system, this signal is 140 ms or more in duration. Then, on receipt of "proceed to send" the initiating exchange spills all digits forward. In the case of R-1 each digit is an MF pulse of 68 ms duration with 68 ms between each pulse. After the last address digit an ST (end-of-pulsing) signal is sent. In the case of R-1 the incoming (far-end) switch register knows the number of digits to expect. Thus there is an explicit acknowledgment that the call setup has proceeded satisfactorily. Thus R-1 is a good example of noncompelled signaling.

A fully compelled signaling system is one in which each signal continues to be sent until an acknowledgment is received. Thus signal duration is not significant and bears no meaning. The R-2 and SOCOTEL are examples of fully compelled signaling systems. Figure 4.6 shows a fully compelled signaling sequence. Note the small overlap of signals, causing that the acknowledging (reverse) signal to start after a fixed time on receipt of the forward signal. This is because of the minimum time required for recognition of the incoming signal. After the initial forward signal, further forward signals are delayed for that short recognition time (see Figure 4.6). Recognition time is normally less than 80 ms.

Fully compelled signaling is advantageous in that signaling receivers do not have to measure duration of each signal, thus making signaling equipment simpler and more economical. Fully compelled signaling adapts automatically to the velocity of propagation, to long circuits, to short circuits, to metallic pairs, or to carrier, and is designed to stand short interruptions in the transmission path. The principal drawback of compelled signaling is its inherent lower speed, thus requiring more time for setup. Setup time over space-satellite circuits with compelled signaling is appreciable and may force the system engineer to seek a compromise signaling system.

There is also a partially compelled type of signaling, where signal duration is fixed in both forward and backward directions according to system specifications; or the forward signal is of indefinite duration and the backward is of fixed duration. The forward signal ceases once the backward signal has been received correctly. The CCITT No. 4 is a variation of partially compelled signaling.

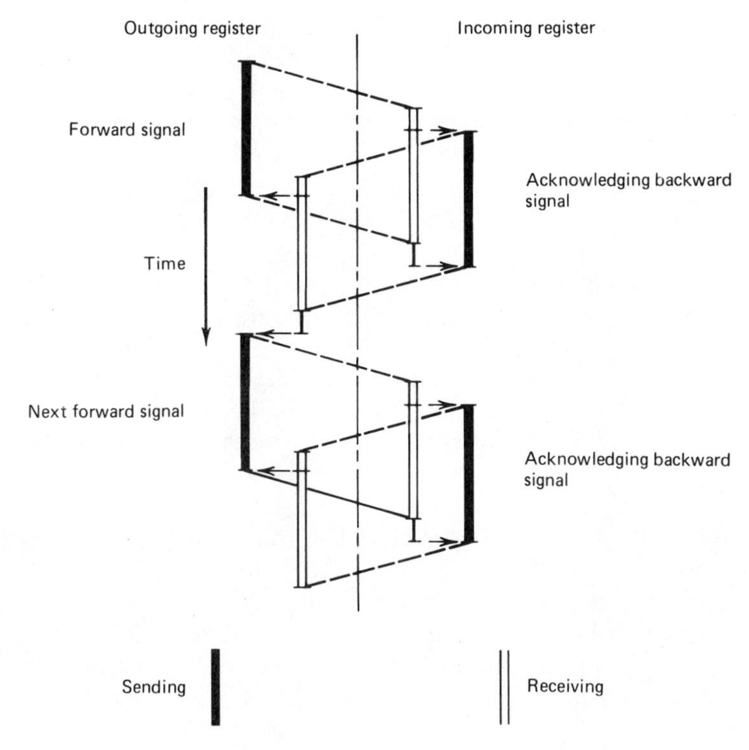

Figure 4.6 Compelled signaling procedure.

6 LINK-BY-LINK VERSUS END-TO-END SIGNALING

An important factor to be considered in switching system design that directly affects both signaling and customer satisfaction is postdialing delay. This is the amount of time it takes after the calling subscriber completes dialing until ring-back is received. Ring-back is a backward signal to the calling subscriber telling him that his dialed number is ringing. Postdialing delay must be made as short as possible.

Another important consideration is register occupancy time for call setup as the setup proceeds from originating exchange to terminating exchange. Call-setup equipment, that equipment used to establish a speech path through a switch and to select the proper outgoing trunk, is expensive. By reducing register occupancy per call, we may be able to reduce the number of registers (and markers) per switch, thus saving money.

Link-by-link and end-to-end signaling each affect register occupancy and postdialing delay, each differently. Of course, we are considering calls involving one or more tandem exchanges in a call setup, as this situation usually occurs on long distance or toll calls. Link-by-link signaling may be defined as a signaling system where *all* interregister address information must be transferred to the subsequent exchange in the call-setup routing. Once this information is received at this exchange, the preceding exchange control unit (register) releases. This same operation is carried on from the originating exchange through each tandem (transit) exchange to the terminating exchange of the call. The R-1 system is an example of link-by-link signaling.

End-to-end signaling abbreviates the process such that tandem (transit) exchanges receive only the minimum information necessary to route the call. For instance, the last four digits of a seven-digit telephone number need be exchanged only between the originating exchange (e.g., the calling subscriber's local exchange or the first toll exchange in the call setup) and the terminating exchange in the call setup. With this type of signaling, fewer digits are required to be sent (and acknowledged) for the overall call-setup sequence. Thus the signaling process may be carried out much more rapidly, decreasing postdialing delay. Intervening exchanges on the call route work much less, handling only the digits necessary to pass the call to the next exchange in the sequence.

The key to end-to-end signaling is the concept of "leading register." This is the register (control unit) in the originating exchange that controls the call routing until a speech path is set up to the terminating exchange before releasing to prepare for another call setup. For example, consider a call from subscriber X to subscriber Y.

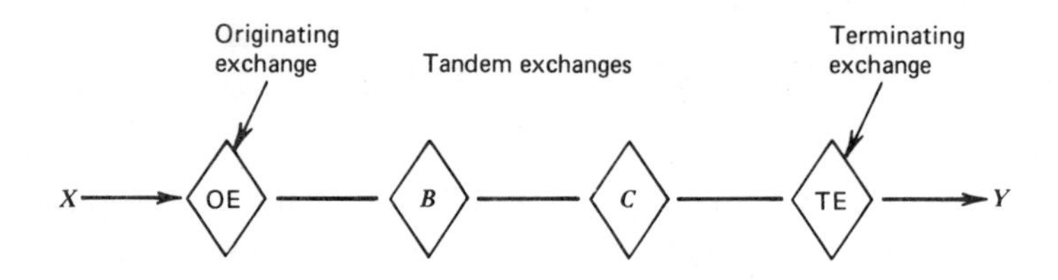

The telephone number of subscriber Y is 345-6789. The sequence of events is as follows using end-to-end signaling:

- A register at exchange OE receives and stores the dialed number 345-6789 from subscriber X.
- Exchange OE analyzes the number and then seizes a trunk (junction) to exchange B. It then receives a "proceed-to-send" signal indicating that the register at B is ready to receive routing information (digits).
- Exchange OE then sends digits 34, which are the minimum necessary to effect correct transit.
- Exchange B analyzes the digits 34 and then seizes a trunk to exchange

C. Exchanges OE and *C* are now in direct contact and exchange *B*'s register releases.

- Exchange OE receives the "proceed-to-send" signal from exchange *C* and then sends digits 45, those required to effect proper transit at *C*.
- Exchange *C* analyzes digits 45 and then seizes a trunk to exchange TE. Direct communication is then established between leading register for this call at OE and the register at TE being used on this call setup. The register at *C* then releases.
- Exchange OE receives the "proceed-to-send" signal from exchange TE, to which it sends digits 5678, the subscriber number.
- Exchange TE selects the correct subscriber line and returns to *A* ring-back, line busy, out of order, or other information after which all registers are released.

Thus we see that a signaling path is opened between the leading register and the terminating exchange. To accomplish this, each exchange in the route must "know" its local routing arrangements and request from the leading register those digits it needs to further route the call along its proper course.

Again, the need for backward information becomes evident, and backward signaling capabilities must be nearly as rich as forward signaling capabilities when such a system is implemented.

R-1 is a system inherently requiring little backward information (interregister). The little information that is needed, such as "proceed to send," is sent via line signaling. The R-2 system has major backward information requirements, and backward information and even congestion and busy signals are sent back by interregister signals.

7 THE EFFECTS OF NUMBERING ON SIGNALING

Numbering, the assignment and use of telephone numbers, affects signaling as well as switching. It is the number or the translated number, as we found out in Chapter 3, that routes the call. There is "uniform" numbering and "nonuniform" numbering. How does each affect signaling? Uniform numbering can simplify a signaling system. Most uniform systems in the nontoll or local-area case are based on seven digits, although some are based on six. The last four digits identify the subscriber. The first three digits (or the first two in the case of a six-digit system) identify the exchange. Thus the local exchange or transit exchanges know when all digits are received. There are two advantages to this sort of scheme:

1. The switch can proceed with the call once all digits are received because it "knows" when the last digit (either the sixth or seventh) has been received.

2. "Knowing" the number of digits to expect provides inherent error control and makes "time out"* simpler.

For nonuniform numbering, particularly on direct distance dialing in the international service, switches require considerably more intelligence built in. It is the initial digit or digits that will tell how many digits are to follow, at least in theory.

However, in local or national systems with nonuniform numbering, the originating register has no way of knowing whether it has received the last digit, with the exception of receiving the maximum total used in the national system. With nonuniform numbering an incompletely dialed call can cause a useless call setup across a network up to the terminating exchange, and the call setup is released only after time out has run its course. It is evident that with nonuniform numbering systems, national (and international) networks are better suited to signaling systems operating end-to-end with good features of backward information such as the R-2 system.

8 COMMON-CHANNEL SIGNALING

8.1 General

There are two types of common-channel signaling (CCS) in existence today, European CCITT No. 6 and the North American CCIS (common-channel interoffice signaling). In our previous discussion all signaling systems covered were associated channel systems, meaning that each voice trunk carried its own signaling, both supervisory (line) and interregister. Signaling was associated with the channel, whether in band, out of band, pulse or MF, or MF pulse.

Common-channel signaling separates signaling from its associated speech path by placing the signaling from a group (or several groups) of voice trunks on a separate path dedicated to signaling only. The signaling information is transmitted by means of serial binary data. Serial binary data transmission is discussed further in Chapter 8.

The basic distinction between conventional associated channel signaling and CCS is shown in Figure 4.7.

From Figure 4.7 we see that CCS is feasible only between processor-controlled exchanges, more commonly known as *stored-program control* (SPC) exchanges. The advantages of CCS become obvious between switch processors. Signaling on the telephone network is essentially digital for both line and interregister signaling. Why, then, translate into a complex analog mode rather than just leaving it digital where it belongs? Common-

*"Time out" is the resetting of call-setup equipment and return of dial tone to subscriber as a result of incomplete signaling procedure, subset left off hook, and so forth.

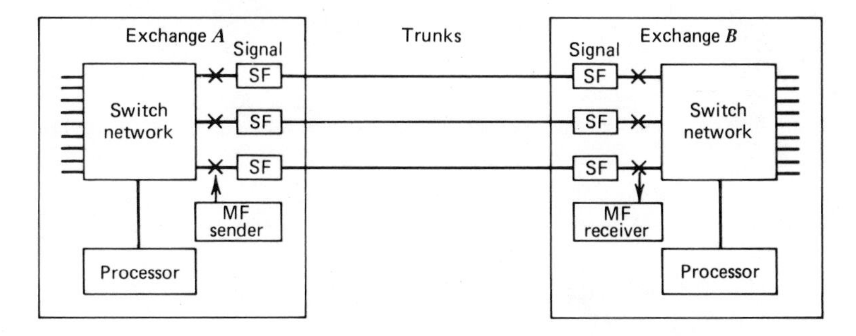

CONVENTIONAL SF-MF

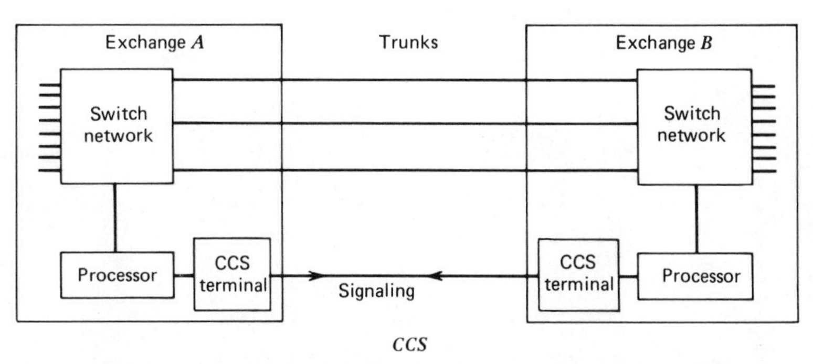

CCS

Figure 4.7 Conventional analog versus CCS signaling techniques. (*Note:* Signaling in upper drawing accompanies voice paths; signaling in lower drawing is conveyed on a separate circuit.)

channel signaling does just this. Of course, the interprocessor digital information has to be conditioned for analog voice channels; this is done by the data modem.

Figure 4.8 illustrates the basic functional components of CCS. The diagram is valid for both CCITT No. 6 and CCIS operation. Essentially, then, leaving aside error control features, a CCS signaling link consists of the voice-frequency channel (four-wire), two signaling terminals, and two modems. The signaling terminals store both incoming signaling information awaiting processing and outgoing signaling information awaiting transmission. The terminals also perform error control in the North American CCIS system.

With conventional signaling, the signal path and the speech path or voice channel occupy the same media; if signaling is effected, there is continuity of the voice or speech path. Because CCS does not pass signaling over the speech trunks that are to be set up and "supervised," call-path continuity must be checked once a call is set up. This is done with tone transceivers that are connected at the time of setup to assure path continuity. With

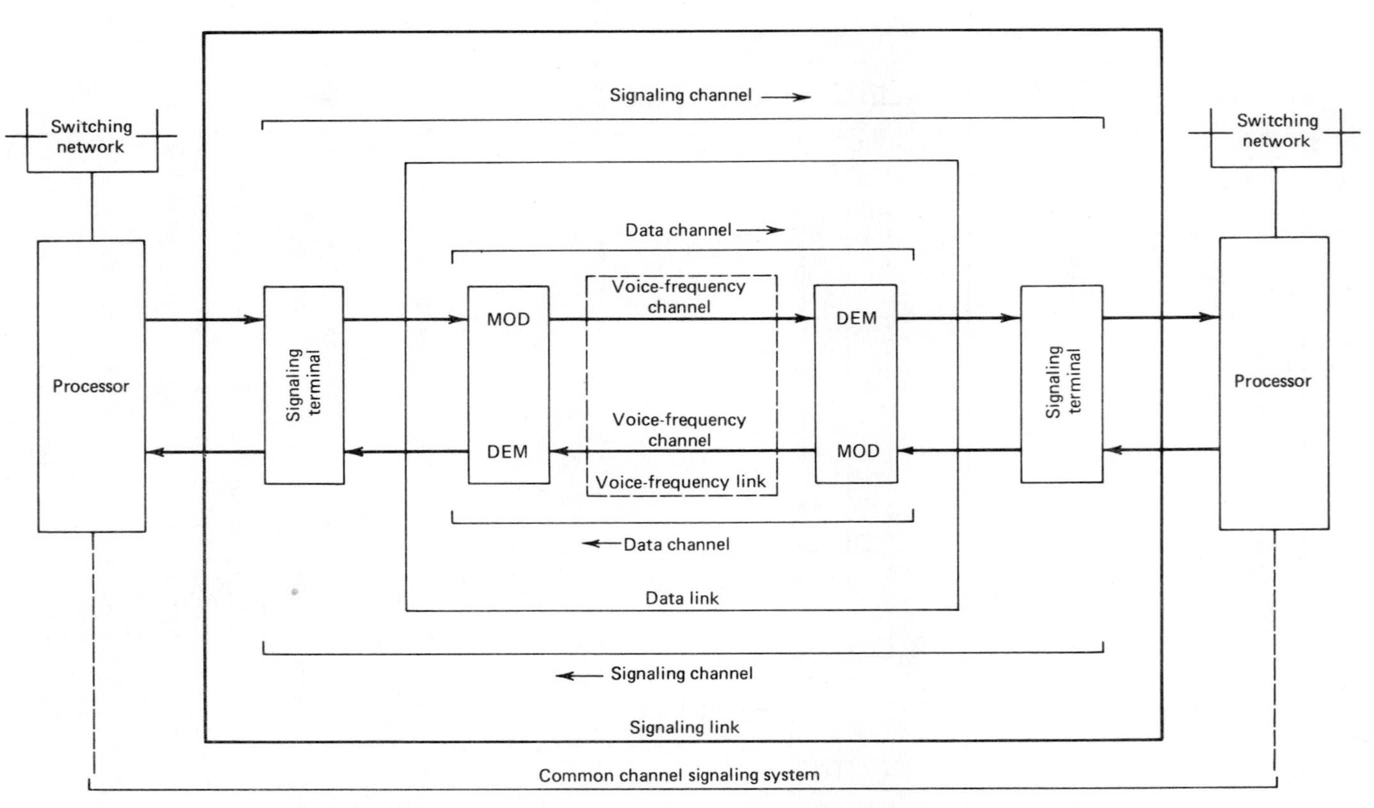

Figure 4.8 Basic functional components of common-channel signaling system. Courtesy CCITT-ITU.

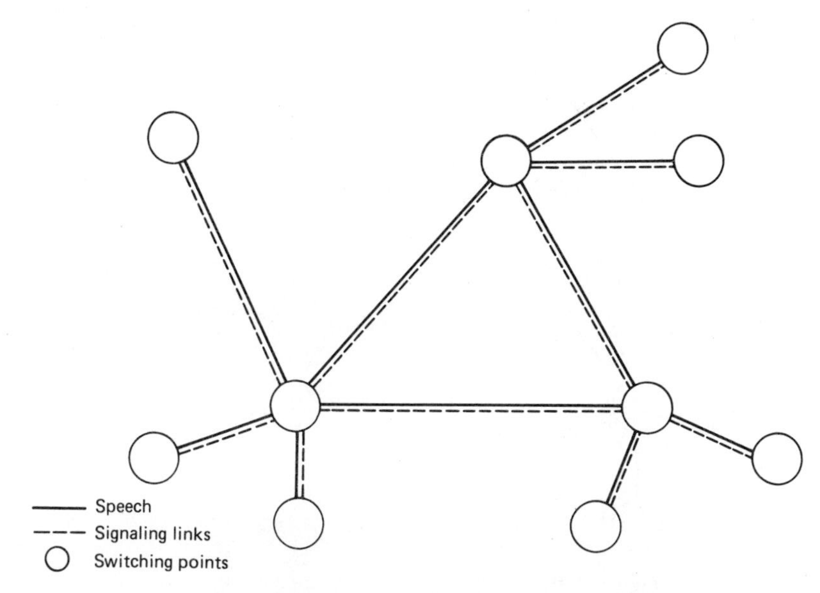

Figure 4.9 Associated CCIS signaling mode.

CCIS the transceivers operate at 2010 Hz on four-wire trunks. On two-wire trunks, 1780 Hz is transmitted by the originating exchange, and a 2010-Hz tone is returned by the terminating exchange. When no continuity is achieved, a second attempt is made and the failed trunk is blocked. The blocked trunk is then retested. If it continues to fail continuity with CCIS, a trunk failure message is printed at the maintenance position.

The CCIS (North America) is designed to operate in two basic signaling modes of operation, referred to as *associated* and *nonassociated.* In the associated mode a separate voice channel or channels carry signaling information, and this channel(s) is (are) routed with the speech channels it (they) serve. Such a concept is represented in Figure 4.9. Compare this to Figure 4.10, which depicts the concept of fully nonassociated signaling, where the signaling information traverses a routing entirely separate from the voice paths that it controls.

Signal-transfer points (STPs)* are used in North America with nonassociated signaling. A signal-transfer point consists of a processor with signaling terminals and data modems (see Chapter 8) on either side. In effect, an STP is a data-message transfer point or switching point, if you will. Signal-transfer points allow the concentration of signaling for a large number of trunks and will also provide for circuit viability by allowing alternative routing of the CCIS signaling path.

*Signal-transfer points are also used in the nonassociated mode with CCITT No. 6; however, North American practice is discussed here.

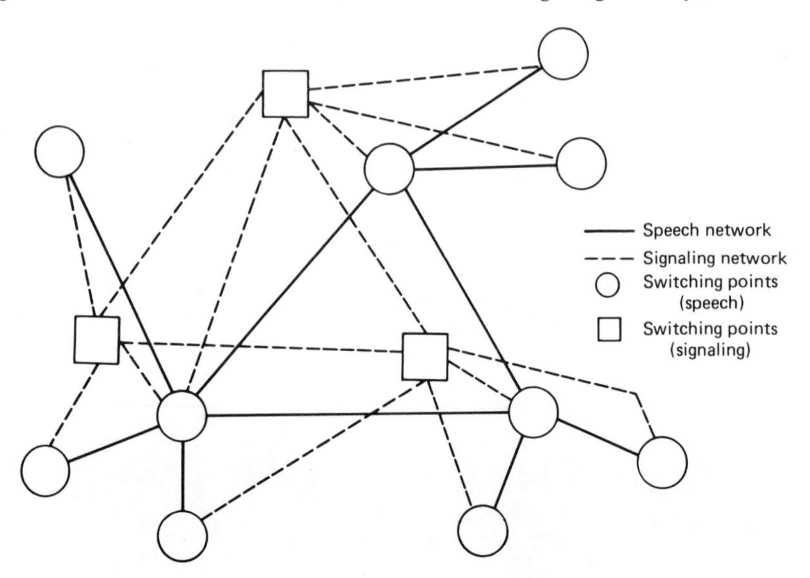

Figure 4.10 Nonassociated CCIS signaling mode.

To improve viability (reliability of communications), all CCIS signaling paths in North America are duplicated and fully redundant. In addition, all STPs are fully interconnected to provide several signaling-path possibilities. The STPs are overdimensioned, so that either STP can carry the full signaling load if one STP fails. Such a network (see Figure 4.11) has a hierarchical structure, with primary and secondary centers. Of course, it is fully nonassociated.

For cost-effectiveness, associated signaling is generally used with large trunk groups. The concentration aspects of nonassociated signaling would make it more attractive for small trunk groups in the long-distance (toll) network. It is also undesirable to have many STPs in tandem because of the added delay. Each STP processes signal units from its input link to its output link, each adding incremental delay. The limit of STPs in tandem is set at two in the normal or primary path by ATT and four under failure conditions.

8.2 Common-Channel Interoffice Signaling Format

Both CCIS and CCITT No. 6 signaling systems carry signaling information in a serial binary format. Thus we are dealing with digital data transmission, which is introduced in Chapter 8.

The basic (data) word in the CCIS system is the signal unit (SU). A signal unit is 28 bits long, with the last 8 bits used for error checking. Thus the signaling information is actually contained in the first 20 bits of the basic "word." Signal units are grouped into blocks of 12 for transmission; thus a

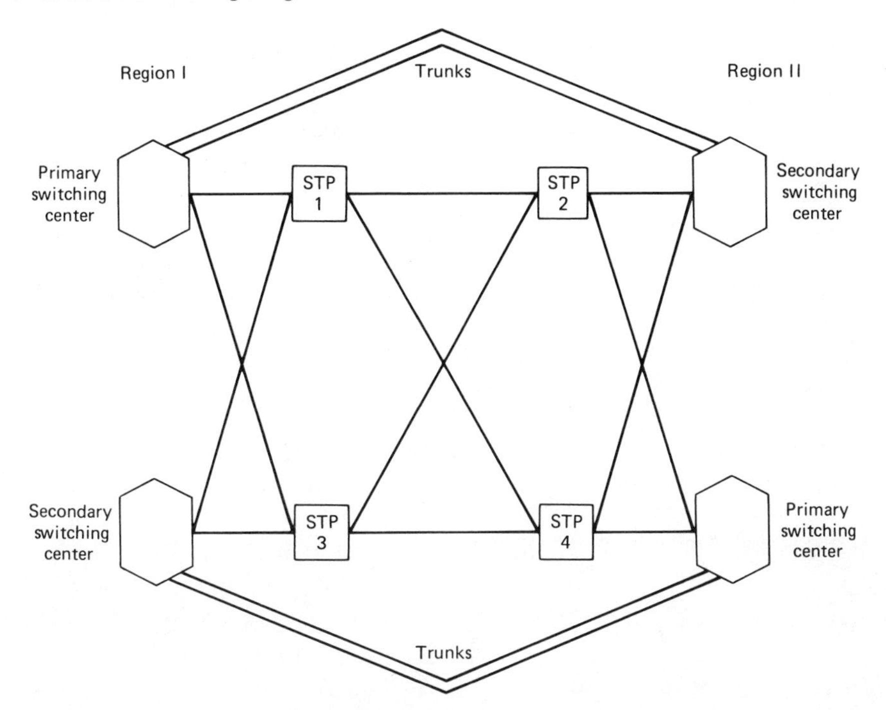

Figure 4.11 Nonassociated CCIS signaling mode, according to concept of signal-transfer point (STP) (CCITT hierarchy) (single solid lines = STP signaling links). Copyright © American Telephone and Telegraph Co., 1975.

block contains 12 × 28 or 336 serial bits. The last signal unit of each block is the acknowledgment signal unit (ACU).

Common-channel interoffice "messages" can be one or more SUs long. Length, as we can imagine, depends on the amount of information to be sent. There are single-unit messages, called *lone signal unit* (LSU) and *multiunit messages* (MUM). The LSU is generally used for specific control information (e.g., "answer"), whereas the MUMs are generally used for passing address information such as digits. Figure 4.12 shows the format of an LSU and a MUM. Note the trunk label, which is used to identify the trunk being served. The trunk label is subdivided into two fields: (1) the band number, one or more of which is associated with a trunk group and that is used to determine the routing of the message in the signaling network, and (2) the trunk number that identifies a specific trunk (see Figure 4.12). The number of trunk labels handled by a link gives a measure of CCS link capacity.

The MUM shown in Figure 4-12 has an ISU (initial signal unit) and a single SSU (subsequent signal unit). The ISU has a unique heading code identifying it as the message start (ISU) of a MUM. The ISU then has a length indicator, (②) in Figure 4.12B, which indicates the number of

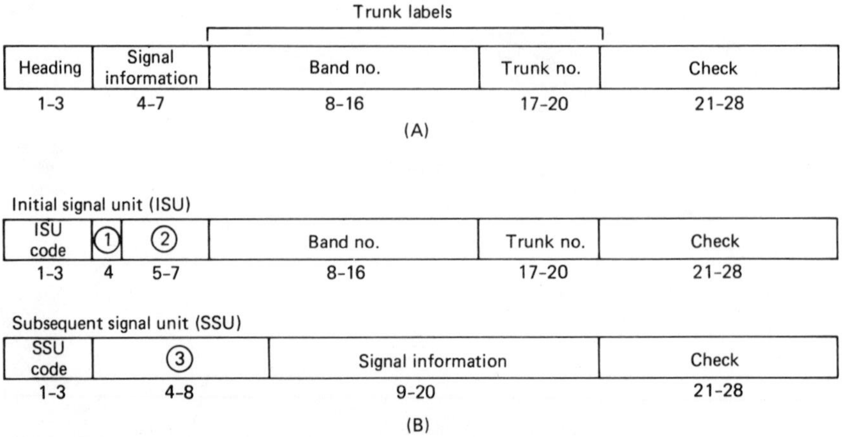

Trunk labels

Heading	Signal information	Band no.	Trunk no.	Check
1-3	4-7	8-16	17-20	21-28

(A)

Initial signal unit (ISU)

ISU code	①	②	Band no.	Trunk no.	Check
1-3	4	5-7	8-16	17-20	21-28

Subsequent signal unit (SSU)

SSU code	③	Signal information	Check
1-3	4-8	9-20	21-28

(B)

Figure 4.12 Message formats of CCIS signals: (A) lone-signal unit (LSU); (B) multiunit message (MUM) [① ISU-type indicator, ② length indicator, ③ message category]. Copyright © American Telephone and Telegraph Co., 1975.

subsequent signal units (SSU) to follow. The SSU starts with two data fields. The first, a unique heading code, is used to identify the SSU (i.e., to differentiate it from LSU, ISU, etc.). The second field gives message category to identify the type of MUM. This is shown in Table 4.12.

The heading field of three bits (binary digits) provide eight different combinations of 1's and 0's (see Chapter 8) to identify signal groups. These combinations of 1's and 0's may be called a *code*. These eight combinations are shown in Table 4.12.

Only the first 11 SUs of a signaling message block carry signaling infor-

Table 4.12 Binary Combinations

Heading Code	Signal Unit Type
000	Lone signal unit (LSU)—telephone signals
001	Lone signal unit (LSU)—telephone signals
010	Lone signal unit (LSU)—telephone signals
011	Acknowledge signal unit (ACU)
100	Lone signal unit (LSU)—telephone signals
101	Initial signal unit (ISU)
110	Subsequent signal unit (SSU)
111	Lone signal units (LSU)
	Telephone signals
	System control signals
	Management signals
	Maintenance signals

mation. The 12th SU of the block is an ACU, which is coded to indicate the number of the blocks being acknowledged, and the acknowledgment bits, indicating whether each of the 11 signal units of the block being acknowledged was received without error.

Figure 4.13 illustrates the IAM (initial address message) and LSUs associated with a routine 10-digit call and the actions performed at the originating and terminating CCIS exchanges. Speech-path continuity checks are also shown in their proper chronological order in Figure 4.13. Error detection of CCIS signaling links is achieved by redundant coding. A data-carrier failure detector complements the bit-error detection and is helpful for longer error bursts. Error-free messages are processed on receipt while a retransmission (a form of ARQ described in Chapter 8) is requested of those found in error.

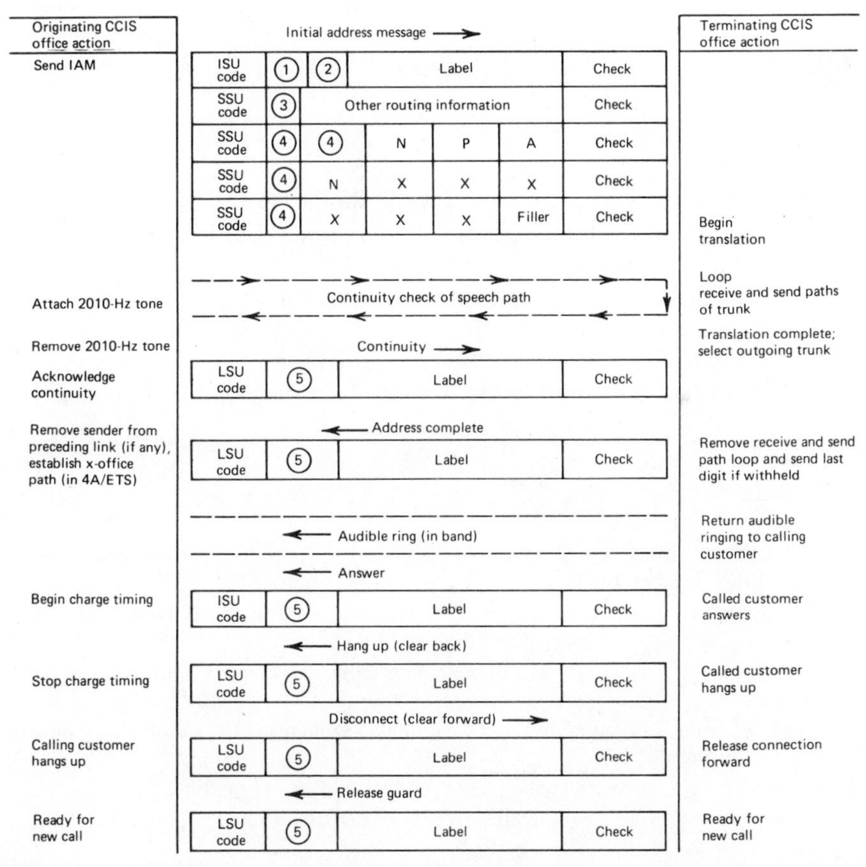

Figure 4.13 Generalized signal sequence for 10-digit CCIS call. [*Note:* All units within dotted lines are in band; ① ICU-type indicator (IAM), ② length indicator, ③ full routing information indicator, ④ no information; all zeros, ⑤ signal information ("answer," "disconnect," etc.)]. Copyright © American Telephone and Telegraph Co., 1975.

Table 4.13 Variance between CCITT No. 6 and CCIS[a]

	CCITT No. 6	CCIS
Heading	5 bits	3 bits
$Signal information	4 bits	4 bits
$Band No.	7 bits	9 bits
Circuit no.–trunk no.	4 bits	4 bits
Check (error)	8 bits	8 bits
	28 bits	28 bits
Label capacity	$2^{11a} = 2048$	$2^{13a} = 8192$

[a]The exponents 11 and 13 are derived from the sum of bits with $ indicator; thus 7 + 4 = 11 and 9 + 4 = 13, respectively. Thus CCIS label capacity is four times greater than CCITT No. 6. There are other format variations as well. Although international and North American signaling requirements differ, spare capacity was allocated in the CCITT No. 6 code to allow for national variants such as the CCIS.

8.3 CCITT No. 6 Code

The CCITT developed a CCS code very similar to the North American CCIS. The No. 6 is embodied in CCITT Rec. Q.251 through Q.267. However, the variances are too great for the two codes to interoperate (see Table 4.13). The CCITT No. 6 format is 28-bit for a signal unit (SU), with the last 8 bits of an SU for error check; however, compare this with an LSU. Both systems use a loop-back technique on four-wire speech paths for continuity using a nominal 2000-Hz tone (CCIS uses a 2010-Hz tone).

REFERENCES

1. C. A. Dahlbom and C. Breen, "Signaling Systems for Control of Telephone Switching," *BSTJ*, **39** (November 1960).
2. J. D. Sipes, "Common Channel Interoffice Signalling—International Field Trial," IEEE International Switching Symposium, June 1972.
3. *Notes on Distance Dialing*, American Telephone and Telegraph Company, New York, 1975.
4. *Signalling*, from Telecommunication Planning Documents, ITT Laboratories (Spain), Madrid, November 1974.
5. *National Networks for the Automatic Service*, CCITT-ITU, Geneva, 1964.
6. M. Den Hertog, "Inter-register Multifrequency Signalling for Telephone Switching in Europe," *Electr. Commun.*, **38** (1) (1972).
7. "USITA Minutes Attachments," May 18, 1972, "CCIS" and "Common Channel Interoffice Signaling" memorandum.
8. International Telephone and Telegraph Corporation, *Reference Data for Radio Engineers*, 6th ed, ITT Howard W. Sams, Indianapolis, 1976.

9. R. Freeman, *Telecommunication Transmission Handbook*, Wiley, New York, 1975.
10. *Lenkurt Demodulator—World-Wide E & M Signalling,* July–August 1977; "*Common Channel Signalling Systems*," April 1973; *A Glossary of Signalling Terms*, April 1974. GTE Lenkurt, San Carlos, Calif.
11. CCITT Recommendations, *Orange Books*, Vol. VI. ITU Geneva 1976 (Sixth Plenary Assembly CCITT).
12. J. Gordon Pearce, "The CCITT No. 6 Signaling System", *Telephony* (May 17, 1971).
13. Karl F. Steinhawer, "International Signalling with CCITT No. 6 System," *Telephony* (May 17, 1971).
14. *Access Area Switching and Signaling: Concepts, Issues, and Alternatives*, NTIA, Boulder, Colo., May 1978.
15. C. A. Dahlbom, "Signaling Systems and Technology," *Proc. IEEE*, **65,** 1349–1353 (1977).

Chapter 5

INTRODUCTION TO TRANSMISSION FOR TELEPHONY

1 GENERAL

The basic building block for transmission is the telephone channel or voice channel. "Voice channel" implies spectral occupancy, whether the voice path is over wire, radio, or optical fiber. If a pair of wires of a simple subscriber loop is extended without loading, we could expect to see the spectral content of the signal deriving from the average talker with frequencies as low as 20 Hz and as high as 20 kHz if the transducer of the telephone set was at all efficient across this band. Our ear, at least in younger people, is sensitive to frequencies from 30 Hz to as high as 30 kHz. However, the primary content of a voice signal (energy plus emotion) will occupy a much narrower band of frequencies (~ 100 Hz to 4000 Hz). Considering these and other factors, we say that the *nominal* voice channel occupies the band of 0 to 4 kHz.

There are essentially four other parameters we can use to define the voice channel:

- Attenuation distortion.
- Envelope delay distortion.
- Level (signal power level).
- Noise and signal-to-noise ratio.

Return loss, singing and stability, echo, and reference equivalent are other important parameters but are more applicable to a voice channel operating in a specific network. Each parameter is treated in its appropriate place further on in the discussion.

2 THE FOUR ESSENTIAL PARAMETERS

2.1 Attenuation Distortion

A signal transmitted over a voice channel suffers various forms of distortion. That is, the output signal from the channel is distorted in some manner such that it is not an exact replica of the input. One form of distortion is called *attenuation distortion* and is the result of imperfect amplitude–frequency response. Attenuation distortion can be avoided if all frequencies within the passband are subjected to exactly the same loss (or gain). Whatever the transmission, however, some frequencies are attenuated more than others. For example, on loaded wire-pair systems higher frequencies are attenuated more than lower ones. On carrier equipment (see Section 4 this chapter), band-pass filters are used on channel units, where, by definition, attenuation increases as the band edges are approached. Figure 5.1 is a good example of the attenuation characteristics of a voice channel operating over carrier equipment.

Attenuation distortion across the voice channel is measured against a reference frequency. The CCITT specifies 800 Hz, as reference, which is universally used in Europe, Africa, and parts of Latin America, whereas 1000 Hz is the common reference frequency in North America. Let us look at some ways attenuation distortion may be stated. For example, one requirement may state that between 600 and 2800 Hz the level will vary no more than −1 +2 dB, where the plus sign means more loss and the minus sign means less loss. Thus if a signal at −10 dBm is placed at the input of the channel, we would expect −10 dBm at the output at 800 Hz (if there were no overall loss or gain), but at other frequencies we could expect a variation between −1 and +2 dB. For instance, we might measure the level at the output at 2500 Hz at −11.9 dBm and at 1000 Hz at −9 dBm.

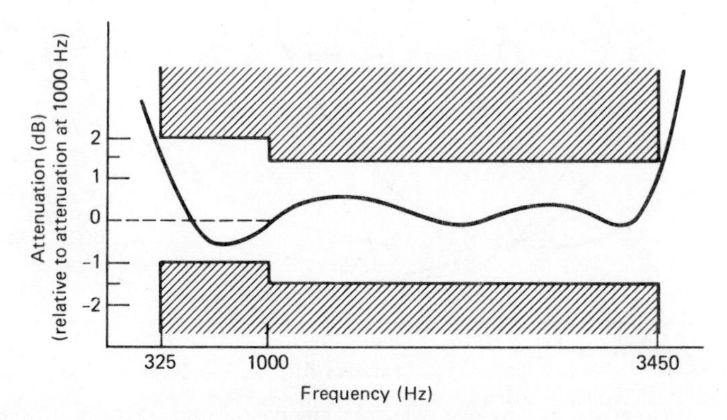

Figure 5.1 Typical attenuation–frequency response (attenuation distortion) for a voice channel. Crosshatched areas show specified limits.

2.2 Envelope Delay Distortion

A voice channel may be regarded as a band-pass filter. A signal takes a finite time to pass through a filter. This time is a function of the velocity of propagation, which varies with the medium involved. This velocity also tends to vary with frequency because of the electrical characteristics associated with it; it tends to increase toward band center and decrease toward band edge, as shown in Figure 5.2.

The finite time during which a signal takes to pass through the total extension of a voice channel or any network is called *delay.* Absolute delay is the delay a signal experiences while passing through the channel at reference frequency. But we see that the propagation time is different for different frequencies with the wavefront of one frequency arriving before the wavefront of another in the passband. Thus we can say that the phase has shifted. A modulated signal will not be distorted on passing through the channel if the phase shift changes uniformly with frequency, whereas if the phase shift is nonlinear with respect to frequency, the output signal is distorted when compared to the input.

In essence, therefore, we are dealing with the phase linearity of a circuit. If the phase–frequency relationship over a passband is not linear, distortion will occur in the transmitted signal. This distortion is best measured by a parameter called *envelope delay distortion.* Mathematically, envelope delay is the derivative of the phase shift with respect to frequency. The maximum

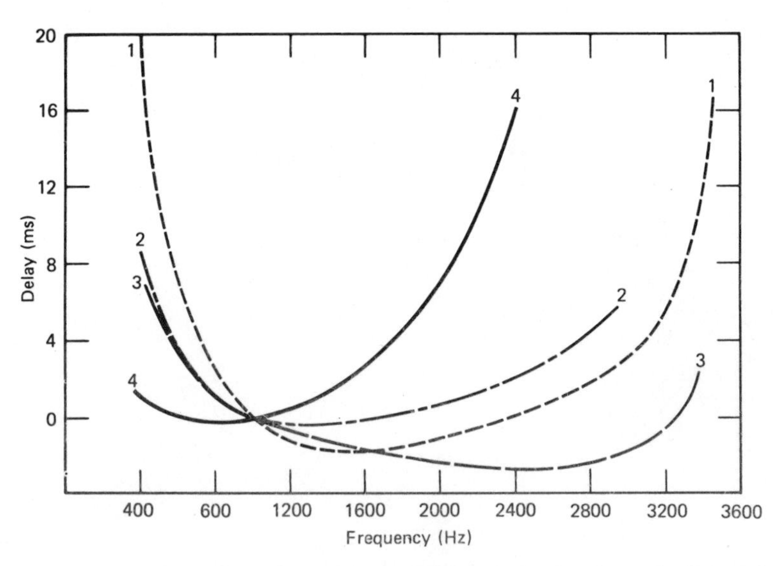

Figure 5.2 Comparison of envelope delay in some typical voice channels. Curves 1 and 3 represent delay of several thousand miles of a toll-quality carrier system. Curve 2 shows delay produced by 100 mi of loaded cable. Curve 4 shows delay in 200 mi of heavily loaded cable. Courtesy of GTE Lenkurt Demodulator, San Carlos, Calif.

difference in the derivative over any frequency interval is called envelope delay distortion (EDD). Therefore, EDD is always a difference between the envelope delay at one frequency and that at another frequency of interest in a passband. The EDD unit is milli- or microseconds. Note that envelope delay is often defined the same as group delay, that is, the rate of change, with angular frequency, of the phase shift between two points in a network [11].

2.3 Level

When referring to level, the connotation is either signal intensity or noise intensity. In most telecommunication systems "level" means power level measured in dBm, dBW, or other power units such as picowatts. One notable exception is video, which uses voltage, usually dBmV. Level is an important system parameter. If levels are maintained at too high a point, amplifiers become overloaded, with resulting increases in intermodulation products or crosstalk. If levels are too low, customer satisfaction may suffer.

System levels are used for engineering a communication system. These are usually taken from a level chart or reference system drawing made by a planning group or as a part of an engineered job. On the chart a 0 TLP (zero test-level point) is established. A test-level point is a location in a circuit or system at which a specified test-tone level is expected during alignment. A 0 TLP is a point at which the test-tone level should be zero dBm.

From the 0 TLP other points may be shown using the unit dBr (decibel reference). A minus sign shows that the level is so many decibels below reference and a positive sign, above. The unit dBm0 is an absolute unit of power in dBm referred to the 0 TLP. The dBm can be related to dBr and dBm0 by the following formula:

$$\text{dBm} = \text{dBm0} + \text{dBr}$$

For instance, a value of -32 dBm at a -22-dBr point corresponds to a referenced level of -10 dBm0. A -10 dBm0 signal introduced at the 0-dBr point (0 TLP) has an absolute signal level of -10 dBm.

2.4 Noise

2.4.1 General

Noise, in its broadest definition, consists of any undesired signal in a communication circuit. The subject of noise and noise reduction is probably the most important single consideration in transmission engineering. It is the major limiting factor in system performance. For this discussion noise has been broken down into four categories: (1) thermal noise (Johnson noise), (2) Intermodulation noise, (3) crosstalk, and (4) impulse noise.

2.4.2 Thermal Noise

Thermal noise occurs in all transmission media and all communication equipment. It arises from random electron motion and is characterized by a uniform distribution of energy over the frequency spectrum with a Gaussian distribution of levels.

Every equipment element and the transmission medium proper contribute thermal noise to a communication system if the temperature of that element of medium is above absolute zero. Thermal noise is the factor that sets the lower limit for sensitivity of a receiving system and is often expressed as a temperature, usually given in degrees referred to absolute zero (degrees Kelvin).

Thermal noise is a general expression referring to noise based on thermal agitations. The term "white noise" refers to the average uniform spectral distribution of energy with respect to frequency. Thermal noise is directly proportional to bandwidth and temperature. The amount of thermal noise to be found in 1 Hz of bandwidth in an actual device is

$$P_n = kT(\text{W/Hz})$$

where k is Boltzmann's constant, equal to $1.3803(10^{-23})$J/°K and T is the absolute temperature (°K) of the circuit (device). At room temperature, T = 17°C or 290°K; thus

$$P_n = 4.00(10^{-21})\text{W/Hz of bandwidth}$$

or

$$= -204 \text{ dBW/Hz of bandwidth}$$
$$= -174 \text{ dBm/Hz of bandwidth.}$$

For a band-limited system (i.e., a system with a specific bandwidth), $P_n = kTB$(W), where B refers to the so-called noise bandwidth in hertz. Thus at 0°K $P_n = -228.6$ dBW/Hz of bandwidth and for a system with a noise bandwidth measured in hertz,

$$P_n = -228.6 \text{ dBW} + 10 \log T + 10 \log B$$

2.4.3 Intermodulation Noise

Intermodulation (IM) noise is the result of the presence of intermodulation products. If two signals with frequencies F_1 and F_2 are passed through a nonlinear device or medium, the result will be IM products that are spurious frequency components. These components may be present either inside or outside the band of interest for the device. Intermodulation products may be produced from harmonics of the signals in question, either as products between harmonics or as one or the other or both signals themselves. The products result when the two (or more) signals beat together or "mix." Look at the mixing possibilities when passing F_1 and F_2 through a nonlinear device. The coefficients indicate first, second, or third harmonics.

- Second-order products $F_1 \pm F_2$.
- Third order products $F_1 \pm 2F_2$; $2F_1 \pm F_2$.
- Fourth-order products $2F_1 \pm 2F_2$; $3F_1 \pm F_2$. . . .

Devices passing multiple signals simultaneously, such as multichannel radio equipment, develop intermodulation products that are so varied that they resemble white noise.

Intermodulation noise may result from a number of causes:

- Improper level setting. If the level of input to a device is too high, the device is driven into its nonlinear operating region (overdrive)
- Improper alignment causing a device to function nonlinearly.
- Nonlinear envelope delay.

To summarize, intermodulation noise results from either a nonlinearity or a malfunction that has the effect of nonlinearity. The cause of intermodulation noise is different from that of thermal noise. However, its detrimental effects and physical nature are identical to those of thermal noise, particularly in multichannel systems carrying complex signals.

2.4.4 Crosstalk

Crosstalk refers to unwanted coupling between signal paths. There are essentially three causes of crosstalk: (1) electrical coupling between transmission media, such as between wire pairs on a VF cable system, (2) poor control of frequency response (i.e., defective filters or poor filter design), and (3) the nonlinear performance in analog (FDM) multiplex systems. There are two types of crosstalk:

- *Intelligible*—where at least four words are intelligible to the listener from extraneous conversation(s) in a 7-s period.
- *Unintelligible*—crosstalk resulting from any other form of disturbing effects of one channel on another.

Intelligible crosstalk presents the greatest impairment because of its distraction to the listener. Distraction is considered to be caused by either fear of loss of privacy or primarily by the user of the primary line consciously or unconsciously trying to understand what is being said on the secondary or interfering circuits; this would be true for any interference that is syllabic in nature.

Received crosstalk varies with the volume of the disturbing talker, the loss from the disturbing talker to the point of crosstalk, the coupling loss between the two circuits under consideration, and the loss from the point of crosstalk to the listener. The most important of these factors for this discussion is the coupling loss between the two circuits under consideration. Talker levels have been discussed elsewhere in this text. Also, we must not lose sight of the fact that the effects of crosstalk are subjective, and other factors have to be considered when crosstalk impairments are to be measured. Among these factors are the type of people who use the channel, the acuity of listeners, traffic patterns, and operating practices.

2.4.5 Impulse Noise

Impulse noise is noncontinuous, consisting of irregular pulses or noise spikes of short duration and of relatively high amplitude. These spikes are often called "hits." Impulse noise degrades voice telephony only marginally, if at all; however, it may seriously degrade error rate on data or other digital circuits. Impulse noise is treated in greater depth when we discuss data and other digital communications.

2.5 Signal-to-Noise Ratio

When dealing with transmission engineering, *possibly* the signal-to-noise ratio is perhaps more frequently used than any other criterion when designing a telecommunication system. Signal-to-noise ratio expresses in decibels the amount by which a signal level exceeds the noise within a specified bandwidth.

As we review several types of material to be transmitted, each will require a minimum signal-to-noise ratio to satisfy the customer or to make the receiving instrument function within certain specified criteria. We might require the following signal-to-noise ratios (S/N) with the corresponding end instruments:

- Voice: 30 dB } based on customer satisfaction
- Video: 45 dB } based on customer satisfaction
- Data: 15 dB based on a specified error rate

In Figure 5.3* a 1000-Hz signal has a signal-to-noise ratio (S/N) of 10 dB. The level of the noise is 5 dBm and the signal, 15 dBm. Thus

$$\left(\frac{S}{N}\right)_{\mathrm{dB}} = \mathrm{level}_{(\mathrm{signal\ in\ dBm})} - \mathrm{level}_{(\mathrm{noise\ in\ dBm})}$$

*Assumes a nominal 4-kHz bandwidth in the example.

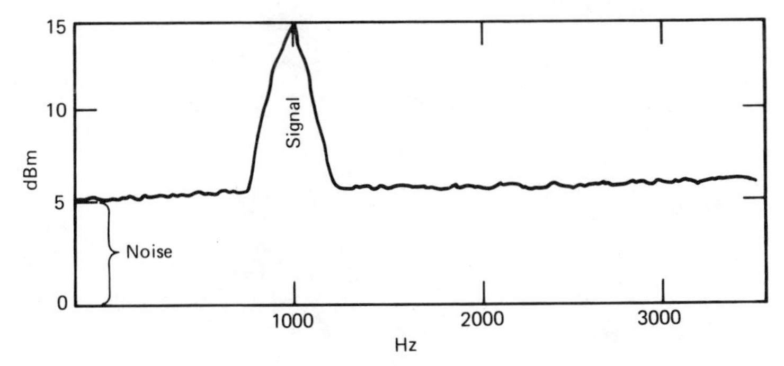

Figure 5.3 Signal-to-noise ratio.

3 TWO-WIRE–FOUR-WIRE TRANSMISSION

3.1 Two-Wire Transmission

A telephone conversation inherently requires transmission in both directions. When both directions are carried on the same pair of wire, it is called *two-wire transmission.* The telephones in our homes and offices are connected to a local switching center (exchange) by means of two-wire circuits. A more proper definition for transmitting and switching purposes is that when oppositely directed portions of a single telephone conversation occur over the same electrical transmission channel or path, we call this *two-wire operation.*

3.2 Four-Wire Transmission

Carrier and radio systems require that oppositely directed portions of a single conversation occur over separate transmission channels or paths (or using mutually exclusive time periods). Thus we have two wires for the transmit path and two wires for the receive path, or a total of four wires for a full-duplex (two-way) telephone conversation. For almost all operational telephone systems, the end instrument (i.e., the telephone subset) is connected to its intervening network on a two-wire basis.

Nearly all long-distance (toll) telephone connections traverse four-wire links. From the near-end user the connection to the long-distance network is two-wire or via a two-wire link. Likewise, the far-end user is also connected to the long-distance (toll) network via a two-wire link. Such a long-distance connection is shown in Figure 5.4. Schematically, the four-wire interconnection is shown as if it were wire line, single channel with amplifiers; however, it would more likely be multichannel carrier on cable and/or multiplex on radio. However, the amplifiers in Figure 5.4 serve to

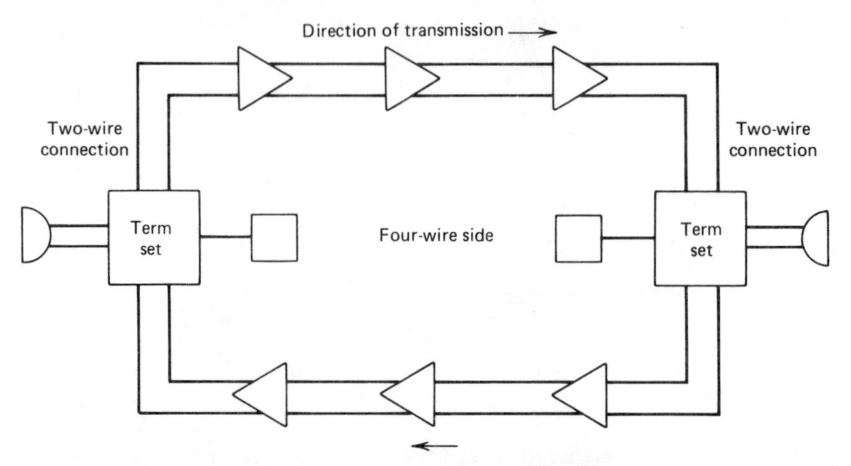

Figure 5.4 A typical long-distance (toll) telephone connection.

convey the ideas that this section considers. As shown in Figure 5.4, conversion from two-wire to four-wire operation is carried out by a terminating set, more commonly referred to in the industry as a *term set,* which contains a three-winding balanced transformer (a hybrid) or a resistive network (the latter is less common).

3.3 Operation of a Hybrid

A hybrid, in terms of telephony (at voice frequency), is a transformer. For a simplified description, a hybrid may be viewed as a power splitter with four sets of wire-pair connections. A functional block diagram of a hybrid device is shown in Figure 5.5. Two of the wire pair connections belong to the four-wire path, which consists of a transmit pair and a receive pair. The third pair is the connection to the two-wire link that is eventually connected to the subscriber subset. The last wire pair of the four connects the hybrid

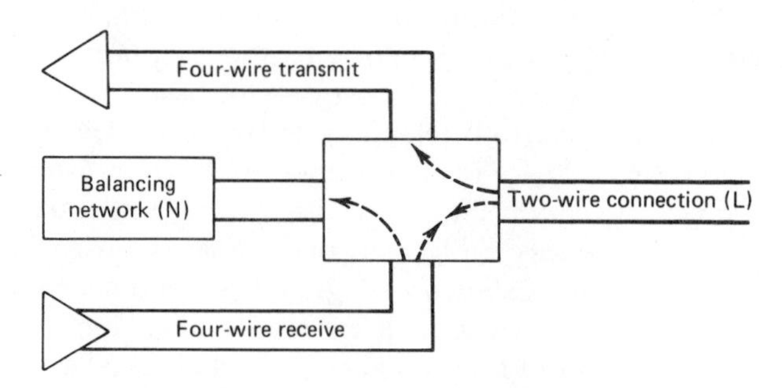

Figure 5.5 Operation of a hybrid transformer.

to a resistance–capacitance balancing network, which electrically balances the hybrid with the two-wire connection to the subscriber subset over the frequency range of the balancing network. An artificial line may be used for this purpose.

Signal energy entering from the two-wire subset connection divides equally, half of it dissipating in the impedance of the four-wire side receive path and the other half going to the four-wire side transmit path as shown in Figure 5.5. Here the *ideal* situation is that no energy is to be dissipated by the balancing network (i.e., there is a perfect balance). The balancing network is supposed to display the characteristic impedance of the two-wire line (subscriber connection) to the hybrid.

The reader notes that in the description of the hybrid, in every case, ideally half of the signal energy entering the hybrid is used to advantage and only half is dissipated or wasted. Also keep in mind that any passive device inserted in a circuit such as a hybrid has an insertion loss. As a rule of thumb, we say that the insertion loss of a hybrid is 0.5 dB. Thus there are two losses here that the reader must not lose sight of:

Hybrid insertion loss	0.5 dB
Hybrid dissipation loss	3.0 dB (half of the power)
	3.5 dB (total)

As far as this section is concerned, any signal passing through a hybrid suffers a 3.5-dB loss. This is a good design number for gross engineering practice. However, some hybrids used on short subscriber connections purposely have higher losses, as do special resistance-type hybrids.

4 FREQUENCY DIVISION MULTIPLEX

4.1 General

Frequency division multiplex (FDM) is a method of allotting a unique band of frequencies in a comparatively wide band frequency spectrum of the transmission medium to each communication channel on a continuous time basis. The communication channel may be a voice channel 4 kHz wide, a 120-Hz telegraph channel (for example), a 15-kHz broadcast channel, a 48-kHz data channel, or a 4.2-MHz television channel.

Before launching into multiplexing, keep in mind that all multiplex systems work on a four-wire basis. The transmit and receive paths are separate (see Figure 5.6). For the discussion that follows, it is assumed that the reader has some background on how an SSB (single sideband) signal is developed as well as how a carrier is suppressed in SSBSC systems.

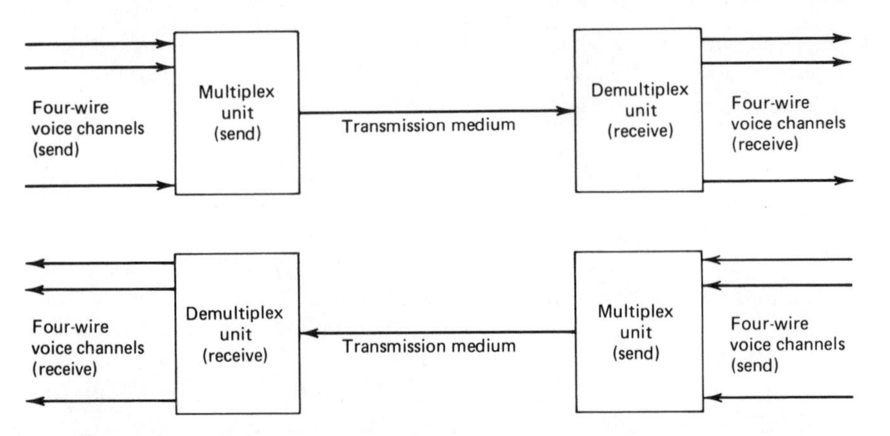

Figure 5.6 Simplified block diagram of a frequency division multiplex link.

4.2 Mixing

The heterodyning or mixing of signals of frequencies A and B is shown in the following diagram. What frequencies may be found at the output of the mixer?

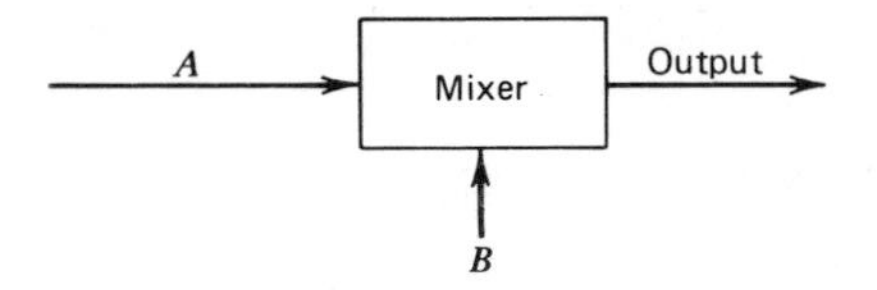

Both of the original signals will be present, as well as signals representing their sum and difference in the frequency domain. Thus signals of frequency $A, B, A + B$, and $A - B$ will be present at the output of the mixer.* Such a mixing process is repeated many times in FDM equipment.

Let us look at the boundaries of the nominal 4-kHz voice channel. These are 300 and 3400 Hz. Let us further consider these frequencies as simple tones of 300 and 3400 Hz. Now consider the mixer below and examine the possibilities at its output.

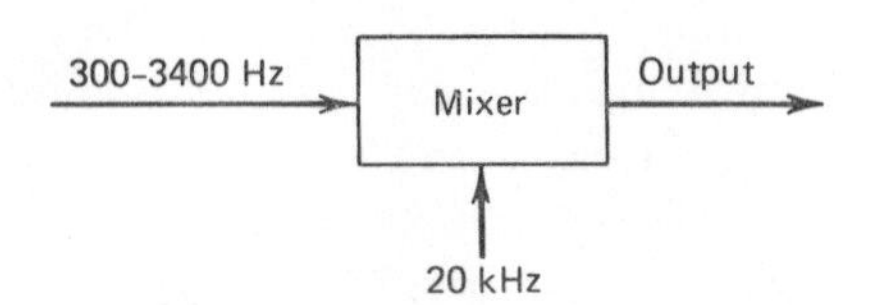

*Of course, if the mixer is a balanced mixer, only signals $A + B$ and $A - B$ will be present.

First, the output may be summed, such that

$$\begin{array}{r} 20{,}000 \text{ Hz} \\ +\ 300 \text{ Hz} \\ \hline 20{,}300 \text{ Hz} \end{array} \qquad \begin{array}{r} 20{,}000 \text{ Hz} \\ +3{,}400 \text{ Hz} \\ \hline 23{,}400 \text{ Hz} \end{array}$$

A simple low-pass filter could filter out all frequencies below 20,300 Hz. Now imagine that instead of two frequencies, we have a continuous spectrum of frequencies between 300 and 3400 Hz (i.e., we have the voice channel). We represent the spectrum as a triangle.

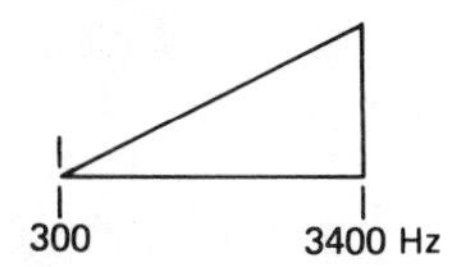

As a result of the mixing process (translation), we have another triangle as follows:

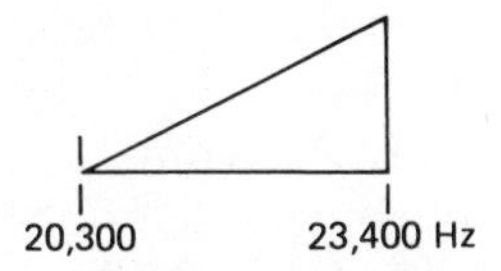

When we take the sum, as we did previously, and filter out all other frequencies, we say we have selected the upper sideband. Thus we have a triangle facing to the right, termed an *upright* or *erect* sideband.

We can also take the difference, such that

$$\begin{array}{r} 20{,}000 \text{ Hz} \\ -\ 300 \text{ Hz} \\ \hline 19{,}700 \text{ Hz} \end{array} \qquad \begin{array}{r} 20{,}000 \text{ Hz} \\ -3{,}400 \text{ Hz} \\ \hline 16{,}000 \text{ Hz} \end{array}$$

and we see that in the translation (mixing process) we have had an inversion of frequencies. The higher frequencies of the voice channel become the lower frequencies of the translated spectrum, and the reverse occurs when the difference is taken. We represent this by a right triangle facing the other direction (left):

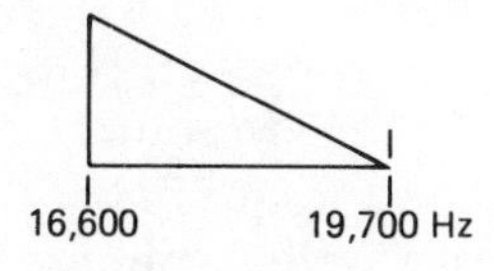

This is called an *inverted sideband.* To review, when we take the sum, we get an erect sideband. When we take the difference in the mixing process, frequencies invert and we have an inverted sideband represented by a triangle facing left.

4.3 CCITT Modulation Plan

4.3.1 Introduction

A modulation plan sets forth the development of a band of frequencies called the *line frequency* (i.e., ready for transmission on the line or other transmission medium). The modulation plan usually is a diagram showing the necessary mixing, local oscillator insertion frequencies, and the sidebands selected by means of the triangles described previously, in a step-by-step process from voice-channel input to line-frequency output. The CCITT has recommended a standardized modulation plan with a common terminology, allowing large telephone networks, in both national and multinational systems, to interconnect. In the paragraphs that follow the reader is advised to be careful with terminology.

4.3.2 Formation of Standard CCITT Group

The standard *group* as defined by CCITT occupies the frequency band 60 to 108 kHz and contains 12 voice channels. Each voice channel is the nominal 4-kHz voice channel occupying the 300-to-3400-Hz spectrum. The group is formed by mixing each of the 12 voice channels with a particular carrier frequency associated with the channel. Lower sidebands are then selected. Figure 5.7 shows the basic (and preferred) approach to the formation of the standard CCITT group. It should be noted that in the 60-to-108-kHz band voice channel 1 occupies the highest frequency segment by convention, between 104 and 108 kHz. The layout of the standard group is shown in Figure 5.8. For more information refer to CCITT Rec. G.232.

Single sideband suppressed carrier (SSBSC) modulation techniques are used in all cases. The CCITT recommends that the carrier leak be down to at least −26 dBm0, referred to the zero relative level point (0 TLP) (see Section 2.3 of this chapter).

4.3.3 Formation of Standard CCITT Supergroup

A supergroup contains five standard CCITT groups, equivalent to 60 voice channels. The standard supergroup before translation occupies the frequency band 312 to 552 kHz. Each group making up the supergroup is translated in frequency to the supergroup band by mixing with the proper carrier frequency. The carrier frequencies are 420 kHz for group 1, 468 kHz for group 2, 516 kHz for group 3, 564 kHz for group 4, and 612 kHz for group 5. In the mixing process the difference is taken (lower sidebands are selected). This translation process is shown in Figure 5.9.

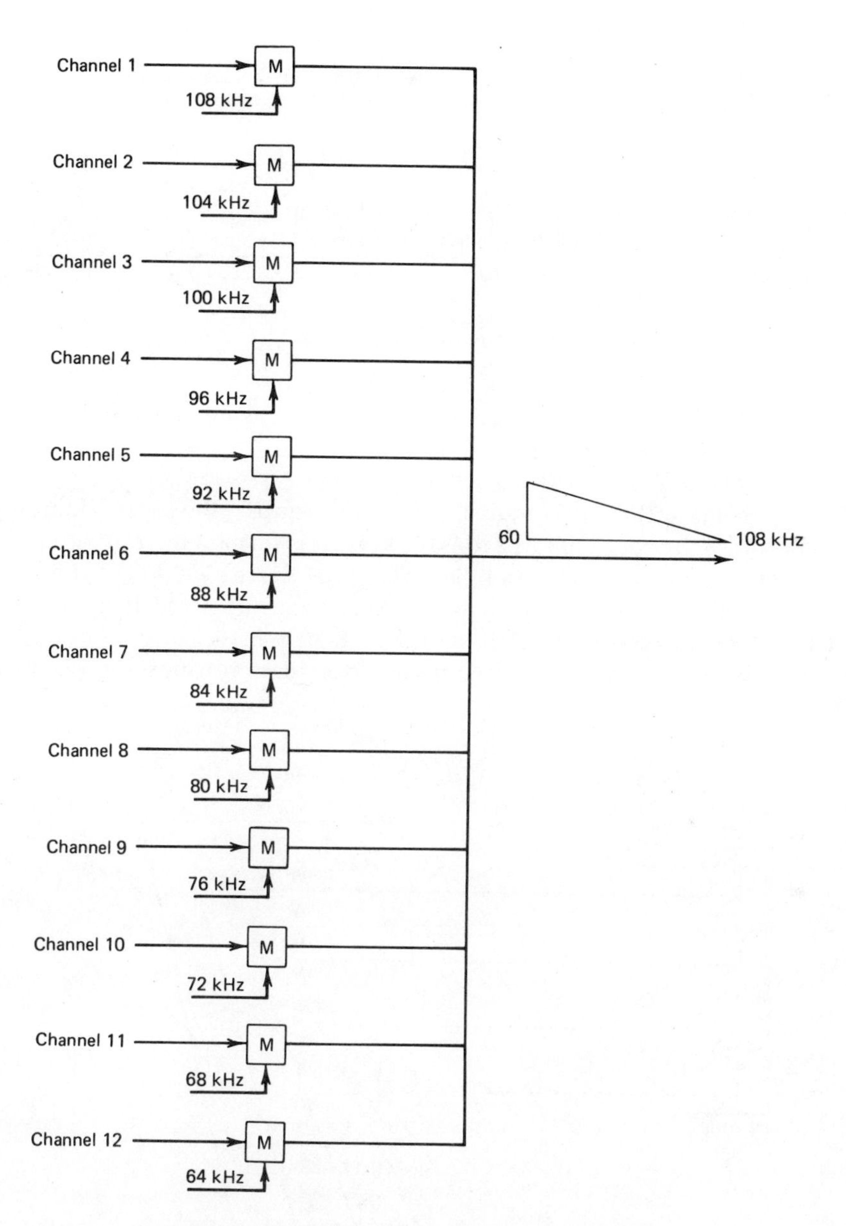

Figure 5.7 Formation of standard CCITT group.

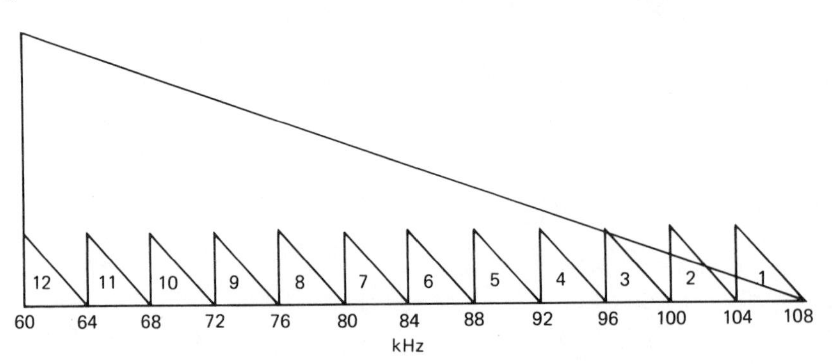

Figure 5.8 Layout of standard CCITT group.

4.3.4 Formation of Standard CCITT Basic Mastergroup and Supermastergroup

The basic mastergroup contains five supergroups (300 voice channels) and occupies the spectrum 812 to 2044 kHz. It is formed by translating the five standard supergroups, each occupying the 312-to-552-kHz band, by a process similar to that used to form the supergroup from five standard CCITT groups. This process is shown in Figure 5.10.

The basic supermastergroup contains three mastergroups and occupies

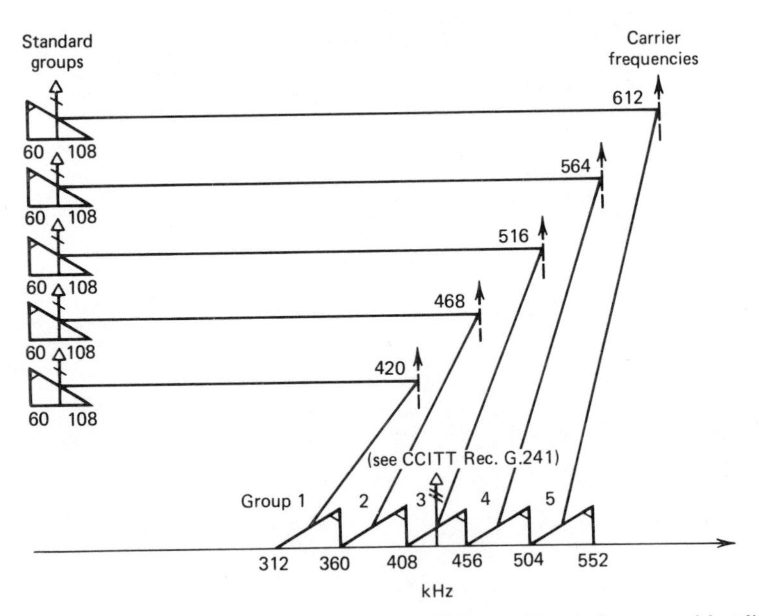

Figure 5.9 Formation of the standard CCITT supergroup. Vertical arrows with solid lines are level-regulating pilot tones; arrows with dashed lines are translation carrier frequencies. Courtesy of the International Telecommunication Union–CCITT.

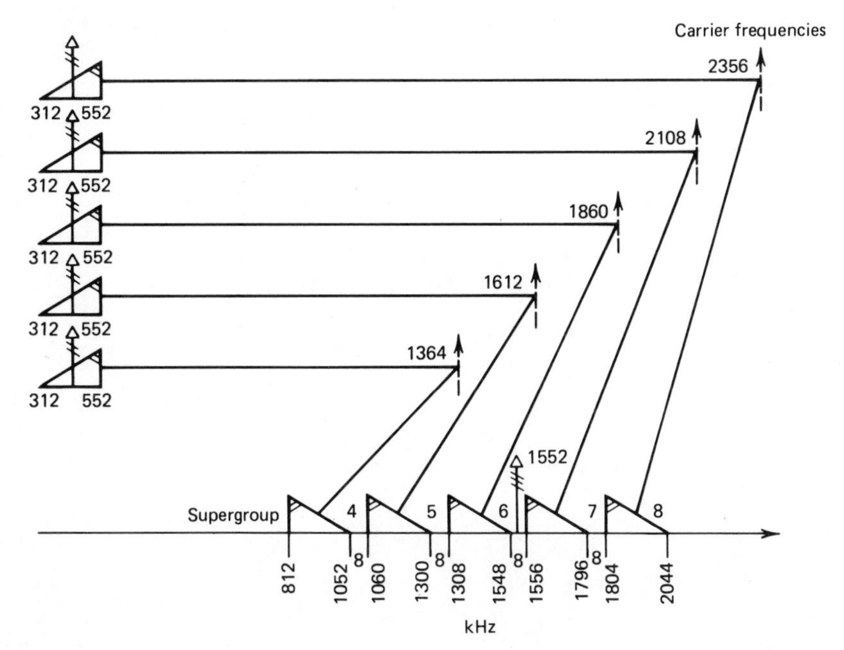

Figure 5.10 Formation of standard CCITT mastergroup. Courtesy of the International Telecommunication Union–CCITT.

the band 8516 to 12,388 kHz. The formation of the supermastergroup is shown in Figure 5.11.

4.3.5 Line Frequency

The band of frequencies that the multiplex applies to the line, whether the line is radiolink, coaxial cable, wire pair, or open wire, is called the *line frequency*. Another expression is HF (or high frequency), which is ambiguous and often confused with HF radio transmission.

The line frequency in any one particular case may just be the direct application of a group or supergroup to the line. However, a final translation stage more commonly occurs, particularly in high-density systems. Two of the recognized CCITT configurations are shown in Figures 5.12 and 5.13. Figure 5.12 shows the makeup of the standard CCITT 15-supergroup assembly and Figure 5.13, the standard 15-supergroup assembly No. 3 deriving from the basic 15-supergroup assembly shown previously.

4.3.6 North American L-Carrier

"L-carrier" is the generic name given by the Bell System of North America to their long-haul SSB carrier system. Its development of the basic group

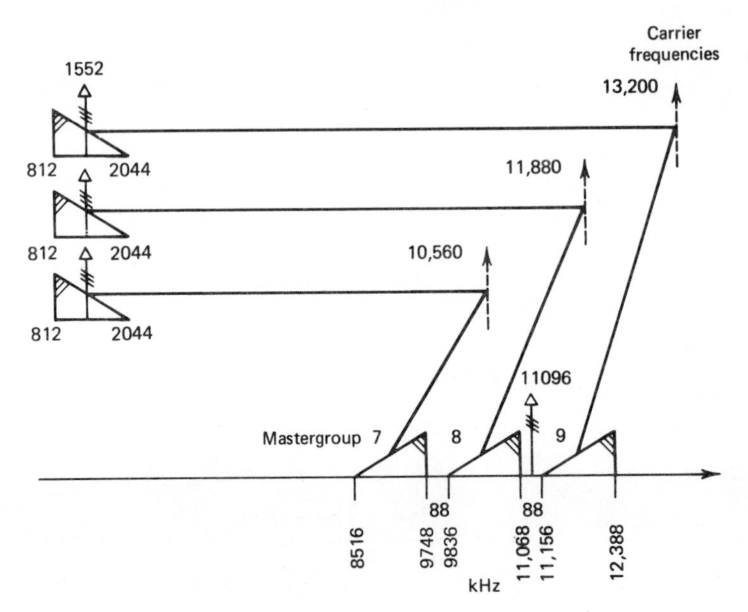

Figure 5.11 Formation of standard CCITT supermastergroup. Courtesy of the International Telecommunication Union–CCITT.

and supergroup assemblies is essentially the same as that of the CCITT described previously. There is a variance in levels and pilot tone frequencies.

The basic mastergroup differs, however. It consists of 600 VF channels (i.e., 10 standard supergroups). The L600 configuration occupies the band 60 to 2788 kHz and the U600 configuration, the band 564 to 2084 kHz. The relevant mastergroup assemblies are shown in Figure 5.14. The Bell System (ATT) identifies specific long-haul line-frequency configurations by adding a simple number after the letter "L." For example, the L3 carrier, which is used on coaxial cable and one type of ATT microwave (TH) has three mastergroups (see Figure 5.14), plus one supergroup, comprising a total of 1800 VF channels on line occupying the band 312 to 8284 kHz. Table 5.1 compares L-carrier with CCITT carrier (FDM) standards. Subsequent paragraphs discuss the parameters and their meanings.

4.4 Loading of Multichannel Frequency-Division-Multiplex Systems

4.4.1 Introduction

Most of the FDM (carrier) equipment in use today carries speech traffic, which is sometimes misnamed "message traffic" in North America. In this context we refer to full-duplex conversations by telephone between two

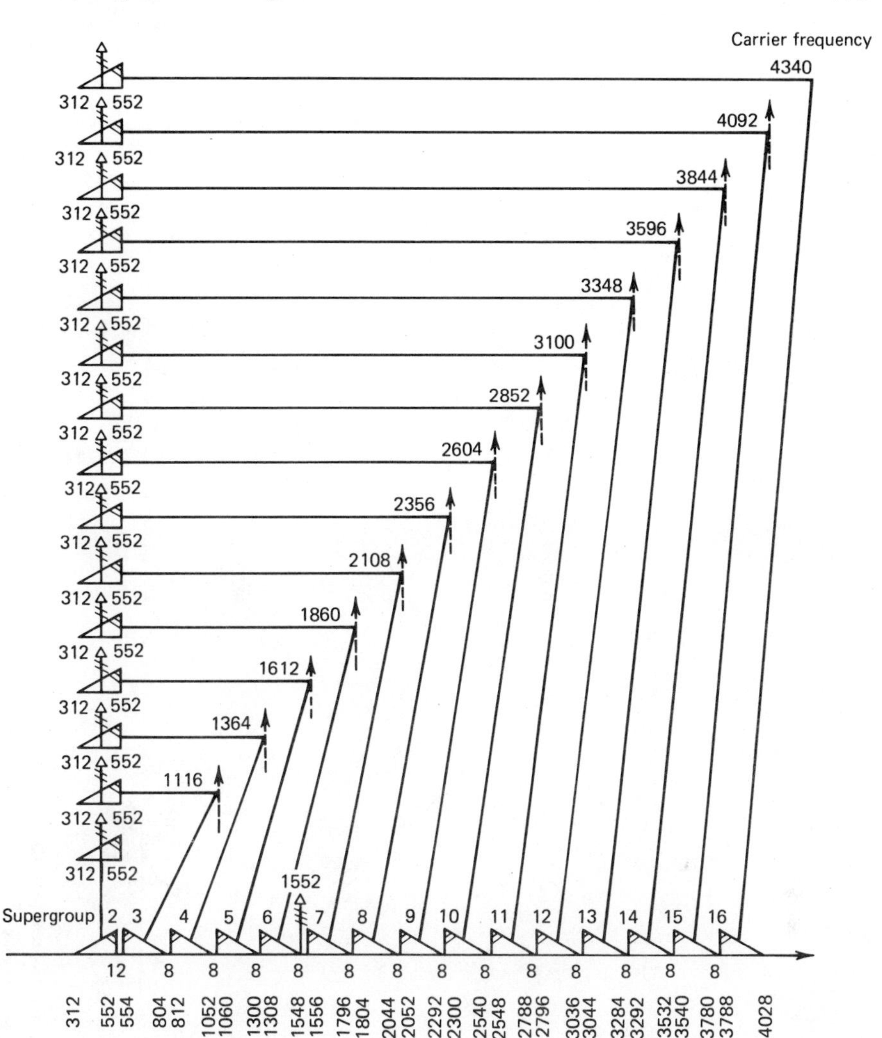

Figure 5.12 Makeup of basic CCITT 15-supergroup assembly. Courtesy of the International Telecommunication Union–CCITT.

"talkers." However, the reader should remember that there is a marked increase in the use of these same intervening talker facilities for the transmission of data and facsimile.

For this discussion, the problem essentially concerns human speech and how multiple users may load a carrier system. If we load a carrier system too heavily—if the input levels are too high—IM noise and crosstalk would become intolerable, eventually leading to the breakdown of the system. If we do not load the system sufficiently, the signal-to-noise ratio will suffer.

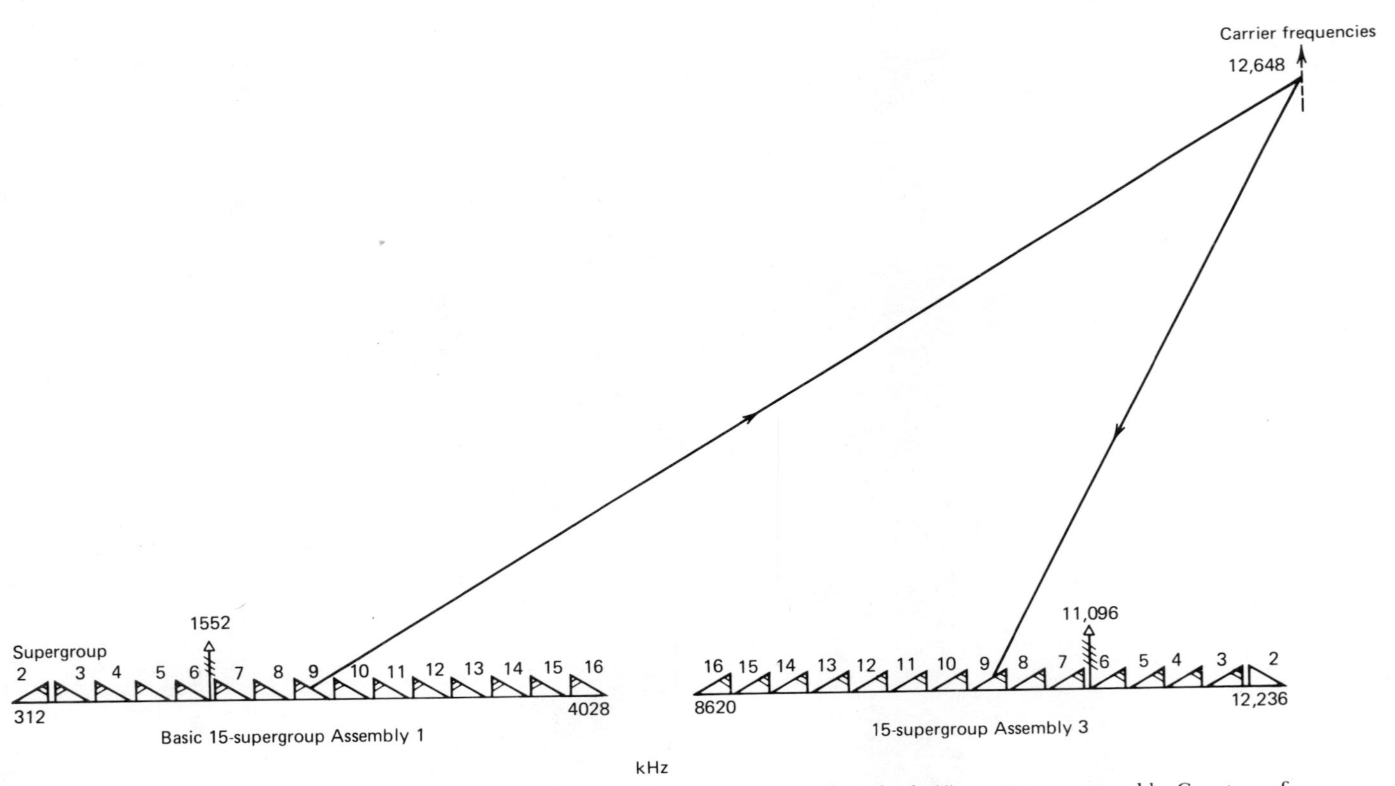

Figure 5.13 Makeup of standard CCITT 15-supergroup assembly No. 3 as derived from basic 15-supergroup assembly. Courtesy of the International Telecommunication Union–CCITT.

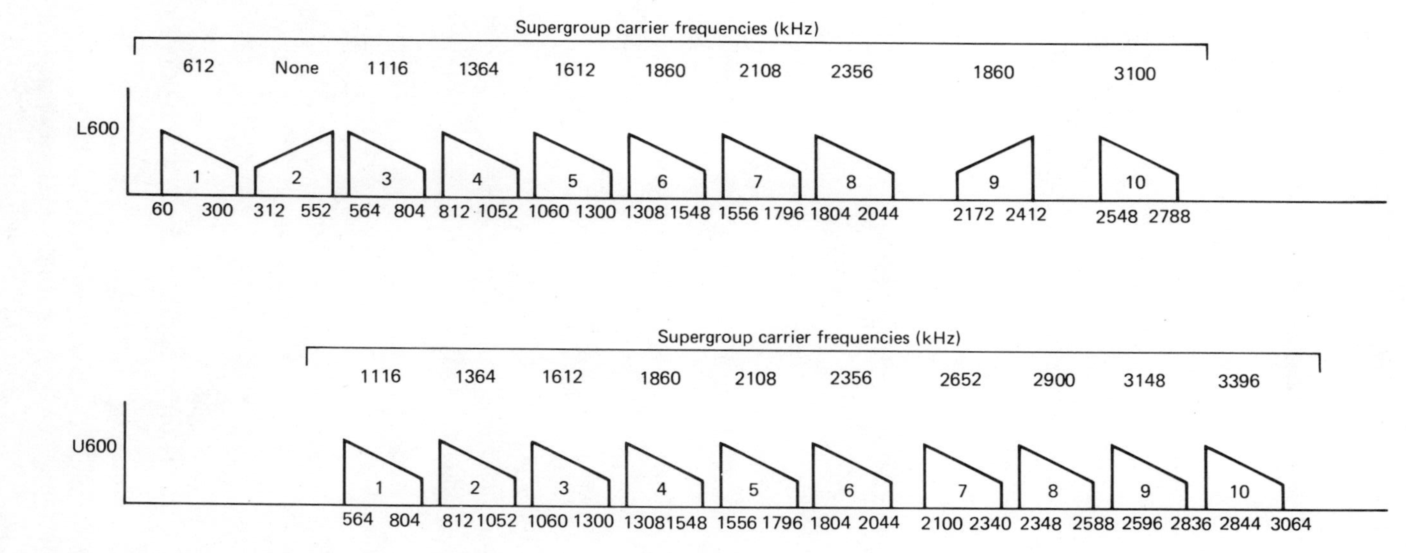

Figure 5.14 L-carrier mastergroup assemblies (all frequencies in kilohertz). Copyright PP© Bell Telephone Laboratories, 1964.

Table 5.1 L-Carrier and CCITT Comparison Table

Item	ATT L-Carrier	CCITT
Level		
Group		
Transmit	−42 dBm	−37 dBm
Receive	−5 dBm	−8 dBm
Supergroup		
Transmit	−25 dBm	−35 dBm
Receive	−28 dBm	−30 dBm
Impedance		
Group	130 Ω balanced	75 Ω unbalanced
Supergroup	75 Ω unbalanced	75 Ω unbalanced
VF channel	200–3350 Hz	300–3400 Hz
Response	+1.0 to −1.0 dB	+0.9 to −3.5 dB
Channel carrier		
Levels	0 dBm	Not specified
Impedances	130 Ω balanced	Not specified
Signaling	2600 Hz in band	3825 Hz out of band
Group pilot		
Frequencies	92 or 104.08 kHz	84.08 kHz
Relative levels	−20 dBm0	−20 dBm0
Supergroup carrier		
Levels	+19.0 dBm per mod or demod	Not specified
Impedances	75 Ω unbalanced	Not specified
Supergroup pilot frequency	315.92 kHz	411.92 kHz
Relative supergroup pilot levels	−20 dBm0	−20 dBm0
Frequency synchronization	Yes, 64 kHz	Not specified
Line pilot frequency	64 kHz	308, 12,435 kHz
Relative line pilot level	−14 dBm0	−10 dBm0
Regulation		
Group	Yes	Yes
Supergroup	Yes	Yes

The problem is fairly complex because speech amplitude varies:

1. With talker volume.
2. At a syllabic rate.
3. At an audio rate.
4. With varying circuit losses as different loops and trunks are switched into the same channel bank voice channel input.

4.4.2 Loading

For the loading of multichannel FDM systems, CCITT Rec. G.223 recommends:

It will be assumed for the calculation of intermodulation below the overload point that the multiplex signal during the busy hour can be represented by a uniform spectrum (of) random noise signal, the mean absolute power level of which, at a zero relative level point—(in dBm0):

$$P_{av} = -15 + 10 \log N$$

when N is equal to or greater than 240

and

$$P_{av} = -1 + 4 \log N$$

When N is equal to or greater than 12 and less than 240. . . .

where N is the number of voice channels.

(all logs to the base 10).

4.4.3 Single-Channel Loading

Many telephone administrations have attempted to standardize on −16 dBm0 for single-channel speech input to multichannel FDM equipment. With this input, peaks in speech level may reach −3 dBm0. Tests indicated that such peaks will not be exceeded more than 1% of the time. However, the conventional value of average power per voice channel allowed by the CCITT is −15 dBm0 (see CCITT Rec. G.223). This assumes a standard deviation of 5.8 dB and the traditional activity factor of 0.25 (see Section 4.4.5). Average talker level is assumed to be at −11.5 VU. We must turn to the use of standard deviation because we are dealing with talker levels that vary with each talker and thus with the mean or average.

4.4.4 Loading with Constant-Amplitude Signals

Speech on multichannel systems has a low duty cycle or activity factor. We exemplify a simplified derivation of the traditional activity factor of 0.25%. Certain other types of signals transmitted over multichannel equipment have an activity factor of 1. This means that they are transmitted continuously, or continuously over fixed time frames. They are also usually characterized by constant amplitude. The following are some examples of these types of signal:

- Telegraph tone or tones.
- Signaling tone or tones.
- Pilot tones.

- Data signals (particularly PSK and FSK modulation).
- Facsimile (in digital or FM mode).

To provide a safety factor on overload the tendency would be to reduce level. Here again, if we reduce level too much, the signal-to-noise ratio, and hence the error rate, will suffer.

Using the loading formulas in Section 4.4.2, may permit the use of several data or telegraph channels without serious consequence. But if any wide use is to be made of the system for data–telegraph–facsimile then other loading criteria must be used. For typical constant-amplitude signals, traditional (CCITT) transmit levels as seen at the input of the channel modulator of FDM carrier equipment are as follows:

- Data: −13 dBm0.
- Signaling (single-frequency supervision, tone on when idle (see Chapter 4): −20 dBm0.
- Composite telegraph: −8.7 dBm0.

For one FDM system now on the market with 75% speech loading and 25% data loading with more than 240 channels (total), the manufacturer recommends the following:

$$P_{\mathrm{rms}} = -11 + 10 \log N$$

(units in dBm0) using −5 dBm0 per channel for data-input levels and −8 dBm0 for the composite telegraph signal level. However, the manufacturer hastens to mention that all VF channels may be loaded with data signals (or telegraph) at −8 dBm0 level per channel. But for the −5-dBm0 signal level, only two channels per group may be assigned this level, with the remainder speech, or the group must be "deloaded," which means that a certain number of channels must remain idle.

4.4.5 Activity Factor

Speech is characterized by large variations in amplitude, ranging from 30 to 50 dB. We often use the VU (volume unit) to measure speech levels. The VU can be equated to the dBm for a simple sine wave in the VF range across 600 ohms. For a complex signal such as our speech signal, the average power measured in dBm of a typical single talker is

$$P_{\mathrm{dBm}} = V_{\mathrm{VU}} - 1.4$$

or that a 0-VU talker has an average power of −1.4 dBm. Empirically, the peak power is about 18.6 dB higher than average power for a typical talker. The peakiness of speech level means that carrier (FDM) equipment must be operated at a low average power to withstand voice peaks to avoid overloading and distortion. These can be related to an activity factor T_a, which is defined as that proportion of the time that the rectified speech envelope exceeds some threshold. If the threshold is about 20 dB below the average power, the activity dependence on threshold is fairly weak. The preceding equation for average talker power can now be rewritten in relation to the activity factor as follows:

$$P_{\text{dBm}} = V_{\text{VU}} + 10 \log T_a$$

If $T_a = 0.725$, the results will be the same for the equation relating VU to dBm.

Now consider adding a second talker at a different frequency segment on the same equipment, but independent of the first talker. Of course, we are describing here the operation of FDM equipment discussed previously. With the second talker added, the system average power will increase 3 dB as expected. If we have N talkers, each on a different frequency segment, the average power developed will be

$$P_{\text{dBm}} = V_{\text{VU}} - 1.4 + 10 \log N$$

where P_{dBm} is the power developed across the frequency band occupied by all talkers.

Empirically, it has been found that the peakiness or peak factor of many talkers over a multichannel analog system reaches the characteristics of random noise peaks when the number of talkers, N, exceeds 64. When $N = 2$, the peaking factor is 18 dB; when $N = 10$, it is 16 dB; when $N = 50$, it is 14 dB; and so forth.

An activity factor—a term analogous to a "duty cycle" of 1—which we have been using in the preceding argument, is quite unrealistic. If it is 1, it means that somebody is talking on each channel all the time. The traditional figure for activity factor accepted by the CCITT and used in North American practice is 0.25. Follow the argument in the next paragraph on how to reach this lower figure.

The multichannel FDM equipment cannot be designed for N callers and no more. If this were true, a new call would have to be initiated every time a call terminated, or calls would have to be turned away because of congestion. Thus the equipment must be "overdimensioned" for BH service. For this dimensioning problem, we drop the activity factor from 1 to 0.70. Other causes will reduce this figure even more. For instance, circuits are essentially inactive during call setup as well as during pauses for thinking during a conversation. The 0.70 figure now must be dropped to 0.50. This latter figure is divided in half because of the talk–listen condition. If we disregard cases of "double-talking," it is obvious that while one end is talking on a full-duplex telephone circuit, the other is listening. Thus a circuit (in one direction) is idle half the time during the "listen" period. The resulting activity factor is 0.25.

4.5 Pilot Tones

In FDM equipment pilot tones are used primarily for level regulation and secondarily for fault alarms. However, some systems still also use pilot tones for end-to-end frequency synchronization of the carrier-generation equipment. The nature of speech, particularly its varying amplitude, makes it a poor prospect as a reference for level control. Ideally, simple, single-sinusoid, constant-amplitude signals with 100% duty cycles provide simple control information for level-regulating equipment. Frequency division multiplex level regulators operate very much in the same manner as automatic-gain control circuits on radio systems, except that their dynamic range is considerably smaller. Modern FDM carrier systems initiate a level-regulating pilot tone on each group at the transmit end of the system. Individual level-regulating pilots are also initiated on all supergroups and mastergroups. The intent is to regulate system level within ±0.5 dB.

Pilots are assigned frequencies that are part of the transmitted spectrum yet do not interfere with voice-channel operation. They usually are assigned a frequency appearing in the guard band between voice channels or are residual carriers (i.e., partially suppressed carriers). The CCITT has assigned the following as group-regulation pilots:

84.080 kHz (at a level of −20 dBm0)

84.140 kHz (at a level of −25 dBm0)

The Defense Communications Agency of the U.S. Department of Defense recommends 104.08 kHz for group regulation and alarm. For CCITT group pilots, the maximum level of interference permissible in the voice channel is −73 dBm0p. CCITT pilot filters have a bandwidth at the 3-dB points of 50 Hz (see CCITT Rec. G.241 for further information on other CCITT pilot frequencies and levels). The operating range of level control equipment activated by pilot tones is usually about ±4 or 5 dB. If the incoming level of a pilot tone in the multiplex receive equipment drops outside the level regulating range, an alarm will be indicated (if such an alarm is included in the system design); CCITT Rec. G.241 suggests such an alarm when the incoming level varies 4 dB up or down from the nominal.

4.6 Noise and Noise Calculations in FDM Carrier Systems

4.6.1 General

Carrier equipment is the principal contributor of noise on coaxial cable systems and other metallic transmission media. On radiolinks (line-of-sight microwave) it makes up about 25% of the total noise on the overall system link. The traditional approach is to consider noise with respect to a hypothetical reference circuit. Several methods are possible, depending on the application. These include the CCITT method, which is based on a 2500-km hypothetical reference circuit, and the U.S. Department of Defense method used in specifying communication systems. Such military systems are based on a 6000-nautical-mile reference circuit with 1000-mi links and 333-mi sections.

4.6.2 CCITT Approach

The mean psophometric* power, which corresponds to the noise produced by all modulating (multiplex) equipment . . . shall not exceed 2500 pW† at a zero relative level point. This value of power refers to the whole of the noise due to various causes (thermal, intermodulation, crosstalk, power supplies, etc.). Its allocation between various equipments can be to a certain extent left to the descretion of design engineers. However, to ensure a measurement agreement in the allocation chosen by different administrations, the following values are given as a guide to the target values:

for 1 pair of channel modulators	200 pW
for 1 pair of group modulators	80 pW
for 1 pair of supergroup modulators	60 pW

The following values are recommended on a provisional basis:

for 1 pair of mastergroup modulators	60 pW
for 1 pair of supermastergroup modulators	60 pW
for 1 pair of 15-supergroup assembly modulators	60 pW

Experience has shown that these target figures can often be improved considerably. The CCITT has purposely loosened the noise value allotted to channel modulators so that other modulation schemes from voice channel to group can be used rather than the direct modulation approach that we have described here. For instance, one solid-state FDM equipment now on the market, when operated with CCITT loading, has the following characteristics:

1 Pair of channel modulators	31 pWp‡
1 Pair of group modulators	50 pWp
1 Pair of supergroup modulators	50 pWp

*See Section 5 of this chapter

†1 pW = 1×10^{-9} W.

‡Picowatts psophometrically weighted. See Section 5 of this Chapter.

Using another solid-state equipment and increasing the loading to 75% voice, 17% telegraph tones, and 8% data, the following noise information is applicable:

1 Pair of channel modulators	322 pWp
1 Pair of group modulators	100 pWp
1 Pair of supergroup modulators	63 pWp

If this equipment is used on a real circuit with heavier loading, the sum for noise for channel modulators, group modulators, and supergroup modulator pairs is 485 pWp. Thus by simple arithmetic we see that such a system would be permitted to demodulate to voice only five times over a 2500-km route if we were to adhere to CCITT recommendations (i.e., $5 \times 485 = 2425$ pWp). This leads to the use of through-group and through-supergroup techniques discussed in Section 4.7. Figure 5.15 shows a typical application of this same equipment using CCITT loading.

4.6.3 U.S. Military Approach

The following values are set forth in U.S. Military Standard 188-100 for loaded noise:

Channel modulator pair	31 pWp0
Group modulator pair	50 pWp0
Supergroup modulator pair	50 pWp0
Through-group equipment	10 pWp0
Through super-group equipment	50 pWp0
Multiplex noise of FDM reference voice bandwidth link	131 pWp0

and it should be noted that the design objective is 100% data loading.

4.7 Through-Group and Through-Supergroup Techniques

In Section 4.6 we indicated that modulation–translation steps in long FDM carrier systems must be limited to avoid excessive noise accumulation. One method used widely is to employ group connectors and through-supergroup devices. Figure 5.16, which illustrates this technique, shows that supergroup 1 is passed directly from point *A* to point *B* while supergroup 2 is dropped at *C*, a new supergroup is inserted for onward transmission to *E*, and so forth. At the same time supergroups 3 to 15 are passed directly from *A* to *E* on the same line frequency (baseband).

The expression "drop and insert" is used in carrier-system terminology

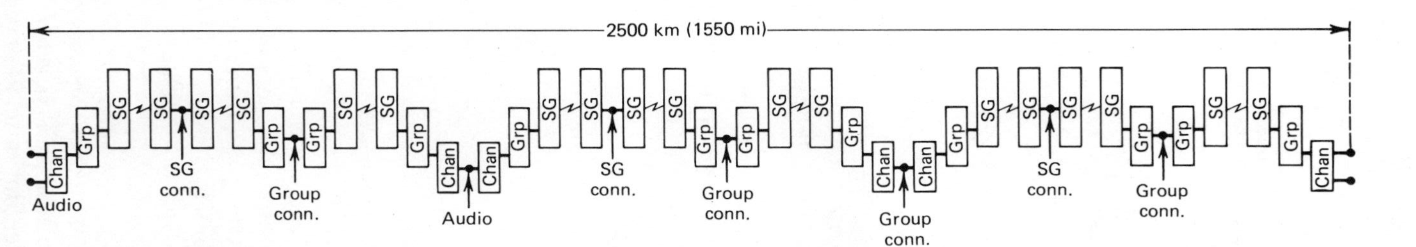

Figure 5.15 Typical CCITT reference-system noise calculations. These calculations are for typical equipment using CCITT loading. Courtesy of GTE Lenkurt, Inc., San Carlos, Calif.

Multiplex		pWp0	dBrnCO
3 complete sets of channel bank equipment (XMT and REC)	at 110 pW	330	26
6 complete sets of group bank equipment	at 45 pW	270	25
9 complete sets of supergroup bank equipment	at 58 pW	522	28
3 sets of group connector equipment	at 5 pW	15	12.6
3 sets of supergroup connector equipment	at 16 pW	48	17.6
Total 46A3 Carrier Noise		1185	31.5
Microwave			
9 complete radio sections of 6 hops each			
78A3 radio (600 ch.) at 500 pWp0/section		4500	37.3
75A3 radio (960 ch.) at 400 pWp0/section		3600	36.4
Overall System Noise			
Using 78A3 radio (600 channels)	(4500 + 1185)	5685	
Using 75A3 radio (960 channels)	(3600 + 1185)	4785	

Note: Conversion from pWp0 to dBrnCO is dBrnCO = (10 log pWp0) + .8

Note: Calculations are for a typical equipment with CCITT loading.
Courtesy GTE Lenkurt Incorporated, San Carlos, Calif.

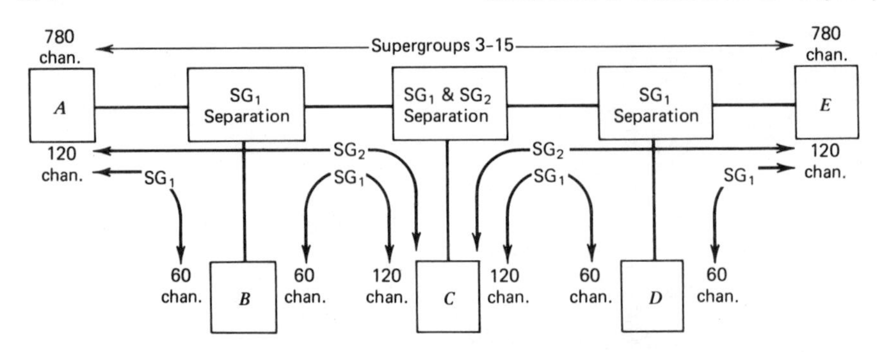

Figure 5.16 Typical drop and insert of supergroups.

to indicate that at some way point a number of channels are "dropped" to voice (if you will) and an equal number are "inserted" for transmission back in the opposite direction. For full-duplex operation, if channels are dropped at *B* from *A*, *B* necessarily must insert the same number of channels going back again to *A*.

Through-group and through-supergroup techniques are used especially on long trunk routes where excessive noise accumulation can be a problem. Such route plans can be very complex. However, the savings on equipment and reduction of noise accumulation should be obvious. When through-supergroup techniques are used, the supergroup pilot may be picked off and used for level regulation. Nearly all FDM carrier equipment manufacturers include level regulators as an option on through-supergroup equipment, whereas through-group equipment may not necessarily have the option.

5 SHAPING OF A VOICE CHANNEL AND ITS MEANING IN NOISE UNITS

The attenuation distortion or frequency response of a voice channel is termed "flat" because when it is tested from input to output, the response from, say, 300 to 3400 Hz may vary by only several dB. Of course, from the input of a voice channel modulator of an FDM carrier link to the output of the companion demodulator flat response, which is indeed what we have, should vary no more than perhaps ± 0.5 dB.

Now connect a telephone handset transmitter (with appropriate talk battery) to the input of the voice channel modulator and handset receiver to the output of the companion voice channel demodulator, and we see a "shaping" effect. The handset "shapes" the channel when the audio level is compared at various frequencies at the input of the acoustical–electrical transducer (handset transmitter) to the audio output from the electrical–acoustical transducer (the handset receiver).*

The frequency response measurement uses a reference frequency located at the point of minimum attenuation in the voice channel. In North America the reference frequency is 1000 Hz, whereas in Europe and many other locations in the world it is 800 Hz. All CCITT recommendations dealing with the voice channel use 800 Hz as reference.

For all voice-channel measurements affecting *speech,* we must note that certain frequencies in the voice channel are attenuated more than others. When transmission systems are tested, weighting networks are used to simulate these effects. Different types of telephone handset have different attenuation–frequency characteristics than others. In the literature we are liable to run into four different types, namely:

144	Line weighting (North America); seldom encountered; noise unit, dBrn
F1A	Line weighting, (North America); being phased out; noise unit, dBa
C–Msg	*C*-Message weighting; presently applicable in North America; noise unit, dBrn*C*
CCIR–1951	Psophometric weighting; European and CCITT–CCIR; noise unit, pWp

Reference frequency was established where the reference signal level was just discernible by the human ear. This level, depending on the handset, is in the range of −85 to −90 dBm. Thus the derived units would be positive numbers. Such units are used for noise measurement, given the zero reference and weighting characteristics referred to one of the four telephone handsets mentioned previously. These weighting characteristics are given in Figure 5.17.

The Western Electric 144 handset was an early model universally used in North America. The noise-measurement unit referred to the 144 was the dBrn, an abbreviation for "decibels reference noise," −90 dBm at 1000 Hz. Note that on the 144 curve in Figure 5.17, a 500-Hz sinusoidal signal would need a 15-dB level increase to have the same interfering effect on the "average" listener over the 1000-Hz reference. A 3000-Hz signal requires an 18-dB increase to have the same interfering effect, 6 dB at 800 Hz, and so on. This curve, as shown in Figure 5.17, is called a *weighting curve.* Noise-measurement instruments have artificial filters made to simulate the

*The acoustical properties of the typical human ear must also be taken into account because the design of a weighting network, using a particular handset, is based on the annoyance factor of a certain single frequency tone in the voice channel compared to the reference frequency.

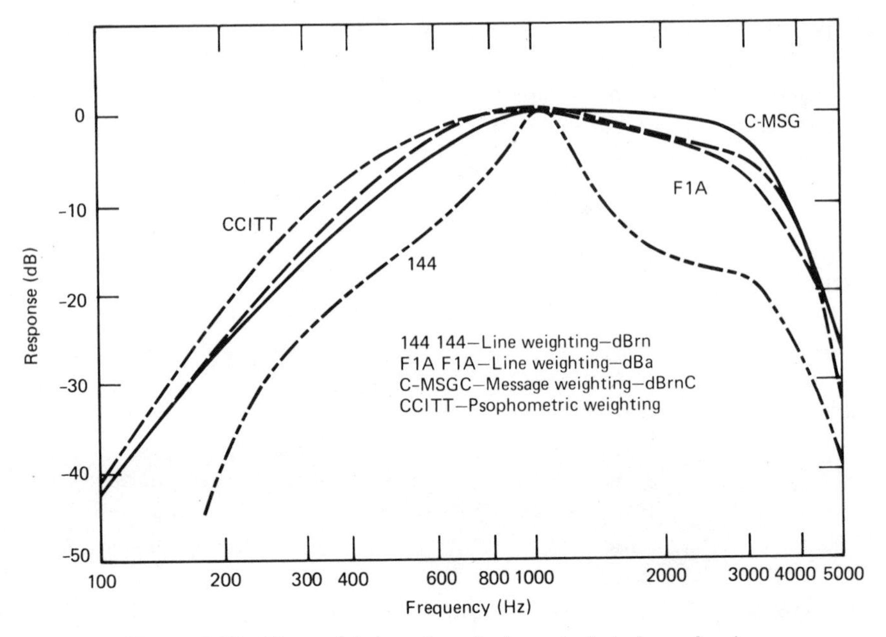

Figure 5.17 Line weightings for telephone (voice) channel noise.

response of the several handsets listed. Such filters are called *weighting networks*.

Subsequent to the 144 handset, the Western Electric Company developed the F1A handset, which had a considerably broader response than the older 144 handset but was 5 dB less sensitive at 1000 Hz (as shown in Figure 5.17). The reference level for this type of handset was −85 dBm. Of course, the new weighting curve and associated weighting network were denoted "F1A." The noise measurement was dBa, meaning dB adjusted.

A third, more sensitive handset is now in use in North America, giving rise to the *C*-message line-weighting curve and its companion noise measurement unit, the dBrn*C*. In Figure 5.17 we see that it is 3.5 dB more sensitive at the 1000-Hz reference frequency than F1A and 1.5 dB less sensitive than the old 144 weighting. Rather than a new reference frequency power level (−88.5 dBm), the reference power level of −90 dBm was maintained.

One important weighting curve is the CCIR or CCITT psophometric weighting curve. The noise-measurement units associated with this curve are the dBmp and the pWp, for dBm psophometrically weighted and picowatts psophometrically weighted, respectively. The reference frequency in this case is 800 Hz rather than 1000 Hz.

The reader must be certain to differentiate between a voice channel where noise measurements are weighted because the eventual user of the channel is the human mouth–ear combination, with the associated handset as the transducer of electrical energy to speech or audio energy. We also talk about a flat channel that has a comparatively flat frequency response across the voice channel, and we generally define the voice channel today as occupying that spectrum from 300 to 3400 Hz. Data–telegraph, facsimile, and video* transmission systems utilize "flat" channels, that is, where the amplitude frequency response varies little between channel edges—say, in the order of ±0.5 or ±0.2 dB. On long networks with numerous points of modulation–demodulation the amplitude frequency response can vary considerably more, but we still refer to the voice channel as being flat.

To convert from flat channel noise measurements in pW or dBm, the following excerpt from CCITT Rec. G.223 may be useful:

If uniform-spectrum random noise is measured in a 3.1 kHz band with flat attenuation frequency characteristic, the noise level must be reduced 2.5 dB to obtain the psophometric power level. For another bandwidth, B, the weighting factor will be equal to:

$$2.5 + \frac{(10 \log B)}{3.1} \text{ (dB)}$$

When $B = 4$ kHz, for example, this formula gives a weighting factor of 3.6 dB.

REFERENCES

1. *Notes on Distance Dialing,* American Telephone and Telegraph Company, New York, 1975.
2. International Telephone and Telegraph Corporation, *Reference Data for Radio Engineers,* 6th ed., Howard W. Sams, Indianapolis, 1976.
3. *Lenkurt Demodulator,* Lenkurt Electric Company, San Carlos, Calif., December 1964, June 1965, September 1965.
4. *Transmission Systems for Communication,* 4th ed., rev., Bell Telephone Laboratories.
5. U.S. Military Standard Mil-Std-188-100, November 15, 1972, Department of Defense, Washington, D.C. 20301.
6. H. H. Smith, *Noise Transmission Level Terms in American and International Practice,* ITT Communication Systems, Paramus, N.J., 1964.
7. CCITT *Green Books,* Fifth Plenary Assembly, Geneva, 1972, G. Recommendations.
8. B. D. Holbrook and J. T. Dixon, "Load Rating Theory for Multichannel Amplifiers," *Bell Syst. Tech. J.,* 624–644 (October 1939).
9. W. Oliver, *White Noise Loading of Multi-channel Communication Systems,* Marconi Instruments, St. Albans, Herts, UK, 1976.
10. Roger L. Freeman, *Telecommunication Transmission Handbook,* Wiley, New York, 1975.
11. *IEEE Standard Dictionary of Electrical & Electronic Terms,* 2nd ed., IEEE, New York, 1977.

*Under certain conditions video signal-to-noise ratios are weighted (i.e., TASO weightings).

Chapter 6

LONG-DISTANCE NETWORKS

1 GENERAL

The design of a long-distance network involves basically three considerations: (1) routing scheme given inlet and outlet points and their traffic intensities, (2) switching scheme and associated signaling, and (3) transmission plan. In the design, each criterion will interact with the other. In addition, the system designer must specify type of traffic, lost-call criteria, or grade of service, a survivability criterion, forecast growth, and quality of service. The trade-off of all these factors with "economy" is probably the most vital part of initial planning and downstream system design.

Consider transcontinental communications in the United States. Service is now available for people in New York to talk to people in San Francisco. From the history of this service, we have some idea of how many people wish to talk, how often, and for how long. These factors are embodied in traffic intensity and calling rate. There are also other cities on the West Coast to be served and other cities on the East Coast. In addition, there are existing traffic nodes at intermediate points such as Chicago and St. Louis. An obvious approach would be to concentrate all traffic into one transcontinental route with drops and inserts at intermediate points.

Again, we must point out that switching enhances the transmission facilities. From an economic point of view, it would be desirable to make transmission facilities (carrier, radio, and cable systems) adaptive to traffic load. These facilities taken alone are inflexible. The property of adaptivity, even when the transmission potential for it has been predesigned through redundancy, cannot be exercised except through the mechanism of switching in some form. It is switching that makes transmission adaptive.

The following requirements for switching ameliorate the weaknesses of transmission systems: concentrate light, discretely offered traffic from a multiplicity of sources and thus enhance the utilization factor of transmission trunks; select and make connections to a statistically described distribution of sinks per source; and restore connections interrupted by internal or external disturbances, thus improving reliabilities (and survivability) from an order of 90% to 99% to levels of the order of 99% to 99.9% or better. Switching cannot carry out this task alone. Constraints have to be iterated or fed back to the transmission systems, even to the local area. The transmission system must not excessively degrade the signal to be transported; it must meet a reliability constraint expressed in MTBF (mean time between failures), and availability and must have an alternative route scheme in case of facility loss, whether switching node or trunk route. This latter may be termed *survivability* and is only partially related to overflow (e.g., alternative routing).

The single transcontinental main traffic route in the United States suggested earlier has the drawback of being highly vulnerable. Its level of survivability is poor. At least one other route would be required. Then, why not route that one south to pick up drops and inserts? Reducing the concentration in the one route would result in a savings. Capital, of course, would be required for the second route. We could examine third and fourth routes to improve reliability–survivability and reduce long feeders for concentration at the expense of less centralization. In fact, with overflow, one to the other, dimensioning can be reduced without reduction of overall grade of service.

2 THE DESIGN PROBLEM

The same factors enter into long-distance network routing decisions as were discussed for the local area in Chapter 2. The first step is exchange placement. Here we follow North American practice and call the exchange in the long distance network a "toll exchange." Rather than base the placement decision on subscriber density distribution and their calling rates, the basic criterion is economy, the most cost-effective optimum. Toll-center placement is discussed in Section 5 of this chapter.

Having chosen toll-center locations, the design procedure is the familiar traffic matrix, where cost ratio studies are carried out to determine whether routing will be direct or tandem. The tendency is tandem (or "transit" exchanges) working and direct routes with overflow. The economic decision arises to balance switching against transmission (considering our arguments in section 1). Compare local versus long-distance networks:

	Switching Cost per Circuit	Transmission Cost per Circuit	Favored Network
Local network	Relatively high	Low	Mesh
Toll network	Relatively low	High	Star

For the long-distance network we can nearly always assume a hierarchical structure with three, four, or even five levels. Ideally, the highest levels should be connected in mesh for survivability.

In local network design, particularly in metropolitan areas, we could generally assume a mesh connection. There might be an exceptional case where tandem working would be economical, where traffic flows were 20 erlangs or less. Because of low traffic flows in long-distance network design, star connections can be assumed at the outset, and we would then proceed to determine cases where direct links may be justified with or without alternative routing.

3 LINK LIMITATION

It is stated in CCITT Rec. Q.13, p. 3, that:

> For reasons of transmission quality and the efficient operation of signalling, it is desirable to limit as much as possible the number of circuits connected in tandem.
>
> The apportionment between national and international circuits in such a chain may vary.
>
> The maximum number of circuits to be used for an international call is 12 with up to a maximum of six of the circuits being international.
>
> In exceptional cases and for a low number of calls, the total number of circuits may be 14, but even in this case the maximum number of international circuits is six.

If 14 circuits in tandem is the absolute maximum and we subtract six for the international portion, eight are left for the national portions. We then allow four maximum for each national network. This limit is crucial in the national network design. The concept is illustrated in Figure 6.1.

4 INTERNATIONAL NETWORK

The national telephone network must be connected to the international network, most probably through a CT (central transit) or transit center. The CCITT has established three hierarchical categories of transit centers—CT1, CT2, and CT3, the highest of which is CT1. Seven CT1 centers 37 CT2s, and 37 CT3s have been established. (See Chapt. 1, Section 9.3) The CCITT assignment of CT has repercussions in national network design. For example, a country with a CT3 to access international service should include no more than three links for the majority of calls, comprising two toll circuits in tandem and one toll-connecting trunk (see Figure 6.1). Of course, the national system has as many links as desired, but since a two-rank hierarchy is necessary for the long-distance service to meet international standards, it would be pointless to introduce more ranks in the hierarchy.

A three-rank hierarchy is acceptable for a nation assigned a CT2. Most nations assigned CT2 have a large geographical area; thus a three-rank hierarchy is necessary.

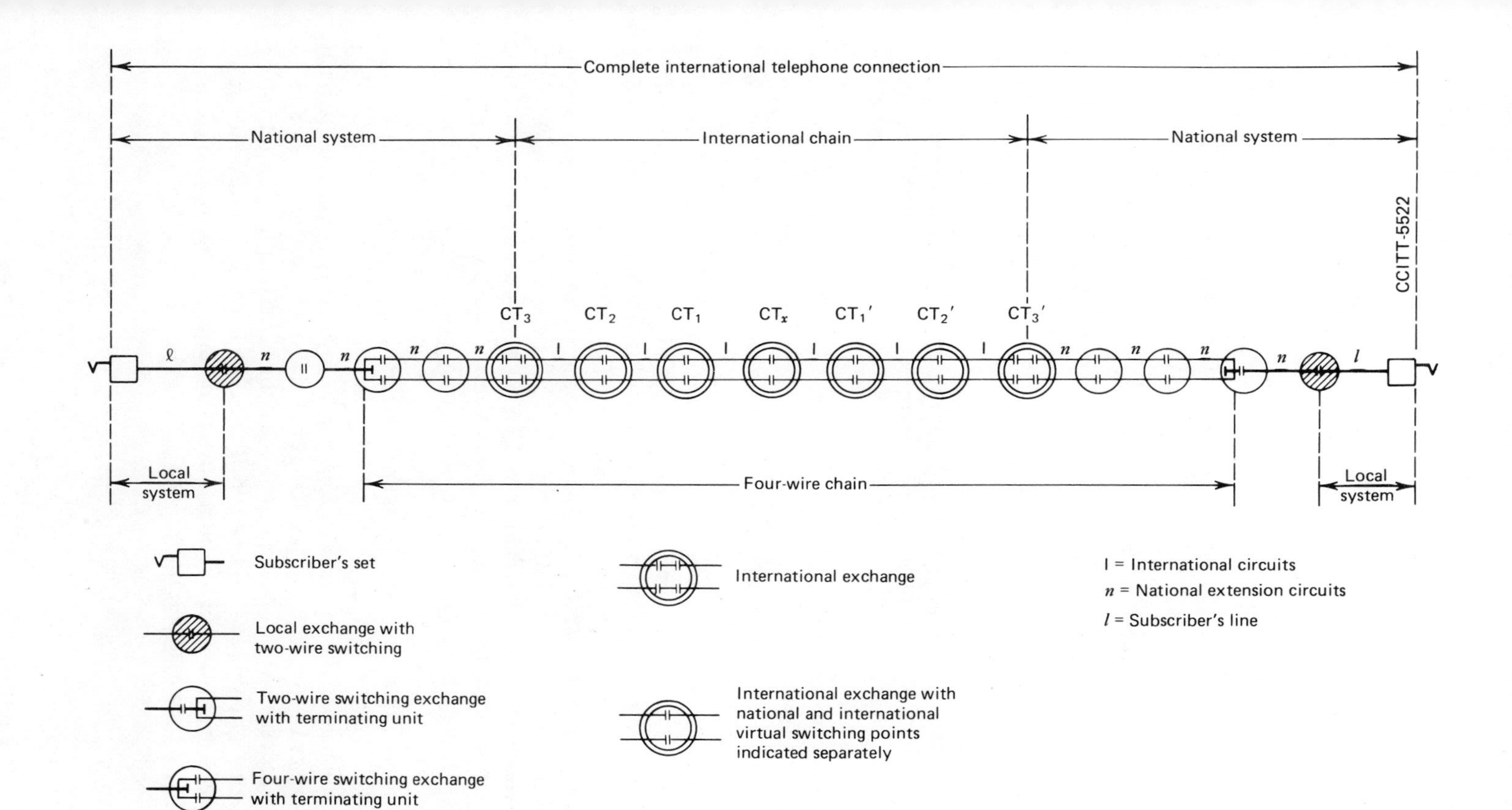

Figure 6.1 The CCITT nomenclature of international transit exchanges (CT) illustrating the limitation of the number of links in tandem on an international connection (see CCITT Rec. Q.41). Courtesy of the International Telecommunication Union—CCITT.

5 EXCHANGE LOCATION

The design of the toll network is closely related to the layout of toll areas; thus the system designer would start by placing a toll exchange in each toll area, probably in or near a large city in the area. Tariffs (tolls) for long-distance calls are usually based on the crow-fly distance traversed by the call. For instance, there may be fixed charges for calls over distances 0 to 20 km, 20 to 50 km, 50 to 100 km, 100 to 200 km, 200 to 300 km, and so on. It is expensive to measure route distance for every call. Hence a country or telephone serving area (in the macro sense) is usually divided into tariff areas and charged the same amount for calls between two different areas, no matter what the exact origin and destination in each of the areas. Tariff areas should not be made too small, as this would result in numerous areas with costly charging and billing equipment. When tariff areas are too large, tariffs may tend to be inequitable to subscribers. An area 50 km in diameter may be used as a guide for desirable size of tariff area. Areas will be considerably larger in sparsely populated regions. After tentatively placing a toll exchange in each toll area, the system designer should then examine adjacent pairs of toll exchanges to determine whether one exchange could serve both areas.

The next step is to examine assignments of toll exchanges regarding numbering. In Chapters 3 and 4 we saw how numbering routes a call and inputs call accounting equipment or call metering. Numbering may entail consideration of more than one toll exchange in geographically large tariff areas. Another consideration is maximum size of a toll exchange. Depending on expected long-distance calling rate and holding times, we might suppose 0.003 erlangs* per subscriber line; thus a 4000-line toll exchange could serve just under a million subscribers. The exchange capacity should be dimensioned to the forecast long-distance traffic load 10 years after installation. If the system undergoes a 15% expansion in long-distance traffic volume per year, it will grow to four times its present size in 10 years. Exchange location in the toll area is not very sensitive to traffic; it is more important to make maximum use of existing plant.

Hierarchy is another essential aspect. One important criterion in establishing the number of hierarchical levels in a national network is the level of CT assigned by CCITT, if any. Once the number of hierarchical levels has been established, the number of fan outs must be considered, to establish the number of long-distance exchanges in the network. Fan outs of six to eight are desirable. Thus with a two-stage hierarchy with a fan out of six and then eight there would be 48 of the lowest-level long-distance (toll) exchanges. A three-level hierarchy, using the same rules, would have $48 \times 8 = 384$, a formidable number.

*There is a marked tendency of toll traffic growth per subscriber in highly developed countries.

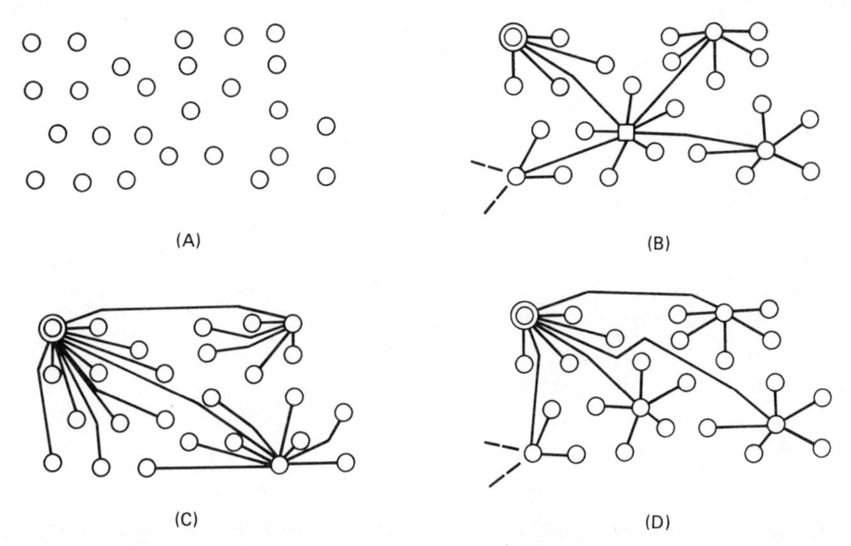

Figure 6.2 Choice of fan outs: (A) basic pattern; (B) three-stage interconnection; (C) two-stage interconnection low initial fan out; (D) two-stage interconnection high initial fan out.

As can be seen, there are many choices open to the system engineer to establish the route-plan hierarchy. For example, if there are 24 long-distance exchanges in an area, the network will initially be a star connection, either three-stage, two-stage with low (initial) fan out, or two-stage with high initial fan out. Here "fan out" refers to the highest level and works downward. Figure 6.2A,B,C,D illustrates these principles. Figure 6.2A shows an area with 26 exchanges and part of a larger area with perhaps 100 or more exchanges, with the principal city in the upper left-hand corner. Three choices of fan outs are shown. Figure 6.2B is a three-level hierarchy with a four-to-five fan out at each stage. For a two-level hierarchy, two possibilities are suggested. Figure 6.2C has low initial fan out and 6.2D, a high one. The choice between 6.2C and 6.2D may depend on traffic intensity between nodes or availability of routes. For national networks, 6.2D may be most economical since traffic is brought to a common point more quickly, leaving the individual branches to be least traffic efficient.

6 NETWORK DESIGN PROCEDURES

The attempt to attain a final design of an optimum national network is a major "cut-and-try" process. It lends itself well to computer techniques such as a program developed by ITT, "Optimization of Telephone Trunking Networks with Alternate Routing." A second program complements the series, namely, "Optimization of Telephone Networks with Hierarchical Structure" [10, 11].

In every case the design must first take into account the existing network. Major changes in that network require large expenditure. The network also represents an existing investment that should be amortized over time. Elements of the system, such as switches, are of varying age; some switches remain in service for up to 40 years, and others have been recently installed. Removal of a switch with only several years of service would not be economical. These switches also have specific signaling characteristics and are interconnected by a trunk network.

To simplify the design process, visualize a group of local areas. That is, the geographical and demographic area of interest in which a national network is to be designed is made up of contiguous local areas (Chapter 2). We assume that there exists a CT as assigned by international convention. There are now three bases to work from:

1. There are existing local areas, each of which has a toll exchange.
2. There is a national CT placed at the top of the network hierarchy.
3. There will be no more than four links in tandem (Section 3) on any connection to reach the CT.*

Point 1 may be redefined as a toll area made up of a grouping of local areas and probably coincide with a numbering (plan) area. This is illustrated in a very simplified manner in Figure 6.3, where T, in European (CCITT) terminology, is a primary center or a class 4 exchange, in North American terminology. Center T, of course, is a tandem exchange with a fan out of four; these are four local exchanges, *A*, *B*, *C*, and *D* homing on T. The entire national geographic area will be made up of small segments as shown in Figure 6.3, and each may be represented by a single exchange such as T.

The next step is to examine traffic flows to and from (originating and terminating at) each T. This information is organized and tabulated on a traffic matrix. A simplified example is shown in Table 6.1. Care must be taken in the preparation and subsequent use of such a table. The convention used here is that values are read *from* the exchange in the left-hand column *to* the exchange in the top row. For example, traffic from exchange 1 to exchange 5 is 23 erlangs, and traffic from exchange 5 to Exchange 1 is 25 Erlangs. It is often useful to set up a companion matrix of distances between exchange pairs. The matrix (Table 6.1) immediately offers candidates for high-usage routes. Nonetheless, this step is carried out after a basic hierarchical structure is established.

The highest level of the structure has been set forth a priori, that is, whether a CT1, CT2, CT3, or CT4. The CCITT hierarchy was shown in Figure 1.14. The lowest level of the network was established with local areas. The fact that Figure 1.14 has a five-level hierarchy does not imply

*There will be four links to a CT2, only three links to a CT3, and so on.

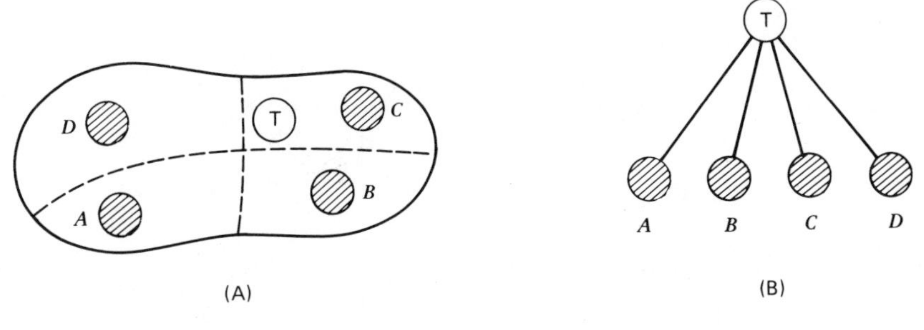

Figure 6.3 Areas and exchange relationships.

the use of five levels. The level of hierarchy is fixed by CCITT. For instance, if the country in question has a CT1, four links are permitted on a five-level hierarchy. A CT2 limits the total links to the top of the hierarchy to three for a four-level hierarchy, a CT3 to two links for a three-level hierarchy, and so forth. We are, of course, working with maximums on final routes.

The outline of a five-level hierarchy is shown in Figure 6.4 with high-usage (HU) routes. Note that the lowest level is not included in the figure, that of the local exchange. High-usage routes ameliorate the problem of excessive links in tandem on the great majority of calls, thereby meeting the intent of CCITT Rec. Q.13.

Suppose, for example, a country had four major population centers and could be divided into four areas around each center. Each of the four major population centers would have a tertiary center assigned, one of which would be the CT. Each tertiary center would have one or several secondary centers homing to it, and a number of primary centers would

Table 6.1 Toll Traffic Matrix (Sample) (in erlangs)

From Exchange	To Exchange 1	2	3	4	5	6	7	8	9	10
1		57	39	73	23	60	17	21	23	5
2	62		19	30	18	26	25	2	9	6
3	42	18		28	17	31	19	8	10	12
4	70	31	23		6	7	5	8	4	3
5	25	19	32	5		22	19	31	13	50
6	62	23	19	8	20		30	27	19	27
7	21	30	17	40	16	32		15	16	17
8	21	5	12	3	25	19	17		18	29
9	25	10	9	1	16	22	18	19		19
10	7	8	7	2	47	25	13	30	17	

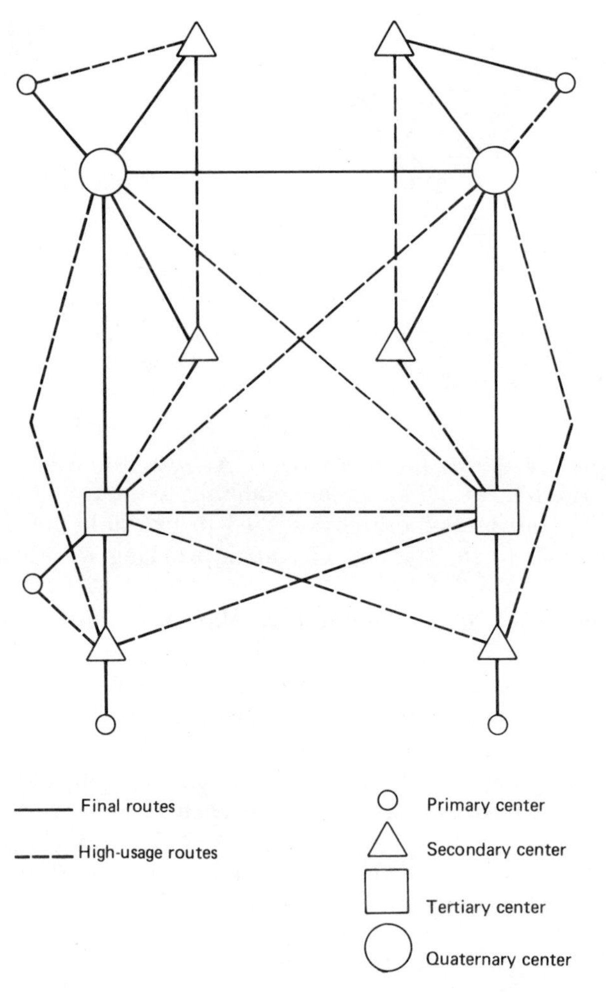

Figure 6.4 A hierarchical network with alternative routing (CCITT).

home to the secondary centers. This procedure is illustrated in Figure 6.5 and is represented systematically in Figure 6.6, thus establishing a hierarchy and setting out the final routes. In this case one of the tertiary exchanges would be a CT3. We define a final route as a route from which no traffic can overflow to an alternative route. It is a route that connects an exchange immediately above or below it in the network hierarchy and there is also connection of the two exchanges at the top level of the network. Final routes are said to make up the "backbone" of a network. Calls that are offered to the backbone but cannot be completed are lost calls.

A *high-usage* route is defined as any route that is not a final route; it may connect exchanges at a level of the network hierarchy *other than* the top

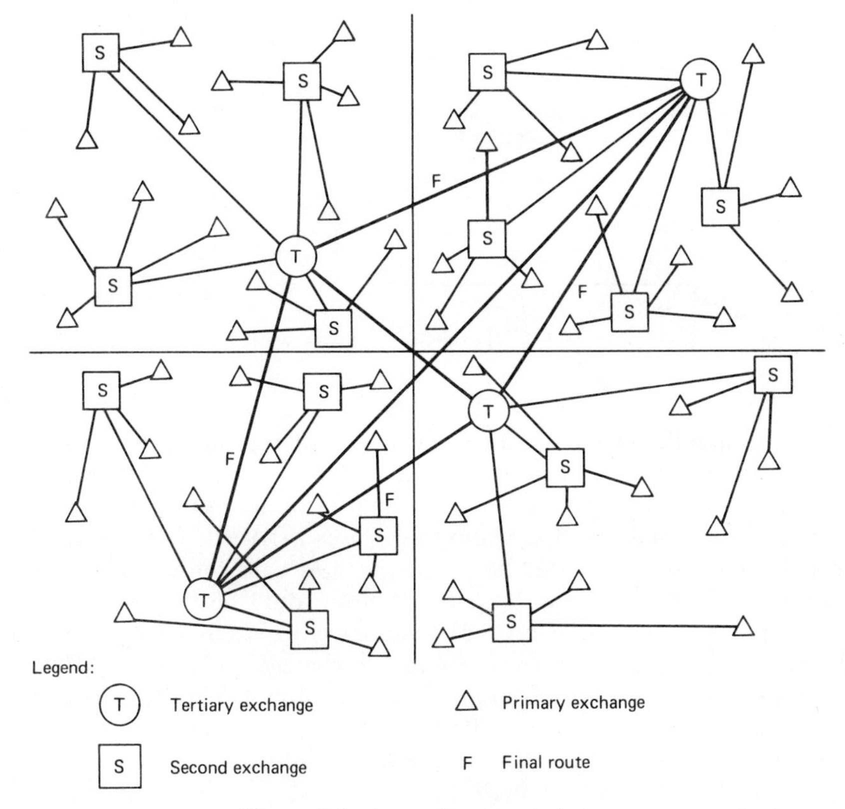

Figure 6.5 A sample network design.

level such as between T_1 and T_2 in Figure 6.6. It may also be a route between exchanges on different hierarchical levels when the lower-level exchange does not home on the higher level. A *direct route* is a special type of high-usage route connecting exchanges of the lowest rank in the hierarchy. Figure 6.7 illustrates these two definitions. High-usage routes are between exchanges 1 and 2 and 3 and 2. The direct route is also between 1 and 2, with exchanges 1 and 2 the lowest level in the hierarchy.

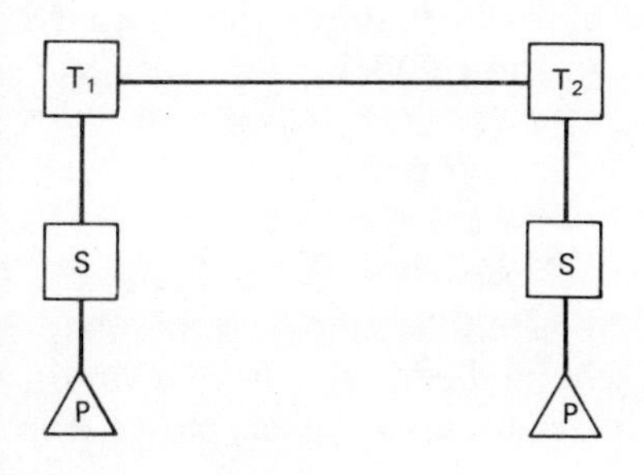

Figure 6.6 Hierarchical representation showing final routes.

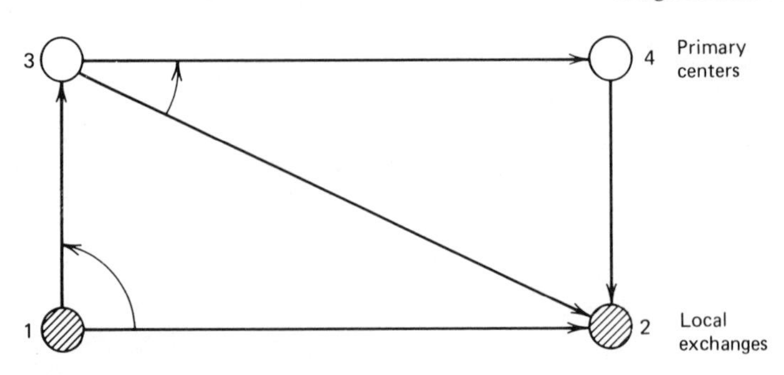

Figure 6.7 Hierarchical network segment.

Before final dimensioning may be carried out, a grade of service p must be established, usually at no greater than 1% per link on the final route during the busy hour (BH). If the maximum number of links in an international call is established at 12 (in some cases 14), the very worst grade of service would be 12 × 1%; however, on most calls the overall grade of service would be significantly better. These would be figures for direct-dialed international calls (see CCITT Recs. Q.13 and Q.95 bis). If there are three links in tandem on final choice routes, there would be up to 3% grade of service under worst conditions and with four links, 4%. These latter figures are for national connections. The use of HU connections reduces tandem operation and tends to improve overall grade of service.

The next step is to lay out high-usage routes. As mentioned earlier, this is done with the aid of the traffic matrix. One method of dimensioning with overflow (alternative routing) was discussed in Section 8.1 of Chapter 1 and the discussion continued in Section 9 of that chapter. Reference should also be made to Section 10 of Chapter 2. Trunks are costly. The exercise is to optimize the number of trunks and maintain a given grade of service. The methodologies given in Chapters 1 and 2 will help carry out this exercise. However, an increasing number of network designers are using computers to carry out this function. Two such computer programs are referenced at the beginning of this subsection. Traffic intensities used in the traffic matrices should be those taken from a 10-year forecast.

In much of this discussion CCITT terminology has been used. It would be helpful if the reader consulted Figure 6.8, which shows the standard structure of a national network according to the CCITT.

7 TRUNKING DIAGRAMS

After toll exchange locations have been established (Section 5 of this chapter), a hierarchical structure and routing scheme been set out (Section 6), and trunks dimensioned, the trunking diagrams must be prepared, for

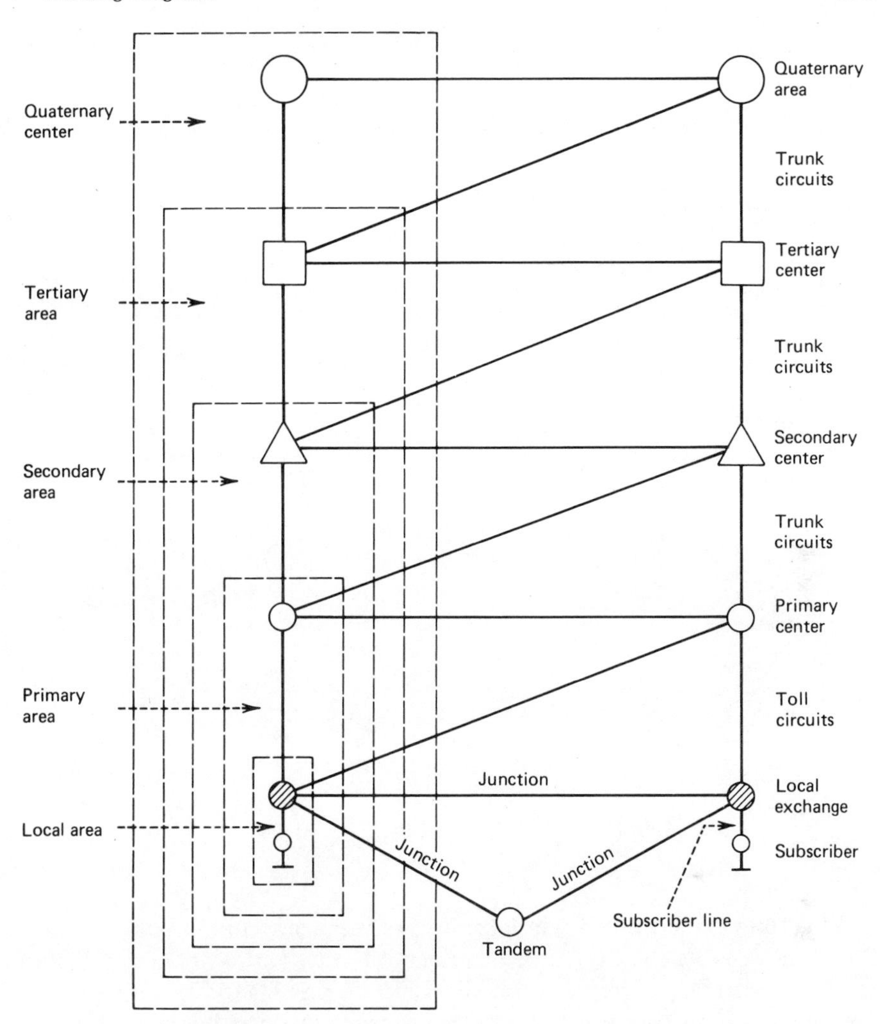

Figure 6.8 Structure of a national network (CCITT).

each toll exchange. A trunking diagram not only assures the network of a dimensioning interface on trunks, but in another context assures him of that all-important interface between switching and transmission. For instance, if there are 12 outgoing trunks at exchange *A* for exchange *B*, then at exchange *B* there must be provision for 12 inlets from *A*. The diagram must also indicate the number of one-way circuits and two-way circuits. Likewise, if exchange *A* is designed for CCITT No. 5 signaling, inlets at exchange *B* proceeding from *A* must be prepared to accept CCITT No. 5 interregister signaling and supervision, or conversion must be made for compatible signaling *A* to *B*.

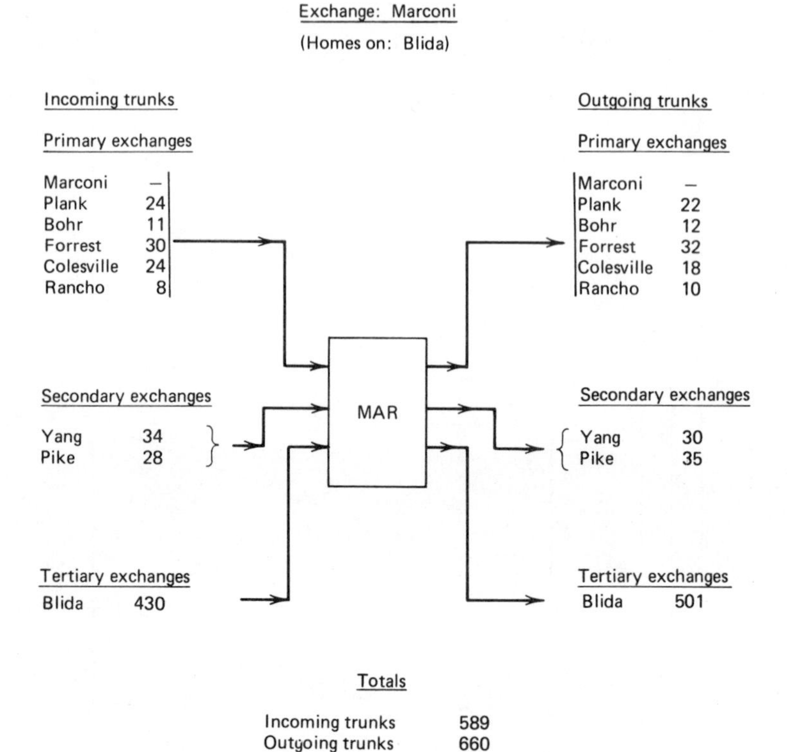

Figure 6.9 A typical trunking diagram for a transit exchange.

A trunking diagram for dimensioning a network contains the name of the exchange and the exchange on which it homes and is dependent within the hierarchy. An example is shown in Figure 6.9, where incoming trunks are on the left and outgoing trunks on the right. All circuits are one-way. In Europe two-way circuits are seldom in the toll plant but are found in rural areas. Two-way circuits in a mix of one- and two-way circuits are quite broadly used in North America. The Americans argue that use of circuit groups is more efficient. Europeans argue that the additional expense for plant mix and additional equipment to ensure no double seizure, are not worth any additional marginal efficiencies. The tendency toward longer-circuit routes in North America may justify attempts to optimize efficiency, particularly when the tendency to traffic peaks in one direction at one time during the day and then peaks in the other direction at another time.

There is still another trunking diagram required to properly engineer trunk groups, that ensuring proper signaling interface. It is common to use a circle associated with each trunk group. The circle is divided into three sections, as shown in the following diagram.

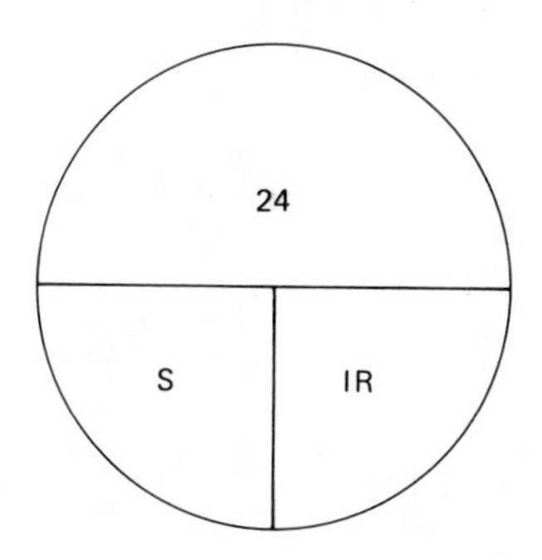

The upper half of the diagram shows the number of trunks, the lower left shows the type of supervision (S), and lower right shows the interregister type (IR) of signaling. One- and two-way trunks must be separated, and a circle such as the preceding one must be used for one-way in each direction and for the two-way. Refer to Chapter 4 for a description of the various signaling and supervision types and methods.

8 TRANSMISSION FACTORS IN LONG-DISTANCE TELEPHONY

8.1 Introduction

Long-distance analog communication systems require some method to overcome losses. As a wire-pair telephone circuit is extended, there is some point where loss accumulates so as to attenuate signals to such a degree that the far-end subscriber is unsatisfied. He cannot hear the near-end talker sufficiently well. Extending the wire connection still further, the signal level can drop below the noise level. For good received level, a 30-dB signal-to-noise ratio is desirable (see Chapter 5, Section 2.5). To overcome the loss, amplifiers can be added; in fact, amplifiers are installed on many wire-pair trunks. Early North American transcontinental circuits were on open-wire lines using amplifiers quite widely spaced. However, as BH demand increased to thousands of circuits, such an approach was not cost-effective.

System designers turned to wideband radio and coaxial cable systems where each bearer or pipe* carried hundreds (and now thousands) of simultaneous telephone conversations. Carrier (frequency division) multiplex techniques made this possible (see Chapter 5). Frequency division multiplex (FDM) requires separation of transmit and receive voice paths. In other words, the circuit must convert from two-wire to four-wire transmission. This is normally carried out by a hybrid transformer, or resistive hybrid. Figure 6.10 is a simplified block diagram of a telephone circuit with transformation from two-wire to four-wire operation at one end and con-

*On a pair of coaxial cables or a pair of radiofrequency carriers, one coming and one going.

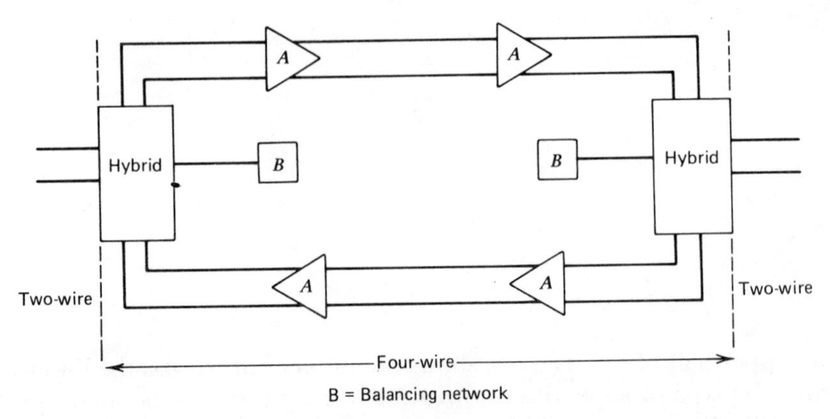

Figure 6.10 Simplified schematic of two-wire/four-wire operation.

version back to two-wire operation at the other end. This concept was introduced in Chapter 5, Section 3.

The two factors which must be considered that greatly affect transmission design, are "echo" and "singing." These are related, as we see in Sections 8.2 and 8.3.

8.2 Echo

As the name implies, echo in telephone systems is the return of a talker's voice. To be an impairment, the returned voice must suffer some noticeable delay. Thus we can say that echo is a reflection of the voice. Analogously, it may be considered as that part of the voice energy that bounces off obstacles in a telephone connection. These obstacles are impedance irregularities, more properly called *impedance mismatches.* Echo is a major annoyance to the telephone user. It affects the talker more than the listener. Two factors determine the degree of annoyance of echo: its loudness and the length of its delay.

8.3 Singing

Singing is the result of sustained oscillations due to positive feedback in telephone amplifiers or amplifying circuits. Circuits that sing are unusable and promptly overload multichannel carrier equipment (FDM—see Chapter 5).

Singing may be regarded as echo that is completely out of control. This can occur at the frequency at which the circuit is resonant. Under such conditions the circuit losses at the singing frequency are so low that oscillation will continue, even after cessation of its original impulse.

8.4 Causes of Echo and Singing

Echo and singing can generally be attributed to the mismatch between the balancing network of the hybrid and its two-wire connection associated with the subscriber loop. It is at this point that the major impedance mismatch usually occurs and an echo path exists. To understand the cause of the mismatch, remember that we always have at least one two-wire switch between the hybrid and the subscriber. Ideally, the hybrid-balancing network must match each subscriber line to which it may be switched. Obviously, the impedances of the four-wire trunks (lines) may be kept fairly uniform. However, the two-wire subscriber lines may vary over a wide range. The subscriber loop may be long or short, may or may not have inductive loading, and may or may not be carrier derived. The hybrid imbalance causes signal reflection or signal "return." The better is the match, the more the return signal is attenuated. The amount that the return signal (or reflected signal) is attenuated is called the *return loss* and is expressed in decibels. The reader should remember that any four-wire circuit may be switched to hundreds or even thousands of different subscribers; if not, it would be a simple matter to match the four-wire circuit to its single subscriber through the hybrid. This is why the hybrid to which we refer has a compromise balancing network rather than a precision network. A compromise network is usually adjusted for a compromise in the expected range of impedance (Z) encountered on the two-wire side.

Let us now consider the problem of match. If the impedance match is between the balancing network (N) and the two-wire line (L) (see Figures 5.4 and 5.5), then

$$\text{Return loss dB} = 20 \log_{10} \frac{Z_N + Z_L}{Z_N - Z_L}$$

If the network perfectly balances the line, then $Z_N = Z_L$, and the return loss would be infinite.

Return loss may also be expressed in terms of reflection coefficient, or:

$$\text{Return loss dB} = 20 \log_{10} \frac{1}{\text{reflection coefficient}}$$

where the reflection coefficient is equal to the reflected signal/incident signal.

The CCITT uses the term "balance return loss" (see CCITT Rec. G.122) and classifies it as two types:

1. Balance return loss from the point of view of echo.* This is the return loss across the band of frequencies from 500 to 2500 Hz.

*Called *echo return loss* (ERL) in via net loss (VNL) (North American practice; Section 8.7) but uses a weighted distribution of level.

2. Balance return loss from the point of view of stability. This is the return loss between 0 and 4000 Hz.

The band of frequencies most important in terms of echo for the voice channel is that from 500 to 2500 Hz. A good value for echo return loss for toll telephone plant is 11 dB, with values on some connections dropping to as low as 6 dB. For the local telephone network, the CCITT recommends better than 6 dB, with a standard deviation of 2.5 dB. For frequencies outside the 500-to-2500-Hz band, return loss values are often below the desired 11 dB. For these frequencies, we are dealing with return loss from the point of view of stability. The CCITT recommends that balance return loss from the point of view of stability (singing) should have a value of not less than 2 db for all terminal conditions encountered during normal operation (CCITT Rec. G.122, p. 3). For further information, the reader should consult CCITT Recs. G.122 and G.131.

Echo and singing may be controlled by:

- Improved return loss at the term set (hybrid).
- Adding loss on the four-wire side (or on the two-wire side).
- Reducing the gain of the individual four-wire amplifiers.

The annoyance of echo to a subscriber is also a function of its delay. Delay is a function of the velocity of propagation of the intervening transmission facility. A telephone signal requires considerably more time to traverse 100 km of a voice-pair cable facility, particularly if it has inductive loading, than it requires to traverse 100 km of radio facility (as low as 22,000 km/s for a loaded cable facility and 240,000 km/s for a carrier facility). Delay is measured in one-way or round-trip propagation time measured in milliseconds. The CCITT recommends that if the mean round-trip propagation time exceeds 50 ms for a particular circuit, an echo suppressor should be used. Bell System practices in North America use 45 ms as a dividing line. In other words, where echo delay is less than that stated previously here, echo can be controlled by adding loss.

An echo suppressor is an electronic device inserted in a four-wire circuit that effectively blocks passage of reflected signal energy. The device is voice operated with a sufficiently fast reaction time to "reverse" the direction of transmission, depending on which subscriber is talking at the moment. The blocking of reflected energy is carried out by simply inserting a high loss in the return four-wire path. Figure 6.11 shows the echo path on a four-wire circuit.

8.5 Transmission Design to Control Echo and Singing

As stated previously, echo is an annoyance to the subscriber. Figure 6.12 relates echo path delay to echo path loss. The curve in Figure 6.12 traces a

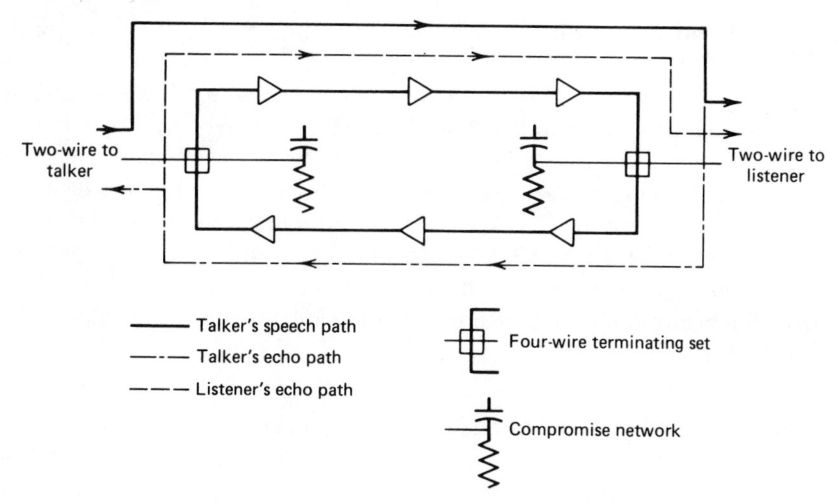

Figure 6.11 Echo paths in a four-wire circuit.

group of points at which the average subscriber will tolerate echo as a function of its delay. Remember that the greater the return signal is delayed, the more annoying it is to the telephone talker (i.e., the more the echo signal must be attenuated). For instance, if the echo path delay on a particular circuit is 20 ms, an 11-dB loss must be inserted to make echo tolerable to the talker. The careful reader will note that the 11 dB designed into the circuit will increase the end-to-end reference equivalent by that amount, which is quite undesirable. The effect of loss design on reference

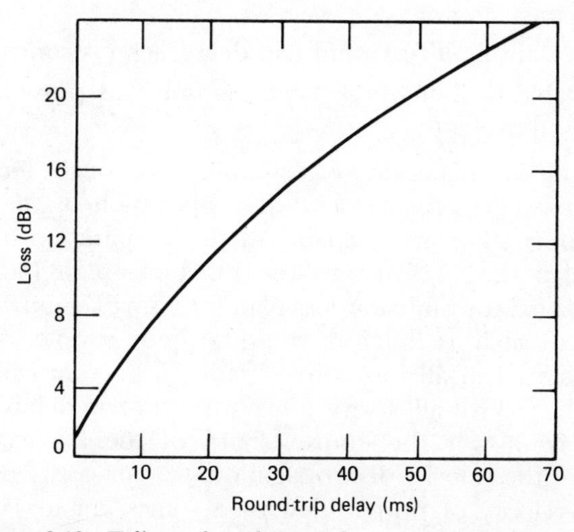

Figure 6.12 Talker echo tolerance for average telephone users.

equivalents and the trade-offs available are discussed in the paragraphs that follow.

If singing is to be controlled, all four-wire paths must have some loss. Once they go into a gain condition, and we refer here to overall circuit gain, positive feedback will result and the amplifiers will begin to oscillate or "sing." North American practice calls for a 4-dB loss on all four-wire circuits to ensure against singing. The CCITT recommends a minimum loss for a national network of 7 dB (CCITT Rec. G.122, p. 2).

Almost all four-wire circuits have some form of amplifier and level control. Such amplifiers are often embodied in the channel banks of the carrier (FDM) equipment.

8.6 An Introduction to Transmission-Loss Engineering

One major aspect of transmission system design for a telephone network is to establish a transmission-loss plan. Such a plan, when implemented, is formulated to accomplish three goals:

- Control singing.
- Keep echo levels within limits tolerable to the subscriber.
- Provide an acceptable overall reference equivalent to the subscriber.

For North America the via net loss (VNL) concept embodies the transmission plan idea (VNL is covered in Section 8.7).

From our preceding discussions we have much of the basic background necessary to develop a transmission loss plan. We know the following:

1. A certain minimum loss must be maintained in four-wire circuits to ensure against singing.
2. Up to a certain limit of round-trip delay, echo is controlled by loss.
3. It is desirable to limit these losses as much as possible to improve reference equivalent.

National transmission plans vary considerably. Obviously, length of circuit is important, as well as the velocity of propagation of the transmission media. Two approaches are available in the preparation of a loss plan, variable-loss plan (i.e., VNL) and the fixed-loss plan (i.e., as used in Europe). A national transmission-loss plan for a small country (i.e., small in geographic area) such as Belgium could be quite simple. Assume that a 4-dB loss is inserted in all four-wire circuits to prevent singing. Consult Figure 6.12, where 4 dB allows for 5 ms of round-trip delay. If we assume carrier transmission for the entire length of the connection and use 105,000 mi/s for the velocity of propagation, we can satisfy Belgium's echo problem. The velocity of propagation used comes out to 105 mi/ms (168 km)/ms. By simple arithmetic, we see that a 4-dB loss on all four-wire

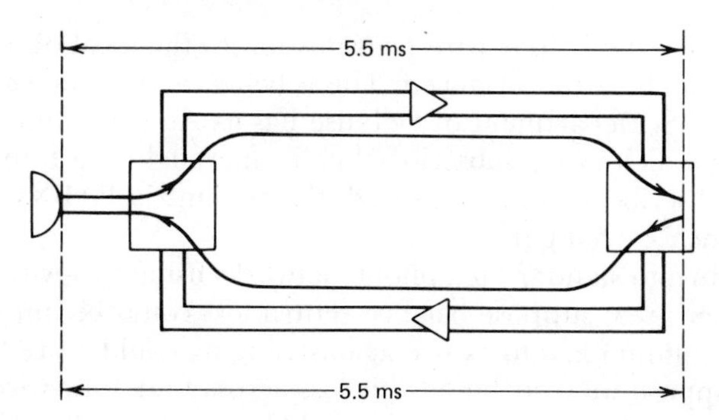

Figure 6.13 Example of echo round-trip delay (5.5 + 5.5 = 11-ms round-trip delay).

circuits will make echo tolerable for all circuits extending 210 mi (336 km). This is an application of the fixed-loss type of transmission plan. In the case of small countries or telephone companies operating over a small geographical extension, the minimum loss inserted to control singing controls echo as well for the entire country.

Let us try another example. Assume that all four-wire connections have a 7-dB loss. Figure 6.12 indicates that 7 dB permits an 11-ms round-trip delay. Assume that the velocity of propagation is 105,000 mi/s. Remember that we are dealing with round-trip delay. The talker's voice reaches the far-end hybrid and is then reflected back. This means that the signal traverses the system twice, as shown in Figure 6.13. Thus 7 dB of loss for the given velocity of propagation allows about 578 mi of extension or, for all intents and purposes, the distance between subscribers and will satisfy the loss requirements for a country with a maximum extension of 578 mi.

It has become evident by now that we cannot continue increasing losses indefinitely to compensate for echo on longer circuits. Most telephone companies and administrations have set the 45- or 50-ms round-trip delay criterion, which sets a top figure above which echo suppressors are to be used. One major goal of the transmission loss plan is to improve overall reference equivalent or to apportion more loss to the subscriber plant so that subscriber loops can be longer or allow the use of less copper (i.e., smaller-diameter conductors). The question arises as to what measures can be taken to reduce losses and still keep echo within tolerable limits. One obvious target is to improve return losses at the hybrids. If all hybrid return losses are improved, the echo tolerance curve shifts; this is because improved return losses reduce the intensity of the echo returned to the talker. Thus he is less annoyed by the echo effect.

One way of improving return loss is to make all two-wire lines out of the hybrid look alike, that is, have the same impedance. The switch at the other end of the hybrid (i.e., on the two-wire side) connects two-wire loops of varying length, thus causing the resulting impedances to vary greatly. One approach is to extend four-wire transmission to the local office such that each hybrid can be better balanced. This is being carried out with success in Japan. The U.S. Department of Defense has its Autovon (automatic voice network), in which every subscriber line is operated on a four-wire basis. Two-wire subscribers connect through the system via PABXs (private automatic branch exchange).

Let us return to standard telephone networks using two-wire switches in the subscriber area; suppose balance return loss could be improved to 27 dB. Thus minimum loss to assure against singing could be reduced to 0.4 dB. Now suppose we distributed this loss across four four-wire circuits in tandem. Thus each four-wire circuit would be assigned a 0.1-dB loss. If we have gain in the network, singing will result. The safety factor between loss and gain is 0.4 dB. The loss in each circuit or link is maintained by amplifiers. It is difficult to adjust the gain of an amplifier to 0.1 dB, much less keep it there over long periods, even with good automatic regulation. *Stability* or *gain stability* is the term used to describe how well a circuit can maintain a desired level. Of course, in this case we refer to a test-tone level. In the preceding example it would take only one amplifier to shift 0.4 dB, two to shift in the positive direction 0.2 dB, and so forth. The importance of stability, then, becomes evident.

The stability of a telephone connection depends on three criteria:

1. The variation of transmission level with time.
2. The attenuation–frequency characteristics of the links in tandem.
3. The distribution of balance return loss.

Each criterion becomes magnified when circuits are switched in tandem. To handle the problem properly, we must talk about statistical methods and standard distributions. In the case of criteria 1 and 2, we refer to the tandeming of four-wire circuits. Criterion 3 refers to switching subscriber loops–hybrid combinations that will give a poorer return loss than will the 11 dB stated earlier. Return losses on some connections can drop to 3 dB or less.

Stability is discussed in CCITT Recs. G.131 and G.151C. In essence, the loss through points a–t–b in Figure 6.14 should have a value not less than $(6 + N)$dB, where N is the number of four-wire circuits in the national chain. Thus the minimum loss is stated (CCITT Rec. G.122), and Rec. G.131 is quoted in part as follows:

> The standard deviation of transmission loss among international circuits routed in groups equipped with automatic regulation is 1 dB. . . . This accords with . . . that the tests . . . indicate that this target is being approached in that 1.1 dB was the standard deviation of the recorded data.
>
> It is also evident that those national networks which can exhibit no better stability balance return loss than 3 dB, 1.5 dB standard deviation, are unlikely to seriously jeopardize the stability of international connections as far as oscillation is con-

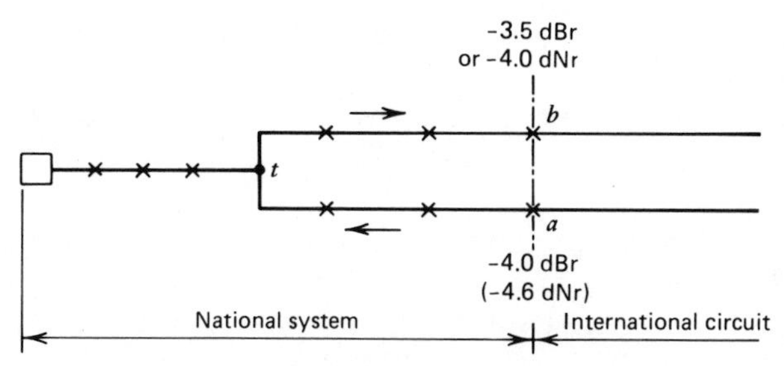

Figure 6.14 Definition of points *a–t–b* (CCITT Rec. G.122); X indicates a switch. Courtesy of the International Telecommunication Union–CCITT.

cerned. However, the near-singing [rain-barrel effect] distortion and echo effects that may result give no grounds for complacency in this matter.

Stability requirements in regard to North American practice are embodied in the VNL concept discussed in the next section.

8.7 Via Net Loss

Via net loss (VNL) is a concept or method of transmission planning that permits a relatively close approach to an overall zero transmision loss in the telephone network (lowest practicably attainable) and maintains singing and echo within specified limits. The two criteria that follow are basic to VNL design:

1. Customer–customer talker echo shall be satisfactorily low on more than 99% of all telephone connections that encounter the maximum delay likely to be experienced.
2. The total amount of overall loss is distributed throughout the trunk segments of the connection by allocation of loss to the echo characteristics of each segment.

One important concept in the development of the discussion on VNL is that of echo return loss (ERL) (see subsection 8.2). For this discussion, we consider ERL as a single-valued, weighted figure of return losses in the frequency band 500 to 2500 Hz. Echo return loss differs from return loss in that it takes into account a weighted distribution of level versus frequency to simulate the nonlinear characteristics of the transmitter and receiver of the telephone instrument. By using ERL measurements, it is possible to arrive at a basic design factor for the development of the VNL formula. This design factor states that the average return loss at class 5 offices (local exchanges) is 11 dB, with a standard deviation of 3 dB. Considering a standard distribution curve and the one, two, or three σ

points on the curve, we could thus expect practically all measurements of ERL to fall between 2 and 20 dB at class 5 offices (local exchanges). Via net loss also considers that reflection occurs at the far end in relation to the talker where the four-wire trunks connect to the two-wire circuits (i.e., at the far-end hybrid).

The next concept in the development of the VNL discussion is overall connection loss (OCL), which is the value of one-way trunk loss between two end (local) offices (not subscribers). Consider that

$$\text{Echo path loss} = 2 \times \text{trunk loss (one-way)} + \text{return loss (hybrid)}$$

where all units are in decibels. Now let us consider the average tolerance for a particular echo path loss. Average echo tolerance is taken from the curve in Figure 6.12. Therefore

$$\text{OCL} = \frac{\text{Average echo tolerance (loss)} - \text{return loss}}{2}$$

where all units are in decibels. Return loss in this case is the average echo return loss that must be maintained at the distant local exchange—the 11 dB given earlier.

An important variability factor not considered in the formula is trunk stability, which determines how close assigned levels are maintained on a trunk. Via-net-loss practice dictates trunk stability to be maintained with a normal distribution of levels and a standard deviation of 1 dB in each direction. For a round-trip echo path the deviation is taken as 2 dB. This variability applies to each trunk in a tandem connection. If there are three trunks in tandem, this deviation must be applied to each.

The reader will recall that the service requirement in VNL practice is satisfactory echo performance for 99% of all connections. This may be considered as a cumulative distribution, or 2.33 standard deviations, summing from negative infinity toward the positive direction. The OCL formula may now be rewritten as follows:

$$\text{OCL} = \frac{\text{Average echo tolerance} - \text{average return loss} + 2.33D}{2}$$

where D is the composite standard deviation of all functions and all units are in decibels. The derivation of D, the composite standard deviation of all functions, is as follows:

$$D = \sqrt{D_t^2 + D_{r1}^2 + ND_1^2}$$

where D_t = standard deviation of distribution of echo tolerance among a large group of observers, given as 2.5 dB

D_{r1} = standard deviation of distribution of return loss, given as 3 dB

D_1 = standard deviation of distribution of the variability of trunk loss for a round-trip echo path, given as 2 dB

N = number of trunks switched in tandem to form a connection class 5 office to class 5 office

Now consider several trunks in tandem; it can be calculated that at just about any given echo path delay, the OCL increases approximately 0.4 dB for each trunk added. With this simplification, once we have the OCL for one trunk, all that is needed to compute the OCL for additional trunks is to add 0.4 dB times the number of trunks added in tandem. This loss may be regarded as an additional constant needed to compensate for variations in trunk loss in the VNL formula.

Figure 6.15 relates echo path delay (round-trip) to overall connection loss (OCL for one trunk, then for a second trunk in tandem, and for four and six trunks in tandem). Although the straight-line curve has been

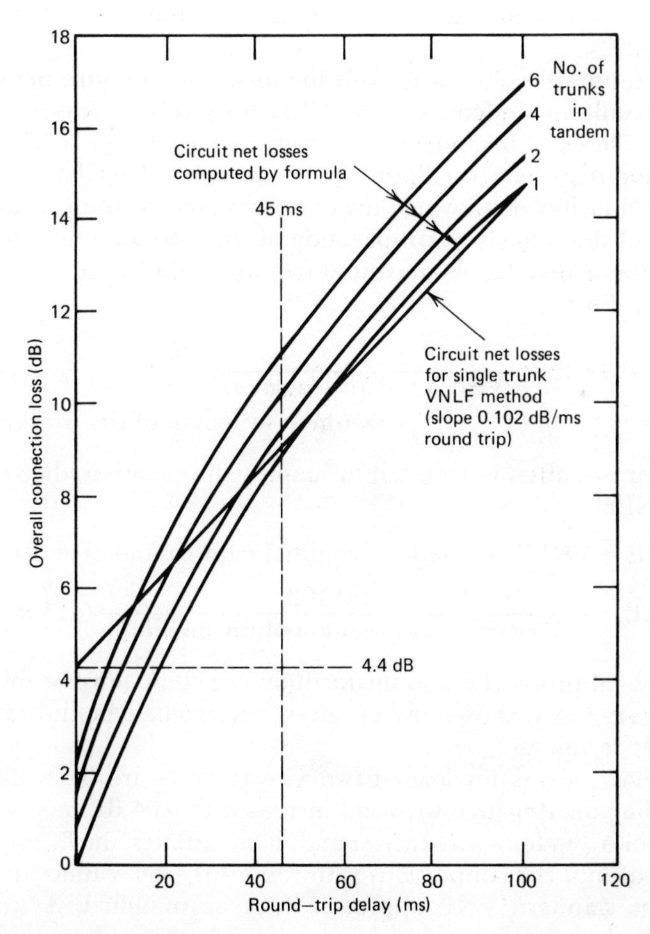

Figure 6.15 Approximate relationship between round-trip echo delay and overall connection loss (OCL).

simplified, the approximation is sufficient for engineering VNL circuits. Note that the straight-line curve in Figure 6.15 cuts the *Y* axis at 4.4 dB, where round-trip delay is 0. This 4.4 dB is based on two conditions, namely, that all trunks have a minimum of 4 dB to control singing and 0.4 dB protection against negative variation of trunk loss. Another important point to be defined on the linear curve in Figure 6.15 is a round-trip delay of 45 ms, which corresponds to an OCL of 9.3 dB. Empirically, it has been determined that echo suppressors must be used for delays greater than 45 ms. From this same linear curve the following formula for OCL may be derived:

$$\text{OCL} = (0.102)\,(\text{path delay in ms}) + (0.4\text{ dB})\,(\text{number of trunks in tandem}) + 4\text{ dB}$$

Usually the 4 dB as shown in the preceding OCL equation is applied to the extremity of each trunk network, namely, to the toll connecting trunks, 2 dB to each.

Overall connection loss deals with the losses of an entire network consisting of trunks in tandem, whereas VNL deals with the losses assigned to one trunk. The VNL formula follows from the OCL formula. The key here is the round-trip delay on the trunk in question. The delay time for a transmission facility employing only one particular medium is equal to the reciprocal of the velocity of propagation of the medium multiplied by the length of the trunk. To obtain round-trip time, this figure must be multiplied by 2; thus

$$\text{VNL} = 0.102 \times 2 \times \frac{1}{\text{Velocity of propagation}} \times (\text{one-way length of the trunk}) + 0.4\text{ dB}$$

Another term is often introduced to simplify the equation, the via-net-loss factor (VNLF):

$$\text{VNL} = \text{VNLF} \times (\text{one-way length of trunk in miles}) + 0.4\text{ dB}$$

$$\text{VNLF} = \frac{2 \times 0.102}{\text{Velocity of propagation of the medium}}\ (\text{dB/mi})$$

The velocity of propagation of the medium used here must be modified by such things as delays caused by repeaters, intermediate modulation points, and facility terminals.

Via-net-loss factors for loaded two-wire facilities are 0.03 dB/mi, with H-88 loading on 19-gauge wire and increased to 0.04 dB/mi on B-88 and H-44 facilities. On four-wire carrier and radio facilities, the factor improves to 0.0015 dB/mi. For connections with round-trip delay times in excess of 45 ms, the standard VNL approach must be modified. As mentioned previously, these circuits use echo suppressors that automatically switch about 50 dB into the echo return path, and the switch actuates when speech is received in the "return" path, thus switching the pad into the "go" path.

Via-net-loss practice in North America treats long delay circuits with up to a maximum of 45 ms of delay in the following manner (see Figure 6.16). Here the total round-trip delay is arbitrarily split into two parts for connections involving regional intertoll trunks. If the regional intertoll delay exceeds 22 ms, echo suppressors are used. If the figure is 22 ms or less, echo is controlled by VNL design. Thus allow 22 ms for the maximum delay for the regional intertoll segments to the connection. This leaves 23-ms maximum delay for the other segments (45 − 22). Now apply the VNL formula for a delay of 22 ms. Thus

$$\text{VNL} = 0.102 \times 22 + 0.4 = 2.6 \text{ B}$$

This loss is equivalent to the maximum length of an intertoll trunk without an echo suppressor. What is that length?

$$\text{Length (one-way)} = \frac{\text{VNL} - 0.4}{\text{VNLF}} \text{ mi}$$

$$= \frac{0.102 \times 22 + 0.4 - 0.4}{0.0015} = 1498 \text{ mi*}$$

In summary, in VNL design we have three types of loss that may be assigned to a trunk:

Type	Loss
Toll-connecting trunk	VNL + 2.5 dB
Intertoll trunk (no echo suppressor)	VNL
Intertoll trunk (with echo suppressor)	0 dB†

8.8 Transmission-Loss Plan for Japan (an Example)

An example of the current transmission-loss plan for Japan is presented in Figure 6.17. It is a fixed-loss plan. The ORE for 97% of national connections is only 26 dB. One way of achieving this level, as suggested previously, is that four-wire switching extend through primary centers and four-wire transmission to local exchanges. The breakdown of the ORE is as follows:

0.5 dB for local switches
12.5 dB TRE for subscriber loop
2.5 dB RRE for subscriber loop
10.0 dB four-wire system, local exchange to local exchange.

26 dB total (=ORE)

The reference equivalents are based on the Japanese 600P telephone sets with a 1200-Ω subscriber loop, and a 0.4-mm nonloaded cable fed with −48

*The VNLF in this equation indicates carrier and/or radio for the whole trunk.
†See discussion on echo suppressor for explanation of the 0-dB figure.

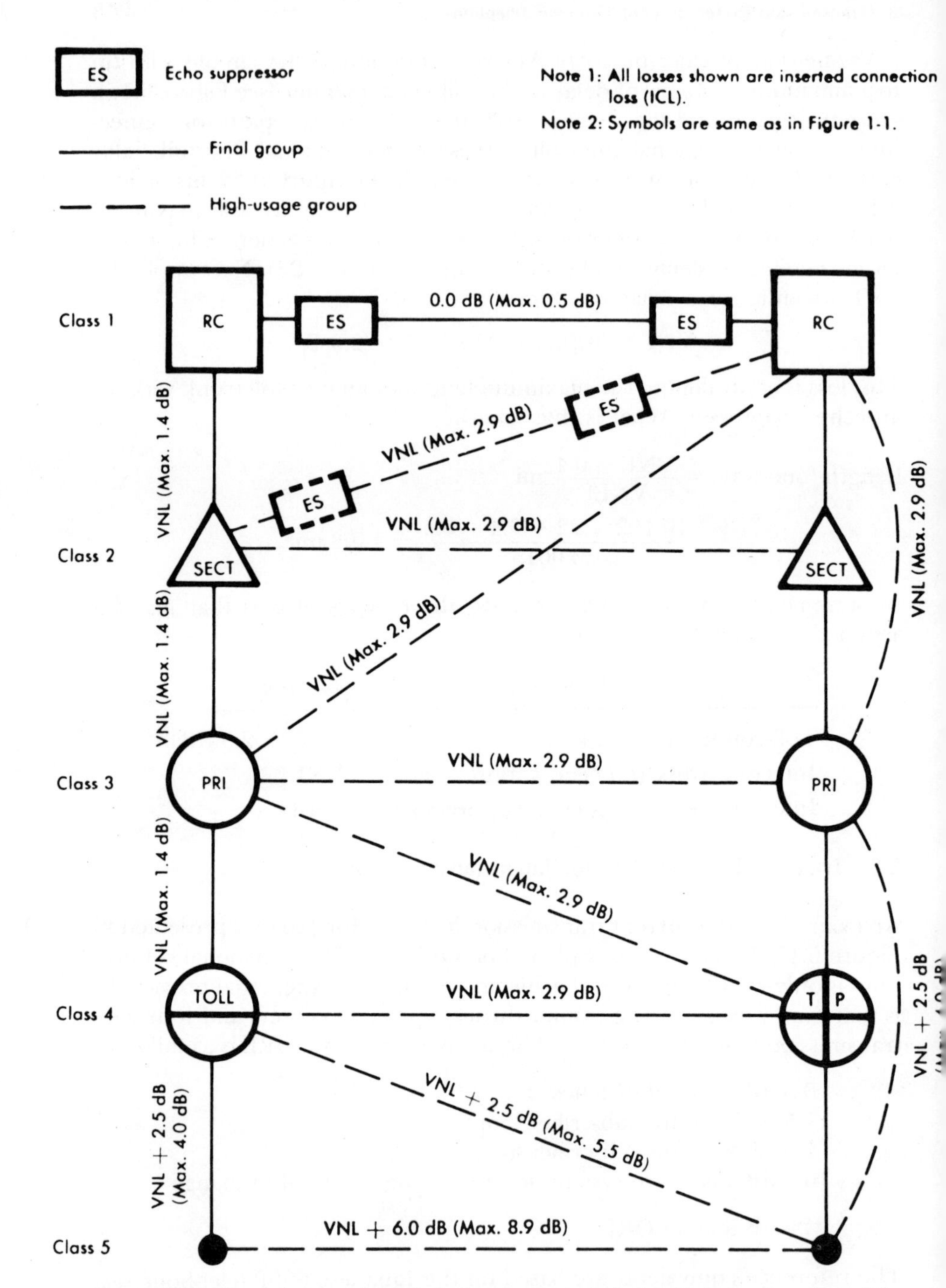

Figure 6.16 Trunk losses with VNL design. Losses include 0.4 dB design loss allowed for maintenance. Copyright © American Telephone and Telegraph Co., 1977.

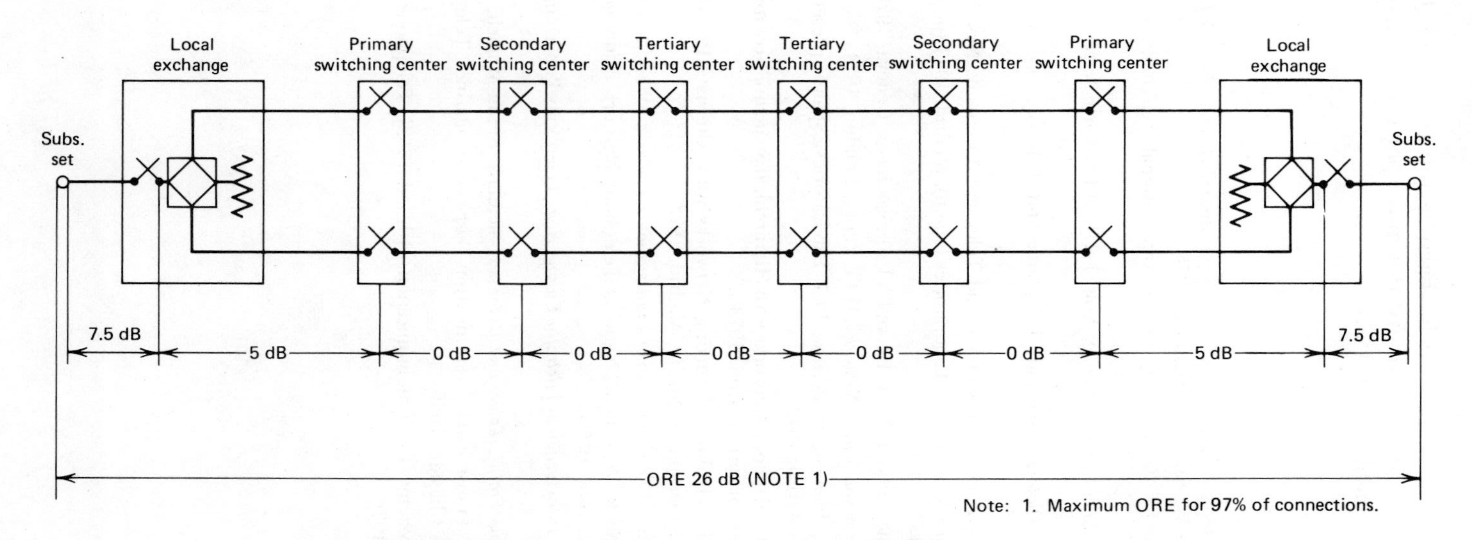

Figure 6.17 National loss distribution for the Japanese network.

V. The plan assumes a mean echo balance return loss of 11 dB. Loss stability is given as 1 dB of standard deviation per each four-wire circuit, with a singing probability of 0.1%.

REFERENCES

1. International Telephone and Telegraph Corporation, *Reference Data for Radio Engineers*, 6th ed., Howard W. Sams, Indianapolis, 1976.
2. *National Networks for the Automatic Service*, International Telecommunications Union, Geneva, 1964.
3. *Overall Communication System Planning*, Vols. I–III, IEEE North Jersey Section Seminar, 1964.
4. *Notes on Distance Dialing*, American Telephone and Telegraph Company, New York, 1975.
5. *Switching Systems*, American Telephone and Telegraph Company, New York, 1961.
6. *Telecommunication Planning*, ITT Laboratories (Spain), Madrid, 1973 (in particular, Section 2, "Networks").
7. CCITT Recommendations, Vols. III and VI, *Orange Books*, Geneva, 1976.
8. J. E. Flood, *Telecommunication Networks*, IEEE Series, London, 1974.
9. *Telecommunication Planning Symposium*, ITT Laboratories (Spain), presented to SAPO, Boksburg, South Africa, 1972.
10. "Optimization of Telephone Networks with Hierarchical Structure" (computer program), ITT Laboratories (Spain), Madrid, 1973.
11. "Optimization of Telephone Trunking Networks with Alternate Routing" (computer program), ITT Laboratories (Spain), Madrid, 1973.
12. Roger L. Freeman, *Telecommunication Transmission Handbook*, Wiley, New York, 1975.
13. F. T. Andrew and R. W. Hatch, "National Telephone Network Planning in the ATT," *IEEE ComTech J.* (June 1971).
14. Ramses R. Mina, *Introduction to Teletraffic Engineering*, Telephony Publishing Corporation, Chicago, 1974.
15. *Theory of Telephone Traffic, Tables and Diagrams*, Siemens, Berlin-Munich, 1971, Part 1.
16. M. A. Clement, "Transmission," reprint from *Telephony* (magazine), Telephony Publishing Corporation, Chicago, 1969.
17. Transmission Systems for Communication, 4th ed., Bell Telephone Laboratories, Holmdel, N.J., 1970.

Chapter 7

THE DESIGN OF LONG-DISTANCE LINKS

1 INTRODUCTION

In Chapter 6 we proposed a methodology for the design of a long-distance network. The network may be defined as a group of switching nodes interconnected by links. We may refer to a link as a transmission highway between switches carrying one or more traffic relations. The link could appear as that in the following diagram, where switches *A*, *B*, and *C* are connected to switches *X*, *Y*, and *Z* over a link as shown. The discussion that follows introduces the essentials of transmission design of such links.

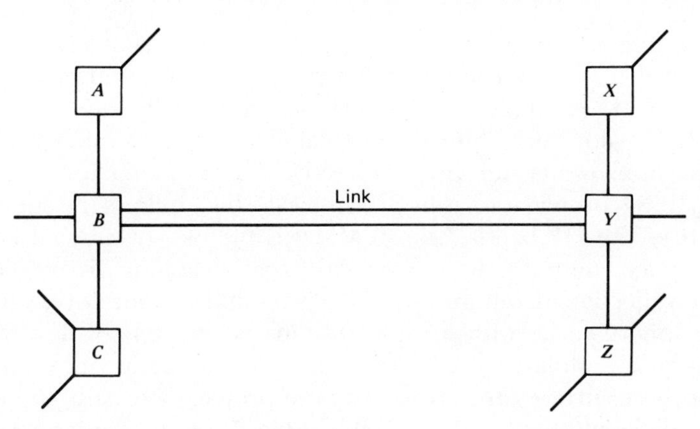

2 THE BEARER

British telephone engineers are fond of the term "bearer," which is quite descriptive. The bearer is what carries the signal(s). It could be a pair of wires or two pairs on a four-wire basis, a radio carrier in each direction, or a coaxial cable in each direction. The wire pair could be open wire line, aerial cable, or buried cable. In the text that follows it is assumed that a number of telephone channels are transmitted by each bearer in an FDM multiplex configuration as discussed in Chapter 5.

Modern long-distance links use one of two types of bearer, either radio or coaxial cable. The decision of which to use is basically economic. For instance, if a link, as defined in Section 1, is to carry 5000 or more circuits (telephone channels) at the end of a 10-year planning period, coaxial cable may be strongly recommended as the bearer. If less than 5000 circuits are to be carried, radio may be the more feasible medium. Other considerations are discussed in Chapter 9, Sections 7 and 8.

3 AN INTRODUCTION TO RADIO TRANSMISSION

Unlike wire and cable as a transmission medium, which are designed for that function, the radio medium is apt to be quite far from optimum, and its characteristics are not always fully understood. Thus much of radio system design is devoted to calculating the probable behavior of a given path and to finding modulation and signal-processing techniques that will overcome defects of the medium. Radio circuits may be characterized by their carrier frequency, which largely determines behavior of the path.

As we look at the electromagnetic spectrum progressing from the lowest to highest frequencies, certain generalities can be made regarding behavior and application. The *lowest frequencies,* below 300 kHz (VLF and LF) are useful for very long range communications but have very limited information bandwidths and require very high power. Propagation in this frequency range is worldwide; thus frequencies can essentially only be assigned once (i.e., they cannot be assigned again in another part of the world). The MF band (300 to 3000 kHz) is traditionally used for broadcast and marine, with fairly limited information capacity, requiring power in the order of kilowatts and with an effective daytime range in the hundreds of miles. Basic propagation is by ground wave (i.e., follows the curvature of the earth). The HF band (3 to 30 MHz) is the old, traditional long-haul point-to-point communication band. Propagation at long distance is by one or more reflections from ionospheric layers and is, therefore, variable as the ionosphere varies with sunspot conditions and time of day. Since the ionosphere is multilayered and irregular in motion, and since fairly large areas of these surfaces are involved in the propagation path, the received signal exhibits multipath effects and is subject to statistical fading. On HF communication paths, communication can be effective for circuits a few hundred miles long, and even worldwide coverage can be obtained. Nearly 90% path reliability on the longer term can be expected for well-designed HF circuits. To accommodate the very high demand for HF frequencies, modulation bandwidths are limited by law to 12 kHz, the equivalent of four 3-kHz telephone channels.

Above approximately 30 to 50 MHz, radio signals tend to pass through the ionosphere, rather than reflect or refract sufficiently for use far beyond the visible horizon. These higher frequencies are useful for line-of-sight communication, troposcatter, diffraction, or satellite relay. All three radio systems offer advantages and disadvantages for the design of telephone transmission links.

Line-of-sight microwaves (radiolinks) in the 150-MHz, 450-MHz, and 900-MHz bands provide multichannel transmission capability of 12 to 120 nominal 4-kHz voice channels in an FDM configuration over line-of-sight paths (FDM is discussed in Chapter 5, Section 4). Above 2 GHz, line-of-sight systems transmit up to 1800 and in some cases, up to 2700 FDM telephone channels per radio carrier frequency. Modulation is usually FM, although some form of digital modulation is being increasingly considered. This point is discussed further in this section and in Chapter 9.

Radio waves in line-of-sight systems travel in a straight line and are limited by the horizon due to the curvature of the earth. Such radio waves propagated line-of-sight (ca. $>$ 50 MHz) are usually bent or defracted beyond the optical horizon, the one that limits our vision beyond a certain point. The optical horizon may be determined by the following formula:

$$d = \sqrt{\frac{3h}{2}} \tag{7.1}$$

where d is the distance to the optical horizon in miles from antenna and h is the height in feet of antenna above the earth's surface.

The distance to the radio horizon varies with the index of refraction. Some designers generalize and say it is $\frac{4}{3}$ the distance to the optical horizon. However, such generalization may be overly optimistic to use in certain circumstances. The concept of optical and radio horizon is shown in Figure 7.1. Microwave radio paths over several miles long may suffer from fading. The longer is the path, the more prone it is to fading. Fading is the variation of a received radio signal level with time. In microwave line-of-sight systems, fading is usually caused by atmospheric changes and ground and water reflections in the propagation path. When using frequencies above 10 GHz, rainfall attenuation must also be taken into account. The most commonly used line-of-sight microwave frequency bands are 2 GHz, 4 GHz, 6 GHz, and 7 GHz. All bands generally will carry up to 1800 FDM

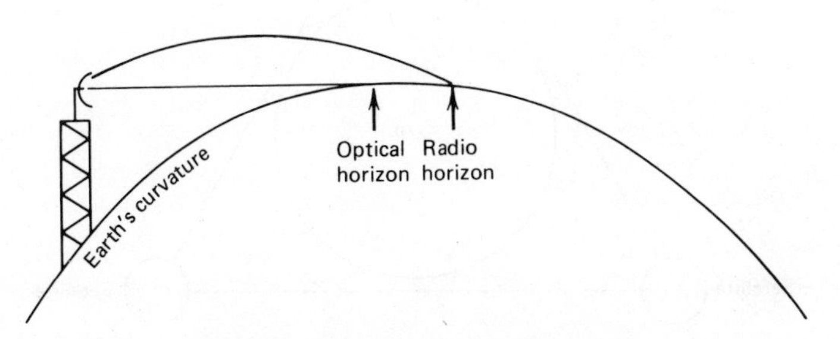

Figure 7.1 Radio and optical horizon.

voice channels by national regulation and CCIR recommendations with portions of the 6-GHz and 7-GHz band where 2700-channel operation is permitted per radio carrier frequency.

Geostationary satellites may be considered as radiofrequency (RF) repeaters for up to 960 or 1200 voice channels (per transponder) in an FDM configuration or, as now developing, in a time-division configuration for digital systems. The 6-GHz band transmits to the satellite that converts and amplifies the received signal to the 4-GHz bands. Both bands are shared with terrestrial line-of-sight microwave services. Due to crowded conditions in the 4 and 6 GHz bands, more and more services are turning to the 11-GHz and 14-GHz bands, especially for domestic and intracontinental communications. Satellite systems are used for very long links and up to about 1972 were almost exclusively used for intercontinental communications. Today there is a veritable rush to what might be familiarly termed "domestic systems." As the technology advances, the cost per voice channel has markedly reduced and can compete or be more economic than line-of-sight terrestrial systems for links over 500 miles (800 km) length or to provide comparatively thin line service to rural or backward areas. Figure 7.2 illustrates a simplified application of satellite communication for intercontinental service where three properly placed geostationary satellites can provide 100% earth coverage. By definition, geostationary satellites are 22,300 statute miles above the earth's surface with a concurrent propagation delay of about 0.25s earth–satellite–earth and a round-trip delay of about 0.5s. This comparatively long propagation time must be taken into account particularly in the design of speech telephone and signaling and certain data circuits.

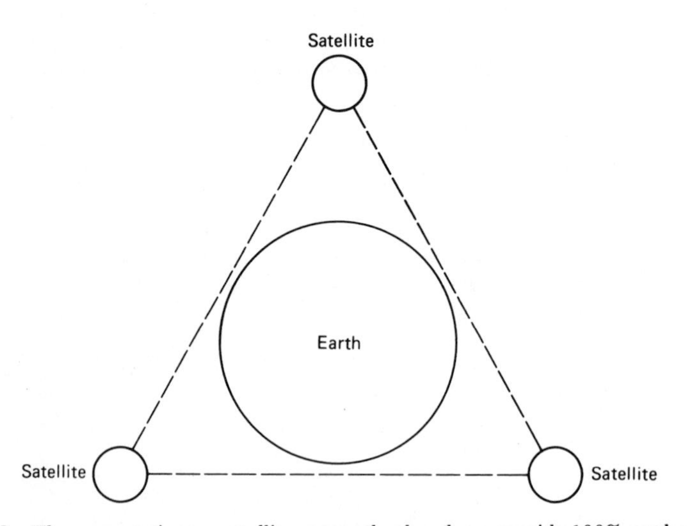

Figure 7.2 Three geostationary satellites properly placed can provide 100% earth coverage.

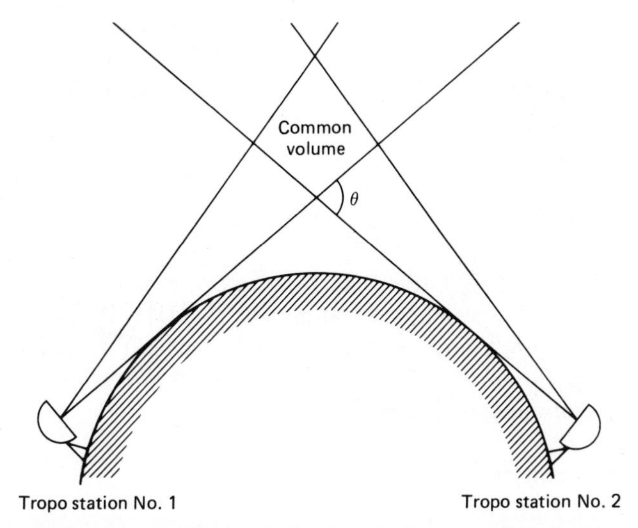

Figure 7.3 Tropospheric scatter model (Θ = scatter angle).

Tropospheric (tropo) scatter is an over-the-horizon microwave communication technique, operating in the 400-MHz and 900-MHz bands and in the 2-GHz and 4-GHz bands, handling 12 to 240 FDM telephone channels. Tropospheric scatter takes advantage of the refraction and reflection phenomena in a section of the earth's atmosphere called the *troposphere*. This is shown functionally in Figure 7.3. With such systems UHF radio signals can be transmitted beyond line-of-sight on single hops up to 640 km (400 miles). Tropo systems are expensive. Transmitters emit 1 kW or 10 kW; antennas use parabolic reflectors 5 m, 10 m, or 20 m in diameter, and receiving systems are quadruple diversity with low noise receivers. Tropo is applied for comparatively thin line links to reach over difficult terrain or over water and into backward regions.

4 DESIGN ESSENTIALS FOR RADIO LINKS

4.1 General

For this discussion, radio link and microwave line-of-sight are considered synonomous. The greater portion of long-distance communication links or "connections" use line-of-sight microwave. A radio link is made up of terminal radios and often one or more repeaters separated approximately 20 to 50 mi (35 to 85 km). Of course, some "hops" may be shorter than 20 mi or longer than 50 mi. Only frequency-modulated systems are considered for this discussion. Digital radio-transmission techniques are discussed in Chapter 9, Section 8.

Frequency modulation is so widely used because of noise-improvement factors, namely, the trade-off of bandwidth for improved signal-to-noise ratio above a certain "noise" threshold. This threshold is called the *FM improvement threshold.* The improvement over amplitude modulation systems is in the order of 20 dB. The bandwidth (Bw) of an FM radio system is defined by the following formula:

$$\text{Bw} = 2(F_{\text{dev}} + BB) \qquad \text{(Carson's rule)}$$

If Bw is in MHz, then F_{dev}, the peak frequency deviation in MHz and BB, maximum baseband frequency, must also be in MHz. Design factors for a radiolink hop include the determination of tower heights by path profiling and path calculations. From the latter, the system designer derives the necessary equipment parameters.

4.2 Site Placement and Tower Heights

As mentioned previously, radio links have terminal sites and repeater sites. At terminal sites all RF carriers are demodulated to baseband; the resulting baseband signal is demultiplexed* to individual voice-frequency channels. Figure 7.4A illustrates the concept pictorially, and Figure 7.4B is a simplified functional block diagram taken from Figure 7.4A. Radiolink repeaters are discussed at length further on. However, it can be seen here that baseband repeating is carried on at the simple repeater site. At exchange *B* there will be straight-through circuit groups from exchanges *A* to *C*. This would most likely be done by using through-group or through-supergroup techniques. This means that not all channels are demultiplexed to voice channels. The "through" channels are bunched in groups of 12 or supergroups of 60. Through-grouping has two advantages: (1) less multiplex equipment is required; and (2) there is less noise accumulation on through-circuits. Figures 7.4A and 7.4B show terminal and relay sites colocated with an exchange. In many cases this may not be feasible. We would want as much concentration as possible of traffic entering (inserts) and leaving (drops) the radio relay system. There are several trade-offs to be considered:

1. Bringing traffic in by wire from several exchanges traded off by elimination of drop and insert at relay site (savings on multiplex equipment).
2. Siting due to propagation constraints (or advantages) versus colocation with exchange or distant from any exchange (savings in land and access problems by colocation).
3. Method of feeding (feeders):† by light-route radio, coaxial cable

*This, of course, refers to frequency division multiplex. See Chapter 5, Section 4.
†These are feeders to the main-line radio system.

(multiplex at exchange), or wire pairs, multiplex on wire (aerial or buried cable).

In gross system design, exchange location, particularly tandem-toll exchanges, must be considered in light of probable radio and cable routes. Another consideration is RF interference or electromagnetic compatibility in general. Midcity relay or terminal sites have the following advantages:

- Colocation with toll exchange.
- Use of tall buildings as natural towers.

And they have the following disadvantages:

- Wave reflections (multipath) off buildings.
- Electromagnetic compatibility (EMC) problems, particularly from other nearby emitters and industrial emission.
- Low-grade labor market.

Sitings in the country have less EMC problems and a usually better labor force available (for operators, other operational personnel, and technicians), and right-of-way for cable is easier.

We are now led to propagation constraints. Terminal sites will be in or near heavily populated areas and preferably colocated with a toll exchange. The tops of modern large office buildings, if properly selected, are natural towers. Relay sites are heavily influenced by intermediate terrain. Accessible hill tops or mountain tops are good prospective locations. Draw a line along the path of the desired route. Sites would zig-zag along the line with optical or "radio" separation distances. If tower costs were $200 per foot ($600 per meter), 300-ft or 100-m towers might be the height limit for economic reasons. If hill-top or mountain-top sites are well selected, towers that high may never have to be considered. High towers are the rule over flat country. The higher the tower, the longer the line-of-sight distance. Thus, on a given link, less repeaters would be required if towers could be higher. Hence there is a trade-off between tower height and number of repeaters.

4.3 Calculation of Tower Heights

Assume now that sites along a microwave radio relay route have been carefully selected. The next step in engineering is the determination of tower heights. The objective is to keep the tower height as low as possible and still maintain effective communication. The towers must just be high enough for the radio beam to surmount obstacles in the path. As the discussion proceeds, the term "high *enough*" is carefully defined. What obstacles might there be in the path? To name some, there are terrain such

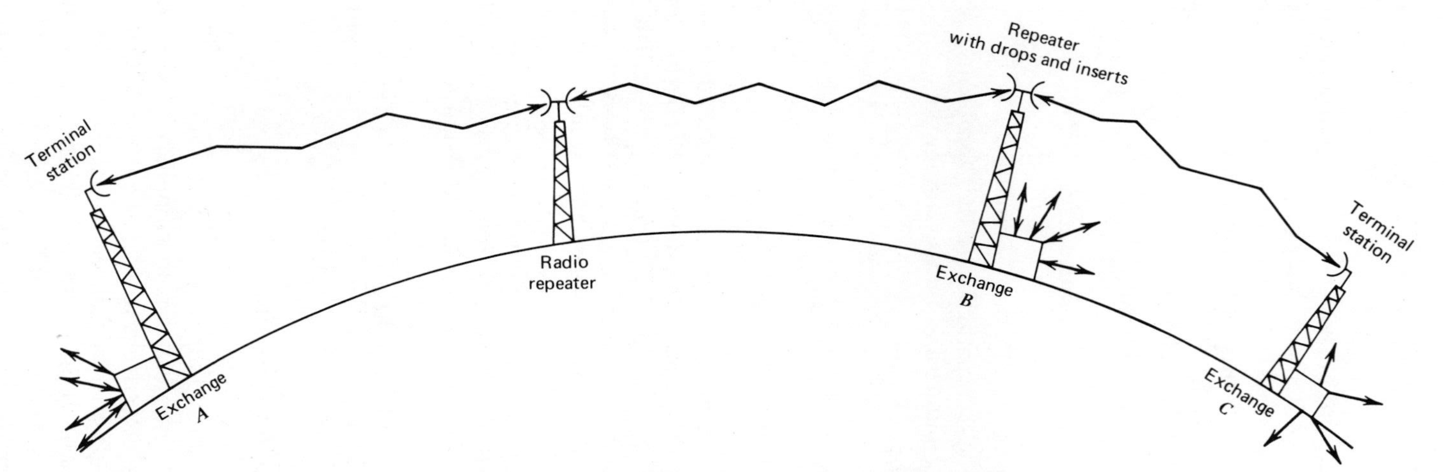

Figure 7.4A Sketch of a microwave (LOS) radio relay system.

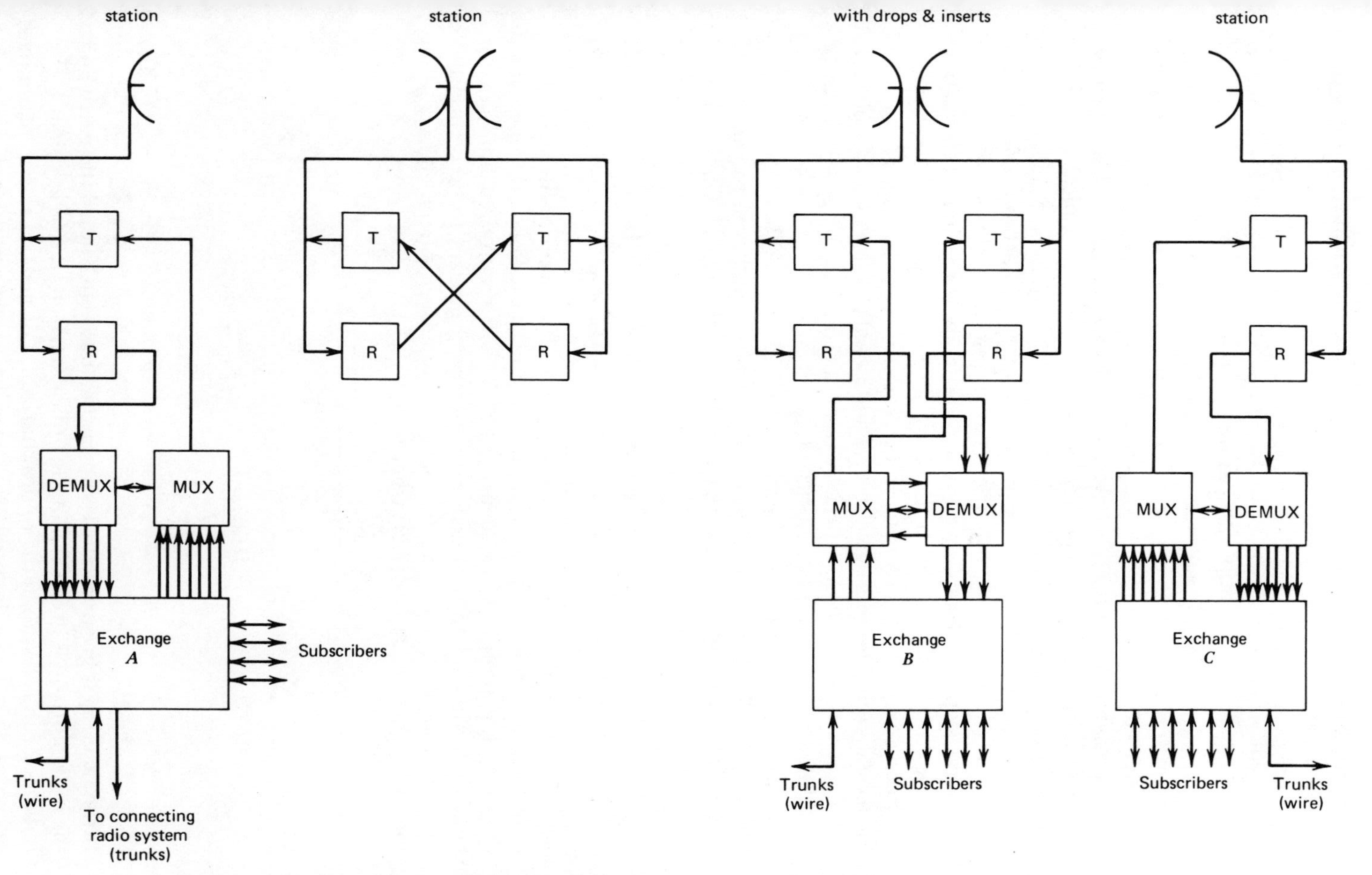

Figure 7.4B Simplified functional block diagram of a microwave (LOS) radio relay system from Figure 7.4A.

as mountains, ridges, hills, and earth curvature—which is highest at midpath—and buildings, towers, grain elevators, and so on.

All obstacles along the path must be scaled on graph paper in an exercise called *path profiling*. Good topographical maps are required of the region. Ideally, such maps should be 1:24,000, although 1:62,500 maps are acceptable. A straight line is drawn between the sites in question and then on linear graph paper scaled to 1 in. for 2 mi on the horizontal or 1 cm, equivalent to 1 or 2 km. Vertical scales depend on the rate of change of elevation along the path. An ideal scaling is 100 ft per inch or 1 cm for 10 m and over hilly country, 1 in. equivalent to 200 ft or 1 cm equivalent to 20 m. In mountainous country the vertical scale may have to be as much as 1 in. equivalent to 1000 ft or 1 cm equivalent to 100 m. Each obstacle encountered must be identified with a letter or number on the horizontal scale. The next step is to establish a point directly above, giving altitude above mean sea level. The bottom of the chart need not be mean sea level; it may be mean sea level plus so many meters. Once the reference altitude has been established, we must give several additional clearances. If the obstacle is terrain with vegetation, especially trees, a clearance for growth must be established; 10 ft is usually sufficient.

To the altitude or height of each obstacle must be added "earth bulge," the number of feet or meters an obstacle is raised higher in elevation (into the path) as a result of earth curvature or "earth bulge." The amount of earth bulge at any point in the path may be calculated by the formula(s):

$$h = 0.677 d_1 d_2 \qquad (h \text{ in feet; } d \text{ in miles}) \tag{7.2A}$$

$$h = 0.078 d_1 d_2 \qquad (h \text{ in meters; } d \text{ in km}) \tag{7.2B}$$

where d_1 is the distance from the near end of the hop to the obstacle in question and d_2 is the distance from the far end of the hop to the obstacle in question. Equation (7.2) is for a ray beam that is a straight line (i.e., no bending). Atmospheric refraction may cause the beam to be bent either toward or away from the earth. This bending effect is handled by adding the factor K to equation (7.2), where

$$K = \frac{\text{Effective earth radius}}{\text{True earth radius}}$$

such that

$$h_{\text{ft}} = \frac{0.667 d_1 d_2}{K} \qquad (d \text{ in miles}) \tag{7.3A}$$

$$h_{\text{m}} = \frac{0.078 d_1 d_2}{K} \qquad (d \text{ in km}) \tag{7.3B}$$

If the factor K is greater than 1, the ray beam is bent toward the earth and the radio horizon is greater than the optical horizon. If K is less than 1, the radio horizon is less than the optical horizon. For general system planning purposes, $K = \frac{4}{3}$ may be used. However, for specific path engineering, K

must be selected with care. The value of h or earth curvature corrected for K from equation (7.3) must be added to obstacle height in the path-profile exercise for each obstacle.

Still another factor must be added to obstacle height, namely, Fresnel zone clearance. This factor derives from the electromagnetic wave theory that a wave front, which our ray beam is, has expanding properties as it travels through space. These expanding properties result in reflections and phase transitions as the wave passes over an obstacle. The outcome is an increase or a decrease in received signal level. The amount of additional clearance over obstacles that must be allowed to avoid problems of Fresnel phenomenon (diffraction) is expressed in Fresnel zones. The first Fresnel zone radius may be calculated from the following formula:

$$R_{\text{ft}} = 72.1\sqrt{\frac{d_1 d_2}{FD}} \tag{7.4A}$$

where F = frequency in GHz
d_1 = distance from transmit antenna to obstacle (statute miles)
d_2 = distance from path obstacle to receive antenna (statute miles)
$D = d_1 + d_2$

For metric units:

$$R_{\text{m}} = 17.3\sqrt{\frac{d_1 d_2}{FD}} \tag{7.4B}$$

where F is the frequency in GHz, and d_1, d_2, and D are the same as in equation (7.4A), but d and D are in kilometers and R in meters.

Previously, a clearance of 0.6 Fresnel zone [0.6 the value of R in equation (7.4)] was considered sufficient. A new rule of thumb is evolving, namely, when $K = \frac{2}{3}$, at least 0.3 Fresnel zone clearance and 1.0 Fresnel zone clearance be allowed when $K = \frac{4}{3}$, whichever is greater. At points near the ends of a path, Fresnel zone clearances should be at least 6 m or 20 ft.

The three basic increment factors that must be added to obstacle heights are now available: vegetation growth, earth bulge corrected for K factor, and Fresnel zone clearance. These are marked as indicated previously on our path-profile chart. A straight line is drawn from right to left, just clearing the obstacle points as corrected for the three factors. Another line is then drawn from left to right. A sample profile is shown in Figure 7.5. Some balance is desirable so at one extreme we have a very tall tower and at the other, a little stubby tower. This is true but for one exception. An exception may occur when a reflection point exists at an inconvenient spot along the path.

4.3.1 Reflection Point

Possible reflection points may be obtained from the profile. The objective is to adjust tower heights such that the reflection point is adjusted to fall on

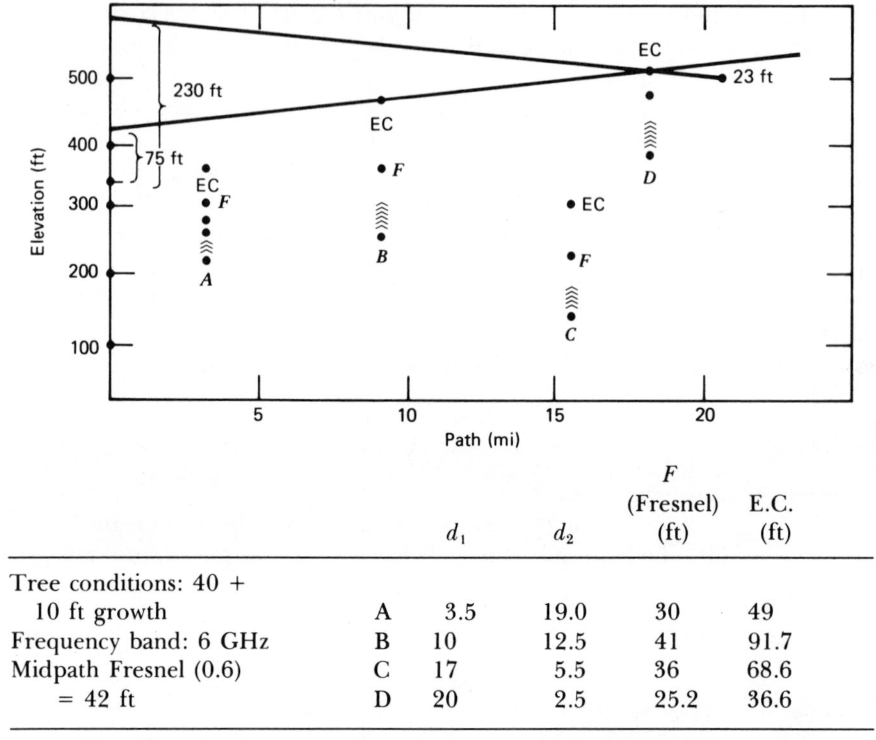

		d_1	d_2	F (Fresnel) (ft)	E.C. (ft)
Tree conditions: 40 + 10 ft growth	A	3.5	19.0	30	49
Frequency band: 6 GHz	B	10	12.5	41	91.7
Midpath Fresnel (0.6)	C	17	5.5	36	68.6
= 42 ft	D	20	2.5	25.2	36.6

Figure 7.5 Practice path profile (x in miles, y in feet; assume that $K = 0.9$).

land area where the reflected energy will be broken up and scattered. Bodies of water and other smooth surfaces cause reflections that are undesirable. Figure 7.6 can facilitate calculations for the adjustment of the reflection point. It uses a ratio of tower heights, h_1/h_2, and the shorter tower height is always h_1. The reflection area lies between a K factor of grazing ($K = 1$) and a K factor of infinity. The distance expressed is always from h_1, the shorter tower. The reflection point can be moved by adjusting the ratio h_1/h_2. The objective is to ensure that the reflection point does not fall on an area of smooth terrain or on water, but rather on land area where the reflected energy will be broken up or scattered (by wooded areas, etc.).

For a path that is highly reflective for much of its length, space-diversity operation may minimize the effects of multipath reception.

4.4 Path Calculations

4.4.1 Introduction

Once the path profile has been completed and checked by a path survey, the next exercise in radio-link design is path calculation. The profile gave us tower heights. Now we want to assign certain parameters to the radio

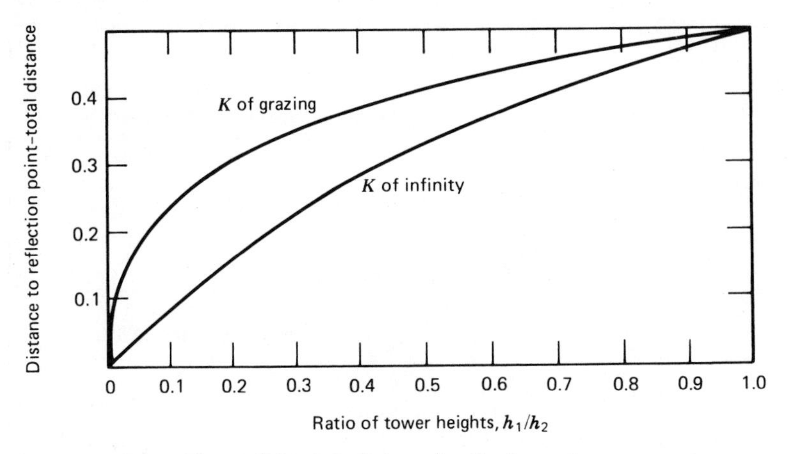

Figure 7.6 Calculation of reflection points.

equipment we wish to install. It was assumed at the outset that the band of frequencies has been selected. Among the other parameters we must work with are:

1. Path loss in decibels.
2. Operating bandwidth and peak deviation (we assume frequency modulation).
3. Receiver noise and FM improvement thresholds
4. A desired, unfaded signal-to-noise ratio in the telephone channel.
5. A fade margin assuring a certain noise specification for 99%, 99.9%, or 99.99% of the time for the worst month or for the year.

From these we will determine parabolic antenna diameters. If we cannot reach the noise objectives in items 4 and 5, the designer may have to resort to:

1. More transmitter power.
2. More sensitive receiver front ends.
3. Use of diversity reception.
4. Reduction of hop length (as a last resort).

As an introduction to the overall problem, Figure 7.7 graphically gives an idea of the gains and losses for a single hop in a radio link system. The gains and losses as shown in Figure 7.7 are summed algebraically. Thence antenna sizes are adjusted to meet a noise objective.

4.4.2 Path Loss

For all intents and purposes, path loss up to about 10 GHz can be considered as "free-space loss." To introduce the reader to the problem, consider

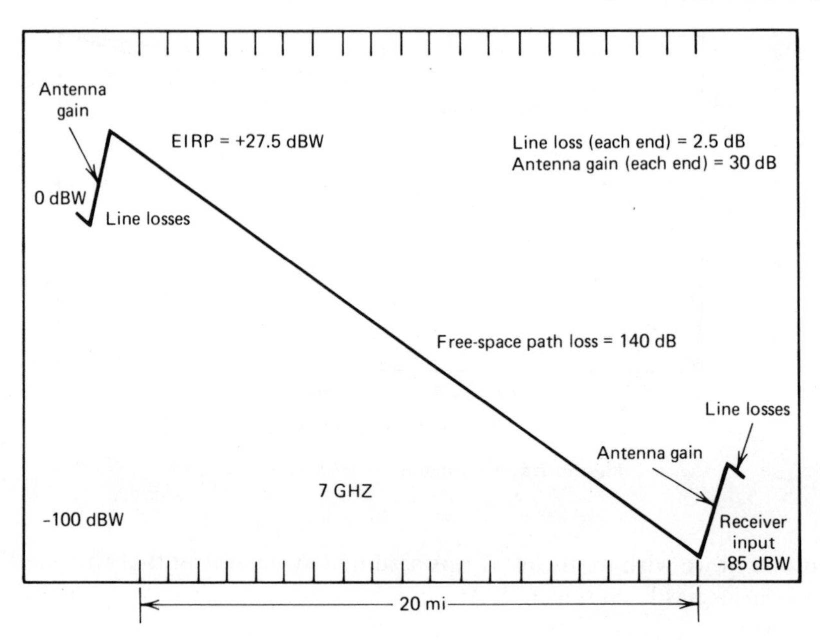

Figure 7.7 Radiolink gains and losses (simplified) (transmitter output = 0 dBW).

an isotropic antenna, that is, an antenna that radiates uniformly in all directions. If the isotropic radiator is fed by a transmitted power P_t, it radiates $P_t/4\pi d^2$ (W/m²) at a distance d, and if a radiator has a gain G_t, the power flow is enhanced by the factor G_t. Finally, the power intercepted by an antenna of effective cross section A (related to the gain by $G_r = 4\pi A/\lambda^2$) is $P_t G_t G_r\ (\lambda/4\pi d)^2$. The term $(\lambda/4\pi d)^2$ is known as the free-space loss and represents the steady decrease of power flow (in W/m²) as the wave propagates. From this we can derive the more common formula for free-space path loss, which reduces to:

$$L = 96.6 + 20 \log_{10} F + 20 \log_{10} D \tag{7.5}$$

where L is the free-space attenuation between isotropic antennas in dB, F is the frequency in GHz, and D is the path distance in statute miles. In the metric system

$$L_{\mathrm{dB}} = 92.4 + 20 \log_{10} F_{\mathrm{GHz}} + 20 \log_{10} D_{\mathrm{km}}$$

Consider the problem from a different aspect. It requires 22 dB to launch a wave to just 1 wavelength (1λ) distant from an antenna. Thus for an antenna emitting +10 dBW, we could expect the signal one wavelength away to be 22 dB down, or −12 dBW. Whenever we double the distance we incur an additional 6 dB of loss. Hence at 2λ from the +10-dBW radiator, we would find −18 dBW; at 4λ, −24 dBW; 8λ, −30 dBW; and so on. Now suppose we have an emitter where $F = 1$ GHz; what is the path loss at one statute mile?

$$L = 96.6 + 20 \log_{10} 1 + 20 \log_{10} 1 = 96.6 \text{ dB}$$

For rough calculations, the 6-dB relationship is worthwhile and also gives insight in that if we have a 20-mi path and shorten or lengthen it by a mile, our signal level will be affected little.

4.4.3 Receiver Threshold—The Starting Point

In this step in path calculations the objective is to calculate the thermal noise level of the receiver to be used at the distant end of the path. If we are given a transmitter with a known output, an attenuation of the signal calculated in Section 4.4.2, and a receiver at the far end, the exercise is to find just the proper signal level input to the receiver to be equal to the thermal noise. The problem is shown diagrammatically as follows:

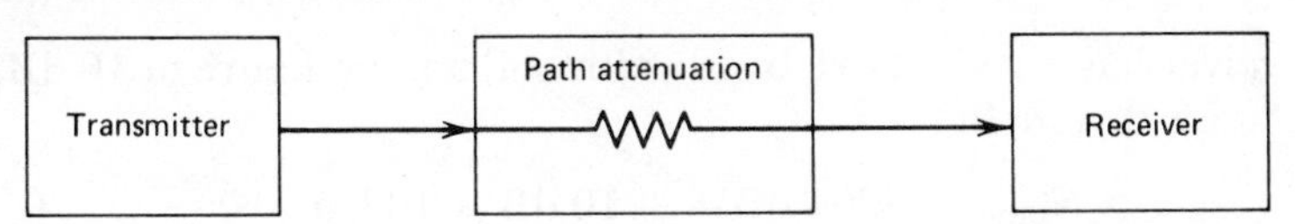

Receiver thermal noise level can be calculated from the following formula using Boltzmann's constant:

$$N_{\text{dBW}} = 10 \log_{10} kTB \tag{7.6}$$

where T is the receiver noise temperature, in degrees kelvin; B is the noise bandwidth, in hertz; and K is Boltzmann's constant (1.3803×10^{-23}J/°K). By converting Boltzmann's constant to dBW, we have

$$N_{\text{dBW}} = -228.6 \text{ dBW} + 10 \log T + 10 \log B_{\text{if}} \tag{7.7}$$

where B_{if} may be calculated according to Carson's rule as $B_{\text{if}} = 2$ (peak FM deviation) + 2(highest modulating frequency). Of course, if we were transmitting only supergroup 2 in an FDM multiplex configuration (Chapter 5), the highest modulating frequency would be 552 kHz.

In equation (7.7) the term −228.6 dBW represents the thermal noise of a receiver with a 1-Hz bandwidth, unadjusted for noise temperature. The term T, or 10 log T, adjusts the receiver to its real noise temperature, and B_{if} adjusts the calculation to its design intermediate-frequency (IF) bandwidth. For the more common radio-link application, noise temperature may be converted to the more familiar noise figure by

$$NF = 10 \log_{10} \left(1 + \frac{T_e}{290}\right) \tag{7.8}$$

where the effective noise temperature of the receiver, T_e, is compared to room temperature, 290°K.

The noise figure is a measure of the noise produced by a practical network compared to an ideal network (i.e., one that is noiseless). For a linear system, noise figure (NF) is expressed by

$$NF = \frac{S/N_{\text{in}}}{S/N_{\text{out}}} \tag{7.9}$$

where S/N is the signal-to-noise ratio.

For NF expressed in dB:

$$NF_{\text{dB}} = 10 \log_{10} NF \tag{7.10}$$

The receiver noise threshold formula may be simplified still further, assuming that the receiver operates at room temperature.

$$N_{\text{dBW}} = -228.6 \text{ dBW} + 10 \log 290°\text{K} + NF_{\text{dB}} + 10 \log B_{\text{if}}$$

or

$$N_{\text{dBW}} = -204 \text{ dBW} + NF_{\text{dB}} + 10 \log B_{\text{if}} \tag{7.11}$$

If a receiver has a 10-MHz IF bandwidth and a noise figure of 10 dB, what is the noise threshold?

$$N_{\text{dBW}} = -204 \text{ dBW} + 10 \text{ dB} + 10 \log_{10} 10^7$$

$$= -124 \text{ dBW}$$

Let us now examine such a receiver on the distant end of a 20-mi hop operating at 6 GHz. First, we would calculate the free-space attenuation of the hop, or

$$\text{Path loss} = 96.6 + 20 \log_{10} 6 + 20 \log_{10} 20$$

$$= 138.24 \text{ dB}$$

Look again at the simplified network diagram and calculate the transmitter output (assuming isotropic antennas and no other losses) so that the incoming signal at the receiver front end just equals the thermal noise level of the receiver.

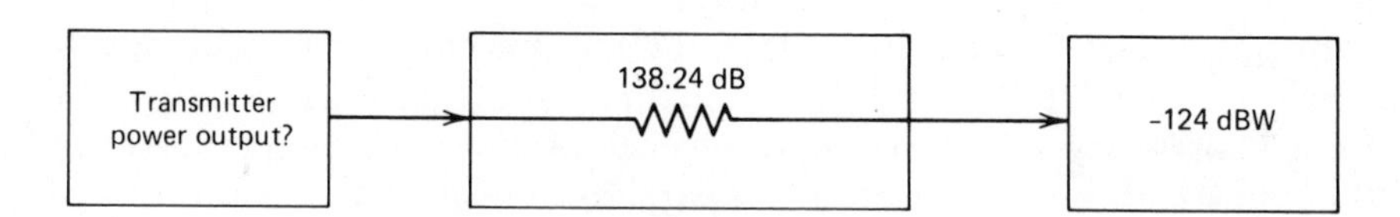

Thus

$$138.24 - 124 \text{ dBW} = +14.24 \text{ dBW}$$

Hence the transmitter power output would have to be at least 14.24 dBW, or 40 W.

To make the exercise more realistic, assume a transmitter with 0 dBW output, 2 dB of loss for transmission lines from the transmitter to its antenna, and 2 dB of loss from the receiver antenna to the receiver front end. What antenna gains will be required to reach a 10-dB carrier-to-noise ratio at the receiver front end?

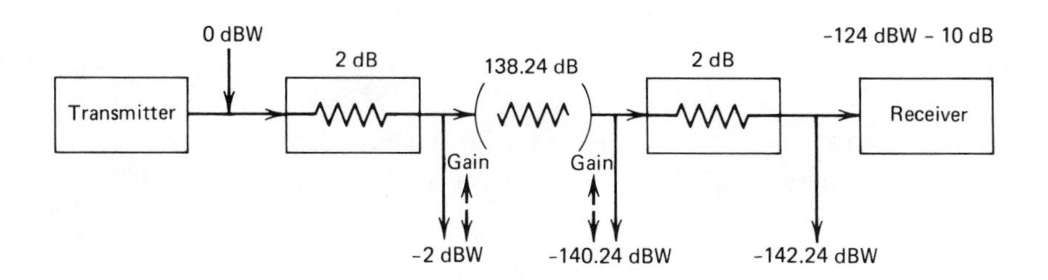

Without any antenna gain, we have a level of -142.24 dBW at the receiver front end, but as we require $(-124 + 10)$ dBW, an antenna gain of $-114 + 142.24$ dB or 28.24 dB is necessary. The antennas (there are two) should have at least 28.24 dB gain between them, or 14.12 dB gain each. We round off upward; thus 15 dB gain is required for each antenna.

4.4.4 Parabolic Antenna Gain

At a given frequency the gain of a parabolic antenna is a function of its effective area and may be expressed by the formula:

$$G = 10 \log_{10} (4\pi A \eta / \lambda^2) \tag{7.12}$$

where G = gain in dB relative to an isotropic antenna
A = area of antenna aperture
η = aperture efficiency
λ = wavelength at the operating frequency

Commercially available parabolic antennas with a conventional horn feed at their focus usually display a 55% efficiency or somewhat better. With such an efficiency, gain (G, in dB) is then

$$G = 20 \log_{10} D + 20 \log_{10} F + 7.5 \tag{7.13}$$

where F is the frequency in gigahertz and D is the parabolic diameter in feet.

In metric units, we have

$$G = 20 \log_{10} D + 20 \log_{10} F + 17.8 \tag{7.14}$$

where D is measured in meters and F in gigahertz.

What size antenna would be required in the preceding example? Let G = 15 dB and F = 6 GHz; then

$$15 = 20 \log_{10} D + 20 \log_{10} 6 + 7.5$$

$$20 \log D = +15 - 20 \times 0.7782 - 7.5 \log_{10} D$$

$$= \frac{8}{20} = 0.4$$

$$D = 2.2 \text{ ft}$$

Parabolic dish antennas, with waveguide (horn) feeds (see Figure 7.8), are probably the most economic antennas for radio links operating from 3 GHz upward. From 900 MHz to about 3 GHz, coaxial feeds are used. Coaxial cable transmission lines deliver the RF energy from/to transmitter/receiver to the antenna in this range. Above 3 GHz coaxial cable becomes too lossy and waveguide is more practical.

To every rule there are exceptions. One method of delivering the RF signal to/from the point of radiation is by the so-called periscopic method. In this case the antenna is mounted on the radio-equipment building or shelter, and a plane reflector is placed at the point of radiation on the tower. If the antenna is directly below the reflector, the reflector would be oriented at 45° to permit the radio beam to be emitted on a straight line parallel to level earth, as in Figure 7.9.

For path calculations, there may be a gain or loss of up to several decibels on each end using periscopic techniques. The gain or loss depends on

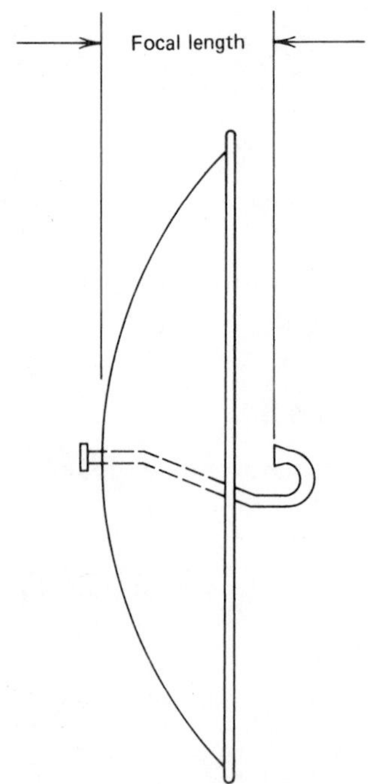

Figure 7.8 Typical parabolic antenna with front feed.

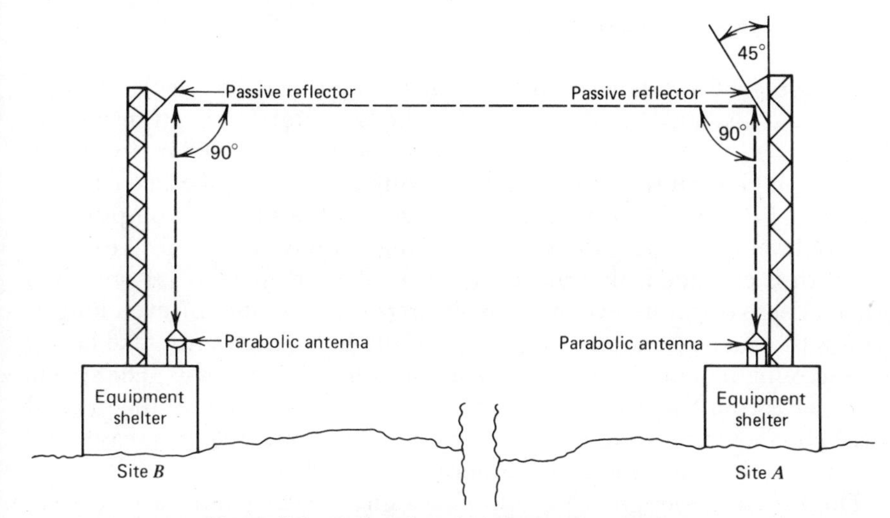

Figure 7.9 Periscopic technique (passive reflectors).

frequency, antenna size, reflector cross area, and distance from antenna to reflector [3, 5–7].

Other types of antennas may also be used such as the "cornucopia", horn, spiral and so forth. Besides cost and gain, other features are front-to-back ratio, side lobes and efficiency. For instance the "cornucopia", called such because it looks like "the horn of plenty", has efficiencies in excess of 60% and improved side-lobe discrimination but is more costly.

4.4.5 Effective Isotropically Radiated Power

Effective isotropically radiated power (EIRP) is a term used by the radio engineer to conveniently describe the power in the radio beam compared to the isotropic antenna. Remember that the isotropic radiator is "a hypothetical antenna radiating or receiving equally in all directions" (IEEE definition). The IEEE dictionary goes on to add that "an isotropically radiating antenna does not exist physically but represents a convenient reference for expressing the directive properties of actual antennas." It has a gain of 1, expressed as 0 dB.

The EIRP is the algebraic sum of the transmitter output (expressed in dBm or dBW) plus the gains and losses of the transmitting antenna system. The system includes all antenna and transmission-line elements from the transmitter output to the antenna feed.

If a transmitter has a +10 dBW output and there are 2 dB of transmission-line losses and the antenna has a 20 dB gain, the EIRP is

$$\begin{aligned} EIRP_{\text{dBW}} &= +10 \text{ dBW} - 2 \text{ dB} + 20 \text{ dB} \\ &= +28 \text{ dBW} \end{aligned}$$

4.4.6 Fades and Fade Margin

In Section 4.4.2 we showed how path attenuation may be calculated. This was a fixed loss and can be simulated in the laboratory with a transmission line attenuator. On short radiolink paths below about 10 GHz, the signal level impinging on the distant end receiving antenna can be calculated to less than 1 dB. If the transmitter continues to give the same output, that receive level will remain the same over long periods of time, for years. As the path is extended, the calculated level will tend to decrease once in a while. These drops in level may last for seconds, minutes, or even longer. This is the phenomenon of fading. The radio-link design must take fading into account, first on a system basis and then on a per-hop basis. The system designer is concerned with accumulated noise contribution in the derived voice channel. As the signal fades, the signal-to-noise ratio decreases and the system noise increases in the derived voice channel.

The system is designed for specified signal levels. Microwave receiver AGC (automatic gain control) and FDM level regulation maintain these levels regardless of the signal-to-noise ratio. Thus in the system design, noise level in the derived voice channel is specified not to exceed a certain value for a percentage of the time, such as 1%, 0.1%, or 0.01%. Or we can say that noise will be *less* than a certain value for 99%, 99.9%, or 99.99% of the time. That is, the system would not meet a predetermined noise criterion for 8.8 hr per year if we assigned a 99.9% path reliability criteria.

If a radio-link system has 10 hops and the system noise criterion is established for 99.9% of the time, the *worst* case per hop criteria must be 99.99%. Fades will not occur at the same time on each hop in the system, so the hop criterion can be relaxed considerably and still maintain the system noise time percentage. But for worst case the designer often assigns such a noise criterion per hop because noise power is the summation of the noise powers of individual hops.

The CCIR recommends 3-pWp noise accumulation per kilometer, adding a fixed noise power quantity of 200 pWp for systems from 50 to 840 km long, 400 pWp from 840 to 1670 km, and 600 pWp from 1670 to 2500 km. For instance, a system 2000 km long would be 600 pWp + 2000 × 3 pWp or 6600 pWp accumulated noise for the system. These, of course, are unfaded figures. The CCIR further states that the 1-min mean power of 47,500-pWp shall not be exceeded 0.1% of the time over a 2500-km reference circuit.

Two approaches may be taken to establish a fade margin. The first, developed by Bullington of Bell Telephone Laboratories [9], is shown in Figure 7.10.

Using Figure 7.10, a 6-GHz path fade margin may be derived from path length for 99%, 99.9%, and 99.99% of the time. For instance, a path 25 mi long with a 99.9% path reliability must have a 30-dB fade margin. The

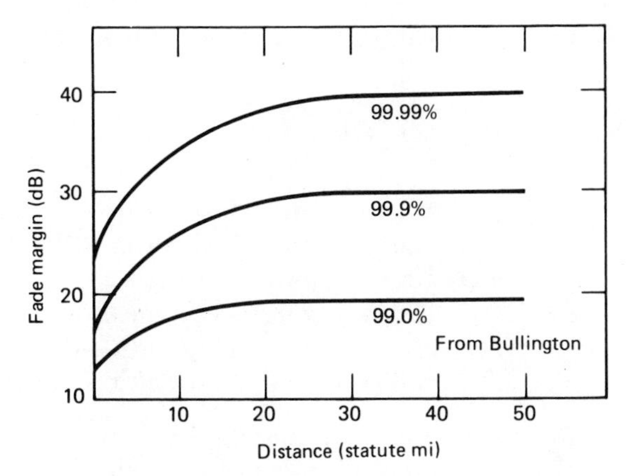

Figure 7.10 Nomogram for calculation of fade margins as a function of path length for the 6-GHz band [3, 6].

other approach is to assume that the fading follows a Rayleigh distribution. Here a 8-dB fade margin would be required for a 90% path reliability.* A Rayleigh distribution has a 10-dB slope for each full order of improvement, or:

Propagation or Path Reliability	Required Fade Margin
90%	8 dB
99%	18 dB
99.9%	28 dB
99.99%	38 dB
99.999%	48 dB

A major cause of fading derives from reflections from a stratified atmosphere or from surface or land conditions along the path. This type of fading is called *multipath* and causes destructive and constructive interference in the level of the incoming signal. Of course, the adjustment of the reflection point as discussed in Section 4.3.1 can reduce or eliminate ground reflections in many cases. Likewise, space diversity can reduce the effects of multipath fading (see Section 4.4.7).

Noise accumulation following North American (ATT) practice states that

*This is often called *time availability*.

intertoll trunks should display no more than for the trunk lengths indicated:

0–50 mi (long) 30 dBrnC0
51–100 mi (long) 31 dBrnC0
101–200 mi (long) 33 dBrnC0
201–400 mi (long) 35 dBrnC0
401–1000 mi (long) 37 dBrnC0
1001–1500 mi (long) 38 dBrnC0
1501–2500 mi (long) 41 dBrnC0
2501–4000 mi (long) 43 dBrnC0

When referring to trunks, in this case total noise accumulation is implied from all sources. It is the sum of radio noise (or cable noise), switch noise, and multiplex noise, among others. The figures are for BH measurement, unfaded. For tandem or toll connecting trunks, the figures may be deteriorated by 2 dB. For instance, on a toll-connecting trunk 300 miles long, 35 + 2 dBrnC0 or 37 dBrnC0 is permitted. On direct trunks about 5 dB of additional noise is permitted, or a 2000-mi direct trunk may have 46 dBrnC0. To relate dBrnC0 to pWp:

$$\text{dBrnC0} = 10 \log \text{pWp} + 0.8 \text{ dB} \tag{7.15}$$

To meet system noise criterion, a per hop fade margin must be established. This means the hop is overbuilt; or, in other words, more gain is engineered in the hop than would be necessary under no-fade conditions. Overbuilding costs money and must be limited to just the proper amount necessary to meet system specifications.

Fading varies with path length and frequency. This is a simplification of a very complex phenomenon. Let us make the assumption first and then discuss the factors that will affect the design. Multipath fading varies not only with path length and frequency as in our simplification, but is also a function of climate and terrain. For example, in dry, windy, mountainous areas the multipath phenomenon may be nonexistent. Flat terrain along a path tends to increase the incidence of fading. In hot, humid coastal regions a very high incidence of fading would be expected.

4.4.7 Diversity

"Diversity" here refers to simultaneous reception of a radio signal over several "paths." The signal "paths" are combined in one way or another in the radio equipment so that the composite signal is less affected by fades. In fact, on well-designed radio diversity systems the frequency of fades (i.e., the number fades per time interval) and the depth of fades (in dB) are notably less. The simplest form of diversity is space diversity. Such a configuration is shown in Figure 7.11.

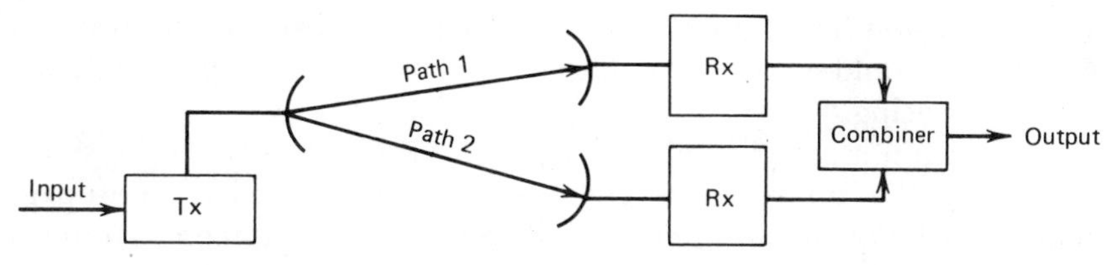

Figure 7.11 Space diversity.

The two diversity paths in space diversity are derived at the receiver end from two separate receivers with a combined output. Each receiver is connected to its own antenna, separated vertically on the same tower. The separation distance should be at least 70 wavelengths and preferably 100 wavelengths. In theory, fading will not occur on both paths simultaneously.

Frequency diversity is more complex and more costly than space diversity. It has advantages as well as disadvantages. Frequency diversity requires two transmitters at the near end of the link. The transmitters are modulated simultaneously by the same signal but transmit on different frequencies. Frequency separation must be at least 2%, but 5% is preferable. Figure 7.12 is an example of a frequency-diversity configuration. The two diversity paths are derived in the frequency domain. When a fade occurs on one frequency, it will probably not occur on the other frequency. The more one frequency is separated from the other, the less is the chance that fades will occur simultaneously on each path.

Frequency diversity is more expensive, but there is greater assurance of path reliability. It provides full and simple equipment redundancy and has the great operational advantage of two complete end-to-end electrical paths. In this case failure of one transmitter or one receiver will not interrupt service and a transmitter, and (or) a receiver can be taken out of service for maintenance. The primary disadvantage of frequency diversity is that it doubles the amount of frequency spectrum required in this day and age when spectrum is at a premium. In many cases it is prohibited by national licensing authorities. For example, the U.S. Federal Communica-

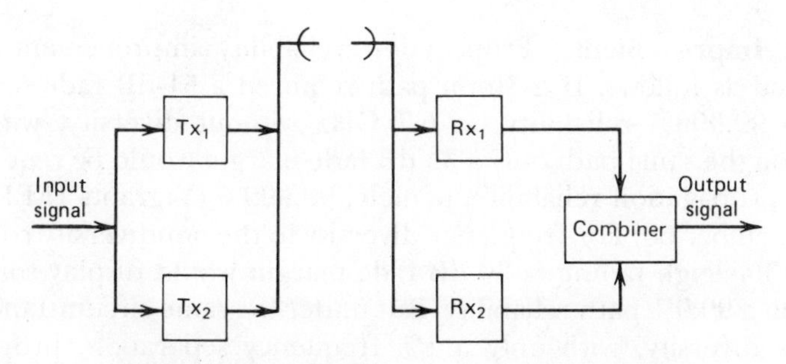

Figure 7.12 Frequency-diversity configuration.

tions Commission (FCC) does not permit frequency diversity for industrial users. It also should be appreciated that it will be difficult to get the desired frequency spacing.

The full equipment redundancy aspect is very attractive to the system designer. Another approach to achieve diversity improvement in propagation plus reliability improvement by fully redundant equipment is to resort to the "hot-standby" technique. On the receive end of the path, a space-diversity configuration is used. On the transmit end a second transmitter is installed as in Figure 7.12, but the second transmitter is on "hot-standby." This means that the second transmitter is on but not radiated by the antenna. On a one-for-one basis the second transmitter is on the same frequency as the first transmitter. On failure of transmitter No. 1, transmitter No. 2 is switched on automatically.

One-for-*N* hot-standby is utilized on large radio-link systems employing several radio carriers, where the cost for duplicate equipment for each channel may be prohibitive. In this case one full set of spare equipment in the "on" condition serves to replace one of several operational channels, and the spare equipment is assigned its own frequency. On the receive side there is just one extra receiver. Relay cutover must be provided, and no space diversity improvement is afforded. Likewise, there is no paralleling of inputs on the transmit side; thus the switching from the operational pair to the standby pair is much more complex. On such multi-RF channel arrangements it is customary to assign some sort of priority arrangement. Often the priority channel enjoys the advantage of frequency diversity, whereas the other RF channels do not. On failure of one of the other channels, the diversity improvement is lost on the priority channel, with the diversity pair switched to carry the traffic on the failed pair. In another arrangement the standby channel carries low-priority traffic and does not operate in a frequency-diversity arrangement, while providing protection for perhaps two or three other channels carrying the higher-priority traffic. On failure of a high-priority channel, the lower-priority channel drops its traffic, replacing one of the RF channels carrying the more important traffic flow. Once this occurs, the remaining channels operate without standby equipment protection.

Diversity Improvement. Propagation reliability improvement can be exemplified as follows. If a 30-mi path required a 51-dB fade margin to achieve a 99.999% reliability on 6.7 GHz without diversity, with space diversity on the same path only a 33 dB fade margin would be required for the same propagation reliability, namely, 99.999% (Vigrants IEEE Trans. Com., December 68) For frequency diversity in the nondiversity condition, assuming Rayleigh fading, a 30-dB fade margin would display something better than a 99.9% path reliability. But under the same circumstances with frequency diversity, with only a 1% frequency separation, propagation reliability on the same path would be improved to 99.995%.

4.4.8 Path Calculations—Conclusions

The exercise of carrying out path calculations basically involves algebraically summing the gains and losses in the system to reach a specified noise criterion that is valid for a time percentage. The basic contributors to loss have been discussed, such as path loss and transmission-line loss. Researching the literature still further will uncover yet other losses that must be considered in any well-engineered system. Gain is achieved by selecting the proper antennas and applying diversity when needed. There are other gains as well such as preemphasis gain, and, when required, the use of low-noise front ends and improved antenna feeds among other tools available to the design engineer to optimize radio-link design.

4.5 Noise Considerations—System Loading

For this discussion, there are two types of noise to be considered in radiolink systems, thermal noise and intermodulation (IM) noise. The calculation of thermal noise was discussed in Section 4.4.3. Once the receiver noise figure is given, thermal noise can be calculated. The receiver noise figure is given by the equipment manufacturer.

Intermodulation noise is also characteristic of the equipment and can be determined either from manufacturer specifications or from actual measurements on the system itself. Under multichannel loading, with more than approximately 60 voice channels, IM noise resembles thermal noise, and both can be considered "white noise." The IEEE dictionary defines white noise as "noise, of either random or impulsive type, that has a flat frequency spectrum at the frequency range of interest." This means that the noise-power distribution is equal throughout the frequency spectrum of the demodulated signal.

The important fact is that the IM noise increases with load (i.e., increased traffic and/or signal level), and when a certain "breakpoint" of load-handling capability is exceeded, IM noise becomes excessively high. On the other hand, residual noise is not affected by the amount of traffic or traffic load of the system. Frequency-modulation radio overcomes much of the residual noise that appears in a radio system by distributing it over a wide radio bandwidth. Such an exchange of bandwidth for lower noise is also proportional to the signal level appearing at the system receiver. At periods of low signal (i.e., during a radio fade) level, thermal or residual noise becomes dominant.

4.5.1 Noise-Power Ratio

Noise-power ratio (NPR) has come into wide use in radio-link systems to describe IM noise performance. This ratio gives an excellent indication of performance of IM noise when measured under standard fixed conditions.

It is measured in decibels, and we can say that a radio-link transceiver combination has a specified NPR under certain load conditions. Noise-power ratio, which is a meaningful measurement for radio-link systems that carry multichannel FDM telephone baseband information, can be defined as the ratio, expressed in decibels, of (1) the noise in a test channel, with all channels loaded with white noise, to (2) noise in the test channel, with all channels except the test channel fully noise loaded.

When NPR is measured on a baseband–baseband basis, a radio-link transmitter is connected back-to-back with a receiver using proper waveguide attenuators to simulate real-path conditions. A white-noise generator is connected to the transmitter baseband input. This generator produces a noise spectrum that approximates a spectrum produced by a multichannel (FDM) multiplex system. The output noise level from the generator is adjusted to a desired *composite noise baseband power.* A notched filter is then switched in to clear a narrow slot in the spectrum of the noise signal, and a noise analyzer is connected at the output of the system. The analyzer is used to measure the ratio of the noise in the illuminated (noise loaded) section of the baseband to the noise power in the cleared slot. The slot noise level is equivalent to the total noise (residual plus intermodulation) that is present in the slot bandwidth. Slot bandwidths are the width of a standard voice channel and are taken at the upper, middle, and lower portions of the baseband.

The composite noise power is taken from one of the following formulas for N telephone channels in an FDM (SSBSC) configuration (see Chapter 5):

$$P(\text{dBm0}) = -1 + 4 \log_{10} N \qquad \text{(CCIR)} \quad (7.16\text{A})$$

when $N < 240$ channels

$$P(\text{dBm0}) = -15 + 10 \log_{10} N \qquad \text{(CCIR)} \quad (7.16\text{B})$$

when $N > 240$ channels

$$P(\text{dBm0}) = -16 + 10 \log_{10} N \qquad \text{(ATT)} \quad (7.16\text{C})$$

$$P(\text{dBm0}) = -10 + 10 \log_{10} N \qquad (7.16\text{D})$$

for certain US military systems with heavy data usage, and

$$NPR = \text{Composite power (dB)} - \text{noise power in slot (dB)} \quad (7.17)$$

A good guide for NPR for high-capacity radio-link systems with a 300-channel capacity (1 hop) should be 55 dB; for 1200 voice channels, it should be in excess of 50 dB.

4.5.2 Derived Signal-to-Noise Ratio

Given the NPR of a system, we can then compute test tone to noise ratio for each voice channel.

$$\frac{S}{N} = NPR + BWR - NLR \tag{7.18}$$

$$BWR \text{ (bandwidth ratio)} = 10 \log \frac{\text{(occupied-baseband bandwidth)}}{\text{(voice-channel bandwidth)}} \tag{7.19}$$

NLR (noise load ratio) = *P* which is taken from the load equation, 7.16. The signal-to-noise ratio as given is unweighted. For F1A or psophometric weighting, add 3 dB (when reference is 1000 Hz).* For an 800-Hz reference,* add 2.5 dB.

4.5.3 Conversion of Signal-to-Noise Ratio to Channel Noise

Using the signal-noise-ratio calculated in equations (7.18) and (7.19),

$$\text{Noise power in dBa0} = 82 - \frac{S}{N} \tag{7.20A}$$

$$\text{Noise-power ratio in dBrnC} = 88.5 - \frac{S}{N} \tag{7.20B}$$

$$\text{Noise-power ratio in pW} = \frac{10^9}{\text{antilog } S/N} \tag{7.20C}$$

$$\text{Noise-power ratio in pWp} = \frac{10^9 \times 0.56}{\text{antilog } S/N} \tag{7.20D}$$

4.6 Radio-link Repeaters

Up to this point the only radio-link repeaters discussed have been baseband repeaters. Such a repeater fully demodulates the incoming RF signal to baseband. In the most simple configuration this demodulated baseband is used to modulate the transmitter used in the next link section. This type of repeater also lends itself to dropping and inserting voice channels, groups, and supergroups. It may also be desirable to demultiplex the entire baseband down to voice channel for switching and insert and drop a new arrangement of voice channels for multiplexing. The new baseband would then modulate the transmitter of the next link section. A simplified block diagram of a baseband repeater is shown in Figure 7.13A. Two other types

*These are voice-channel test-tone reference signals (tones); 1000 Hz is used in North America, whereas 800 Hz is used in Europe and is recommended by the CCITT (see Chapter 5, Section 5).

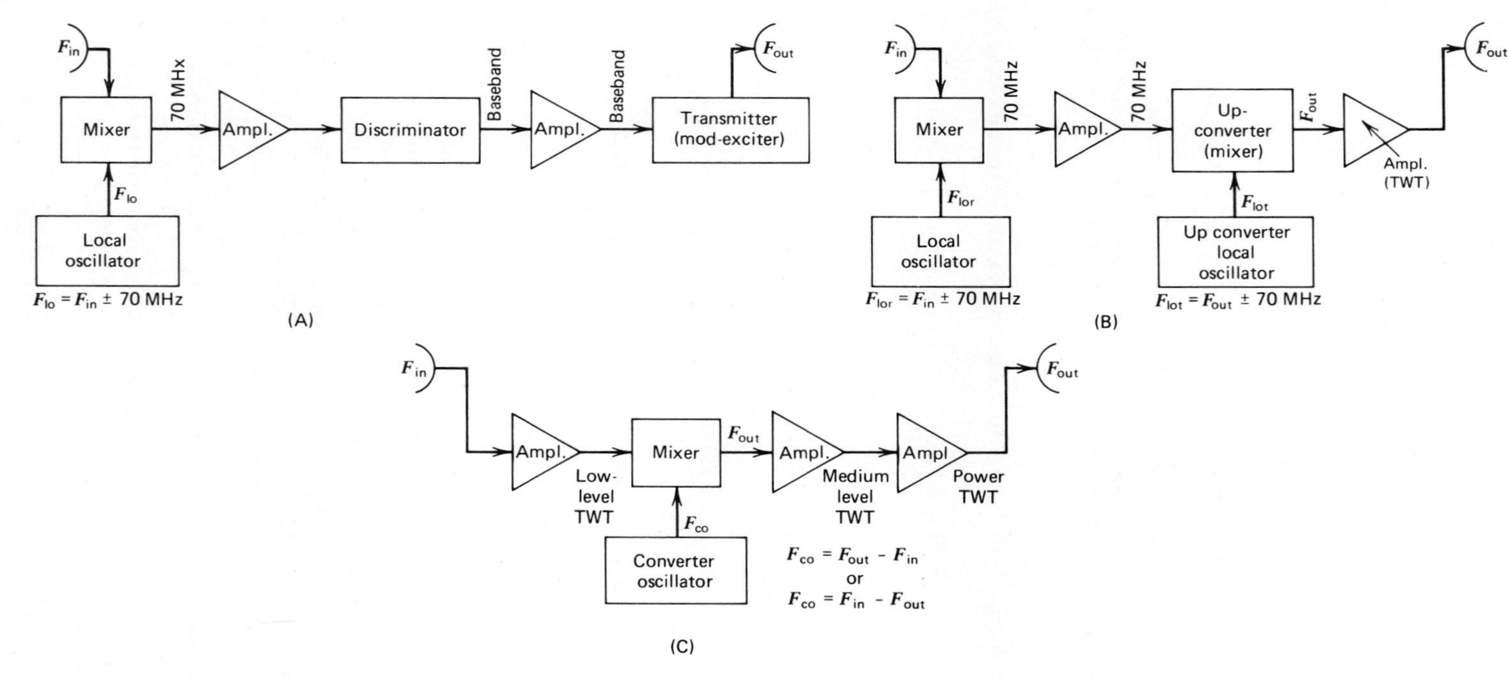

Figure 7.13 Radiolink repeaters: (A) baseband repeater; (B) IF heterodyne repeater; (C) RF heterodyne repeater; F_{in}, input frequency to receiver; F_{out}, output frequency of transmitter; F_{co}, output frequency of the converter local oscillator; F_{lor}, frequency of receiver local oscillator; F_{lot}, frequency of transmitter local oscillator; F_{lo}, frequency of the local oscillator; twt, traveling-wave tube.

of repeater are also available: the IF heterodyne repeater (Figure 7.13B) and the RF heterodyne repeater (Figure 7.13C). The IF repeater is attractive for use on long, backbone systems where noise and/or differential phase and gain should be minimized.

Generally, a system with fewer modulation–demodulation stages or steps is less noisy. The IF repeater eliminates two modulation steps. The repeater simply translates the incoming signal to IF with appropriate local oscillator and a mixer, amplifies the derived IF, and then up-converts it to a new RF frequency. The up-converted frequency may then be amplified by a traveling-wave tube amplifier (TWT). An RF heterodyne repeater is shown in Figure 7.13C. With this type of repeater amplification is carried out directly at RF frequencies. The incoming signal is translated to a different frequency amplified, usually by a TWT, and then reradiated. Radiofrequency repeaters are troublesome in design in things such as sufficient selectivity, limiting and automatic gain control, and methods to correct envelope delay. However, some RF repeaters are now available, particularly for operation below 6 GHz.

4.7 Frequency Planning

4.7.1 General

To derive maximum performance from a radio-link system, the system engineer must set out a frequency-usage plan that may or may not have to be approved by the local administration.

The problem has many aspects. First, the useful RF spectrum is limited from above dc to about 150 GHz. The frequency range of discussion for radio links is essentially from the VHF band at 150 MHz (overlapping) to the millimeter region of 15 GHz. Second, the spectrum from 150 MHz to 15 GHz often must be shared with other services such as radar, navigational aids, research (i.e., space), meteorological, broadcast, and so on. Some radio-link frequency assignments are shown in Table 7.1.

Although many of the allocated bands are wide, some up to 500 MHz in width, FM inherently is a wide-band form of emission. It is not uncommon

Table 7.1 Some Radio-link Frequency Assignments

450– 470 MHz	5,925– 6,425 MHz
890– 960 MHz	7,300– 8,400 MHz
1,710–2,290 MHz	10,550–12,700 MHz
2,550–2,690 MHz	14,400–15,250 MHz
3,700–4,200 MHz	

Source: Ref. 1.

to have B_{rf} = 25 (B_{rf} = RF bandwidth) or 30 MHz for just one emission. Guard bands must also be provided. These are a function of frequency drift of transmitters as well as "splatter" or out-of-band emission, which in some areas is not well specified.

Occupied bandwidth has been specified in Section 4.4.3, according to Carson's Rule. This same rule is followed by the U.S. regulatory agency, FCC.

4.7.2 Radiofrequency Interference

On planning a new radio-link system or on adding RF carriers to an existing installation, RFI of the existing (or planned) emitters in the area must be carefully considered. Usually the governmental authorizing agency has information on these and their stated radiation limits. Limits are established by national authorities. Equally important is antenna directivity and side-lobe radiation. Not only must the radiation of other emitters be examined from this point of view, but also the capability of the planned antenna to reject unwanted signals. The radiation pattern of all licensed emitters should be known. The side-lobe level should be converted in the direction of the planned installation to EIRP in dBW. This should be done for all interference candidates within interference frequency range. For each emitter's EIRP, a path loss should be run to the planned installation to determine interference. Such a study could well affect a frequency plan or antenna design.

Nonlicensed emitters should also be considered. Many such emitters may be classified as industrial noise sources such as heating devices, electronic ovens, electric motors, unwanted radiation from privately owned and other microwave installations (i.e., radar harmonics). In the 6-GHz band a coordination contour should be carried out to verify interference from earth stations (see CCIR Rep. 448 and Rec. 359, p. 2). For general discussion on the techniques for calculating interference noise in radio-link systems, see CCIR Rep. 388, p. 1.

4.7.3 Overshoot

Overshoot interference may occur when radio-link hops in tandem are in a straight line. Consider stations A, B, C, and D in a straight line, or that a straight line on a map drawn between A and C also passes through B and D. Link A–B has frequency F_1 from A to B, and F_1 is reused in direction C–D. Care must be taken that some of the emission F_1 on the A–B hop does not spill into the receiver at D. Reuse may even occur on an A, B, and C combination, so F_1 at A–B may spill into a receiver at C tuned to F_1. This can be avoided if stations are removed from the straight line. In this case the station at B should be moved to the north of a line A–C, for example.

4.7.4 Transmit–Receive Separation

If a transmitter and receiver are operated on the same frequency at a radio-link station, the loss between them must be at least 120 dB. One way to assure the 120 figure is to place all "go" channels in one-half of an assigned band and all "return" channels in the other. The terms "go" and "return" are used to distinguish between the two directions of transmission.

4.7.5 Basis of Frequency Assignment

"Go" and "return" channels are assigned as in the preceding section. For adjacent RF channels in the same half of the band, horizontal and vertical polarizations are used alternately; thus we may assign, as an example, horizontal polarization (H) to the odd numbered channels in both directions on a given section and vertical polarization (V) to the even-numbered channels. The order of isolation between polarizations is on the order of 35 dB.

5 COAXIAL CABLE TRANSMISSION LINKS

5.1 Introduction

A coaxial cable is simply a transmission line consisting of an unbalanced pair made up of an inner conductor surrounded by a grounded outer conductor, which is held in a concentric configuration by a dielectric. The dielectric can be of many different types, such as solid "poly" (polyethylene or polyvinyl chloride), foam, Spirafil, air, or gas. In the case of air–gas dielectric, the center conductor is kept in place by spacers or disks.

Systems have been designed to use coaxial cable as a transmission medium with a capability of transmitting a frequency division multiplex configuration ranging from 120 voice channels to 10,800. Community antenna television (CATV) systems use single cables for transmitted bandwidths on the order of 300 MHz.

Frequency division multiplex was developed originally as a means to increase the voice-channel capacity of wire systems. At a later date the same techniques were applied to radio. Then for the 20 years after World War II radio systems became the primary means for transmitting long-haul, toll-telephone traffic. Lately, coaxial cable has been making a strong comeback in this area.

One advantage of coaxial cable systems is reduced noise accumulation when compared to radio links. For point-to-point multichannel telephony the FDM line frequency (see Chapter 5) configuration can be applied directly to the cable without further modulation steps as required in radio links, thus substantially reducing system noise. In most cases radiolinks will prove more economical than coaxial cable. Nevertheless, because of the congestion of centimetric radio wave (radio links) systems, coaxial cable is making a new debut. Coaxial cable should be considered in lieu of radio links using the following general guidelines:

- In areas of heavy-microwave (including radio link) RFI.
- On high-density routes where coaxial cable may be more economical than radio links. (Consider a system that will require 5000 or more circuits at the end of 10 years.)
- On long national or international backbone routes where the system designer is concerned with noise accumulation.

Coaxial cable systems may be attractive for the transmission of television or other video applications. Some activity has been noted in the joint use of TV and FDM telephone channels on the same conductor. Another advantage in some circumstances is that system maintenance costs may prove to be less than for equal-capacity radio links.

One deterrent to the implementation of coaxial cable systems, as with any cable installation, is the problem of getting right-of-way for installation, and its subsequent maintenance (gaining access), particularly in urban areas. Another consideration is the possibility of damage to the cable once it is installed. Construction crews may unintentionally dig up or cut the cable. For more details on the choice between coaxial cables and radio links, see Section 6 of this chapter.

5.2 Basic Construction Design

Each coaxial line is called a "tube." A pair of these tubes is required for full-duplex, long-haul application. One exception is the CCITT small bore coaxial cable system where 120 voice channels, both "go" and "return", are accommodated in one tube. For long-haul systems, more than one tube is carried in a sheath. In the same sheath filler pairs or quads are included, sometimes placed in the interstices, depending on the size and lay-up of the cable. The pairs and quads are used for order wire and control purposes as well as for local communication. Some typical cable lay-ups are shown in Figure 7.14. Coaxial cable is usually placed at a depth of 90 to 120 cm, depending on frost penetration, along the right-of-way. Tractor-drawn trenchers or plows normally are used to open the ditch where the cable is placed, using fully automated procedures.

Cable repeaters are spaced uniformly along the route. Secondary or "dependent" repeaters are often buried. Primary power feeding or "main" repeaters are installed in surface housing. Cable lengths are factory-cut so that the splice occurs right at repeater locations.

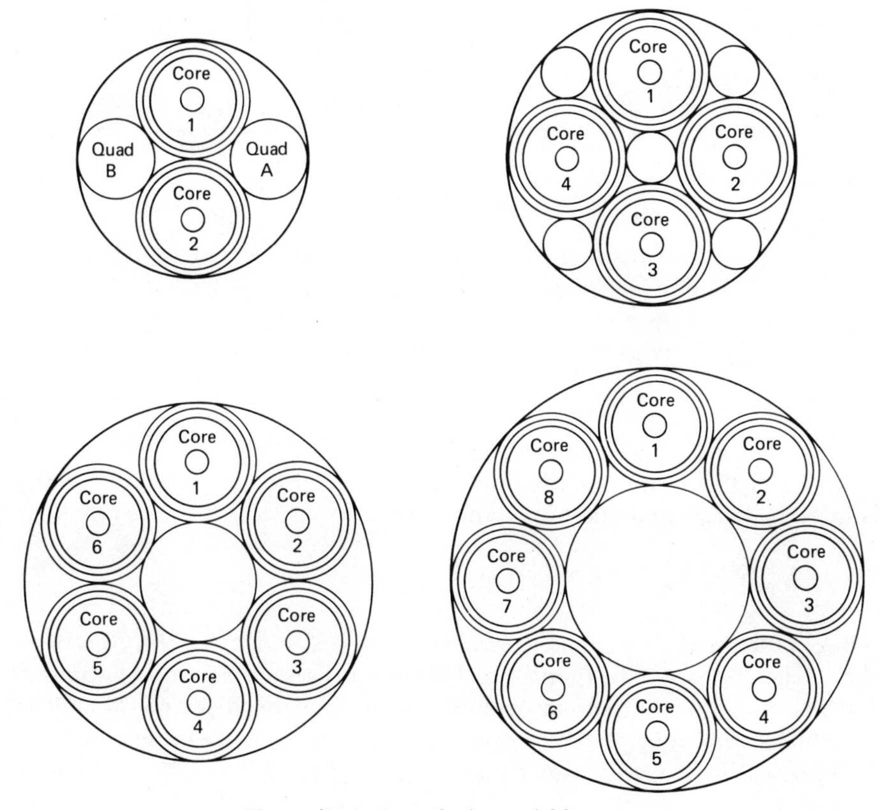

Figure 7.14 Some basic coaxial lay-ups.

5.3 Cable Characteristics

For long-haul transmission, standard cable sizes are as follows:

Inches	Millimeters
0.047/0.174	1.2/4.4 (small diameter)
0.104/0.375	2.6/9.5

The fractions express the outside diameter of the inner conductor over the inside diameter of the outer conductor. For instance, for the large bore cable, the outside diameter of the inner conductor is 0.104 in. and the inside diameter of the outer conductor is 0.375 in. This is shown in Figure 7.15. As can be seen from the equations (7.21) and (7.22), the ratio of the diameters of the inner and outer conductors has an important bearing on attenuation. If we can achieve a ratio of $(b/a) = 3.6$, a minimum attenuation per unit length will result.

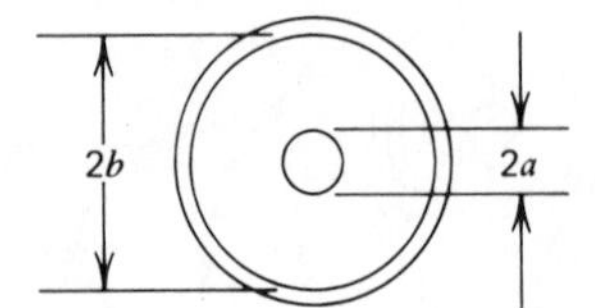

Figure 7.15 Basic electrical characteristics of coaxial cable. See equations (7.21) and (7.22).

ϵ is the dielectric constant and
for air dielectric cable pair, ϵ = 1.0
Outside diameter of inner conductor = $2a$
Inside diameter of outer conductor = $2b$
Attenuation constant (dB)/mi,

$$\alpha = 2.12 \times 10^{-5} \frac{\sqrt{f}[(1/a) + (1/b)]}{\log b/a} \tag{7.21}$$

where a is the radius of inner conductor and b is the radius of outer conductor. Characteristic impedances (Ω) are

$$Z = \left(\frac{138}{\sqrt{\epsilon}}\right) \log \frac{b}{a} = 138 \log \frac{b}{a} \text{ in air} \tag{7.22}$$

The characteristic impedance of coaxial cable is $Z_0 = 138 \log (b/a)$ for an air dielectric. If $b/a = 3.6$, then $Z_0 = 77\Omega$. Using a dielectric, other than air reduces the characteristic impedance. If we use the disks mentioned in Section 5.1 to support the center conductor, the impedance lowers to 75Ω.

Figure 7.16 is a curve giving attenuation per unit length in decibels

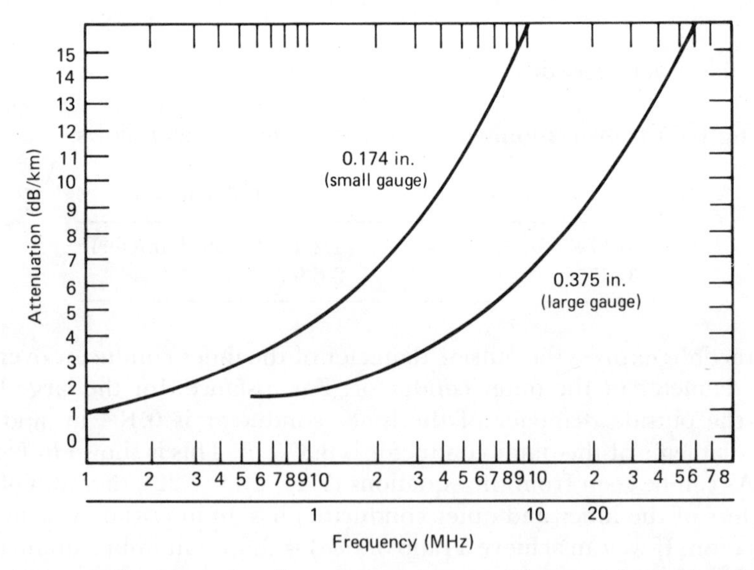

Figure 7.16 Attenuation–frequency response per kilometer of coaxial cable.

versus frequency for the two most common types of coaxial cable discussed in this chapter. Attenuation increases rapidly as a function of frequency and is a function of the square root of frequency as shown in equation 7.21. The transmission system engineer is basically interested in how much bandwidth is available to transmit an FDM line-frequency configuration. For instance, the 0.375-in cable has an attenuation of about 5.8 dB/mi at 2.5 MHz and the 0.174-in cable, 12.8 dB/mi. At 5 MHz the 0.174-in. cable has about 19 dB/mi and the 0.375-in. cable, 10 dB/mi. Attenuation is specified for the highest frequency of interest.

Coaxial cable can transmit signals down to dc, but in practice, frequencies below 60 kHz are not used because of difficulties of equalization and shielding. Some engineers raise the lower limit to 312 kHz. The high frequency limit of the system is a function of the type and spacing of repeaters as well as cable dimensions and the dielectric constant of the insulating material. It will be appreciated from equation 7.21 that the gain-frequency characteristics of the cable follow a root-frequency law, and equalization and "preemphasis" should be designed accordingly.

5.4 System Design

Figure 7.17 is a simplified application diagram of a coaxial cable system in long-haul, point-to-point multichannel telephone service. To summarize system operation, an FDM line frequency (Chapter 5) is applied to the coaxial cable system via a line-terminal unit. Dependent repeaters are spaced uniformly along the length of the cable system. These repeaters are fed power from the cable itself. In the ITT design [14] the dependent repeater has a plug-in automatic level-control unit. In temperate zones

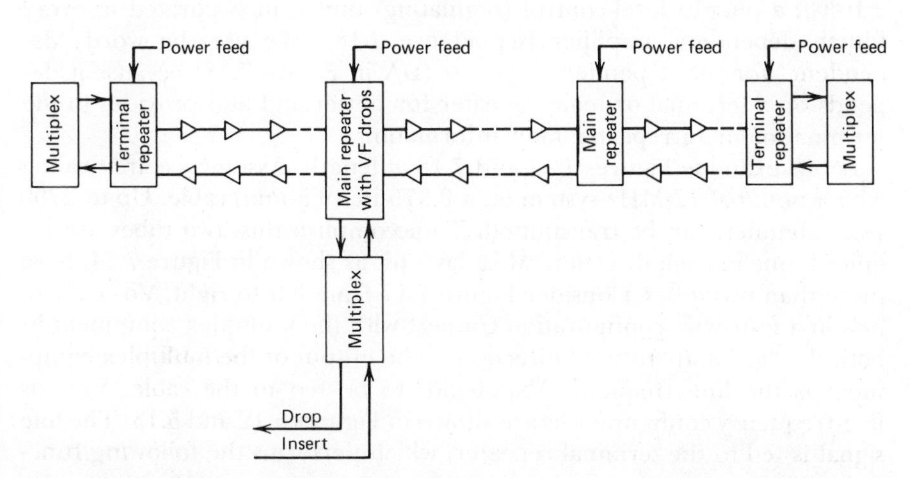

Figure 7.17 Simplified application diagram of a long-haul coaxial cable system for multichannel telephony.

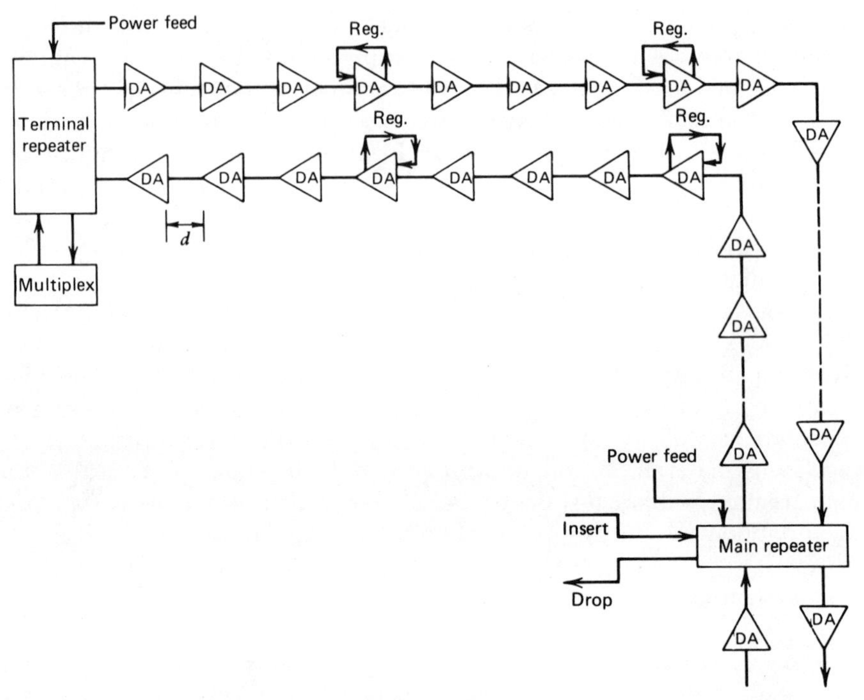

Figure 7.18 Detail of application diagram: DA, dependent amplifier (repeater); Reg, regulation circuitry; *d*, distance between repeaters.

where cable is layed to sufficient depth and where diurnal and seasonal temperature variations are within the "normal" (a seasonal swing of ±10°C), a plug-in level-control (regulating) unit is incorporated in every fourth dependent amplifier (see Figure 7.18). We use the word "dependent" for the dependent repeater (DA in Figure 7.18) because it depends on a terminal or main repeater for power and also provides to the terminal or main repeater fault information.

Let us examine Figures 7.17 and 7.18 at length. Assume we are dealing with a nominal 12-MHz system on a 0.375-in. (9.5-mm) cable. Up to 2700 voice channels can be transmitted. To accomplish this, two tubes are required, one in each direction. Most lay-ups, as shown in Figure 7.14, have more than two tubes. Consider Figure 7.17 from left to right. Voice channels in a four-wire configuration connect with the multiplex equipment in both the "go" and "return" directions. The output of the multiplex equipment is the line frequency (baseband) to be fed to the cable. Various line-frequency configurations are shown in Figures 5.12 and 5.13. The line signal is fed to the terminal repeater, which performs the following functions:

- Combines the line-control pilots with the multiplex line frequency.
- Provides "preemphasis" to the transmitted signal, distorting the output signal such that the higher frequencies get more gain than the lower frequencies (see Figure 7.16).
- Equalizes the incoming wide-band signal.
- Feeds power to dependent repeaters.

The output of the terminal repeater is a preemphasized signal with required pilots along with power feed. In the ITT design this is a dc voltage up to 650 V with a stabilized current of 110 mA. A main (terminal) repeater feeds, in this design, up to 15 dependent repeaters in each direction. Thus a maximum of 30 dependent repeaters appear in a chain for every main or terminal repeater. Other functions of a main repeater are to equalize the wide-band signal and to provide access for drop and insert of telephone channels by means of through-group filters. Figure 7.18 is an enlargement of a section of Figure 7.17 showing each fourth repeater with its automatic level regulation circuitry. Distance d between "DA" repeaters (dependent amplifier) is 4.5 km or 2.8 mi for a nominal 12-MHz system (0.375-in. cable). Amplifiers have gain adjustments of ± 6 dB, equivalent to varying repeater spacing ±570 m.

As can be seen from the preceding discussion, the design of coaxial cable systems for both long-haul multichannel telephone service as well as CATV systems has become, to a degree, a "cookbook" design. Basically, system design involves the following:

- Repeater spacing as a function of cable type and bandwidth.
- Regulation of signal level.
- Temperature effects on regulation.
- Equalization.
- Cable impedance irregularities.
- Fault location or the so-called supervision.
- Power feed.

Other factors are, of course, right-of-way for the cable route with access for maintenance and the laying of the cable. With these factors in mind, consult Tables 7.2 and 7.3, which review the basic parameters of the CCITT approach (Table 7.2) and the Bell System approach (Table 7.3) to standard coaxial cable systems. For the 0.375-in. coaxial cable systems, practical noise accumulation is less than 1 pWp/km, whereas radio links allocate 3 pWp/km. These are good guideline numbers to remember for gross system considerations. Noise in coaxial cable systems derives from the active devices in the line (e.g., the repeaters) as well as the terminal equip-

Table 7.2 Characteristics of CCITT Specified Coaxial Cable[a] Systems (Large-diameter Cable)

Item	Nominal Top Modulation Frequency				
	2.6 MHz	4 MHz	6 MHz	12 MHz	60 MHz
CCITT Rec.	G.337A	G.338	G.337B	G.332	G.333
Repeater type	Tube	Tube	Tube	Transistor	Transistor
Video capability	No	Yes	Yes	Yes	Not stated
Video + FDM capability	No	No	No	Yes	Not stated
Nominal repeater spacing	6 mi–9 km	6 mi–9 km	6 mi–9 km	3 mi–4.5 km	1 mi–1.55 km
Main line reg. pilot	2604 kHz	4092 kHz	See CCITT Rec. J.72	12,435 kHz	12,435/4287 kHz
Auxiliary reg. pilot(s)		308, 60 kHz	See Rec. J.72	4287, 308 kHz	61,160, 40,920, and 22,372 kHz

[a]Cable type for all systems, 0.104/0.375 in. = 2.6/9.5 mm.

Table 7.3 Characteristics of "L" Coaxial Cable Systems

Item	"L" System Identifier[a]			
	L_1	L_3	L_4	L_5
Maximum design line length	4000 mi	4000 mi	4000 mi	4000 mi
Number of 4-kHz FDM VF channels[b]	600	1860	3600	10,800[c]
TV NTSC	Yes	Yes plus 600 VF	No	Not stated
Line frequency	60–2788 kHz	312–8284 kHz	564–17,548 kHz	1590–68,780 kHz
Nominal repeater spacing	8 mi	4 mi	2 mi	1 mi
Power feed points	160 mi or every 20 rptrs	160 mi or 42 rptrs	160 mi or every 80 rptrs	75 mi or every 75 rptrs

[a]Cable type of all "L" systems, 0.375 in.
[b]Number of VF channels expressed per pair of tubes, one tube "go" and one tube "return."
[c]L5E ~ 13,200 VF channels.

ment, both line conditioning and multiplex. Noise design of these devices is a trade-off between thermal and intermodulation (IM) noise. Intermodulation noise is the principal limiting parameter forcing the designer to install more repeaters per unit length with less gain per repeater.

Refer to Chapter 5 for CCITT-recommended FDM line-frequency configurations, in particular Figures 5.12 and 5.13, which are valid for 12 MHz systems.

5.5 Repeater Considerations—Gain and Noise

Consider a coaxial cable system 100 km long using 0.375-in. cable capable of transmitting up to 2700 voice (VF) channels in an FDM configuration, in this case a 12-MHz system. At the highest modulation frequency, 12 MHz, the cable attenuation per kilometer is approximately 8.3 dB (from Figure 7.16). The total loss at 12 MHz for the 100-km cable section is $8.3 \times 100 = 830$ dB. Thus one approach the system design engineer might take would be to install a 830-dB amplifier at the front end of the 100-km section. This approach is rejected out of hand.

Another approach would be to install a 415-dB amplifier at the front end and another at the 50-km point. Suppose the signal level was −15 dBm composite at the originating end. Thus −15 dBm + 415 dB = +400 dBm or +370 dBW. Remember that +60 dBW is equivalent to a megawatt; otherwise, we would have an amplifier with an output of 10^{37} W or 10^{31} MW. Still other approaches would be to have 10 amplifiers with 83-dB gain, each spaced at 10-km intervals, or to install 20 amplifiers or (830/20) = 41.5 dB each, or to install 30 amplifiers at (830/30) = 27.67 dB, each spaced at 3.33-km intervals. As we see later, the latter approach begins to reach an optimum from a noise standpoint keeping in mind that the upper limit for noise accumulation is 3 pWp/km. The gain most usually encountered in coaxial cable amplifiers is 30 to 35 dB.

If we remain with the 3-pWp/km criterion, in nearly all cases radio links will be installed because of their economic advantage. Assuming 10 full-duplex RF channels per radio system at 1800 VF channels per RF channel, the radio link can transmit 18,000 full-duplex channels, which is probably more economical on an installed cost basis. On the other hand, if we can show noise accumulation less on coaxial cable systems, these systems will prove in at some number of channels less than 18,000 if the reduced cumulative noise is included as an economic factor. There are other considerations, such as maintenance and reliability, but let us discuss noise further. In the basic design of coaxial cable repeaters noise consists of two major components, thermal noise (white noise) and intermodulation (IM) noise. To reach a goal of 1 pWp/km of noise accumulation coaxial cable amplifier design, must walk a "tightrope" between thermal and IM noise. It is also very sensitive to overload, with its consequent impact on intermodulation noise. For a deeper analysis of repeater design, the reader should consult Refs. 3 and 4.

5.6 Powering the System

Power feeding of buried repeaters in a typical coaxial cable link such as the ITT system permits the operation of 15 dependent repeaters from each end of a feed point (12-MHz cable). Thus up to 30 dependent repeaters can be supplied power between power feed points. A power feed unit at the power feed point provides up to 650 V dc of voltage between center conductor and ground using 110-mA stabilized direct current. Power feed points may be spaced as far apart as 140 km (87 mi) on large-diameter cable.

5.7 60-MHz Coaxial Cable Systems

Wide-band coaxial cable systems are presently being implemented because of the ever-increasing demand for long-haul, toll-quality telephone channels. Such systems are designed to carry 10,800 nominal 4-kHz FDM channels per pair of tubes. The line-frequency configuration for such a system, as recommended in CCITT Rec. G.333, is shown in Figure 7.19. For long-haul noise objectives, the large diameter cable is recommended (e.g., 2.6 to 9.5 mm).

When expanding a coaxial cable system, a desirable objective is to use the same repeater locations as with the old cable and to add additional repeaters at intervening locations. For instance, if we have 4.5-km spacing for a 12-MHz system and our design shows that we need three times the number of repeaters for an equal-length 60-MHz system, repeater spacing should be at 1.5-km intervals.

The ITT 12-MHz system uses 4.65-km spacing. Thus its 60-MHz system will use 1.55-km (0.95-mi) spacing with a mean cable temperature of 10°C. The attenuation characteristic of the large-gauge cable is shown in Figure 7.20, which is an extension of Figure 7.16. Repeater gain for the ITT system is nominally 28.5 dB at 60 MHz and can be varied ±1.5 dB. Line build-out networks allow still greater tolerance. The overload point, following CCITT Rec. 223, is taken at +20 dBm with a transmit level of −18 dBm. System pilot frequency is 61.160 MHz for regulation. A second pilot frequency of 4.287 MHz corrects the level of the lower-frequency range. Pilot regulation repeaters are installed at from seven to 10 nonregulated repeaters, with deviation equalization at every 24th repeater. All repeaters have temperature control (controlled by the buried ambient). Power feeding is carried out at every 100 km (63 mi). Thus 64 repeaters will be fed remotely using constant direct current feed over the conductors. Each repeater will tap off about 15 V, requiring 2 W. Thirty-two repeaters at 2 × 15 V each will require 960 V. An additional 120-V dc is required for

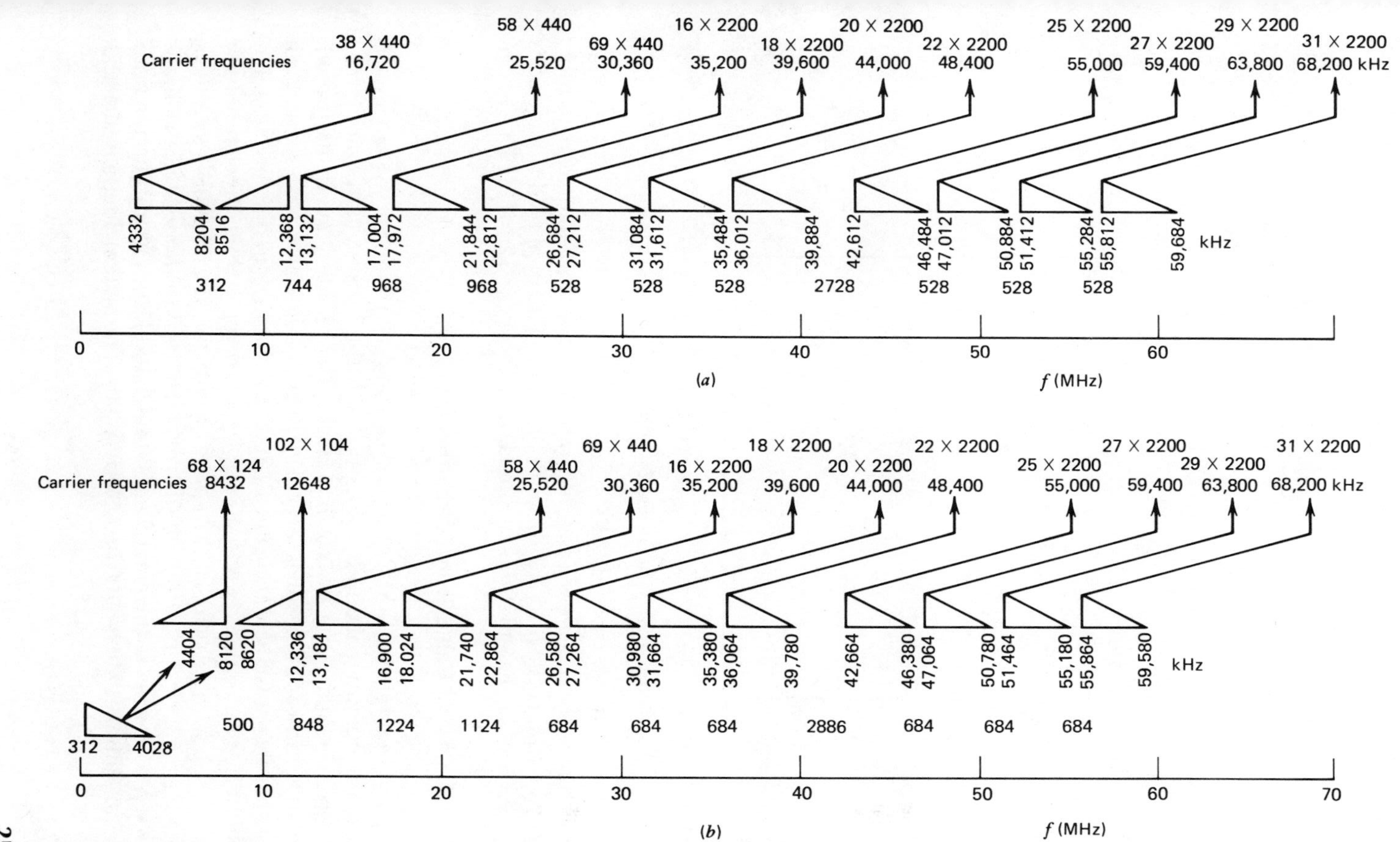

Figure 7.19 Line-frequency allocation recommended for 40-MHz and 60-MHz systems on 2.6/9.5 mm coaxial cable pairs using (A) plan 1 and (B) plan 2 from CCITT Rec. G.333). Courtesy of the International Telecommunication Union–CCITT.

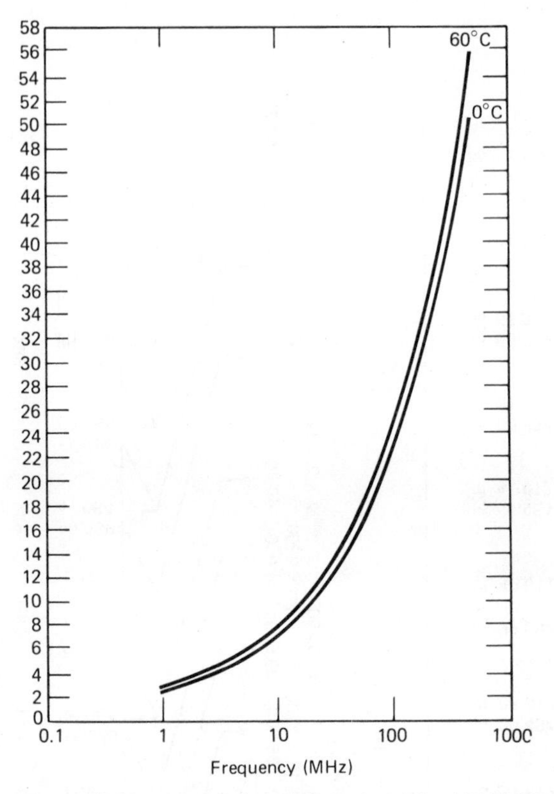

Figure 7.20 Attenuation of large coaxial cable (0.375 in.).

pilot-regulated repeaters plus one repeater with deviation equalization. Added to this is the 50-V IR drop on the cable. The total feed voltage adds to 1226 V of dc. Fault location is similar to that for the 12-MHz ITT system.

5.8 L5 Coaxial Cable Transmission System

A good example of a 60-MHz coaxial cable transmission system that is presently operational and carrying traffic is the L5 system operating on a transcontinental route in North America (see Table 7.3). In its present lay-up it consists of 22 tubes, of which 20 are on-line and 2 are spare. Each tube has the capacity to transmit 10,800* VF channels in one direction. For full-duplex operation, two tubes are required for 10,800 VF channels, or the total system capacity is [(22 − 2)/2] × 10,800, or 108,000 VF channels. The system is designed for a 40-dBrnC0 (8000-pWp) noise objective in the worst VF channel at the end of a 4000-mi (6400-km) system. This is a noise accumulation of 8000/6400 or 1.25 pWp/km. The system design is such that

*The ATT L5E system has a channel capacity of 13,200 voice channels.

it is second-order intermodulation and thermal-noise limited. Repeater overload is at the 24-dBm point.

The modulation plan is an extension of that shown in Figure 5.14. The basis of the plan is the development of the "jumbo group" (JG) made up of six mastergroups (Bell System FDM hierarchy). Keep in mind that the basic mastergroup consists of 600 voice channels (in this case) or 10 standard supergroups and occupies the band 564 to 3084 kHz. The basic jumbo group occupies the band 564 to 17,548 kHz with a level-control pilot at 5888 kHz. The three jumbo groups are assigned the following line frequencies:

JG1	3,124 to 20,108 kHz
JG2	22,068 to 39,056 kHz
JG3	43,572 to 60,556 kHz

Equalizing pilots are at 2976 kHz, 20,992 kHz, and 66,048 kHz as transmitted to the line. There is a temperature pilot at 42,880 kHz. The basic jumbo-group frequency generator is built around an oscillator that has an output of 5.12 MHz. This oscillator has a drift rate of less than 1 part in 10^{10}/day after aging and a short-term stability of better than 1 part in 10^{8}/ms. Excessive frequency offset is indicated by an alarm. Automatic protection of the 10 operating systems is afforded by the LPSS (line-protection switching system) on a 1:10 basis. A maximum length of switching span is 150 mi. Power feeds are at 150-mi intervals, feeding power in both directions. Thus a power span is 75 mi long, or has 75 repeaters. Power is 910 mA on each cable, +1150 V and −1150 V operating against ground. The basic repeater is a fixed-gain amplifier, spaced at 1-mi intervals. Typically, every fifth repeater is a regulating repeater, and this regulation is primarily for temperature compensation.

6 COAXIAL CABLE OR RADIO LINK—THE DECISION

6.1 General

One major decision that the transmission system engineer often faces is whether to install a radio link or coaxial cable on a particular point-to-point circuit. The factors determining the choice fall into two categories, technical and economic. Table 7.4 compares the two media from a technical viewpoint. These comparisons can serve as a fundamental guide for making a technical recommendation in the selection of facility. System mixes may also be of interest.

The discussion that follows is an expansion of some of the points covered in Sections 4 and 5 of this chapter. Table 7.4 summarizes the factors, which are discussed subsequently.

Table 7.4 Comparison of Coaxial Cable versus Radio Link

Item	Cable	Radio
Land acquisition	Requires land easements or right-of-way along entire route and recurring maintenance access later	Repeater site acquisition every 30–50 km with building, tower, access road at each site
Insert and drop	Insert and drop at any repeater. Should be kept to minimum. Land buys, building required at each insert location	Insert and drop at more widely spaced repeater sites
Fading	None except temperature variations	Important engineering parameter
Noise accumulation	Less, 1 pWp/km	More, 3 pWp/km
Radiofrequency interference (RFI)	None	A major consideration
Limitation on number of carriers or basebands transmitted	None	Strict, band-limited plus RFI ambient limitations
Repeater spacing	1.5, 4.5, 9 km	30–50 km
Comparative cost of repeaters	Considerably lower	Considerably higher
Power considerations	High-voltage dc in milliampere range	48-V dc static no-break at each site in ampere range
Cost versus traffic load	Full load proves more economical than radio link	Less load proves more economical than cable
Multiplex	FDM—CCITT	FDM—CCITT
Maintenance and engineering	Lower level, lower cost	Higher level, higher cost
Terrain	Important consideration in cable laying	Can jump over, and even take advantage of difficult terrain

6.2 Land Acquisition as a Limitation to Coaxial Cable Systems

Acquisition of land detracts more from the attractiveness of the use of coaxial cable than does any other consideration; it adds equally to the attractiveness of selecting radio links (LOS microwave). With a radio-link system large land areas are jumped and the system engineer is not concerned with what goes on between. One danger that many engineers tend to overlook is that of the chance building of a structure in the path of the radio beam after installation on the routes has been completed.

Cable, on the other hand, must physically traverse the land area that intervenes. Access is necessary after the cable is laid, particularly at repeater locations. This may not be as difficult as it first appears. One method is to follow parallel to public highways, keeping the cable lay on public land. Otherwise, with a good public-relations campaign, easement or rights-of-way often are not difficult to obtain. This leads to another point. The radio-link relay sites are fenced. Cable lays are marked, but the chances of damage by the farmer's plow or construction activity are fairly high.

6.3 Fading

Radio links are susceptible to fading. Fades of 40 dB on long hops are not unknown. Overbuilding a radio-link system tends to keep the effect of fades on system noise within specified limits.

On coaxial cable systems, signal level variation is mainly a function of temperature variation. Level variations are well maintained by regulators controlled by pilot tones and, in some cases, auxiliary regulators controlled by ground ambient.

6.4 Noise Accumulation

Noise accumulation is a most important consideration and has been discussed in Section 5.5. Either system will serve for long-haul backbone routes and meet the minimum specific noise criteria established by CCITT–CCIR. However, in practice, the engineering and installation of a radio system may require more thought and care to meet those noise requirements. Modern coaxial cable systems have a design target of 1 pWp/km of noise accumulation. With care, using IF repeaters, radio systems can meet the 3-pWp criterion. Besides the disadvantage of fading, radio links, by definition, have more modulation steps and thus are noisier.

6.5 Group Delay—Attenuation Distortion

Group delay is less of a problem with radio links. Figure 7.16 shows amplitude response of a cable section before amplitude equalization. The cable plus amplifiers plus amplitude equalizers add to the group delay problem.

It should be noted that for video transmission on cable, an additional modulation step is required to translate the video to the higher frequencies and invert the band. While on radio links, video can be transmitted directly without additional translation or inversion besides RF modulation.

6.6 Radiofrequency Interference

There is no question as to the attractiveness of broad-band buried coaxial cable systems over equivalent radio-link systems when the area to be traversed by the transmission medium is one of dense RFI. Usually these areas are built-up metropolitan centers with high industrial and commercial activity. Unfortunately, as a transmission route enters a dense RFI area, land values increase disproportionately, as do construction costs for cable laying. Yet the trade-off is there.

6.7 Maximum VF Channel Capacity

In heavily populated areas of highly developed nations, frequency assignments are becoming severely limited or unavailable. Although some of the burden on assignment will be removed as the tendency toward usage of the millimeter region of the spectrum is increased, coaxial cable remains the most attractive of the two for high-density FDM configurations.

If it is assumed that there are no RFI or frequency assignment problems, a radio link can accommodate up to eight carriers in each direction (CCIR Rec. 384-1) with 2700 VF channels per carrier. Thus the maximum capacity of such a system is 8 × 2700 = 21,600 VF channels. Now assume a 12-MHz coaxial cable system with 22 tubes, 20 operative or 10 "go" and 10 "return." Each coaxial tube has a capacity of 2700 channels. Thus the maximum capacity is 10 × 2700 = 27,000 VF channels.

It should be noted that the radio system with a full load of 2700 VF channels may suffer from some multipath problems. Coaxial cable systems have no similar interference problems; However, cable impedance must be controlled carefully when splicing cable sections. Such splices usually are carried out at repeater locations.

Consider now 60-MHz cable systems with 20 active tubes, 10 "go" and 10 "return." Assume 10,000 channels capacity per tube; thus 10,000 × 10 = 100,000 VF channels or equivalent to five full radio-link systems.

6.8 Repeater Spacing

As discussed in Chapter 5, a high average for radiolink repeater spacing is 50 km (30 mi), depending on drop and insert requirements as well as an economic trade-off between tower height and hop distance. For coaxial cable systems, repeater separation depends on the highest frequency to be transmitted, which ranges from 9.0 km for 4-MHz systems to 1.5 km for 60-MHz systems. A radio-link repeater is much more complex than a cable repeater. Coaxial cable repeaters are much cheaper than radio-link repeater, considering tower, land, and access roads for radio links. However, much of this advantage for coaxial cables is offset because radio links require much fewer repeaters. It should also be kept in mind that a radio-link system is more adaptable to difficult terrain.

6.9 Power Considerations

The 12-MHz ITT coaxial cable system can have power feed points separated by as much as 140 km (87 mi) using 650 V dc at 150 mA. In a 140-km section of a radio-link route at least four power feed points would be required, one at each repeater site. About 2 A is required for each transmitter receiver combination using standard 48-V dc battery, usually with static no-break power. Power will also be required for tower lights and perhaps for climatizing equipment enclosures.

6.10 Engineering and Maintenance

Cable systems are of "cookbook" design. Radio-link systems require a greater engineering effort prior to and during installation. Likewise, the level of maintenance of radio systems is higher than cable.

6.11 Multiplex Modulation Plans

Interworking, tandem working of radio and coaxial cable systems, is made easier because both broad-band media use the same standard CCITT or "L"-system modulation plans (see Chapter 5).

7 AN INTRODUCTION TO EARTH-STATION TECHNOLOGY

7.1 General Application

Satellite communication is another method of establishing a communication link. For network planning, satellite trunks for telephony may be economically the optimum for a variety of applications, including the following:

1. International high-usage trunks country-to-country.
2. On national trunks, over 500 miles between switching nodes in highly developed countries. Again, the tendency is high-usage to serve in addition to radio link–coaxial cable.
3. In areas under development where satellite links replace HF and a high growth is expected to be eventually supplemented by radio link and coaxial cable.
4. In sparsely populated, highly rural, "out-back" areas where it may be the only form of communication. Northern Canada and Alaska are good examples.
5. Final routes for overflow on a demand-assignment basis. Route length again is a major consideration.
6. In many cases reduces the international connection to one link.

7.2 Definition

A number of world bodies, including the CCIR and the U.S. Federal Communications Commission (FCC), have now accepted the term "earth station" as a radio facility located on the earth's surface that communicates with satellites. A "terrestrial station" is a radio facility on the earth's surface that communicates with other similar facilities on the earth's surface. Section 4 of this chapter dealt with one form of terrestrial stations. The term "earth station" as used today has come more to mean a radio station operating with other stations on the earth via an orbiting satellite relay.

Nearly all commercial communication satellites are geostationary. Such satellites orbit the earth in a 24-hr period. Thus they appear stationary over a particular geographic location on earth. For a 24-hr synchronous orbit, the altitude of a geostationary satellite is 22,300 statute miles or 35,900 km above the earth.

7.3 The Satellite

A communication satellite is an RF repeater, which may be represented in its most simple configuration as that shown in Figure 7.21. Theoretically, as mentioned earlier, three such satellites properly placed in equatorial synchronous orbit could provide communication from one earth station to any other located anywhere on the earth's surface (see Figure 7.2).

7.4 Three Basic Technical Problems

As the reader can appreciate, satellite communication is nothing more than radio-link (microwave line-of-sight) communication using one or two RF repeaters located at great distances from the terminal earth stations as shown in Figure 7.22. Because of the distance involved, consider the slant range from earth antenna to satellite to be the same as the satellite altitude. This would be true if the antenna were pointing at zenith to the satellite. Distance increases as the pointing angle to the satellite decreases (elevation angles).

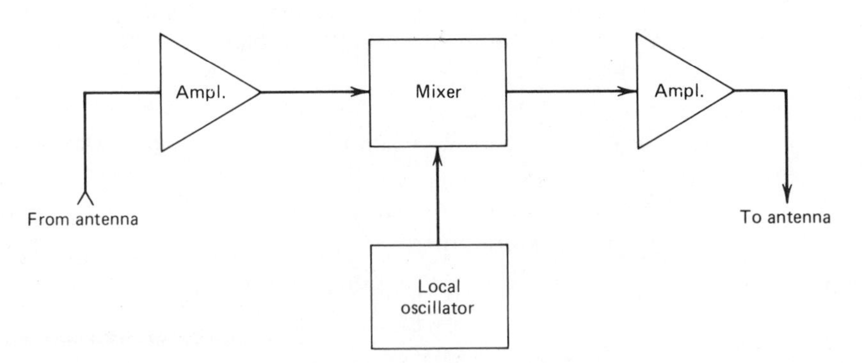

Figure 7.21 Simplified functional block diagram of radio relay portion of a typical communication satellite.

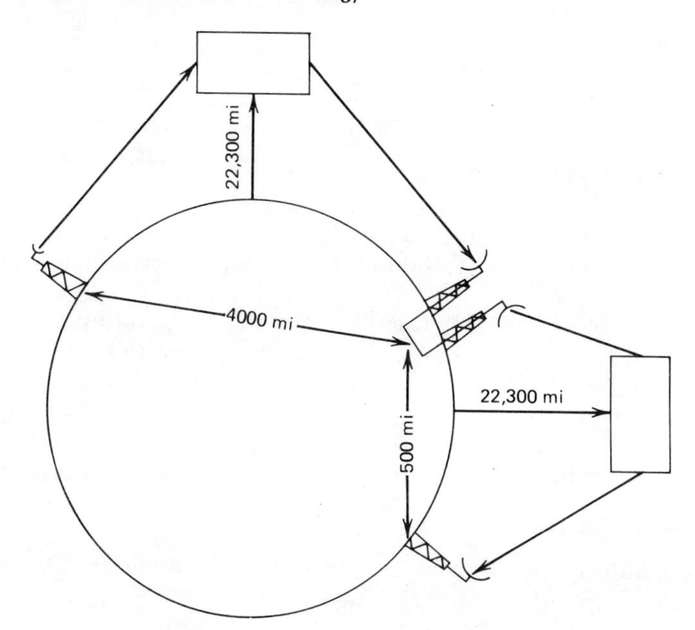

Figure 7.22 Distances involved in SatCom.

We thus are dealing with very long distances. The time required to traverse these distances, namely earth station to satellite to another earth station, is in the order of 240 ms. Round-trip delay will be 2 × 240 on 480 ms. These propagation times are much greater than those encountered on conventional terrestrial systems. So one major problem is propagation time and resulting echo on telephone circuits. It influences certain data circuits in delay to reply for block or packet transmission systems and requires careful selection of telephone signaling systems, or call-setup time may become excessive.

Naturally, there are far greater losses. For radio links, we encounter free-space losses possibly as high as 145 dB. At 22,300 mi at 4.2 GHz, the free-space loss is 196 dB at and 6 GHz, 199 dB. At 14 GHz the loss is about 207 dB. This presents no insurmountable problem from earth to satellite, where comparatively high power transmitters and very high gain antennas may be used. On the contrary, from satellite to earth, the link is power limited for two reasons: (1) in bands shared with terrestrial services such as the popular 4-GHz band to assure noninterference with those services and (2) in the satellite itself, which can derive power only from solar cells. It takes a great number of solar cells to produce the RF power necessary; thus the down-link, from satellite to earth, is critical, and received signal levels will be much lower than on comparative radio links, as low as −150 dBW. A third problem is crowding. The equatorial orbit is filling with geostationary satellites. Radiofrequency interference from one satellite system to another is increasing. This is particularly true for systems employing smaller antennas at earth stations with their inherent wider beamwidths. It all boils down to a frequency "crowding."

7.5 Frequency Bands—Desirable and Available

The most desirable frequency bands for commercial satellite communication are in the spectrum of 1000 to 10,000 MHz. These bands are:

3700–4200 MHz (satellite-to-earth or down-link)
5925–6425 MHz (earth-to-satellite or up-link)
7250–7750 MHz* (down-link)
7900–8400 MHz* (up-link)

These bands are preferred by design engineers for the following primary reasons:

- Less atmospheric absorption than higher frequencies.
- Less noise, both galactic and man-made.
- A well-developed technology.
- Less free-space loss compared to the higher frequencies.

There are two factors contraindicating application of these bands and pushing for the use of higher frequencies:

- The bands are shared with terrestrial services.
- There is orbital crowding (discussed earlier).

Higher-frequency bands for commercial satellite service are:

11.7–12.2 GHz (down-link)
14.0–14.5 GHz (up-link)
17.7–21.26 GHz (down-link)
27.5–30.0 GHz (up-link)

Above 10-GHz rainfall attenuation and scattering, other moisture and gas absorption must be taken into account. The satellite link must meet a 10,000-pWp total VF channel noise criterion at least 99.9% of the time. One solution is a space-diversity scheme where we can be fairly well assured that one of the two antenna installations will not be seriously affected by the heavy rainfall cell affecting the other installation. Antenna separations of 1 to 10 km are being studied. Another advantage with the higher frequencies

*These two bands are intended mainly for military application.

is that requirements for down-link interference are less; thus satellites may radiate more power. This is often carried out on the satellite using spot-beam antennas rather than general-coverage antennas.

7.6 Multiple Access of a Satellite

Multiple access is defined as the ability of a large number of earth stations to interconnect their respective communication links through a common satellite. Satellite access is classified (1) by assignment, whether quasipermanent or temporary, namely, (a) preassigned multiple access or (b) demand-assigned multiple access (DAMA) and (2) according to whether the assignment is in the frequency domain or the time domain, namely, (a) frequency-division multiple access (FDMA) or (b) time-division multiple access (TDMA). On comparatively heavy routes, ($\geq$10 to 20 erlangs) preassigned multiple access may become economical. Other factors, of course, must be considered, such as whether the earth station is "INTELSAT" standard as well as the space-segment charge that is levied for use of the satellite. In telephone terminology, "preassigned" means dedicated circuits. Demand-assigned multiple access is useful for low-traffic, multipoint routes where it becomes interesting from an economic standpoint. Also, an earth station may resort to DAMA as a remedy to overflow for its FDMA circuits.

7.6.1 Frequency-Division Multiple Access

Historically, FDMA has the highest usage and application of the various access techniques. The several RF bands available from Section 7.5 have a 500-MHz bandwidth. A satellite contains a number of transponders, each of which cover a frequency segment of the 500-MHz bandwidth. One method of segmenting the 500 MHz is by utilizing 12 transponders each with a 36-MHz bandwidth. Sophisticated satellites such as INTELSAT V segment the 500 MHz available with transponders up to 77-MHz bandwidth at 4 to 6 GHz and at 11 to 14 GHz have one transponder with a 241-MHz bandwidth.

With FDMA operation, each earth station is assigned a segment or a portion of a segment. For a nominal 36-MHz transponder, 14 earth stations may access in an FDMA format, each with 24 voice channels in a standard CCITT modulation plan (Chapter 5). The INTELSAT IV assignments for a 36-MHz transponder are shown in Table 7.5, where it can be seen that when larger channel groups are used, less earth stations can access the same transponder.

Consider the following hypothetical example of how FDMA works. Frequency translation is via a 2225-MHz local oscillator in the satellite, and the difference mode is used in the mixer; for example:

	Satellite Receive Frequency (MHz)		Mixer Frequency (MHz)		Satellite Transmit Frequency (MHz)
	5925	–	2225	=	3700
	6425	–	2225	=	4200
If a transponder had the 6262–6298-MHz segment, then:					
	6262	–	2225	=	4037
	6298	–	2225	=	4073

In the segment, assign three RF carriers from three locations in common view of an Atlantic satellite. Each carrier will be frequency modulated with two supergroups as in Table 7.5. Subgroup A, the remaining 12 channels of the 132 VF channel total, are spare and are disregarded. Let us assign Spain (Buitrago) the transponder subsegment 6262 to 6272 MHz, Etam, West Virginia (USA), 6272 to 6282 MHz; and Longoville, Chile, 6282 to 6292 MHz. These are the up-link frequencies. When converted in the satellite, they are as follows:

	Up-link (MHz)	Down-link (MHz)
Buitrago	6262–6272 MHz	4037–4047 MHz
Etam	6272–6282 MHz	4047–4057 MHz
Longoville	6282–6292 MHz	4057–4067 MHz

The United States would transmit one carrier at Etam to communicate with both Buitrago and Longoville. The carrier contains two supergroups, as shown in the following diagram.

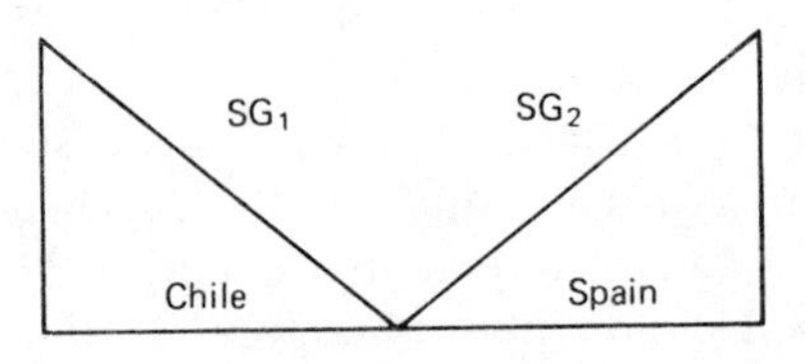

The United States would receive two carriers at Etam for the receive end, one from Buitrago and one from Longoville, shown in the following diagram, and would pick off supergroup 1 (as in our example) from the Chile carrier and supergroup 2 from the Spanish carrier. A similar exercise can be carried out for the Buitrago and Longoville situations. Note that one transmitter can access many locations, but to receive multiple locations, a

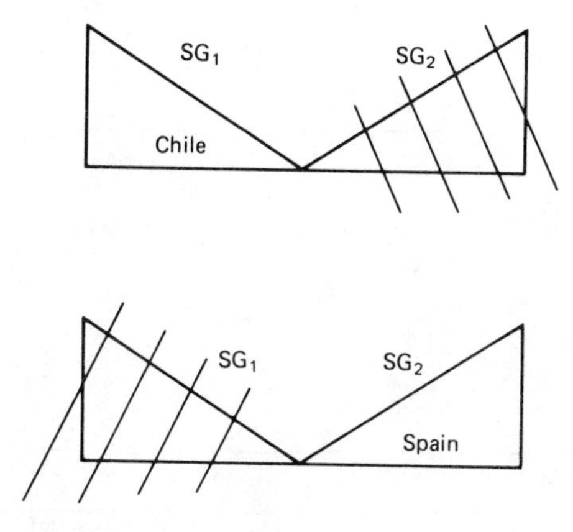

receiver chain is required for each distant location to be received. If a single uplink carrier occupied an entire transponder, (Table 7.5), it could serve (36/2.5) or 14 down-link separate locations, provided that each location required no more than 24 VF channels.

7.6.2 Time-Division Multiple Access

Time-division multiple access (TDMA) operates in the time domain. Use of the satellite transponder is on a time-sharing basis. Individual time slots are assigned to earth stations in a sequential order. Each earth station has full and exclusive use of the transponder bandwidth during its time-assigned segment. Depending on the bandwidth of the transponder, bit rates of 10 to 100 Mbps (megabits per second) are used.

With TDMA operation earth stations use digital modulation and transmit with bursts of information. The duration of a burst lasts for the time period of the slot assigned. Timing synchronization is a major problem.

A frame, in digital format, may be defined as a repeating cycle of events. It occurs in a time period containing a single digital burst from each earth

Table 7.5 INTELSAT IV Global Beam: Voice-channel Capacity versus Frequency Assignments

Carrier capacity (number of voice channels)	24.0	60.0	96.0	132.0	252.0	432.0	972.0
Top baseband frequency (kHz)	108.0	252.0	408.0	552.0	1052.0	796.0	4028.0
Allocated satellite bandwidth (MHz)	2.5	5.0	7.0	10.0	15.0	25.0	36.0
Occupied bandwidth (MHz)	2.0	4.0	5.9	7.5	12.4	20.7	36.0

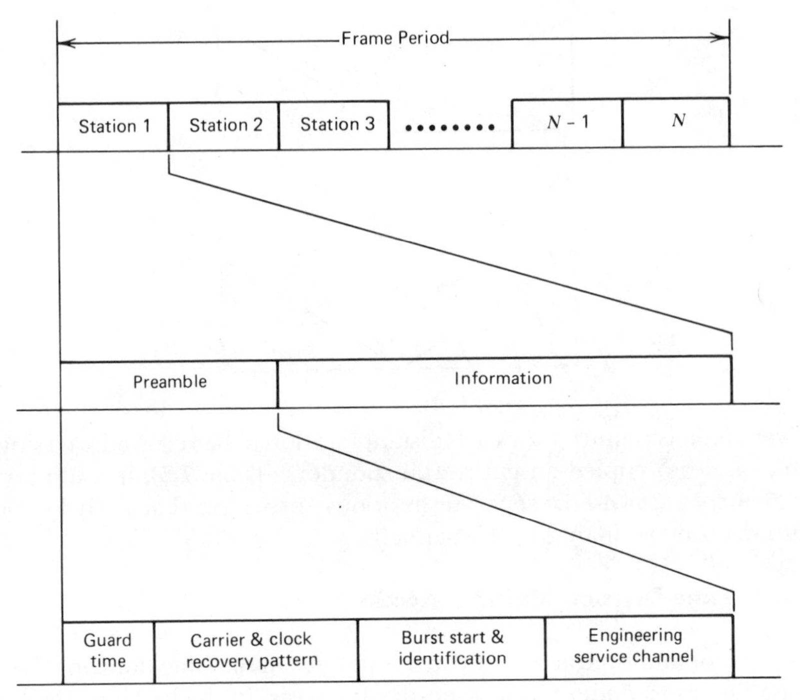

Figure 7.23 Example of TDMA burst format.

station and the guard periods or guard times between each burst. A sample frame is shown in Figure 7.23 for earth stations 1, 2, and 3 to N. Typical frame periods are 750 μsec for INTELSAT and 250 μsec for the Canadian Telesat.

The reader will appreciate that timing is crucial to effective TDMA operation. The greater N becomes (i.e., the more stations operating in the frame period), the more clock timing affects the system. The secret lies in the "carrier and clock (timing) recovery pattern" as shown in Figure 7.23. One way to assure that all stations synchronize to a master clock is to place a sync burst as the first element in the format frame. The INTELSAT (V) does just this. The burst carries 44 bits, starting with 30 bits carrier and bit timing recovery, 10 bits for the "unique word," and 4 bits for the station identification code.

Why use TDMA in the first place? It lies in a major detraction of FDMA. Satellites use traveling-wave tubes (TWTA) in their transmitter final amplifiers. A TWTA has the undesirable property of nonlinearity in its input–output (I/O) characteristics when operated at full power. When there is more than one carrier accessing the transponder simultaneously, high levels of intermodulation products are produced, thus increasing noise and crosstalk. When a transponder is operated at full power output, such noise can be excessive and intolerable. Thus input must be backed off (i.e., level reduced) by $\geq$ 3 dB. This, of course, reduces the EIRP and results in reduced efficiency and reduced information capacity. Consequently, each earth station's up-link power must be carefully coordinated to assure proper loading of the satellite. The complexity of the problem increases when a large number of earth stations access a transponder, each with varying traffic loads.

On the other hand, TDMA allows the transponder's TWTA to operate at full power because only one earth-station carrier is providing input to the satellite transponder at any one instant.

To summarize, consider the following advantages and disadvantages of FDMA and TDMA. The major advantages of FDMA are:

- No network timing is required.
- Channel assignment is simple and straightforward.

The major disadvantages of FDMA are:

- Up-link power levels must be closely coordinated to obtain efficient use of transponder RF output power.
- Intermodulation difficulties require power back-off as the number of RF carriers increase with inherent loss of efficiency.

The major advantages of TDMA are:

- There is no power sharing and IM product problems do not occur
- The system is flexible with respect to user differences in up-link EIRP and data rates.

The major disadvantages of TDMA are:

- Accurate network timing is required.
- There is some loss of throughput due to guard times and preambles.
- Large buffer storage may be required if frame lengths are long.

7.6.3 Demand-Assignment Multiple Access

The demand-assignment multiple-access (DAMA) method where single voice channels are allocated to an earth station on demand. A pool of idle channels is available, and assignments from the pool are made on request. When a call has been completed on the channel, the channel is returned to the idle pool for reassignment.

The DAMA method is analogous to a telephone switch. When a subscriber goes "off-hook," a line is seized; on dialing, a connection is made; and when the call is completed and there is an "on-hook" condition, the voice path through the switch is returned to "idle" and is ready for use by another subscriber. There are three methods available for handling DAMA in a satellite system:

- The polling method.
- The random access–central control.
- Random access–distributed control.

The polling method is fairly self-explanatory. A master station "polls" all other stations in the system sequentially. When a positive reply is received, a channel is assigned accordingly. As the number of stations increases, the polling interval becomes longer and the system tends to become unwieldy. With the random access–central control method, status of channels is coordinated by a central control computer, which is usually located at a "master" earth station. Call requests (call attempts in switching) are passed to the central processor via digital order wire (digitally over the radio service channel) and a channel is assigned if available. Once the call is completed and the subscriber goes on-hook, the speech path is taken down and the channel used is returned to the demand-access pool. According to the system design, there are various methods to handle blocked calls ["all trunks busy" (ATB)] such as queuing and other repeat attempts.

The distributed-control random-access method utilizes a processor control at each earth station in the system. All earth stations in the network monitor the status of all channels where channel status is continuously updated via a digital order-wire circuit. When an idle circuit is seized, all users are informed and the circuit is removed from the pool. Similar information is transmitted to all users when the circuit returns to the idle condition. The same problems arise regarding blockage (ATB) as in the central control system. Distributed control is more costly, particularly in large systems with many users. It is attractive in the international environment as it eliminates the "politics" of a master station.

Most systems use a mix of preassigned channels on an FDMA basis and of DAMA channels. The DAMA concept only uses transponder "space" when in use and all DAMA stations in the system can *directly* access each other. On low-usage routes and at earth stations with a low traffic volume, DAMA is attractive, as proven on low space-segment-usage costs. The DAMA employs the technique of single channel per carrier (SCPC), whereas TDMA and FDMA systems are multichannel. SCPC may use FM or phase-shift keying (PSK) modulation. Frequency-modulation systems operate on an analog basis, usually with preemphasis, threshold extension of the carrier, and syllabic companding. With PSK systems the voice is digitized in a PCM or Δ-modulation format. Modulation rates for PCM are 64 kbps per channel and with δ modulation, 32 or 40 kbps (see Chapter 9).

A typical distributed control DAMA system (SPADE) used by INTELSAT is shown in Figure 7.24. This system uses PCM on the voice circuit and four-phase modulation. A typical SCPC FM system is illustrated in Figure 7.25.

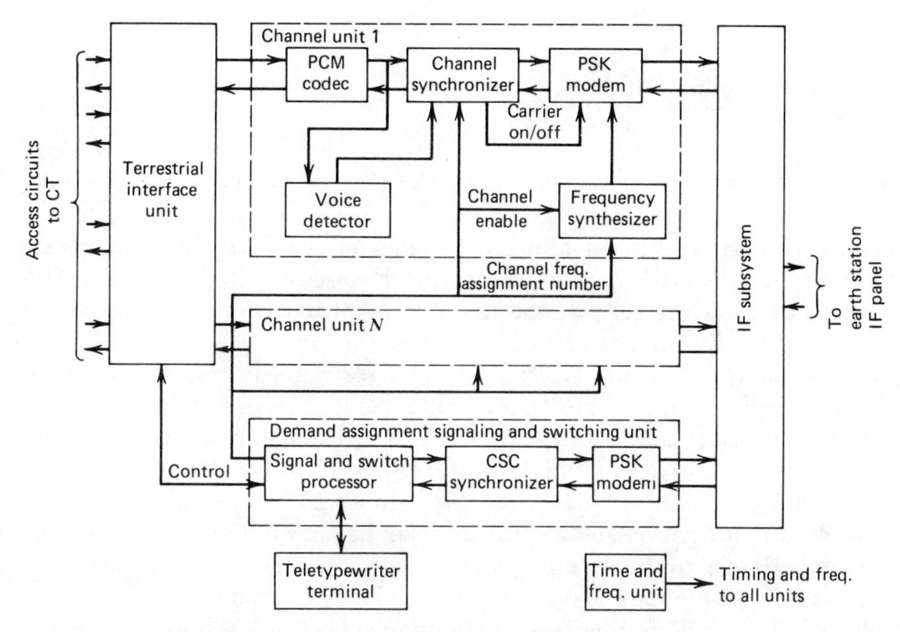

Figure 7.24 Block diagram of a SPADE terminal.

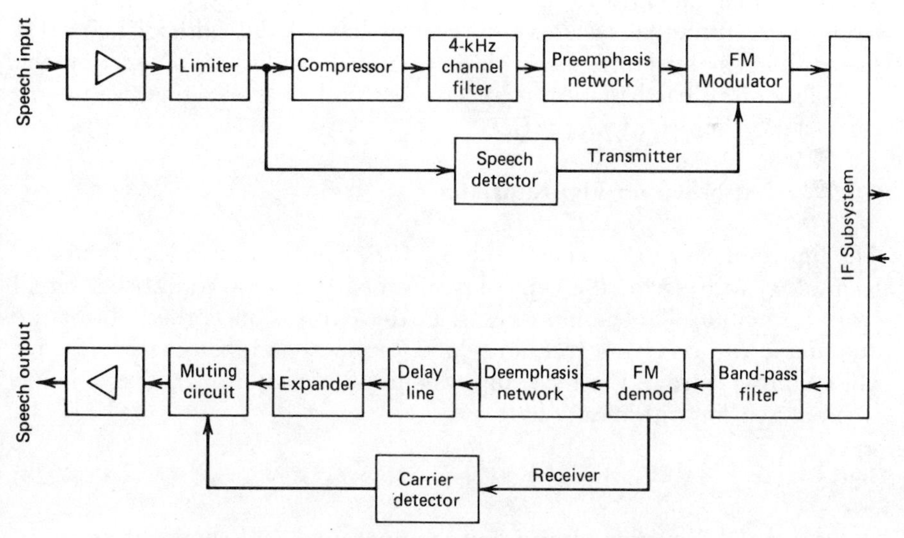

Figure 7.25 Block diagram of a typical SCPC FM station unit.

7.7 Earth-station Link Engineering

7.7.1 Introduction

Up to this point we have discussed basic earth-communication problems such as access and coverage. This section reviews some of the link-engineering problems associated with earth-station system engineering. The approach used to introduce the reader to essential path engineering expands on the basic principles previously discussed (Section 4) in this chapter on radio links. As we saw in Section 7.3, an earth station is a distant RF repeater. By international agreement, the repeater's EIRP is limited because nearly all bands are shared by terrestrial services. The limit for satellites transmitting in the 4-GHz band is -142 dBW/m^2 (flux density) in a 4-kHz bandwidth and -140 dBW/m^2 for the 11-GHz band. Given a free-space loss in the order of 196 dB in the 4-GHz band (satellite to earth) and about 207 dB for the 11-GHz band, how do we go about engineering such links? Assume that a satellite transponder has a +6-dBW power output, about 1 dB loss to the antenna, and 25 dB gain; thus the EIRP would be +30 dBW. With a path loss of 196 dB, we have a signal (+30 dBW − 196 dB) of -166 dBW impinging on an antenna. The trick is how to make this broad-band signal (36 MHz) with such a low signal level usable. Assume frequency modulation. A very large antenna is used at the earth station, with very small line losses and very low noise receiver front-end. In fact, on large intercontinental systems the receiver has a bandwidth of 500 MHz that amplifies the entire bandwidth of the down-link. To describe the capability of an earth station to receive such a low level signal, we use an earth-station figure of merit, G/T.

7.7.2 Earth-Station Figure of Merit

The figure of merit of an earth station, G/T, has been introduced into the technology to describe the capability of an earth station to *receive* a signal from a satellite. The numerator, G, in the expression is the gain of the antenna at the receiving frequency and T, the denominator, the effective noise temperature of the receiving system (often written T_e). This relation is expressed logarithmically as

$$\frac{G}{T} = G_{\text{dB}} - 10 \log T_e \tag{7.23}$$

In Section 4.4.3 we introduced noise temperature and showed how it can be related to the noise figure (dB) of a device. The kTB equation (7.6) is the basic equation for thermal noise. Because the earth station is thermal noise limited, we find it more useful to keep all receiving noise calculations in noise temperature units (°K) rather than the more familiar noise figure (dB). A "standard" INTELSAT earth station must display a G/T of 40.7 dB/°K. The antenna must also display a gain (G) of 57 dB at 4 GHz;

however, most standard earth stations have a 60-dB gain or greater. Substitute these numbers in the G/T equation, or

$$40.7 = 60 - 10 \log T_e$$

$$10 \log T_e = 60 - 40.7$$

$$10 \log T_e = 19.3 \text{ dB}$$

$$T_e = 81.5°\text{K}$$

Many system designers want a certain field margin and design to a G/T of 41.5 dB/°K; therefore:

$$10 \log T_e = 60 - 41.5$$

$$= 18.5 \text{ dB}$$

$$T_e = 70°\text{K}$$

Consider point-to-point (radio-link) microwave at 4 GHz and allow a 10-dB noise figure for a typical FM receiver operating in that band. The receiver noise temperature turns out to be approximately 2600°K. Noise temperature can be related to noise figure by the formula

$$F_{\text{dB}} = 10 \log \left(1 + \frac{T}{290}\right)$$

The T value that we have been using is more commonly referred to as *equivalent noise temperature* and is the summation of all noise components in the receiving system. This concept has been with us throughout this chapter (Section 4); we have simply shunted it aside because receiver noise temperatures were so large (e.g., the 2600°K) that the other noise components seemed insignificant. In earth station technology, the receiver noise temperature has been lowered to such an extent that other contributors now become important.

The noise components comprising T may be broken down into four categories:

- Antenna noise, usually taken at the 5° elevation angle of the antenna, which is the worst allowable case. This noise includes antenna "spillover," galactic noise, and atmospheric noise.
- Passive component noise, the summation of the equivalent noise from the passive components before the incoming signal reaches the first active component.
- High-power amplifier (HPA) leakage noise.
- Sum of the excess noise contributions of the various active amplifying stages of the receiving system.

At an angle of 5°, sky noise temperature reaches the order of 25°K. Minimum antenna noise occurs when the antenna is at zenith (i.e., an elevation angle of 90°). (*Note:* Elevation angles are referred to the horizon; thus elevation angle would be 0° when the antenna is pointed directly at the horizon.) "Antenna spillover" refers to radiated energy from the antenna to the ground and scatter off antenna spars. In both cases noise generators are formed and must be considered in a "noise budget." "Spars" refer here to those metal elements used to support the antenna feeding device. The sum total of antenna noise may reach 39 or 40°K, 25° of which is sky noise.

The next group of noise contributors to the total effective noise temperature (at 4 GHz) includes the feed, for which 10°K is a typical noise temperature; the directional coupler, 1.45°K; the waveguide switch, 0.58°K; and a short piece of flexible waveguide at 2.92°K. Noise temperature of passive, two-port devices (i.e., where G, the active gain, equals 1) in series may be added directly. The noise of these components is a function of their insertion loss; thus the effective noise temperature of all the passive components is

$$T_{ep} = T_{ant} + T_f + T_{dc} + T_{wg} \tag{7.24A}$$

where ant = antenna
f = feed
dc = directional coupler
wg = waveguide

$$= 39 + 10 + 1.45 + 2.92$$
$$= 53.37°K$$

Now let us connect this equation to several stages of a cooled parametric amplifier and a TWT (traveling-wave tube) or a similar driver. Then the effective noise temperature is

$$T_e = T_{ep} + T_{A_1} + \frac{T_{A_2}}{G_1} + \frac{T_{A_3}}{G_1G_2} + \frac{T_{twt}}{G_1G_2G_3}^{*} \tag{7.24B}$$

where T_{ep} = effective noise temperature of passive components and antenna
T_{A_1} = effective noise temperature of the first stage of the parametric amplifier
T_{A_2} = effective noise temperature of the second stage of the parametric amplifier
G_1, G_2, G_3 = gains of the active amplification stages 1, 2, 3, and so on

If we simplify this equation by considering the parametric amplifier as one amplifier, then

$$T_e = T_{ep} + T_a + \frac{T_{twt}}{G_a} \tag{7.24C}$$

*Valid when the contribution to T_{ep} from ohmic losses is comparatively low.

where T_a is the noise temperature of the parametric amplifier and G_a is the gain of the parametric amplifier equal to 40 dB or 10,000.

A good helium-cooled parametric amplifier should display an equivalent noise temperature of no more than 15 to 20°K at any frequency in the band 3700 to 4200 MHz. For this discussion, 18°K is used. The parametric amplifier is often followed by a TWT that drives longer lengths of transmission line connecting the antenna facility to the receiving radio equipment (i.e., the down-converters and demodulators). Typically, this device may have a 7-dB noise figure or a noise temperature of approximately 1165°K; thus

$$T_e = 53.37°\text{K} + 18°\text{K} + \frac{1165°\text{K}}{10{,}000}$$

$$= 53.37 + 18 + 0.11 = 71.48°\text{K}$$

Again, assuming G to be 60 dB,

$$\frac{G}{T} = 60 - 10 \log T_e$$

$$= 60 - 10 \log 71.48$$

$$= 60 - 18.54$$

$$= 41.46 \text{ dB/°K for this example.}$$

7.7.3 Station Margin

One major consideration in the design of radio-link and tropo systems is the fade margin, which is the additional signal level added in the system calculations to allow for fading. This value is often on the order of 20 to 50 dB. In other words, the receive signal level system in question was overbuilt to provide 20 and 50 dB above threshold to overcome most fading conditions or to ensure that noise would not exceed a certain norm for a fixed time frame.

As we saw in Section 4, fading is caused by anomalies in the intervening medium between stations or by the reflected signal, thus causing interference to the direct ray signal. There would be no fading phenomenon on a radio signal being transmitted through a vacuum. Thus satellite earth-station signals are subject to fade only during the time they traverse the atmosphere. For this case, most fades, if any, may be attributed to rainfall.

Margin or station margin is an additional design advantage that compensates for deteriorated propagation conditions or fading. The margin designed into an LOS system is large and is achieved by increasing antenna size, improving the receiver noise figure, or increasing transmitter output power. The station margin of a satellite earth station in comparison is small, on the order of 4 to 6 dB. Typical rainfall attenuation exceeding 0.01% of a year may be from 1 to 2 dB (in the 4-GHz band) without a radome on the antenna and when the antenna is at 5° elevation angle. As the antenna elevation increases to zenith, the rain attenuation notably decreases because the signal passes through less atmosphere. The addition of a radome could increase the attenuation to 6 dB or greater during precipitation. Receive station margin for an earth station, those extra decibels on the down-link, is sometimes achieved by use of threshold-extension demodulation techniques. Up-link margin is provided by using larger transmitters and by increasing power output when necessary. A G/T ratio in excess of 40.7 dB at the 5° elevation angle will also provide margin but may prove expensive to provide.

7.7.4 UP-Link Considerations

A typical specification for INTELSAT states that the EIRP per voice channel must be +61 dBW (example); thus to determine the EIRP for a specific number of voice channels to be transmitted on a carrier, we take the required output per voice channel in dBW (the above) and add logarithmically 10 log N, where N is the number of voice channels to be transmitted.

For example, consider the case for INTELSTAT III for an up-carrier transmitting 60 voice channels; thus

$$+61 \text{ dBW} + 10 \log 60 = 61 + 17.78 = +78.78 \text{ dBW}$$

If the nominal 100-ft (30-m) antenna has a gain of 63 dB (at 6 GHz) and losses typically of 3 dB, the transmitter output power, P_t required is

$$EIRP_{\text{dBW}} = P_t + G_{\text{ant}} - \text{line losses}_{\text{dB}}$$

where P_t is the output power of the transmitter (in dBW) and G_{ant} is the antenna gain (in dB) (up-link). Then in the example:

$$\begin{aligned} +78.78 \text{ dBW} &= P_t + 63 - 3 \\ P_t &= +18.78 \text{ dBW} \\ &= 75.6 \text{ W} \end{aligned}$$

7.8 Domestic Earth Stations

7.8.1 Introduction

"Standard" INTELSAT earth stations have installed costs well in excess of \$1 million. Nominal 30-m (100-ft) steerable antennas with monopulse tracking* make up much of the burden of money outlay. Cryogenically cooled low-noise amplifiers with total T_e of 75°K for the receiving subsystem looking at 500 MHz bandwidth is also costly. Transmitters with up to

*Step-tracking is proving effective and is more economic.

Table 7.6 INTELSAT Utilization Charges

Effective Date	Annual Charge
January 1965	$32,000
January 1966	$20,000
January 1971	$15,000
January 1972	$12,960
January 1973	$11,160
January 1974	$ 9,000
January 1975	$ 8,460

6-kW outputs make up another major segment of cost. The extra power is needed for simultaneous TV transmission with 5-MHz baseband.

Space-segment costs were initially high, indeed. Consider Table 7.6, which shows charges for ½ of a nominal 4-kHz telephone channel. This describes what a user pays for his end of a telephone connection. Two ends are required, thus the term "½" of a telephone circuit or voice channel. The costs given in Table 7.6 were based on standard earth-station usage. Substandard earth stations (i.e., those with a G/T of less than 40.7 dB) paid a hefty space-segment penalty because such smaller earth stations required full satellite power in the down-link, thus often requiring cutback in "standard" users (refer to our discussion of TWTA back-off).

Single-channel per carrier DAMA systems are charged only for time utilization, whereas FDMA operations require full-period payment of a circuit, regardless of whether it is used. Compare the charges shown in Table 7.6 (1975) with those of SPADE 1976, which were $0.10 per minute of satellite circuit holding time applicable on the ½ circuit basis. Even with space-segment penalties, SCPC DAMA operations began to be feasible with utilization of smaller earth stations. Further, if earth stations looked at smaller bandwidths, at only one transponder, (36 MHz) for example, the received signal-noise-ratio met CCIR requirements and penalties were not invoked. Satellite power did not require increase.

As TWTA technology advanced, IM products decreased at full power on multi-RF channel usage. "Frequency reuse" put more available channels on the satellite, thus amortizing space-segment costs over more users. For instance, there were only 1049 leased ½ circuits in the INTELSAT system in 1967 and nearly 13,000 in 1975. INTELSAT IV-A had a 6000-VF channel capacity whereas INTELSAT V has over 12,000 full telephone-circuit capacity per satellite. Military developments were equally as advanced and showed that very small earth stations with less than 1-m dishes could support several voice channels if they looked at small bandwidths (e.g., <100 kHz). Satellite station-keeping improved, and with smaller antennas—under 10 to 15 m in diameter with their inherent wider beamwidths—it was possible to remove automatic tracking features. Small earth stations were now feasible, with installed costs of less than $400,000. The scene was set for domestic satellite service.

7.8.2 Application of DOMSAT

Small earth stations do not need the heavy traffic-load revenues to amortize investment (e.g. the investment is much smaller). Stations with busy-hour (BH) loads of under 20 erlangs or a single video channel or a combination of video, data, and voice traffic become attractive for such uses as:

1. Industrial private networks.
2. National and regional systems over fairly wide geographic areas.
3. Serving remote or "bush" areas, particularly in harsh environments.
4. Certain mobile applications such as ships at sea.
5. Video and other broadcast applications.

Industrial usage is most applicable in free-enterprise situations where the only measure of feasibility is to prove-in economically. This is so in countries where corporations may set up private networks with minimal governmental control. If telecommunications, in all its ramifications, is government owned and operated—not for convenience, but for political control—such DOMSAT facilities are not likely to be accepted or considered. On the contrary, where such control is absolute application 3 in the preceding list may get added impetus, so said governments can spread their "good word" to their people. Another use of application 3 is the educational benefits. India, Brazil, and Indonesia are combining applications 2 and 3, as is northern Canada and Alaska.

7.8.3 Sample Applications

A good example of industrial application (item 1 in the preceding list) is the joint venture of (USA) COMSAT, IBM, and a large insurance company, Aetna Casualty and Surety Company—the SBS (Satellite Business Systems). This system will cover the continental United States with two spot beams. One spot beam is superimposed on the other; the narrower one will cover the east central states where rainfall will affect path reliability more, and the wider one will cover nearly all the contiguous states. The narrow beam has an EIRP of +42 dBW and the wide beam, +38 dBW. The satellite will keep station with a total excursion of 0.07°, as seen from an optimally pointed 7-m dish. At maximum satellite excursion, up-link additional loss will be about 1.2 dB and for that critical down-link, 0.8 dB. Operation for the system is in the 12 and 14 GHz bands (see Section 7.5); thus rainfall is a major consideration. The system availability goal is 99.5%. The down-link margin is 5 dB. Margin, as discussed in Section 7.7.2, is the additional gain added to the system to assure the specified reliability–availability.

Earth stations will have 5-m antennas in the region of coverage for the narrow beam and 7-m antennas for the remainder of the coverage area. Autotracking will not be required. The estimated system noise temperature (T_e) is 225°K; G/T for the 5-m antenna is 30.4 dB/°K and for the 7-m antenna, 33.3 dB/°K. Some of the rain margin in the narrow beam region is made up from decreased losses due to satellite excursion. This is because smaller-diameter parabolic antennas (5 m) have wider beamwidths. The two different antenna configurations have beamwidths as follows:

	Receive	Transmit
5 m	0.37°	0.31°
7 m	0.27°	0.22°

Full redundancy in the satellite will help ensure the 99.95% total-availability goal figures. With a bit error rate (Chapter 8) of 1×10^{-4}, the availability is 99.5%. The operational mode will be TDMA and demand access, with burst transmissions with bit rates as high as 50 Mbits. Although we indicated previously that satellite communications was down-link limited, the SBS system is partially up-link limited. A greater noise contribution is allowed the up-link so that earth-station transmitter size (and resulting cost) may be reduced. Maximum transmitter output for any earth station is 4 kW. In other systems the transmitter available power is sufficient for the up-link to have little effect as a noise contributor when compared to the down-link. Individual transmitter output will be a function of the bit rate to be transmitted. A 38-Mbps burst will require about 300 W; 45 Mbps, 1 kW; and 50 Mbps, about 4 kW.

Considering that SBS has ventured into the Ku band (11 to 12 GHz down-link), a down-link analysis would be revealing. The signal level impinging on the antennas will be:

SAT EIRP—Free-space Loss (12 GHz)

5 m	+42 dBW − 206 dB = −164 dBW
7 m	+38 dBW − 206 dB = −168 dBW

Signal Level at Receiver Front End

	Level	Gain (dB)
5 m	−164 dBW	+ 52.5 = −111.5 dBW
7 m	−168 dBW	+ 56.5 = −111.5 dBW

With a system noise temperature of 225°K and an IF of approximately 60 MHz, the bandwidth noise threshold is N_{th}:

$$\begin{aligned} N_{th} &= -228.6 \text{ dBW} + 10 \log K + 10 \log B_{if} \\ &= -228.6 \text{ dBW} + 10 \log 225 + 10 \log 60 \times 10^6 \\ &= -127.18 \text{ dBW} \end{aligned}$$

$$\begin{aligned} C/N &= 111.5 - (-127.18) \\ &= 15.68 \text{ dB} \end{aligned}$$

where 15.68 dB for C/N (carrier-to-noise ratio) provides a good margin to achieve the minimum acceptable bit-error rate (BER) for digital modulation. This minimum BER is specified as 1×10^{-4} 99.6% of the time (per year). With the use of forward error correction (FEC) data transmitted over an SBS link will have a BER better than of 1×10^{-7}. Modulation has been assumed as QPSK, or quaternary phase shift keying. (*Note:* Digital transmission (PCM) is discussed in Chapter 9, data transmission in Chapter 8, and data network operation in Chapter 10.)

REFERENCES

1. International Telephone and Telegraph Corporation, *Reference Data for Radio Engineers,* 6th ed., Howard W. Sams, Indianapolis, 1976.
2. *Overall Communication Systems Planning,* Vol. 3, IEEE North Jersey Section, 1964.
3. Roger L. Freeman, *Telecommunication Transmission Handbook,* Wiley, New York, 1975.
4. *Transmission Systems for Communications,* 4th ed., Bell Telephone Laboratories, Holmdel, N.J., 1973.
5. *Engineering Considerations for Microwave Communication Systems,* GTE Lenkurt, San Carlos, Calif., 1975.
6. *Jerrold Path Calculations,* Jerrold Electronics Corporation, Philadelphia, 1967.
7. Philip F. Panter, *Communication System Design—Line-of-Sight and Tropo-Scatter Systems,* McGraw-Hill, New York, 1972.
8. J. Jasik, *Antenna Engineering Handbook,* McGraw-Hill, New York, 1961.
9. K. Bullington, "Radio Propagation Fundamentals", *Bell Syst. Tech. J.* (May 1957).
10. *A Survey of Microwave Fading Mechanisms, Remedies and Applications,* ESSA Technical Report ERL-69-WPL4 Boulder, Colo., March 1968.
11. R. G. Medhurst, "Rainfall Attenuation of Centimeter Waves: Comparison of Theory and Measurement," *IEEE Transact. Antennas Propag.*, July 1965.
12. *Technical Note 100,* National Bureau of Standards, January 1962.
13. M. J. Tant, *Multichannel Communication Systems and White Noise Testing,* Marconi Instruments, St. Albans, Herts (UK), 1974.
14. P. J. Howard, M. F. Alarcon, and S. Tronsli, "12-Megahertz Line Equipment," *Electr. Commun.,* **48,** (1973).
15. William A. Rheinfelder, *CATV System Engineering,* TAB Books, Blue Ridge Summit, Pa., 1970.
16. CCITT, *Orange Books,* Fifth Plenary Assembly, Geneva, 1976, particularly Vol. III.
17. CCIR, *Green Books,* Geneva, 1974.

18. J. A. Lawlor, *Coaxial Cable Communication Systems, Management Overview,* ITT, New York, February 1972 (technical memorandum).
19. F. J. Herr, "The L5 Coaxial System—Transmission System Analysis," *IEEE Trans. Commun.*, (February 1974).
20. L. Becker, "60-Megahertz Line Equipment," *Electr. Commun.,* **48** (1973).
21. *Satellite Communication Reference Data Handbook,* Defense Communication Agency (NTIS), Washington, D. C., July 1972.
22. "Space Communications" (course notes), Dr. J. Neal Birch, Marbella, Spain, 1975.
23. B. Edelson, "Cost Effectiveness in Global Satellite Communications," *IEEE Commun. Soc. Mag.* (January 1977).
24. R. B. Marsten, "Service Needs and Systems Architecture in Satellite Communications," *IEEE Commun. Soc. Mag.* (May 1977).
25. *Communication Satellite Systems Worldwide,* Horizon House, Dedham, Mass., 1975.
26. *Small Earth Stations 1976–1986,* ComQuest Corporation, Palo Alto, Calif., 1977.
27. *Edited Lectures, US Seminar on Communication Satellite Earth Station Technology,* ComSat, Washington, D. C., 1966.
28. *INTELSAT V Satellite Specification, Exhibit A,* ComSat, Washington, D. C., BG-23-10E W/9/76, 1976.
29. Roger L. Freeman, "An Approach to Earth Station Technology", *Telecommun. J. (ITU),* (June 1971).
30. R. G. Gould and Y. F. Lum, *Communication Satellite Systems: An Overview of the Technology,* IEEE Press, New York, 1976.
31. *Determining Site Coordinates,* GTE Lenkurt Demodulator, San Carlos, Calif., January–February 1978.
32. Madhu S. Gupta, Ed., *Electrical Noise: Fundamentals & Sources,* IEEE Press, New York, 1977.

Chapter 8

THE TRANSMISSION OF OTHER INFORMATION OVER THE TELEPHONE NETWORK

1 INTRODUCTION

Up to this point we have discussed voice communications over telephone systems. The word "telephone" implies sound, particularly "articulate speech." The existing telephone network covers the entire world. It is gigantic, with an annual investment in a new plant in the many thousands of million dollars. This worldwide network, in part or in whole, can be used to transmit information other than speech. Such information includes data, graphics such as written messages, and video. The bottleneck in the telephone network would appear to be the limitations of a wire pair. There are wire pairs connecting subscribers to switches; the conventional analog switch is built around wire pairs, and wire pairs make up the greater portion of local trunk connections. We find that even with the limitations of bandwidth and envelope delay, wire pairs may be used as is for the transmission of data, telegraph, and facsimile and may even be used over short runs for video. To improve transmission characteristics, wire pairs can be conditioned for wider bandwidths and improved delay characteristics. The real bottleneck that restricts transmission capabilities and places a firm top on data rates or facsimile speed in the conventional telephone network is carrier [frequency-division multiplex (FDM)] equipment.

The objective of this chapter is to introduce the principles of data, telegraph, and facsimile over the telephone network. No distinction is made between data and telegraph communication from the transmission viewpoint. Both are binary, and often the codes used for one serve equally well for the other. The transmission problems of data apply equally to telegraph communication, which is usually in message format, transmitted at rates less than 110 words per minute (wpm), and asynchronous. Data transmission is most often synchronous (but not necessarily so) and usually is an alphanumeric mix of information primarily destined for computers or other EDP (electronic data processing). Data are often transmitted at rates in excess of 110 wpm.

Besides data, there are other forms of information that can be transmitted over the telephone network, such as telemetry, video, and facsimile. Because of its forecast rapid growth and our references to it in Chapter 11, we elected to provide a short description of current facsimile technology at the end of this chapter.

2 ANALOG–DIGITAL

Up to this point the text has dealt essentially with analog signals. An analog transmission system has an output at the far end that is a continuously variable quantity representative of the input. With analog transmission, the signal containing the information is continuous; with digital transmission, the signal is discrete. The simplest form of digital transmission is binary, where an information element is assigned one of two states. There are many binary situations in real life where only one of two possible values can exist; for example, a light may be either on or off, an engine is running or not, and a person alive or dead.

An entire number system has been based on two values, which by convention have been assigned the symbols 1 and 0. This is the binary system, and its number base is 2. Our everyday number system has a base of 10 and is called the *decimal system*. Another system has a base of 8 and is called the *octal system*, and still another is the *hexadecimal* representation, with a number base of 16. The basic information element of the binary system is called the *bit*, which is an acronym for "binary digit." The bit may have the value of either 1 or 0. A number of discrete bits can be encoded to identify a larger piece of information, which we may call a *character*. A code is defined by the IEEE as "a plan for representing each of a finite number of values or symbols as a particular arrangement or sequence of discrete conditions or events."

Binary coding of written information has been in existence for a long time. An example is teleprinter service (i.e., that used in the transmission of a telegram). The majority of computers now in operation operate in binary languages; thus binary transmission fits in well with computer communication, whether terminal-to-computer or computer-to-computer. The facilities used to effect this communications make up the elements of a data transmission system (see Chapter 10).

3 THE BIT–BINARY CONVENTION

In a binary transmission system the smallest unit of information is the bit, which we know to be either one of two states. We call one state a *mark* and

Table 8.1 Equivalent Binary Designations

Active Condition	Passive Condition
Mark or marking	Space or spacing
Current on	Current off
Positive voltage	Negative voltage
Hole (in paper tape)	No hole (in paper tape)
Binary "1"	Binary "0"
Condition Z	Condition A
Tone on (amplitude modulation)	Tone off
Low frequency (frequency shift keying)	High frequency
No phase inversion (differential phase shift keying)	Inversion of phase
Reference phase	Opposite to reference phase

Source: CCITT Rec. V.1.

the other, a *space*. These states may be indicated electrically by the presence or absence of current flow. Unless some rules are established, an ambiguous situation would exist. Is the "1" condition a mark or a space? Does the "no-current" condition mean transmission of a "0" or a "1"? To avoid confusion and to establish a unique identity to binary conditions, CCITT Rec. V.1 recommends equivalent binary designations. These are shown in Table 8.1. If this table is adhered to universally, no confusion should exist as to which is a mark, which is a space, which is the active condition, which is the passive condition, which is "1," and which is "0." Table 8.1 defines the sense of transmission so that the mark and space, the "1" and "0," respectively, will not be inverted. Data-transmission engineers often refer to such a table as a "table of mark–space convention."

4 CODING

4.1 Introduction to Binary-Coding Techniques

Written information must be coded before it can be transmitted over a digital system. The discussion that follows covers binary codes only. Not only does this simplify our argument, but certainly binary codes are by far the most widely used codes in data and telegraph networks.

One of the first questions that arises is in regard to the size or extent of a binary code. The answer involves yet another question, specifically, how much information is to be transmitted. One binary digit (bit) carries little information; it has only two possibilities. If two binary digits are transmitted in sequence, there are four possibilities:

00	10
01	11

or four pieces of information. Suppose 3 bits are transmitted in sequence. Now there are eight possibilities:

000	100
001	101
010	110
011	111

We can now see that for a binary code, the number of distinct information characters available is equal to two raised to a power equal to the number of elements or bits per character. For instance, the last example was based on a three-element code giving eight possibilities or information characters, or 2^3.

Another more practical example is the Baudot teleprinter code, which has 5 bits or information elements per character. Therefore, the number of different graphics* and characters available are $2^5 = 32$. The American Standard Code for Information Interchange (ASCII) has seven information elements per character, or $2^7 = 128$; thus it has 128 distinct combinations of marks and spaces that are available for assignment as characters or graphics.

The number of distinct characters for a specific code may be extended by establishing a bit sequence (a special character assignment) to shift the system or machine to uppercase (as is done with a conventional typewriter). Uppercase is a new character grouping. A second distinct bit sequence is then assigned to revert to lowercase. For example, the CCITT ITA No. 2 code (Figure 8.1) is a five-unit code with 58 letters, numbers, graphics, and operator sequences. The additional characters and graphics (additional above $2^5 = 32$) originate from the use of uppercase. Operator sequences appear on a keyboard as "space" (spacing bar), "figures" (uppercase), "letters" (lowercase), "carriage return," "line feed" (spacing vertically), and so on. When we refer to a five-unit, six-unit, or 12-unit code, we refer to the number of information units or elements that make up a single character or symbol. That is, we refer to those elements assigned to each character that carry information and that distinguish from all other characters or symbols of the code.

*In this context a graphic is a symbol, other than a letter or number. Typical graphics are asterisks, punctuation, parentheses, dollar signs, and so forth.

Characters				Code Elements[a]						
Letters Case	Communications	Weather	CCITT #2[b]	START	1	2	3	4	5	STOP
A	-	↑			X	X				X
B	?	⊕			X			X	X	X
C	:	○				X	X	X		X
D	$	↗	WRU		X			X		X
E	3	3			X					X
F	1	→	Unassigned		X		X	X		X
G	&	↘	Unassigned			X		X	X	X
H	STOP[c]	↓	Unassigned				X		X	X
I	8	8				X	X			X
J	'	↙	Audible signal		X	X		X		X
K	(	←			X	X	X	X		X
L	)	↖				X			X	X
M	.	.					X	X	X	X
N	,	⦶					X	X		X
O	9	9						X	X	X
P	θ	θ				X	X		X	X
Q	1	1			X	X	X		X	X
R	4	4				X		X		X
S	BELL	BELL	'		X		X			X
T	5	5							X	X
U	7	7			X	X	X			X
V	;	⦶	=			X	X	X	X	X
W	2	2			X	X			X	X
X	/	/			X		X	X	X	X
Y	6	6			X		X		X	X
Z	"	+	+		X				X	X
BLANK		-								X
SPACE							X			X
CAR. RET.								X		X
LINE FEED						X				X
FIGURE					X	X		X	X	X
LETTERS					X	X	X	X	X	X

[a] Blank, spacing element; crosshatched, marking element.
[b] This column shows only those characters that differ from the American "communications" version
[c] Figures case H(COMM) may be STOP or +.

Figure 8.1 Communication and weather codes, CCITT International Alphabet No. 2 (ITA2).

4.2 Hexadecimal Representation and BCD Code

The hexadecimal system is a numeric representation in the number base 16. This number base uses "0" through "9" as in the decimal number base, and the letters "A" through "F" to represent the decimal numbers 10 through 15. The "hex" numbers can be translated to the binary base as follows:

Hex	Binary	Hex	Binary
0	0000	8	1000
1	0001	9	1001
2	0010	A	1010
3	0011	B	1011
4	0100	C	1100
5	0101	D	1101
6	0110	E	1110
7	0111	F	1111

Two examples of the hexadecimal notation are:

Number Base 10	Number Base 16
21	15
64	40

The BCD (binary-coded decimal) is a compromise code assigning 4-bit binary numbers to the digits between 0 and 9. The BCD equivalents to decimal digits appear as follows:

Decimal Digit	BCD Digit	Decimal Digit	BCD Digit
0	1010	5	0101
1	0001	6	0110
2	0010	7	0111
3	0011	8	1000
4	0100	9	1001

To cite some examples, consider the number 16; it is broken down into 1 and 6. Thus its BCD equivalent is 0001 0110. If it were written in straight binary notation, it would appear as 10000. The number 25 in BCD combines the digits 2 and 5 above as 0010 0101.

4.3 Some Specific Binary Codes for Information Interchange

In addition to the ITA No. 2 code (Figure 8.1), some of the more commonly used binary codes are the American Standard Code for Information Interchange (ASCII) (Figure 8.3), the CCITT No. 5 (Figure 8.4) Code, the EBCDIC (Figure 8.5), and the Hollerith Code (Figure 8.6). The ASCII, CCITT No. 5, and EBCDIC codes are referred to as "eight-level" code sets because a character is expressed with 8 binary digits or bits. The levels are numbered from 1 to 8, starting from the right or least significant binary digit.

The term "level" originates from perforated-paper-tape technology and implies a channel on the tape. On paper tape a "1" is expressed with a punched hole (Table 8.1), and a "0" is expressed with no hole punched in the appropriate channel. With magnetic tape, "1"s and "0"s are represented by changes in magnetic flux. An example of punched paper tape is shown in Figure 8.2. The small holes offset just to the right of tape center are the

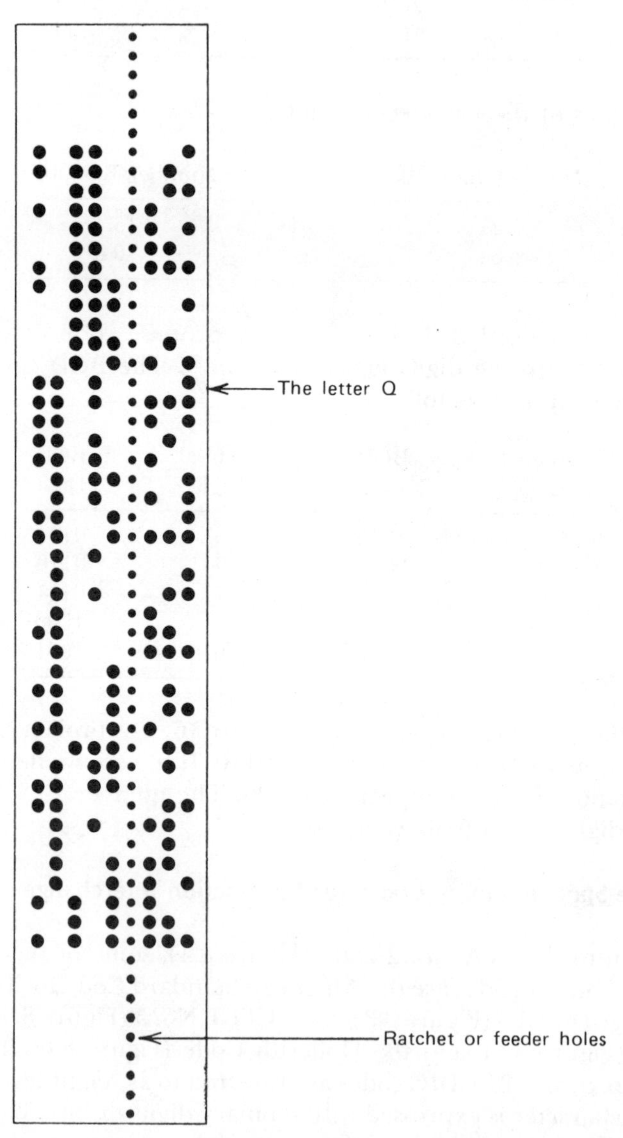

Figure 8.2 Perforated paper tape with the ASCII code. (*Note:* Even parity; parity bits are in far-left column, characters are top to bottom: 1234567890: QWERTYUIOPASD-FGHJKL;ZXCVBNM, etc.)

feed sprocket holes. The ASCII, CCITT No. 5 and EBCDIC are stated as eight-level codes. However, only EBCDIC* is a true eight-level code; ASCII and CCITT No. 5 codes are actually 7-bit codes with an extra bit added to each character for parity. Here we mean that the 7 bits (levels) are used for information, whereas EBCDIC uses all 8 bits for information.

Parity or parity checks provide a means for determining whether a character contains an error after transmission. We speak of "even parity" and "odd parity." On a system using an odd-parity check, the total count of "1"s or marks has to be an odd number per character (e.g., it carries 1, 3, 5, or 7 marks or "1"s). To explain parity and parity checks a little more clearly, let us look at some examples. Consider a seven-level code with an extra parity bit. By system convention, even parity has been established; suppose a character is transmitted as 1111111. There are 7 marks, so to maintain even parity we would need an even number of marks. Thus an eighth bit is added, which must be a mark ("1"). Now consider another bit pattern, 1011111. Here there are 6 marks, even; thus the eighth (parity) bit must be a space. Still another example would be 0001000. To obtain even parity, a mark must be added on transmission and the character transmitted would be 00010001, maintaining even parity. Suppose that, as a result of some sort of signal interference, one signal element was changed on reception. No matter which element was changed, the receiver would indicate an error because we would no longer have even parity. If two elements were changed, however, the error could be masked. This would happen in the case of even or odd parity if 2 marks were substituted for two spaces or vice versa at any element location in the character.

The American Standard Code for Information Interchange (see Figure 8.3), better known as the ASCII, is the latest effort on the part of the North American industry and common-carrier systems, backed by the U.S. Standards Institute (now the American National Standards Institute), to produce a universal common language code. The ASCII is a seven-unit code with all 128 combinations available for assignment. Here again, the 128 bit patterns are divided into two groups of 64. One group is assigned to a subset of graphic printing characters. The second subset of 64 is assigned to control characters. An eighth bit is added to each character for parity check. The ASCII is widely used in North America and has received considerable acceptance in Europe and Latin America.

A seven-level code is recommended in CCITT Rec. V.3 as an international standard for information interchange. It is not intended as a substitute for the CCITT No. 2 code. The CCITT No. 5, or the new alphabet No. 5, as the seven-level code is more commonly referred to, is basically intended for data transmission. Although the CCITT No. 5 is considered a seven-level code, CCITT Rec. V.4 advises that an eighth bit may be added for parity. Under certain circumstances odd parity is recommended; otherwise, even parity.

*EBCDIC—Extended Binary Coded Decimal Interchange Code.

Bit Number									
b_7 →		0	0	0	0	1	1	1	1
b_6 →		0	0	1	1	0	0	1	1
b_5 →		0	1	0	1	0	1	0	1
$b_4 b_3 b_2 b_1$	Row \ Column	0	1	2	3	4	5	6	7
0 0 0 0	0	NUL	DLE	SP	∅	‘	P	@	p
0 0 0 1	1	SOH	DC1	!	1	A	Q	a	q
0 0 1 0	2	STX	DC2	”	2	B	R	b	r
0 0 1 1	3	ETX	DC3	#	3	C	S	c	s
0 1 0 0	4	EOT	DC4	$	4	D	T	d	t
0 1 0 1	5	ENQ	NAK	%	5	E	U	e	u
0 1 1 0	6	ACK	SYN	&	6	F	V	f	v
0 1 1 1	7	BEL	ETB	’	7	G	W	g	w
1 0 0 0	8	BS	CAN	(	8	H	X	h	x
1 0 0 1	9	HT	EM	)	9	I	Y	i	y
1 0 1 0	10	LF	SS	*	:	J	Z	j	z
1 0 1 1	11	VT	ESC	+	;	K	[	k	{
1 1 0 0	12	FF	FS	,	<	L	~	l	¬
1 1 0 1	13	CR	GS	–	=	M	]	m	}
1 1 1 0	14	SO	RS	.	>	N	^	n	|
1 1 1 1	15	SI	US	/	?	O	-	o	DEL

Figure 8.3 American Standard Code for Information Interchange (ASCII). Columns 2, 3, 4, and 5 indicate printable characters in DCS Autodin (U.S. Defense Communications System Automatic Digital Network); columns 6 and 7 fold over into columns 4 and 5, respectively, except DEL.

Figure 8.4 shows the CCITT No. 5 code, where b_1 is the first signal element in serial transmission, and b_7 is the last element of a character. Like the ASCII code, the CCITT No. 5 does not normally need to shift out (i.e., uppercase or lowercase, as in CCITT No. 2). However, like ASCII, it is provided with an escape,* 1101100. Few differences exist between the ASCII and CCITT No. 5 codes.

The ASCII and CCITT No. 5 codes are known as "computable codes," where the letters of the alphabet and all other characters and graphics are assigned values in continuous binary sequence; thus these codes are in the native binary language of today's common digital computers. The CCITT No. 2 (ITA No. 2) is not a computable code and when used with a computer, often requires special processing.

The EBCDIC (extended binary-coded decimal-interchange code) is similar to the ASCII but is a true 8-bit code. The eighth bit is used as an added bit to "extend" the code, providing 256 distinct code combinations for assignment. Figure 8.5 illustrates the EBCDIC code.

The Hollerith Code was specifically designed for use with perforated (punched) cards. It has attained wide acceptance in the business machine and computer fields. Hollerith is a 12-unit character code wherein a character is represented on a card by one or more holes perforated in one column having 12 potential hole positions. It is most commonly used with the standard 80-column punched cards. The theoretical capacity of a 12-unit binary code is very great and by our definition, using all hole patterns available, is 2^{12} or 4096 bit combinations. In the modern version of the Hollerith code only 64 of these, with none using more than three holes, are assigned to graphic characters. Because of its unwieldiness, the Hollerith code is seldom used directly for transmission. Most often it is converted to one of the more conventional transmission codes such as ASCII, CCITT No. 5, the BCD interchange code, or EBCDIC. Figure 8.6 shows the Hollerith Code and its BCD interchange equivalents.

5 ERROR DETECTION AND ERROR CORRECTION

5.1 Introduction

In data transmission one of the most important design goals is to minimize the error rate. Error rate may be defined as the ratio of the number of bits incorrectly received to the total number of bits transmitted. On many data circuits the design objective is an error rate better than one error in 1×10^5 (often expressed 1×10^{-5}), and for telegraph circuits, one error in 1×10^4.

*An "escape" is a code sequence indicating that the succeeding sequences have an interpretation that differs from the conventional meanings of the code in use. In other words, an escape has been made from the code.

Bit Number b_7		0	0	0	0	1	1	1	1
b_6		0	0	1	1	0	0	1	1
b_5		0	1	0	1	0	1	0	1
$b_4\,b_3\,b_2\,b_1$	Row \ Column	0	1	2	3	4	5	6	7
0 0 0 0	0	NUL	(TC$_r$)DLE	SP	0	(@)	P	‘	p
0 0 0 1	1	(TC$_1$)SOH	DC$_1$	!	1	A	Q	a	q
0 0 1 0	2	(TC$_2$)STX	DC$_2$	”	2	B	R	b	r
0 0 1 1	3	(TC$_3$)ETX	DC$_3$	£	3	C	S	c	s
0 1 0 0	4	(TC$_4$)EOT	DC$_4$	$	4	D	T	d	t
0 1 0 1	5	(TC$_5$)ENQ	(TC$_8$)NAK	%	5	E	U	e	u
0 1 1 0	6	(TC$_6$)ACK	(TC$_9$)SYN	&	6	F	V	f	v
0 1 1 1	7	BEL	(TC$_{10}$)ETB	’	7	G	W	g	w
1 0 0 0	8	FE$_0$(BS)	CAN	(	8	H	X	h	x
1 0 0 1	9	FE$_1$(HT)	EM	)	9	I	Y	i	y
1 0 1 0	10	FE$_2$(LF)	SUB	*	:	J	Z	j	z
1 0 1 1	11	FE$_3$(VT)	ESC	+	;	K	([)	k	•
1 1 0 0	12	FE$_4$(FF)	IS$_4$(FS)	,	<	L	•	l	•
1 1 0 1	13	FE$_5$(CR)	IS$_3$(GS)	–	=	M	(])	m	•
1 1 ! 0	14	SO	IS$_2$(RS)	.	>	N	^	n	‾
1 1 1 1	15	SI	IS$_1$(US)	/	?	O	—	o	DEL

Figure 8.4 CCITT No. 5 Code for Information Interchange (CCITT Rec. V.3).

One method for minimizing the error rate would be to provide a "perfect" transmission channel, one that will introduce no errors in the transmitted information at the output of the receiver. However, that perfect channel can never be achieved. Besides improvement of the channel transmission parameters themselves, error rate can be reduced by forms of systematic redundancy. In old-time Morse code, words on a bad circuit were often sent twice; this is redundancy in its simplest form. Of course, it took twice as long to send a message; this is not very economical if the number of useful words per minute received is compared to channel occupancy.

This illustrates the trade-off between redundancy and channel efficiency. Redundancy can be increased such that the error rate could approach zero. Meanwhile, the information transfer or efficiency across the channel would also approach zero. Thus unsystematic redundancy is wasteful and merely lowers the rate of useful communication. On the other hand, maximum efficiency could be obtained in a digital transmission system if all redundancy and other code elements, such as "start" and "stop" elements, parity bits, and other "overhead" bits were removed from the transmitted bit stream. In other words, the channel would be 100% efficient if all bits transmitted were information bits. Obviously, there is a trade-off of cost–benefits somewhere between maximum efficiency on a data circuit and systematically added redundancy (see Chapter 10, Sections 6 and 7).

5.2 Throughput

Throughput of a data channel is the expression of how much data are put through. In other words, throughput is an expression of channel efficiency. The term gives a measure of *useful* data put through the communication link. These data are directly useful to the computer or DTE (data-terminal equipment).

Therefore, on a specific circuit throughput varies with the raw-data rate; is related to the error rate and the type of error encountered (whether burst or random); and varies according to the type of error detection and correction system used, the message-handling time, and the block length from which we must subtract overhead bits such as parity, flags, cyclic redundancy checks, and so forth. Throughput and the operational features of data circuits are described in detail in Chapter 10.

5.3 The Nature of Errors

In data transmission an error is a bit that is incorrectly received; for instance, "1" is transmitted in a particular time slot and the element received in that slot is interpreted as a "0." Bit errors occur either as single random errors or bursts of error. For instance, lightning or other forms of

Bit Levels 4567 ↓	0→ 0 1→ 0 2→ 0 3→ 0	0 0 0 1	0 0 1 0	0 0 1 1	0 1 0 0	0 1 0 1	0 1 1 0	0 1 1 1
0000	NULL		DS		SP	&	−(Minus)	
0001			SOS				/(Slash)	
0011			FS					
0100	PF	RES	BYP	PN				
0101	HT	NL	LF	RS				
0110	LC	BS	EOB	US				
0111	DEL	IL	PRE	EOT				
1000								
1001								
1010			SM			!(Exclam. pt.)		:(Colon)
1011					.(Period)	&	,(Comma)	#
1100					<	*	%	
1101						)	−(Undesc.)	'(Prime)
1110					(	;(Semicolon)	>	=(Equals)
1111					(Vert. Bar)	(not)	?	"(Quote)

Figure 8.5 Extended binary coded decimal interchange code (EBCDIC).

Bit Levels 4567 ↓	0 ⟶ 1	1	1	1	1	1	1	1
	1 ⟶ 0	0	0	1	1	1	1	1
	2 ⟶ 0	0	1	0	0	0	1	1
	3 ⟶ 0	1	0	1	0	1	0	1
0000					PZ	MZ		0
0001	a	j			A	J		1
0010	b	k	s		B	K	S	2
0011	c	l	t		C	L	T	3
0100	d	m	u		D	M	U	4
0101	e	n	v		E	N	V	5
0110	f	o	w		F	O	W	6
0111	g	p	x		G	P	X	7
1000	h	q	y		H	Q	Y	8
1001	i	'	z		I	R	Z	9
1010								
1011								
1100								
1101								
1110								
1111								

Figure 8.5 Continued

BCD Code		Hollerith Code	BCD Code		Hollerith Code
Bit Nos: B A 8 4 2 1	Graphics	Rows Punched	Bit Nos: B A 8 4 2 1	Graphics	Rows Punched
0 0 0 0 0 0	Blank	No holes	1 0 0 0 0 0	—	11
0 0 0 0 0 1	1	1	1 0 0 0 0 1	J	11–1
0 0 0 0 1 0	2	2	1 0 0 0 1 0	K	11–2
0 0 0 0 1 1	3	3	1 0 0 0 1 1	L	11–3
0 0 0 1 0 0	4	4	1 0 0 1 0 0	M	11–4
0 0 0 1 0 1	5	5	1 0 0 1 0 1	N	11–5
0 0 0 1 1 0	6	6	1 0 0 1 1 0	O	11–6
0 0 0 1 1 1	7	7	1 0 0 1 1 1	P	11–7
0 0 1 0 0 0	8	8	1 0 1 0 0 0	Q	11–8
0 0 1 0 0 1	9	9	1 0 1 0 0 1	R	11–9
0 0 1 0 1 0	0	0	1 0 1 0 1 0	!	11–0
0 0 1 0 1 1	① # or =	3–8	1 0 1 0 1 1	$	11–3–8
0 0 1 1 0 0	① @ or '	4–8	1 0 1 1 0 0	*	11–4–8
0 0 1 1 0 1	:	5–8	1 0 1 1 0 1	]	11–5–8
0 0 1 1 1 0	>	6–8	1 0 1 1 1 0	;	11–6–8
0 0 1 1 1 1	√	7–8	1 0 1 1 1 1	△	11–7–8
0 1 0 0 0 0	ƀ	2–8	1 1 0 0 0 0	① & or +	12

0 1 0 0 0 1	/	0–1	1 1 0 0 0 1	A	12–1
0 1 0 0 1 0	S	0–2	1 1 0 0 1 0	B	12–2
0 1 0 0 1 1	T	0–3	1 1 0 0 1 1	C	12–3
0 1 0 1 0 0	U	0–4	1 1 0 1 0 0	D	12–4
0 1 0 1 0 1	V	0–5	1 1 0 1 0 1	E	12–5
0 1 0 1 1 0	W	0–6	1 1 0 1 1 0	F	12–6
0 1 0 1 1 1	X	0–7	1 1 0 1 1 1	G	12–7
0 1 1 0 0 0	Y	0–8	1 1 1 0 0 0	H	12–8
0 1 1 0 0 1	Z	0–9	1 1 1 0 0 1	I	12–9
0 1 1 0 1 0	‡	0–2–8	1 1 1 0 1 0	?	12–0
0 1 1 0 1 1	,	0–3–8	1 1 1 0 1 1	.	12–3–8
0 1 1 1 0 0	① % or (	0–4–8	1 1 1 1 0 0	① □ or)	12–4–8
0 1 1 1 0 1	^	0–5–8	1 1 1 1 0 1	·[	12–5–8
0 1 1 1 1 0	/	0–6–8	1 1 1 1 1 0	<	12–6–8
0 1 1 1 1 1	###	0–7–8	1 1 1 1 1 1	≢	12–7–8

Figure 8.6 Hollerith Card Code and BCD Interchange Code.

① "Arrangement A" (for reports) or "Arrangement H" (for programming).

Control characters: √TM (tape mark); b̸ SB (substitute blank); ‡ RM (record mark); ∨ WS (word separator); ### SM (segment mark); △ MC (mode change); ≢ GM (group mark); □ CW (clear word mark)

Special characters: ! MZ (minus zero); ? PZ (plus zero)

impulse noise often cause bursts of error, where many contiguous bits show a very high number of bits in error. The IEEE defines error burst as "a group of bits in which two successive erroneous bits are always separated by less than a given number of correct bits."

5.4 Error Detection and Correction Defined

The data system engineer differentiates between error detection and error correction. Error detection identified that a symbol has been received in error. As discussed earlier, parity is primarily used for error detection. However, parity bits add redundancy and thus decrease channel efficiency or throughput.

Error correction corrects the detected error. Basically, there are two types of error-correction technique: forward acting (FEC) and two-way error correction [automatic repeat request (ARQ)]. This latter system uses a return channel (backward channel). When an error is detected, the receiver signals this fact to the transmitter over the backward channel, and the block* of information containing the error is transmitted again. Forward-acting error correction (FEC) utilizes a type of coding that permits a limited number of errors to be corrected at the receiving end by means of software (or hardware) implemented at both ends.

5.4.1 Error Detection

There are various arrangements or techniques available for the detection of errors. All error-detection methods involve some form of redundancy, those additional bits or sequences that can inform the system of the presence of error or errors. The parity discussed in section 4.3 was character parity, and its weaknesses were presented. Commonly the data system engineer refers to such parity as *vertical redundancy checking* (VRC).

Another form of error detection utilizes longitudinal redundancy checking (LRC), which is used in block transmission where a data message consists of one or more blocks. Remember that a block is a specific group of digits or data characters sent as a "package." An LRC character, often called a BCC or block check character, is appended at the end of each block. The BCC verifies the total number of "1"s and "0"s in the columns of the block (vertically). The receiving end sums the "1"s (or the "0"s) in the block, depending on the parity convention for the system. If that sum does not correspond to the BCC, an error (or errors) exists in the block. The LRC ameliorates much of the problem of undetected errors that could slip through with VRC, if used alone. The LRC method is not foolproof, however, as it uses the same thinking of VRC. Suppose errors occur such that two "1"s are replaced by two "0"s in the second and third bit po-

*A "block" is a group of digits (data characters) transmitted as a unit over which a coding procedure is usually applied for synchronization and error-control purposes.

sitions of characters "1" and "3" in a certain block. In this case the BCC would read correctly at the receive end and the VRC would pass over the errors as well. A system using both LRC and VRC is obviously more immune to undetected error than either system implemented alone. A more effective method of error detection is CRC (cyclic redundancy check), which is based on a cyclic code and is used in block transmission with a BCC. In this case the transmitted BCC represents the remainder of a division of the message block by a "generating polynomial."

Mathematically, a message block can be treated as a function such as:

$$a_n X^n + a_{n-1} X^{n-1} + a_{n-2} X^{n-2} + \cdots + a_1 X + a_0$$

where coefficients a are set to represent a binary number. Consider the binary number 11011, which is represented by the polynomial

1	1	0	1	1
a_4	a_3	a_2	a_1	a_0

and then becomes

$$X^4 + X^3 + X + 1$$

or consider another example

0	1	1	0	1
a_4	a_3	a_2	a_a	a_0

which becomes

$$X^3 + X^2 + 1$$

The CRC character used as the BCC is the remainder of a data polynomial divided by the generating polynomial.

More specifically, if a data polynomial $D(X)$ is divided by the generating polynomial $G(X)$, the result is a quotient polynomial $Q(X)$ and a remainder polynomial $R(X)$ or

$$\frac{D(X)}{G(X)} = Q(X) + \frac{R(X)}{G(X)}$$

The CRC character in most applications is 16 bits in length or two 8-bit bytes. At present there are three standard generating polynomials commonly in use:

1. CRC-16 (ANSI): $X^{16} + X^{15} + X^5 + 1$
2. CRC (CCITT): $X^{16} + X^5 + 1$
3. CRC-12: $X^{12} + X^{11} + X^3 + 1$

Of course, again, if the computed BCC at the receive end differed from the BCC received from the transmit end, there would be an error (or errors) in the received block.

Reference 34 states that CRC-12 provides error detection of bursts of up to 12 bits in length. Additionally, 99.955% of error bursts up to 16 bits in length. The CRC-16 provides detection of bursts up to 16 bits in length. Additionally, 99.955% of error bursts greater than 16 bits can be detected.

5.5 Forward-Acting Error Correction

Forward-acting error correction (FEC) uses certain binary codes that are designed to be self-correcting for errors introduced by the intervening transmission media. In this form of error correction the receiving station has the ability to reconstitute messages containing errors.

The codes used in FEC can be divided into two broad classes: block codes and convolutional codes. In block codes information bits are taken k at a time, and c parity bits are added, checking combinations of the k information bits. A block consists of $n = k + c$ digits. The code consists of $2k$ words, each n digits long. When used for the transmission of data, block codes may be systematic. A systematic code is one in which the information bits occupy the first k positions in a block and are followed by the $(n - k)$ check digits.

Still another block code is the group code, where the modulo 2 sum of any two n-bit code words is another code word. Modulo 2 addition is denoted by the symbol $\oplus$. It is a binary addition without the "carry" or $1 + 1 = 0$, and we do *not* carry the 1. Summing 10011 and 11001 in modulo 2, we get 01010.

The minimum Hamming distance is a measure of the error detection and correction capability of a code. This "distance" is the minimum number of digits in which two encoded words differ. For example, to detect E digits in error, a code of a minimum Hamming distance of $(E + 1)$ is required. To correct E errors, a code must display a minimum Hamming distance of $(2E + 1)$. A code with a minimum Hamming distance of 4 can correct a single error *and* detect two digits in error.

A convolution(al) code is another form of coding used for error detection and correction. As the word "convolution" implies, this is one code wrapped around or convoluted on another. It is the convolution of an input-data stream and the response function of an encoder. The encoder is usually made up of shift registers. Modulo 2 adders are used to form check digits, each of which is a binary function of a particular subset of the informations digits in the shift register.

Error performance can also be improved by the assistance of a microprocessor in the decoder. Mark or space decisions are not hard or irrevocable decisions; rather, these are called "soft" decisions. In this case a tag of 3 bits is attached to each received digit to indicate the confidence level of the decision in the demodulator before processing. After processing, when errors are indicated, the bits with the lowest confidence level are changed from 0 to 1 or 1 to 0, as the case may be.

The codes discussed up to this point are effective at detecting and correcting errors that are random, where the cause of error is due exclusively to perturbations by additive white Gaussian noise and limited signal power. On many transmission circuits burst noise is encountered where noise hits are in excess of 1-bit or 2-bit duration. Of course, one form of combatting such noise is by extending the duration of a bit. Such bursts, however, may have a duration from 10 to 100 ms. Typically, these are "hits" due to impulse noise.

One forward-acting error correcting code that can combat error bursts quite efficiently is the Hagelbarger code. Two requirements must be fulfilled with this code: the burst must be no longer than 8 digits in length, and there must be at least 91 correct digits between bursts. To pay for this capability, efficiency is reduced to 75% or, in other words, redundancy is added; 1 digit in 4 are check digits.

5.6 Error Correction with Feedback Channel

Two-way or feedback error correction is used very widely today on data and some telegraph circuits. Such a form of error correction is called ARQ. The letter sequence ARQ derives from the old Morse and telegraph signal, "automatic repeat request."

In most modern data systems block transmission is used, and the block is of a convenient length of characters sent as an entity. That "convenient" length is an important consideration, as we see in Chapter 10. One "convenient" number relates to the standard "IBM" card with 80 columns. With 8 bits per column, a block of 8 × 80 or 640 bits would be desirable as data text so we could transmit an IBM card in each block. In fact, one such operating system, Autodin, bases block length on that criterion, with blocks 672 bits long. The remaining bits, those in excess of 640, are overhead and check bits.

Optimal block length is a trade-off between block length and error rate, or the number of block repeats that may be expected on a particular circuit. Longer blocks tend to amortize overhead bits (see Chapter 10) better but are inefficient regarding throughput when an error rate is high. Under these conditions, long blocks tend to tie up a circuit with longer retransmission periods.

ARQ is based on the block-transmission concept. When a receiving station detects an error, it requests a repeat of the block in question from the transmitting station. That request is made on a "feedback" channel, which may be a channel especially dedicated for that purpose or may be the return side of a full duplex link. Such an especially dedicated return channel is generally slow speed, often 75 bps, whereas the forward channel may be 2400 bps or better. For further discussion of ARQ, see Chapter 10, Section 8.3.

6 THE DC NATURE OF DATA TRANSMISSION

6.1 Loops

Binary data are transmitted on a dc loop. More correctly, the binary data end instrument delivers to the line and receives from the line one or several dc loops. In its most basic form a dc loop consists of a switch, a dc voltage, and a termination. A pair of wires interconnects the switch and termination. The voltage source in data and telegraph work is called the *battery*, although the device is usually electronic, deriving the dc voltage from an ac power line source. The battery is placed in the line so as to provide voltage(s) consistent with the type of transmission desired. A simplified dc loop is shown in Figure 8.7.

6.2 Neutral and Polar DC Transmission Systems

Nearly all dc data and telegraph systems functioning today are operated in either a neutral or a polar mode. The words "neutral" and "polar" describe the manner in which battery is applied to the dc loop. On a "neutral" loop, following the mark–space convention of Table 8.1, battery is applied during marking (1) conditions and is switched off during spacing (0). Current therefore flows in the loop when a mark is sent and the loop is closed.

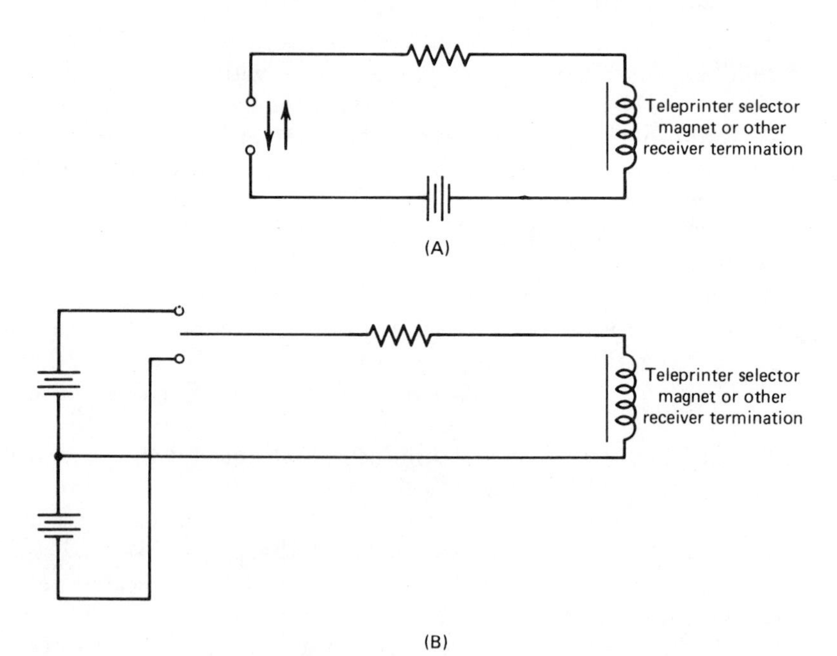

Figure 8.7 Simplified diagram of a dc loop with (A) neutral and (B) polar keying.

Spacing is indicated on the loop by a condition of no current. Thus we have the two conditions for binary transmission, an open loop (no current flowing) and a closed loop (current flowing). Keep in mind that we could reverse this, namely, change the convention (Table 8.1) and assign spacing to a condition of current flowing or closed loop and marking to a condition of no current or an open loop. This is sometimes done in practice and is called "changing the sense." Either way, a neutral loop is a dc loop circuit where one binary condition is represented by the presence of voltage and the flow of current, and the other, by the absence of voltage–current. Figure 8.7A illustrates a neutral loop.

Polar transmission approaches the problem a little differently. Two batteries are provided, one "negative" and the other "positive." During a condition of marking, a positive battery is applied to the loop, following the convention of Table 8.1, and a negative battery is applied to the loop during the spacing condition. In a polar loop, current is always flowing. For a mark or binary "one," it flows in one direction and for a space or binary "zero," it flows in the opposite direction. Figure 8.7B shows a simplified polar loop.

7 BINARY TRANSMISSION AND THE CONCEPT OF TIME

7.1 Introduction

Time and timing are most important factors in digital transmission. For this discussion, consider a binary end instrument sending out in series a continuous run of marks and spaces. Those readers who have some familiarity with the Morse code will recall that the spaces between dots and dashes told the operator where letters ended and where words ended. With the sending device or transmitter delivering a continuous series of characters to the line, each consisting of five, six, seven, eight, or nine elements (bits) per character, a receiving device that starts its print cycle when the transmitter starts sending, and subsequently is perfectly in step with the transmitter, can be expected to provide good printed copy and few, if any, errors at the receiving end.

It is obvious that when signals are generated by one machine and received by another, the speed of the receiving machine must be the same or very close to that of the transmitting machine. When the receiver is a motor-driven device, timing stability and accuracy are dependent on the accuracy and stability of the speed of rotation of the motors used. Most simple data–telegraph receivers sample at the presumed center of the signal element. It follows, therefore, that whenever a receiving device accumulates timing error of more than 50% of the period of one bit, it will print in error.

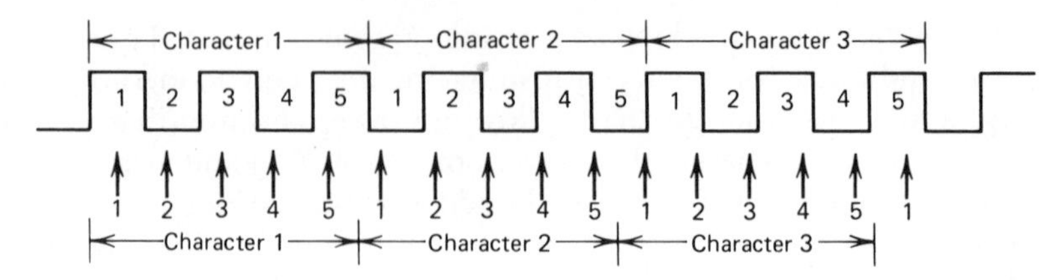

Figure 8.8 Five-unit synchronous bit stream with timing error.

The need for some sort of synchronization is shown in Figure 8.8. A five-unit code is employed, and three characters transmitted sequentially are shown. Sampling points are shown in Figure 8.8 as vertical arrows. Receiving timing begins when the first pulse is received. If there is a 5% timing difference between the transmitter and receiver, the first sampling at the receiver will be 5% away from the center of the transmitted pulse. At the end of the 10th pulse or signal element the receiver may sample in error. The 11th signal element will, indeed, be sampled in error, and all subsequent elements will be errors. If the timing error between transmitting machine and receiving machine is 2%, the cumulative error in timing would cause the receiving device to print all characters in error after the 25th bit.

7.2 Asynchronous and Synchronous Transmission

In the earlier days of printing telegraphy, "start–stop" transmission, or asynchronous operation, was developed to overcome the problem of synchronism. Here timing starts at the beginning of a character and stops at the end. Two signal elements are added to each character to signal the receiving device that a character has begun and ended.

For example, consider a five-element code such as CCITT No. 2 (see Figure 8.1). In the front of a character an element called a "start space" is added, and a stop mark is inserted at the end of each character. To send the letter "A" in Figure 8.1, the receiving device starts its timing sequence on signal element "1," which is a space or "0," followed by 11000, which is the code sequence for the character "A," followed by a stop mark, which terminates the timing sequence. In such an operation, timing errors can accumulate only inside each character. Suppose the receiving device is again 5% slower or faster than its transmitting counterpart; now the fifth information element will be no more than 30% displaced in time from the transmitted pulse and well inside the 50% or halfway point for correct sampling to take place.

In start–stop transmission information signal elements are each of the same duration, which is the duration or pulse width of the start element. The stop element has an indefinite length or pulse width beyond a certain minimum. If a steady series of characters is sent, the stop element is always of the same width or has the same number of unit intervals. Consider the transmission of two As, 01100010110001111 1——————→11111. The start space (0) starts the timing sequence for six additional elements, which are the five code elements in the letter A and the stop mark. Timing starts again on the mark-to-space transition between the stop mark of the first A and the start of the second. Sampling is carried out at pulse center for most asynchronous systems. Note that a continuous series of marks is sent at the end of the second A; thus the signal is a continuation of the stop element or just a continuous mark. It is the mark-to-space transition of the start element that tells the receiving device to start timing a character.

Minimum lengths of stop elements vary. The preceding example above shows a stop element of one-unit interval duration (1 bit). Some are 1.42-unit intervals, others are of 1.5 and 2-unit interval duration. The proper semantics of data–telegraph transmission would describe the code of the previous paragraph as a five-unit start–stop code with a one-unit stop element.

A primary objective in the design of telegraph and data systems is to minimize errors received or to minimize the error rate. There are two prime causes of errors, noise and improper timing relationships. With start–stop systems a character begins with a mark-to-space transition at the beginning of the start space. Then 1.5-unit intervals later the timing causes the receiving device to sample the first information element, which simply is a mark or space decision. The receiver continues to sample at one-bit intervals until the stop mark is received. In start–stop systems the last information bit is most susceptible to cumulative timing errors. Figure 8.9 is an example of a five-unit start–stop bit stream with a 1.5-unit stop element.

Another problem in start–stop systems is the mutilation of the start element. Once this happens, the receiver starts a timing sequence on the next mark-to-space transition it sees and then continues to print in error until, by chance, it cycles back properly on a proper start element.

Synchronous data–telegraph systems do not have start and stop elements but consist of a continuous stream of information elements or bits (see Figure 8.8). The cumulative timing problems eliminated in asynchronous

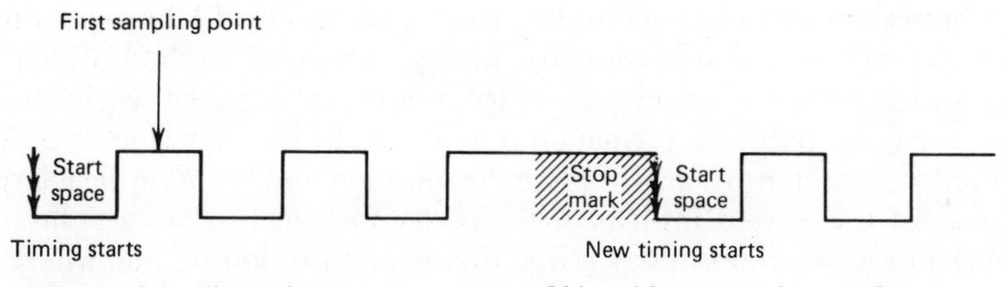

Figure 8.9 Five-unit start–stop stream of bits with a 1.5 unit stop element.

(start–stop) systems are present in synchronous systems. Codes used on synchronous systems are often seven-unit codes with an extra unit added for parity, such as the ASCII or CCITT No. 5 codes. Timing errors tend to be eliminated because the exact rate at which the bits of information are transmitted is known.

If a timing error of 1% were to exist between transmitter and receiver, not more than 100 bits could be transmitted until the synchronous receiving device would be off in timing by the duration of 1 bit from the transmitter, and all bits received thereafter would be in error. Even if timing accuracy were improved to 0.05%, the correct timing relationship between transmitter and receiver would exist for only the first 2000 bits transmitted. It follows, therefore, that no timing error whatsoever can be permitted to accumulate since anything but absolute accuracy in timing would cause eventual malfunctioning. In practice, the receiver is provided with an accurate clock that is corrected by small adjustments, as explained in Section 7.3.

7.3 Timing

All currently used data-transmission systems are synchronized in some manner. Start–stop synchronization has already been discussed. All fully synchronous transmission systems have timing generators or clocks to maintain stability. The transmitting device and its companion receiver at the far end of the circuit must maintain a timing system. In normal practice, the transmitter is the master clock of the system. The receiver also has a clock that in every case is corrected by some means to its transmitter's master clock equivalent at the far end.

Another important timing factor is the time it takes a signal to travel from the transmitter to the receiver. This is called *propagation time.* With velocities of propagation as low as 20,000 mi/s, consider a circuit 200 mi in length. The propagation time would then be 200/20,000 s or 10 ms. Ten milliseconds is the time duration of 1 bit at a data rate of 100 bps; thus the receiver in this case must delay its clock by 10 ms to be in step with its incoming signal. Temperature and other variations in the medium may also affect this delay, as well as variations in the transmitter master clock.

There are basically three methods of overcoming these problems. One is to provide a separate synchronizing circuit to slave the receiver to the transmitter's master clock. However, this wastes bandwidth by expending a voice channel or subcarrier just for timing. A second method, which was quite widely used until several years ago, was to add a special synchronizing pulse for groupings of information pulses, usually for each character. This method was similar to start–stop synchronization and lost its appeal largely because of the wasted information capacity for synchronizing. The most prevalent system in use today is one that uses transition timing, where the receiving device is automatically adjusted to the signalling rate of the transmitter by sampling the transitions of the incoming pulses. This type of timing offers many advantages, particularly automatic compensation for variations in propagation time. With this type of synchronization the receiver determines the average repetition rate and phase of the incoming signal transition and adjusts its own clock accordingly.

In digital transmission the concept of a transition is very important. The transition is what really carries the information. In binary systems the space-to-mark and mark-to-space transitions (or lack of transitions) placed in a time reference contain the information. In sophisticated systems, decision circuits regenerate and retime the pulses on the occurrence of a transition. Unlike decision circuits, timing circuits that reshape a pulse when a transition takes place must have a memory in case a long series of marks or spaces is received. Although such periods have no transitions, they carry meaningful information. Likewise, the memory must maintain timing for reasonable periods in case of circuit outage. Note that synchronism pertains to both frequency and phase, and that the usual error in high-stability systems is a phase error (i.e., the leading edges of the received pulses are slightly advanced or retarded from the equivalent clock pulses of the receiving device). Once synchronized, high stability systems need only a small amount of correction in timing (phase). Modem internal timing systems may have a long-term stability of 1×10^{-8} or better at both the transmitter and receiver. At 2400 bps, before a significant timing error can build up, the accumulated time difference between transmitter and receiver must exceed approximately 2×10^{-4} s. Whenever the circuit of a synchronized transmitter and receiver is shut down, their clocks must differ by at least 2×10^{-4}s before significant errors take place. This means that the leading edge of the receiver-clock equivalent timing pulse is 2×10^{-4} in advance or retarded from the leading edge of the pulse received from the distant end. Often an idling signal is sent on synchronous data circuits during periods of no traffic to maintain the timing. Some high-stability systems need resynchronization only once a day.

Note that thus far in our discussion we have considered dedicated data circuits only. With switched (dial-up) synchronous circuits, the following problems exist:

- No two master clocks are in perfect phase synchronization.
- The propagation time on any two paths may not be the same.

Thus such circuits will need a time interval for synchronization for each call setup before traffic can be passed.

To summarize, synchronous data systems use high-stability clocks, and the clock at the receiving device is undergoing constant but minuscule corrections to maintain an in-step condition with the received pulse train from the distant transmitter, which is accomplished by responding to mark-to-space and space-to-mark transitions. The important considerations of digital network timing are also discussed in Chapter 9.

7.4 Distortion

It has been shown that the key factor in data transmission is timing. Although the signal must be either a mark or space, this alone is not sufficient. The marks and spaces (or ones and zeros) must be in a meaningful sequence based on a time reference.

In the broadest sense, distortion may be defined as any deviation of a signal in any parameter, such as time, amplitude, or wave shape, from that of the ideal signal. For data and telegraph binary transmission, distortion is defined as a displacement in time of a signal transition from the time that the receiver expects to be correct. In other words, the receiving device must make a decision as to whether a received signal element is a mark or a space. It makes the decision during the sampling interval, which is usually at the center of where the received pulse or bit should be; thus it is necessary for the transitions to occur between sampling times and preferably halfway between them. Any displacement of the transition instants is called "distortion." The degree of distortion suffered by a data signal as it traverses the transmission medium is a major contributor in determining the error rate that can be realized.

Telegraph and data distortion is broken down into two basic types, systematic and fortuitous. Systematic distortion is repetitious and is broken down into bias distortion, cyclic distortion, and end distortion, which is more common in start–stop systems. Fortuitous distortion is random and characterized by a displacement of a transition from the time interval in which it should have occurred. Distortion caused by noise spikes or other transients in the transmission medium may be included in this category. Characteristic distortion is still another type and is caused by transients in the modulation process that then appear in the demodulated signal.

Figure 8.10 shows some examples of distortion. Figure 8.10A is a binary signal without distortion, and Figure 8.10B shows the sampling instants, which should ideally occur in the center of the pulse to be sampled. From this we can see that the displacement tolerance is nearly 50%; namely, the point of sample could be displaced by up to 50% of a pulse width and still record the mark or space condition present without error. However, the sampling interval does require a finite amount of time thus in actual practice, the permissible displacement is somewhat less than 50%. Figure 8.10C, D show the two typical types of bias distortion. Spacing bias is shown in Figure 8.10C, where all the spacing impulses are lengthened at the expense of the marking impulses. Figure 8.10D shows marking bias, which is the reverse, the marking impulses are lengthened at the expense of the spaces. Figure 8.10E shows fortuitious distortion, which is random.

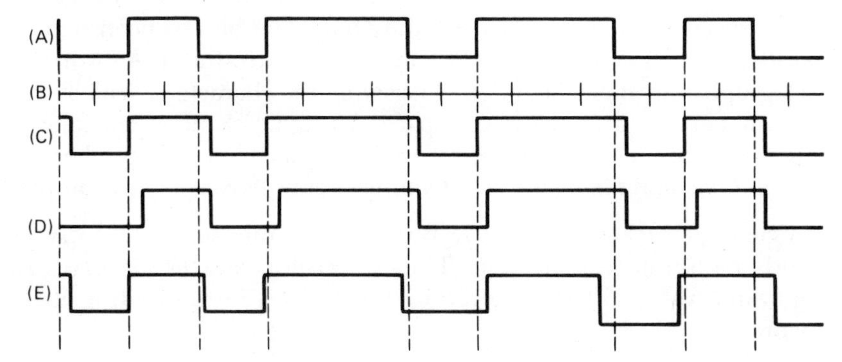

Figure 8.10 Three typical distorted data signals.

Figure 8.11 shows distortion that is more typical of start–stop transmission. Figure 8.11A is an undistorted start–stop signal. Figure 8.11B shows cyclic or repetitive distortion typical of mechanical transmitters. In this type of distortion the marking elements may increase in length for a period of time and then the spacing elements will increase in length. Figure 8.11C shows peak distortion. Identifying the type of distortion present on a signal often gives a clue to the source or cause of distortion. Distortion-measurement equipment measures the displacement of the mark-to-space transition from the ideal of a digital signal. If a transition occurs too near to the sampling point, the signal element is liable to be in error.

7.5 Bits, Bauds, and Words per Minute

There is much confusion among transmission engineers in handling some of the semantics and simple arithmetic used in data and telegraph trans-

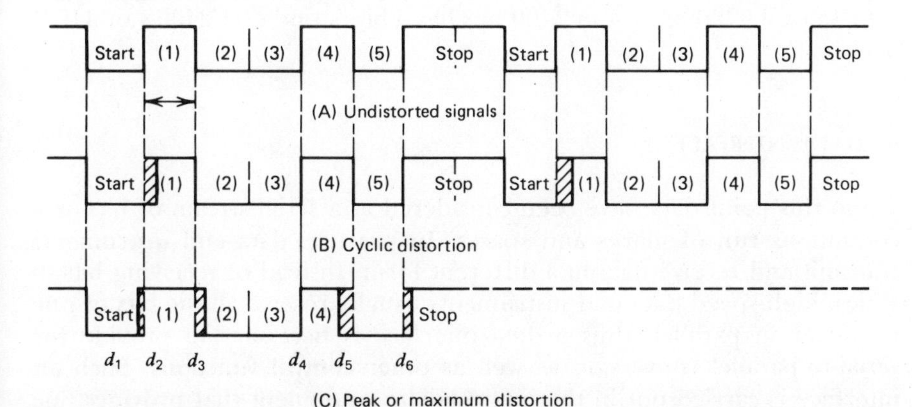

Figure 8.11 Distorted telegraph signals illustrating cyclic and peak distortion. Peak distortion appears at transition d_5.

mission, especially the terms "baud" and "bit." The bit has been defined previously. Now the term "words per minute" is introduced. A "word" in our telegraph and data language consists of six characters, usually five letters, numbers, or graphics and a space. All signal elements transmitted must be counted, such as "carriage return" and "line feed." Remember that the unit interval and the bit are synonymous. Let us look at some examples:

1. A channel is transmitting at 75 bps using a 5-unit start and stop code with a 1.5-unit stop element. Thus for each character there are 7.5 unit intervals (7.5 bits). Therefore, the channel is transmitting at 100 wpm.

$$\frac{75 \times 60}{6 \times 7.5} = 100 \text{ wpm}$$

2. A system transmits in CCITT No. 5 code at 1500 wpm with parity. How many bits per second are being transmitted?

$$\frac{1500 \times 8 \times 6}{60} = 1200 \text{ bps}$$

The baud is the unit of modulation rate. In binary transmission systems, the number of bauds is equal to the number of bits per second. Thus a modem in a binary system transmitting to the line 110 bps has a modulation rate of 110 bauds. In multilevel (M-ary systems), the number of bauds is indicative of the number of transitions per second; it is synonymous with symbol rate. The baud is more meaningful to the transmission engineers concerned with the line side of a modem. This concept is discussed at greater length in Section 10.5.

The period or time duration of 1 bit is another parameter of interest. This is the inverse of the data rate in bits per second. A 75-bps system has a bit period of 1/75 or 0.01333 s. A 45.45-bps system has a bit period of 1/45.45 or 0.0220022 s, and 2400 bps has a bit period of 1/2400 s or 416.7 μsec.

8 DATA INTERFACE

Up to this point data have been considered as a serial stream of bits or a continuous run of marks and spaces. In practice, data end instruments transmit and receive data in a different form. Instead of receiving bits in series, high-speed data end instruments usually receive all the bits of one character in parallel; thus a data interface is necessary to provide for serial-to-parallel conversion as well as other control functions. Such an interface is carried out in the intermediate equipment that provides line buffering. This equipment is usually mounted with the data end instrument such as a card punch or a card reader. The intermediate equipment receives the serial data stream from the data-communication equipment (modem) as well as timing and other control signals for reception. It stores the data, converts it to parallel data, and provides the data to the end instrument on command in parallel form. The reverse holds true for transmission.

Typical intermediate equipment interfaces with input/output (I/O) and user equipment are shown in Figure 8.12A, B. The intermediate equipment interfaces with the line or data set, modem, or what is generically called the DCE, data communication equipment. The interface and differentiation between DTE (data terminal equipment) and DCE is shown in the following diagram.

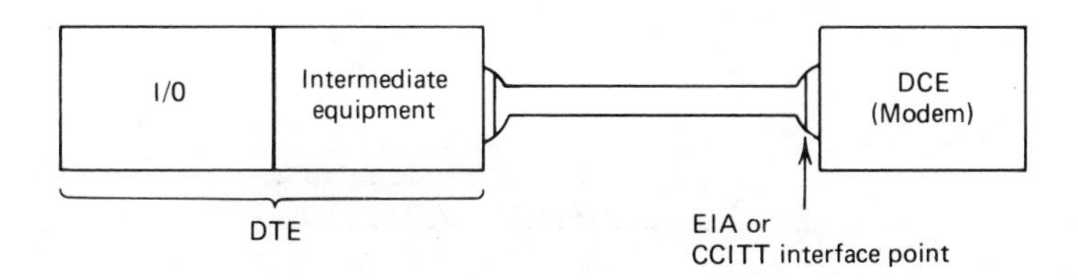

The interface is a 25-pin connector. For the purpose of this discussion, the actual pin assignments are not important and are readily available in the literature [10, 23]. This interface, as described in the standard, is called "low level." To abstract from the RS-232C standard:

For data interchange circuits, the signal shall be considered in the marking condition when the voltage on the interchange circuit, measured at the interface point, is more negative than minus three volts with respect to signal ground. The signal shall be considered in the spacing condition when the voltage is more positive than plus three volts with respect to signal ground. The region between plus 3 volts and minus 3 volts is defined as the transition region. The signal state is not uniquely defined when the (signal) voltage is in this transition region.

9 DATA INPUT–OUTPUT DEVICES

The following paragraphs are intended to give the reader a general familiarity with data-subscriber equipment, which we refer to as input–output (I/O) devices.* Such equipment converts user information (data or messages) into electrical signals and vice versa. The working human interface of a communication system is the I/O device. Electrically, a data subscriber terminal consists of the end instrument, the intermediate equipment (buf-

*With other disciplines there may be an ambiguity here. For instance a computer can be a source or sink which also may be considered I/O equipment.

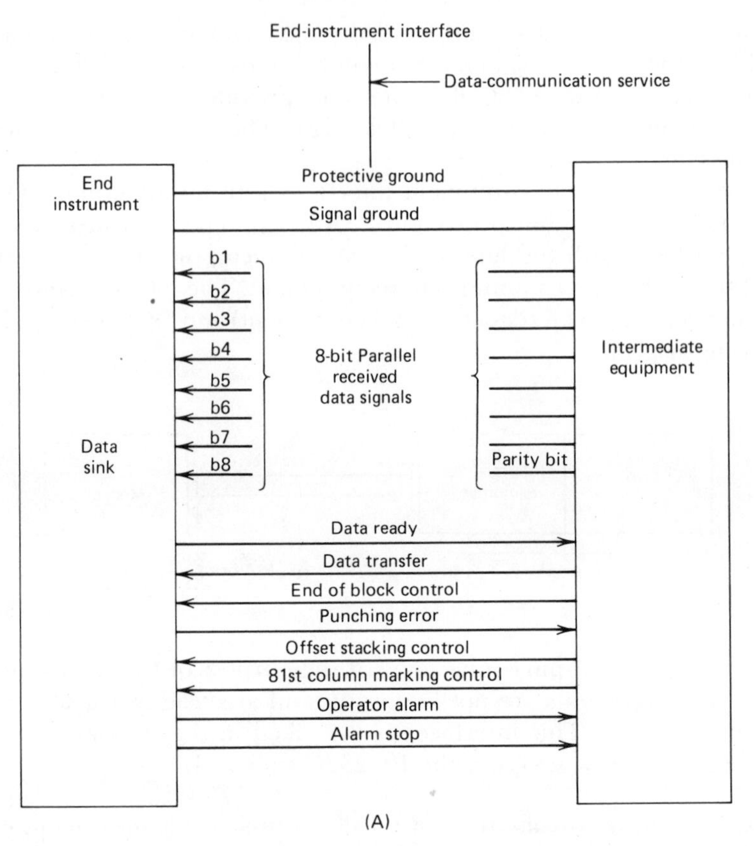

Figure 8.12 End-instrument interface interchange circuits for (A) card punches and (B) card readers.

fering equipment), and a device called a *modem* or *communication set.* The intermediate equipment has been discussed in Section 8. Modems are described in Section 10. The data source is the input device, and the data sink is the output device.

Data input–output devices handle paper tape, punched cards, magnetic tape, drums, disks, visual displays, and printed page copy. Input devices may be broken down into the following categories:

Keyboard sending units.
Card readers.
Paper tape readers.
Magnetic tape (disk and drum) readers.
Optical character readers.

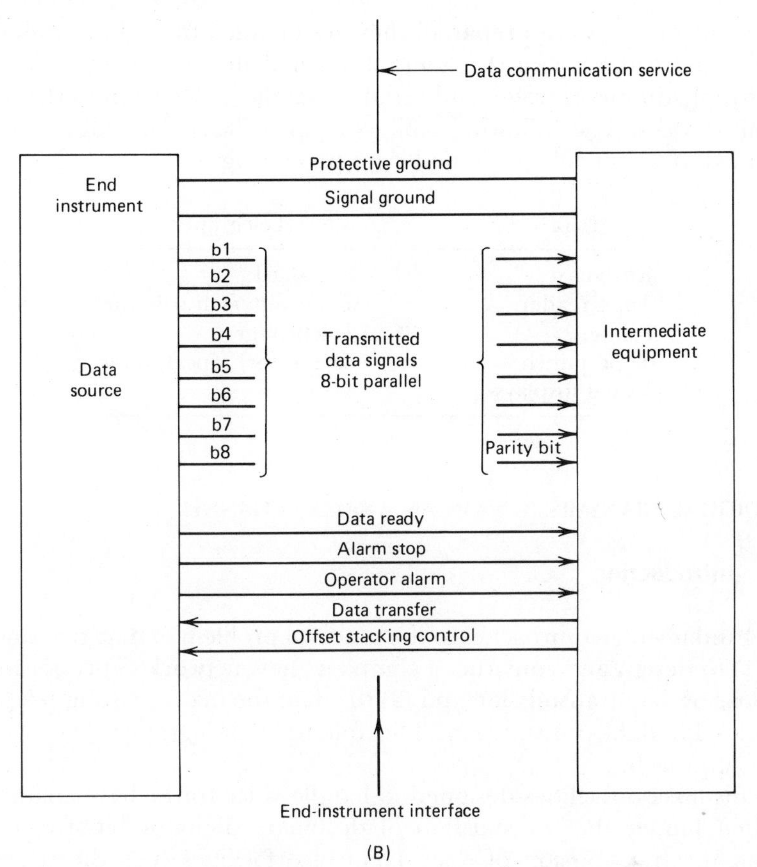

Figure 8.12 Continued

Output devices are as follows:

Printers (paper and hard copy).
Card punches.
Paper tape punches.
Magnetic tape recorders, magnetic cores, and disks.
Visual display units (VDU) (CRT and plasma).

Further, these devices may be used as on-line or off-line devices. The latter are not connected directly to the communication system but serve as auxiliary equipment.

Off-line devices are used for tape or card preparation for eventual transmission. In this case a keyboard is connected to a card punch for card preparation. Also, the keyboard may be connected to a paper tape punch.

Once tape or cards are prepared, they are handled by on-line equipment, either tape readers or card readers. Intermediate equipment or line buffers supply timing, storage, and serial-to-parallel conversion to the input–output devices. The following table compares the terminology of input–output devices for telegraph and data processing.

Data	Telegraph
Keyboard	Keyboard
Tape reader	Transmitter–distributor*
Printer	Teleprinter
Tape punch	Perforator, reperforator
Visual displays	—

10 DIGITAL TRANSMISSION ON AN ANALOG CHANNEL

10.1 Introduction

Two fundamental approaches to the practical problem of data transmission are (1) to design and construct a complete, new network expressly for the purpose of data transmission and (2) to adapt the many existing telephone facilities for data transmission. The following paragraphs deal with the latter approach.

Transmission facilities designed to handle voice traffic have characteristics that hinder the transmission of dc binary digits or bit streams. To permit the transmission of data over voice facilities (i.e., the telephone network), it is necessary to convert the dc data into a signal within the voice frequency range. The equipment that performs the necessary conversion to the signal is generally called a *data modem*, an acronym for MOdulator–DEModulator.

10.2 Modulation–Demodulation Schemes

A modem modulates and demodulates a carrier signal with digital data signals. The types of modulation used by present-day modems may be one or a combination of the following:

Amplitude modulation, double sideband (DSB).

Amplitude modulation, vestigial sideband (VSB).

Frequency shift modulation, commonly called frequency shift keying (FSK).

Phase shift modulation, commonly called phase shift keying (PSK).

*The distributor performs the parallel-to-serial equivalent conversion of data transmission.

10.2.1 Amplitude Modulation–Double Sideband

With the double-sideband (DSB) modulation technique, binary states are represented by the presence or absence of an audio tone or carrier. More often it is referred to as "on–off telegraphy." For data rates up to 1200 bps, one such system uses a carrier frequency centered at 1600 Hz. For binary transmission, amplitude modulation has significant disadvantages, which include (1) susceptibility to sudden gain change and (2) inefficiency in modulation and spectrum utilization, particularly at higher modulation rates (see CCITT Rec. R.70).

10.2.2 Amplitude Modulation—Vestigial Sideband

An improvement in the amplitude modulation, the double sideband (DSB) technique results from the removal of one of the information-carrying sidebands. Since the essential information is present in each of the sidebands, there is no loss of content in the process. The carrier frequency must be preserved to recover the dc component of the information envelope. Therefore, digital systems of this type use VSB modulation in which one sideband, a portion of the carrier, and a "vestige" of the other sideband are retained. This is accomplished by producing a DSB signal and filtering out the unwanted sideband components. As a result, the signal takes only about 75% of the bandwidth required for a DSB system. Typical VSB data modems are operable up to 2400 bps in a telephone channel. Data rates up to 4800 bps are achieved using multilevel (*M*-ary) techniques. The carrier frequency is usually located between 2200 and 2700 Hz.

10.2.3 Frequency-shift Modulation

Many data-transmission systems utilize frequency-shift modulation (FSK). The two binary states are represented by two different frequencies and are detected by using two frequency-tuned sections, one tuned to each of the 2 bit-frequencies. The demodulated signal is then integrated over the duration of 1 bit, and a binary decision is based on the result.

Digital transmission using FSK modulation has the following advantages: (1) the implementation is not much more complex than an AM system; and (2) since the received signals can be amplified and limited at the receiver, a simple limiting amplifier can be used, whereas the AM system requires sophisticated automatic gain control for operation over a wide level range. Another advantage is that FSK can show a 3-to-4-dB improvement over AM in most types of noise environment, particularly at distortion threshold (i.e., at the point where the distortion is such that good printing is about to cease). As the frequency shift becomes greater, the advantage over AM improves in a noisy environment.

Another advantage of FSK is its immunity from the effects of nonselective level variations, even when they occur extremely rapidly. Thus a major application is on worldwide high-frequency radio transmission where rapid fades are a common occurrence. In the United States FSK has nearly universal application for the transmission of data at the lower data rates (i.e., ≤ 1200 bps).

10.2.4 Phase Modulation

For systems using higher data rates, phase modulation becomes more attractive. Various forms are used, such as two-phase, relative phase, and quadrature phase. A two-phase system uses one phase of the carrier frequency for one binary state and the other phase for the other binary state. The two phases are 180° apart and are detected by a synchronous detector using a reference signal at the receiver that is of known phase with respect to the incoming signal. This known signal operates at the same frequency as the incoming signal carrier and is arranged to be in phase with one of the binary signals. In the relative-phase system a binary "1" is represented by sending a signal burst of the same phase as that of the previous signal burst sent. A binary "0" is represented by a signal burst of a phase opposite to that of the previous signal transmitted. The signals are demodulated at the receiver by integrating and storing each signal burst of 1-bit period for comparison in phase with the next signal burst. In the quadrature phase system (QPSK), two binary channels (2 bits) are phase multiplexed onto one tone by placing them in phase quadrature, as shown in the following sketch. An extension of this technique places two binary channels on each of several tones spaced across the voice channel of a typical telephone circuit.

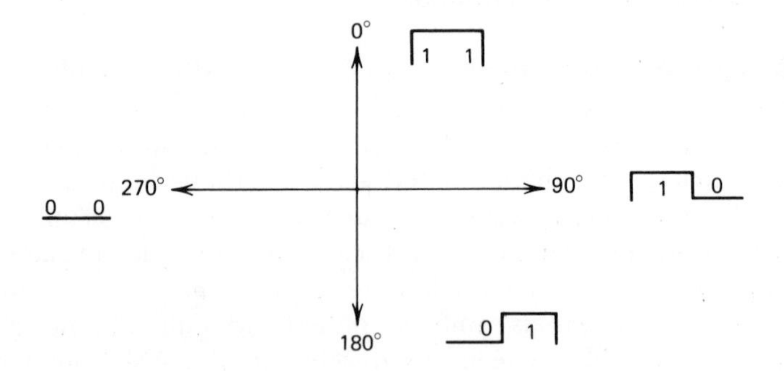

Some of the advantages of phase modulation are as follows:

1. All available power is utilized for intelligence conveyance.
2. The demodulation scheme has good noise-rejection capability.
3. The system yields a smaller noise bandwidth.

A disadvantage of such a system is the complexity of equipment required.

10.3 Critical parameters

The effect of the various telephone-circuit parameters on the capability of a circuit to transmit data is a most important consideration. The following discussion is intended to familiarize the reader with the problems most likely to be encountered in the transmission of data over analog circuits (e.g., the telephone network) and to make certain generalizations in some cases, which can be used to facilitate planning the implementation of data systems.

10.3.1 Delay Distortion

Delay distortion "constitutes the most limiting impairment to data transmission, particularly over telephone voice channels" [3]. When specifying delay distortion, the terms "envelope delay distortion" (EDD) and "group delay" are often used. The IEEE Standard Dictionary states that "Envelope delay is often defined the same as group delay, that is the rate of change, with angular frequency, of the phase shift between two points in a network." (See Chapter 5, Section 2.2.)

The problem is that in a band-limited analog system such as the typical telephone voice channel, not all frequency components of the input signal will propagate to the receiving end in exactly the same elapsed time, particularly on loaded cable circuits and carrier systems. In carrier systems it is the cumulative effect of the many filters used in the FDM equipment. On long-haul circuits the magnitude of delay distortion is generally dependent on the number of carrier modulation stages that the circuit must traverse rather than the length of the circuit. Figure 8.13 shows a typical frequency–delay response curve in milliseconds of a voice channel due to FDM equipment only. For the voice channel (or any symmetrical passband,

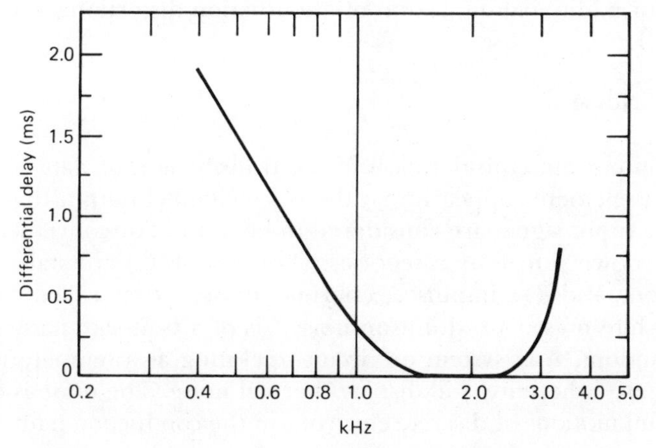

Figure 8.13 Typical differential delay across a voice channel, FDM equipment back-to-back.

for that matter), delay increases toward band edge and is minimum around the center portion.

Delay distortion is the major limitation to modulation rate. The shorter is the pulse width (the width of 1 bit in binary systems), the more critical will be the EDD parameters. As we discuss in Section 10.5, it is desirable to keep the delay distortion in the band of interest below the period of 1 bit.

10.3.2 Amplitude Response (Attenuation Distortion)

Another parameter that seriously affects the transmission of data and can place very definite limits on the modulation rate is amplitude response, also called "attenuation distortion." Ideally, all frequencies across the passband of a channel of interest should undergo the same attenuation. For example, let a −10-dBm signal enter a channel at any frequency between 300 and 3400 Hz. If the channel has 13 dB of flat attenuation, we would expect an output of −23 dBm at any and all frequencies in the band. This type of channel is ideal but unrealistic in a real, working system.

In Rec. G.132, the CCITT recommends no more than 9 dB of attenuation distortion relative to 800 Hz between 400 and 3000 Hz. This figure, 9 dB, describes the maximum variation that may be expected from the reference level at 800 Hz. This variation of amplitude response is often called attenuation distortion. A conditioned channel, such as a Bell System C-4 channel, will maintain a response of −2 to +3 dB from 500 to 3000 Hz and −2 to +6 dB from 300 to 3200 Hz. Channel conditioning is discussed in Section 10.6.

Considering tandem operation, the deterioration of amplitude response is arithmetically cumulative when sections are added. This is particularly true at band edge in view of channel unit transformers and filters that account for the upper and lower cutoff characteristics. Figure 8.14 illustrates a typical example of amplitude response across FDM carrier equipment (see Chapter 5) connected back-to-back at the voice channel input–output. For additional discussion of attentuation distortion see Chapter 5, Section 2.1.

10.3.3 Noise

Another important consideration in the transmission of data is noise. All extraneous elements appearing at the voice channel output that were not due to the input signal are considered to be noise. For convenience, noise is broken down into four categories: (1) thermal, (2) crosstalk, (3) intermodulation, and (4) impulse. Thermal noise, often called "resistance noise," "white noise," or "Johnson noise," is of a Gaussian nature or completely random. Any system or circuit operating at a temperature above absolute zero inherently will display thermal noise. The noise is caused by the random motions of discrete electrons in the conduction path. Crosstalk

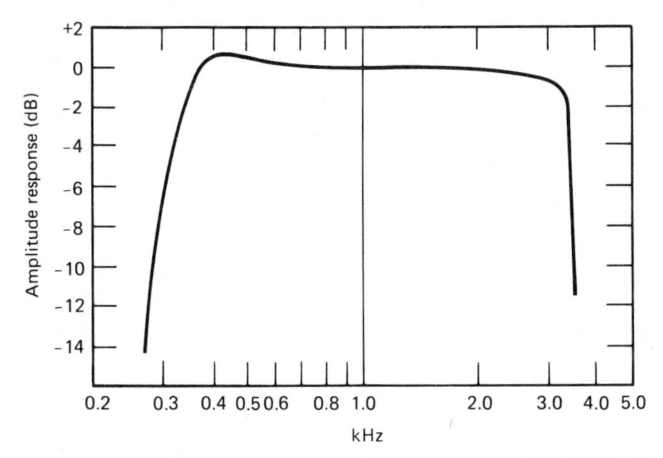

Figure 8.14 Typical amplitude–frequency response across a voice channel; channel modulator, demodulator back-to-back, FDM equipment.

is a form of noise caused by unwanted coupling from one signal path into another. It may be caused by direct inductive or capacitive coupling between conductors or between radio antennas. Intermodulation noise is another form of unwanted coupling, usually caused by signals mixing in nonlinear elements of a system. Carrier and radio systems are highly susceptible to intermodulation noise, particularly when overloaded. Impulse noise is a primary source of errors in the transmission of data over telephone networks. It is sporadic and may occur in bursts or discrete impulses called "hits." Some types of impulse noise are natural, such as that from lightning. However, man-made impulse noise is everincreasing, such as that from automobile ignition systems and power lines. Impulse noise may be of a high level in conventional telephone switching centers as a result of dialing, supervision, and switching impulses that may be induced or otherwise coupled into the data-transmission channel. The worst offender in the switching area is the step-by-step exchange.

For our discussion of data transmission, two types of noise are considered, random (or Gaussian) noise and impulse noise. Random noise measured with a typical transmission measuring set appears to have a relatively constant value. However, the instantaneous value of the noise fluctuates over a wide range of amplitude levels. If the instantaneous noise voltage is of the same magnitude as the received signal, the receiving detection equipment may yield an improper interpretation of the received signal and an error or errors will occur. Thus we need some way of predicting the behavior of data transmission in the presence of noise. Random noise or white noise has a Gaussian distribution and is considered representative of the noise encountered on the analog telephone channel (i.e., the voice channel). From the probability distribution curve for Gaussian noise shown

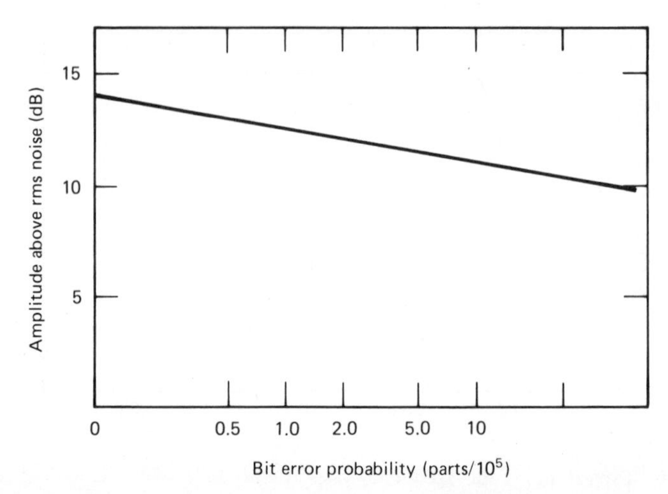

Figure 8.15 Probability of bit error in Gaussian noise; binary polar transmission with a Nyquist bandwidth.

in Figure 8.15, we can make some statistical predictions. It may be noted from this curve that the probability of occurrence of noise peaks that have amplitudes 12.5 dB above the rms level is 1 in 10^5. Hence if we wish to ensure an error rate of 10^{-5} in a particular system using binary polar modulation, the rms noise should be at least 12.5 dB below the signal level [3, p. 114]. This simple analysis is valid for the type of modulation used, assuming that no other factors are degrading the operation of the system and that a cosine-shaped receiving filter is used. If we were to interject distortion such as EDD into the system, we could translate the degradation into an equivalent signal-to-noise-ratio improvement necessary to restore the desired error rate. For example, if the delay distortion were the equivalent of one pulse width, the signal-to-noise-ratio improvement required for the same error rate would be about 5 dB, or the required signal-to-noise-ratio would now be 17.5 dB.

Let us assume a telephone system where the signal level is −10 dBm at the zero transmission-level point of the system; the rms noise measured at the same point would be −27.5 dBm to retain the error rate of 1 in 10^5. This figure can be significant only if it is related to the actual noise found in a channel. The CCITT recommends no more than 50,000 pW of noise psophometrically weighted (−43 dBmp) on an international connection made up of six circuits in a chain. However, the CCITT states (Recs. G.142A and 142D) that for data transmission at as high a modulation rate as possible without significant error rate, a reasonable circuit objective for maximum random noise would be −40 dBm0p for leased circuits (impulse noise not included) and −36 dBm0p for switched circuits without compandors. This figure obviously appears quite favorable when compared to the

−27 dBm0 (−29.5 dBm0p) required in the preceding example. However, other factors developed later will consume much of the noise margin that appears to be available.

Unlike random noise, which is measured by its rms value when we measure level, impulse noise is measured by the number of "hits" or "spikes" per interval of time above a certain threshold. In other words, it is a measurement of the recurrence rate of noise peaks over a specified level. The word "rate" should not mislead the reader. The recurrence is not uniform per unit time, as the word "rate" may indicate, but we can consider a sampling and convert it to an average. The Bell System circuit objective on leased lines is 15 counts in 15 min at −69 dBm with equivalent *C*-message weighting. The CCITT (Rec. Q.45) states that "In any four-wire international exchange the busy hour impulsive noise counts should not exceed 5 counts in 5 min at a threshold level of −35 dBm0."

Remember that random noise has a Gaussian distribution and will produce peaks at 12.5 dB over the rms value (unweighted) 0.001% of the time on a data-bit stream for an equivalent error rate of 1×10^{-5}. It should be noted that some references use 12 dB, some use 12.5 dB, and others use 13 dB. The 12.5 dB above the rms random noise floor should establish the impulse-noise threshold for measurement purposes. We should assume that in a well-designed data-transmission system traversing the telephone network, the signal-to-noise ratio of the data signal will be well in excess of 12.5 dB. Thus impulse noise may well be the major contributor to degradation of the error rate.

When an unduly high error rate has been traced to impulse noise, there are some methods for improving conditions. Noisy areas may be bypassed, repeaters may be added near the noise source to improve signal-to-impulse-noise ratio, or in special cases pulse smearing techniques may be used. This latter approach uses two delay-distortion networks that complement each other such that the net delay distortion is zero. By installing the networks at opposite ends of the circuit, impulse noise passes through only one network* and hence is smeared because of the delay distortion. The signal is unaffected because it passes through both networks.

Signal-to-noise ratio may be traded for the implementation of forward-acting error correction (FEC). This trade-off may be economically feasible or even mandatory, such as in certain digital satellite circuits, to reduce power output of a transmitter or reduce level out of a modem. Reducing power by 3 dB (Figure 8.15) on a circuit displaying a bit error rate of 1×10^{-5} would deteriorate error rate to some 5×10^{-2}. The error rate could be recovered by selecting the proper method of FEC, decoding algorithm with soft mark–space decisions. Soft decision decoding alone can improve error performance by an equivalent 2 dB. For additional discussion of FEC see Section 5 of this chapter.

*This assumes that impulse noise enters the circuit at some point beyond the first network.

10.3.4 Levels and Level Variations

The design signal levels of telephone networks traversing FDM carrier systems are determined by average talker levels, average channel occupancy, permissible overloads during busy hours, and so on. Applying constant amplitude digital data tone(s) over such an equipment of 0 dBm0 on each channel would result in severe overload and intermodulation within the system.

Loading does not affect (metallic) wire systems except by increasing crosstalk. However, once the data signal enters carrier multiplex (voice) equipment, levels must be carefully considered, and the resulting levels most probably have more impact on the final signal-to-noise ratio at the far end than anything else. The CCITT (Recs. G.151C, H.51, H.23, and V.2) recommends −10 dBm0 in some cases, and −13 dBm0 when the portion of nonspeech circuits on an international carrier circuit exceeds 10 or 20%. For multichannel telegraphy −8.7 dBm0 for the composite level, or for 24 channels, each individual telegraph channel would be adjusted for −22.5 dBm0. Even this loading may be too heavy if a large portion of the voice channels is loaded with data. Depending on the design of the carrier equipment, cutbacks to −13 dBm0 or less may be advisable.

In a properly designed transmission system the standard deviation of the variation in level should not exceed 1.0 dB/circuit. However, data-communication equipment should be able to withstand level variations in excess of 4 dB.

10.3.5 Frequency-Translation Errors

Total end-to-end frequency-translation error on a voice channel being used for data or telegraph transmission must be limited to 2 Hz (CCITT Rec. G.135); this is an end-to-end requirement. Frequency translations occur largely because of carrier equipment modulation and demodulation steps. Frequency-division-multiplex carrier equipment widely uses single-sideband, suppressed-carrier techniques. Nearly every case of error can be traced to errors in frequency translation (we refer here to deriving the group, supergroup, mastergroup, and its reverse process; see also Chapter 5) and carrier reinsertion frequency offset, where the frequency error is exactly equal to the error in translation and offset or the sum of several such errors. Frequency-locked (e.g., synchronized) or high-stability master carrier generators (1×10^{-7} or 1×10^{-8}, depending on the system), with all derived frequency sources slaved to the master source, are usually employed to maintain the required stability.

Although 2 Hz seems to be a very rigid specification, when added to the possible back-to-back error of the modems themselves, the error becomes more appreciable. Much of the trouble arises with modems that employ sharply tuned filters. This is particularly true of telegraph equipment. But for the more general case, high-speed data modems occupying the whole voice channel can be designed to withstand greater carrier shifts than can their slower speed counterparts used for asynchronous data or printing telegraphy.

10.3.6 Phase Jitter

The unwanted change in phase or frequency of a transmitted signal caused by modulation by another signal during transmission is defined as "phase jitter." If a simple sinusoid is frequency or phase modulated during transmission, the received signal will have sidebands. The amplitude of these sidebands compared to the received signal is a measure of the phase jitter imparted to it during transmission.

Phase jitter is measured in degrees of variation peak-to-peak for each hertz of transmitted signal. Phase jitter is manifest as unwanted variations in zero crossings of a received signal. It is the zero crossings that data modems use to distinguish marks and spaces. Thus the higher is the data rate, the more jitter can affect error rate on the receive bit stream.

The greatest cause of phase jitter in the telephone network is FDM carrier equipment, where it is manifest as undesired incidental phase modulation. Modern FDM equipment derives all translation frequencies from one master frequency source by multiplying and dividing its output. To maintain stability, phase-lock techniques are used; thus the low-jitter content of the master oscillator may be multiplied many times. It follows, then, that there will be more phase jitter in the voice channels occupying the higher baseband frequencies.

Jitter most commonly appears on long-haul systems at rates related to the power line frequency (e.g., 60 Hz and its harmonics and submultiples) or is derived from 20-Hz ringing frequency. Modulation components that we define as "jitter" usually occur close to the carrier ± 300 Hz maximum.

10.4 Channel Capacity

A leased or switched voice channel represents a financial investment. Therefore, one goal of the system engineer is to derive as much benefit as possible from the money invested. For the case of digital transmission, this is done by maximizing the information transfer across the system. This section discusses how much information in bits can be transmitted, relating information to bandwidth, signal-to-noise ratio, and error rate. These matters are discussed empirically in Section 10.5.

First, looking at very basic information theory, Shannon stated in his classic paper [13] that if input information rate to a band-limited channel is less than C (bps), a code exists for which the error rate approaches zero as the message length becomes infinite. Conversely, if the input rate exceeds C, the error rate cannot be reduced below some finite positive number.

The usual voice channel is approximated by a Gaussian band-limited channel (GBLC) with additive Gaussian noise. For such a channel, consider a signal wave of mean power of S watts applied at the input of an ideal low-pass filter that has a bandwidth of W (Hz) and contains an internal source of mean Gaussian noise with a mean power of N watts uniformly distributed over the passband. The capacity in bits per second is given by:

$$C = W \log_2 \left(1 + \frac{S}{N}\right)$$

Applying Shannon's "capacity" formula to an ordinary voice channel (GBLC) of bandwidth (W) 3000 Hz and a signal-to-noise (S/N) ratio of 1023, the capacity of the channel is 30,000 bps. (Remember that bits per second and bauds are interchangeable in binary systems.) Neither S/N nor W is an unreasonable value. Seldom, however, can we achieve a modulation rate greater than 3000 bauds. The big question in advanced design is how to increase the data rate and keep the error rate reasonable.

One important item not accounted for in Shannon's formula is intersymbol interference. A major problem of a pulse in a band-limited channel is that the pulse tends not to die out immediately, and a subsequent pulse is interfered with by "tails" from the preceding pulse (see Figure 8.16).

Nyquist provided another approach to the data-rate problem, this time using intersymbol interference (the tails in Figure 8.16) as a limit [12]. This resulted in the definition of the so-called Nyquist rate = $2W$ symbols/s, where W is the bandwidth (Hz) of a band-limited channel. In binary transmission we are limited to $2W$ bps, where a symbol is 1 bit. If we let W = 3000 Hz, the maximum data rate attainable is 6000 bps. Some refer to this as "the Nyquist 2-bit rule."

The key here is that we have restricted ourselves to binary transmission and are limited to $2W$ bps no matter how much we increase the signal-to-noise ratio. The Shannon GBLC equation indicates that we should be able to increase the information rate indefinitely by increasing the signal-to-noise ratio. The way to attain a higher C value is to replace the binary transmission system with a multilevel system, often termed an M-ary

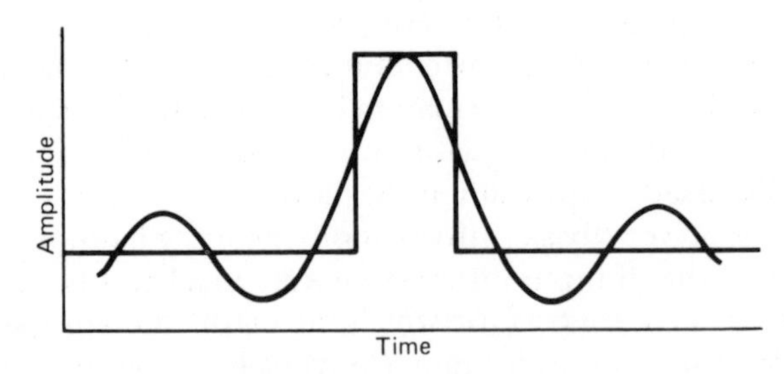

Figure 8.16 Pulse response through a Gaussian band-limited channel (GBLC).

transmission system, with $M > 2$. An M-ary channel can pass $2W \log_2 M$ bps with an acceptable error rate. This is done at the expense of signal-to-noise ratio. As M increases (as the number of levels increases), so must S/N increase to maintain a fixed error rate.

10.5 Some Modem Selection Considerations

The critical parameters that affect data transmission have been discussed; these are amplitude–frequency response (sometimes called "amplitude distortion"), envelope delay distortion, and noise. Now we relate these parameters to the design of data modems to establish some general limits or "boundaries" for equipment of this type. The discussion that follows purposely avoids HF radio considerations.

As stated earlier in the discussion of envelope delay distortion, it is desirable to keep the transmitted pulse (bit) length equal to or greater than the residual differential EDD. Since about 1.0 ms is assumed to be reasonable residual delay after equalization (conditioning), the pulse length should be no less than approximately 1 ms. This corresponds to a modulation rate of 1000 pulses per second (binary). In the interest of standardization (CCITT Rec. V.22), this figure is modified to 1200 bps.

The next consideration is the usable bandwidth required for the transmission of 1200 bps. This figure is approximately 1800 Hz, using modulation methods such as phase shift (PSK), frequency shift (FSK) or double-sideband AM (DSB–AM), and somewhat less for vestigial-sideband AM (VSB–AM). Since delay distortion of a typical voice channel is at its minimum between 1700 and 1900 Hz, the required band, when centered about these points, extends from 800 to 2600 Hz or 1000 to 2800 Hz. From the previous discussion and with reference to Figure 8.13, we can see that the EDD requirement is met easily over the range of 800 to 2800 Hz.

Bandwidth limits modulation rate. However, the modulation rate in bauds and the data rate in bits per second may not necessarily be the same. This is a very important concept. Suppose a modulator looked at the incoming serial bits stream 2 bits at a time rather than the conventional 1 bit at a time. Now let four discrete signal amplitudes be used to define each of four possible combinations of two consecutive bits such that

$$A_1 = 00$$
$$A_2 = 01$$
$$A_3 = 11$$
$$A_4 = 10$$

where A_1, A_2, A_3, and A_4 represent the four pulse amplitudes. This form of treating 2 bits at a time is called "di-bit coding" (see Section 10.2).

Similarly, we could let eight amplitude levels cover all the possible combinations of three consecutive bits so that with a modulation rate of 1200 bauds it is possible to transmit information at a rate of 3600 bps. Rather than vary amplitude to four or eight levels, phase can be varied. A four-phase system (PSK) could be coded as follows.

$$F_1 = 0° = 00$$
$$F_2 = 90° = 01$$
$$F_3 = 180° = 11$$
$$F_4 = 270° = 10$$

Again, with a four-phase system using di-bit coding, a tone with a modulation rate of 1200 bauds PSK can be transmitting 2400 bps. An eight-phase PSK system at 1200 bauds could produce 3600 bps of information transfer. Obviously, this process cannot be extended indefinitely. The limitation comes from channel noise. Each time the number of levels or phases is increased, it is necessary to increase the signal-to-noise ratio to maintain a given error rate. Consider the case of a signal voltage S and a noise voltage N. The maximum number of increments of signal (amplitude) that can be discerned is S/N (since N is the smallest discernible increment). Add to this the no-signal case, and the number of discernible levels now becomes $(S/N) + 1$ or $\left(\frac{S+N}{N}\right)$. Where S and N are expressed as power, the number of levels becomes the square root of this expression. This formula shows that in going from two to four levels or from four to eight levels, approximately 6-dB noise penalty is incurred each time the number of levels is doubled. If a similar analysis is carried out for the multiphase case, the penalty in going from two to four phases is 3 dB, and to eight phases, 6 dB.

Sufficient background has been developed to appraise the data modem for the voice channel. Now consider a data modem for a data rate of 2400 bps. By using quaternary phase shift keying (QPSK) as described earlier, 2400 bps is transmitted with a modulation rate of 1200 bauds. Assume that the modem uses differential phase detection wherein the detector decisions are based on the change in phase between the last transition and the preceding one. Assume the bandwidth of the data modem under consideration to be 1800 Hz. It is now possible to determine whether the noise requirements can be satisfied. Figure 8.15 shows that a 12.5-dB signal-to-noise ratio (Gaussian noise) is required to maintain an error rate 1×10^{-5} for a binary polar (AM) system. It is well established that FSK or PSK systems have about 3-dB improvement. In this case only a 9.5-dB signal-to-noise ratio would be needed, if all other factors were held constant (no other contributing factors). Assume the input from the line to be -10 dBm0 to satisfy loading conditions. To maintain the proper signal-to-noise ratio, the channel noise must be down to -19.5 dBm0. To improve the modulation rate without expense of increased bandwidth, quaternary phase-shift keying (four-phase) is used. This introduces a 3-dB noise-degradation factor, bringing the required noise level down to -22.5 dBm0.

Now consider the effects of EDD. It has been found that for a four-phase differential system, this degradation will amount to 6 dB if the permissible delay distortion is one pulse length. This impairment brings the noise requirement down to −28.5 dBm0 of average noise power in the voice channel. Allow 1 dB for frequency-translation error or other factors, and the noise requirement is now down to −29 dBm0. If the transmission level were reduced by 3 dB to −13 dBm0 (see Section 10.3.4), the noise level must be reduced downward another 3 dB to −32.5 dBm0 (or 19.5 dB signal-to-noise ratio). Thus it can be seen that to achieve a certain error rate for a given modulation rate, several modulation schemes should be considered. It is safe to say that the noise requirement will fall somewhere between −25 and −40 dBm0. This is well inside the CCITT figure of −43 dBmp referenced in Section 10.3.3.

10.6 Circuit Conditioning (Equalization)

Of the critical circuit parameters mentioned in Section 10.3, two that have severely deleterious effects on data transmission can be reduced to tolerable limits by circuit conditioning, often called *equalization*. These two are amplitude–frequency response (amplitude distortion) and EDD.

The most common method of performing equalization or conditioning is the use of several networks in tandem. Such networks tend to flatten response and in the case of amplitude response, add attenuation increasingly toward channel center and less toward its edges. The overall effect is one of making the amplitude response flatter. The delay equalizer operates in a similar manner. Delay increases toward channel edges parabolically from the center. To compensate, delay is added in the center much like an inverted parabola, with less and less delay added as the band edge is approached. Thus the delay response is flattened at some small cost to absolute delay, which has no effect in most data systems. However, care must be taken with the effect of a delay equalizer on an amplitude equalizer and conversely, of an amplitude equalizer on the delay equalizer. Their design and adjustment must be such that the flattening of the channel for one parameter does not entirely distort the channel for the other.

Another type of equalizer is the transversal type of filter, which is useful where it is necessary to select among, or to adjust, several attenuation (amplitude) and phase characteristics. The basis of the filter is a tapped delay line to which the input is presented. The output is taken from a summing network that adds or sums the outputs of the taps. Such a filter is adjusted to the desired response (equalization of both phase and amplitude) by adjusting the contributions for each tap.

If the characteristics of a line are known, a common method of equalization is predistortion of the output signal of the data set. Some devices use a shift register and a summing network. If the equalization needs to be varied, a feedback circuit from the receiver to the transmitter would be required to control the shift register. This type of dynamic predistortion is practical for binary transmission only.

A major drawback of all the equalizers discussed (with the exception of the latter one with the feedback circuit) is that they are useful only on dedicated or leased circuits where the circuit characteristics are known and remain fixed. Obviously, a switched circuit would require a variable automatic equalizer, or conditioning would be required on every circuit in the switched system that would be transmitting data.

Circuits are usually equalized on the receiving end. This is called *postequalization.* An equalizer must be balanced and must present the proper impedance to the line. Administrations may choose to condition (equalize) trunks and attempt to eliminate the need to equalize station lines; the economy of considerably fewer equalizers is obvious. In addition, each circuit that would possibly carry high-speed data in the system would have to be equalized, and the equalization must be sufficient for any possible combination to meet the overall requirements. If equalization requirements become greater (i.e., parameters more stringent), the maximum number of circuits (trunks) in tandem may have to be restricted.

Conditioning to meet amplitude-frequency response requirements is less exacting in the overall system than is envelope delay. Equalization for envelope delay and its associated measurements are time consuming and expensive. In general, envelope delay is arithmetically cumulative. If there is a requirement of overall envelope delay distortion of 1 ms for a circuit between 1000 and 2600 Hz, then in three links in tandem, each link must be better than 333 μs between the same frequency limits. For four links in tandem, each link would have to be at least 250 μs. In practice, accumulation of delay distortion is not entirely arithmetical, as it results in a reduction of requirements by about 10%. Delay distortion tends to be inversely proportional to the velocity of propagation. Loaded cables display greater delay distortion than do nonloaded cables. Likewise, with sharp filters a greater delay is experienced for frequencies approaching band edge than for filters with a more gradual cutoff.

In carrier multiplex systems, channel banks contribute more to the overall EDD than does any other part of the system. Because channels 1 and 12 of the standard CCITT modulation plan (those nearest the group band edge) suffer additional delay distortion because of group filter effects and, in some cases, supergroup filters, the system engineer should allocate channels for data transmission near group and supergroup center. On long-haul critical-data systems, the data channels should be allocated to through-groups and through-supergroups, minimizing as much as possible the steps of demodulation back to voice frequencies (channel demodulation).

Automatic equalization for both amplitude and delay shows promise, particularly for switched data systems. Such devices are self-adaptive and require a short adaptation period after switching, on the order of 1 to 2 s. This can be carried out during synchronization. Not only is the modem clock being "averaged" for the new circuit on transmission of a synchronous idle signal, but the self-adaptive equalizer adjusts for optimum equalization as well. The major drawback of adaptive equalizers is cost.

10.7 Practical Modem Applications

10.7.1 Voice-Frequency Carrier Telegraph

Narrow-shifted FSK transmission of digital data are commonly referred to as voice-frequency telegraph (VFTG) and voice-frequency carrier telegraph (VFCT).

In practice, VFCT techniques handle data rates of up to 1200 bps by a simple application of FSK modulation. The voice channel is divided into segments or frequency bounded zones or bands. Each segment represents a data or telegraph channel, each with a frequency-shifted subcarrier.

For proper end-to-end system interface, it is convenient to use standard modulation plans, particularly on international circuits. For the far-end demodulator to operate with the near-end modulator, the former must be tuned to the same center frequency and accept the same shift. Center frequency is that frequency in the center of the passband of the modulator–demodulator. The shift is the number of hertz that the center frequency is shifted up and down in frequency for the mark–space condition. From Table 8.1, by convention, the mark condition is the center frequency shifted downward and the space, upward. For modulation rates below 80 bps, bandpasses have either 170-Hz (CCITT Rec. R.39) or 120-Hz bandwidths, with frequency shifts of ±42.5 or ±30 Hz, respectively. The CCITT recommends (R.31) the 120-Hz channels for operating at 50 bps and below; however, some administrations operate these channels at higher modulation rates.

The number of tone telegraph or data channels that can be accommodated on a voice channel depends partly on the usable voice-channel bandwidth. For high-frequency radio with a voice-channel limit on the order of 3 kHz, 16 channels may be accommodated using 170-Hz spacing (170 Hz between center frequencies). On the nominal 4-kHz voice channel twenty-four VFCT channels may be accommodated between 390 and 3210 Hz with 120-Hz spacing, or 12 channels with 240-Hz spacing. This can easily meet standard telephone frequency division multiplex carrier channels of 300 to 3400 Hz (see Chapter 10, Section 6).

Some administrations use a combination of voice and telegraph data simultaneously on a telephone channel. This technique is commonly referred to as "voice plus" or $S + D$ (speed plus derived). There are two approaches to this technique. The first is recommended by CCITT and is used widely by INTELSAT order wires. It places five telegraph channels (channels 20 through 24) above a restricted voice band with a roofing filter*

*This is a low-pass filter.

near 2500 Hz. Speech occupies a band between 300 and 2500 Hz. Up to five 50-bps telegraph channels appear above 2500 Hz. The second approach removes a slot from the center of the voice channel into which up to two telegraph channels may be inserted. The slot is a 500-Hz band centered on 1275 Hz.

However, some administrations use a slot for telegraphy of frequencies 1680 Hz and 1860 Hz by either amplitude or frequency modulation (FSK) (see CCITT Rec. R.43). The use of speech plus should be avoided on trunks in large networks because it causes degradation to speech and also precludes the use of the channel for higher-speed data. In addition, the telegraph channels have to be removed before going into two-wire telephone service (i.e., at the hybrid or term set); otherwise, service drops to half-duplex on telegraph.

10.7.2 Medium Data-Rate Modems

In normal practice, FSK is used for the transmission of data rates up to 1200 bps. The 120-Hz channel is nominally modified such that one 240-Hz channel replaces two 120-Hz channels. Administrations use the 240-Hz channel for modulation rates of up to 150 bps. The same process can continue using 480-Hz channels for 300 bps FSK and 960-Hz channels for 600 bps. The CCITT (Rec. in *White Books,* V.23 Vol. III) specifies 600/1200 bps operation in the nominal 4-kHz voice band. In reference to "2000 bps Modem Standardized for Use in the General Switched Telephone Networks," CCITT Rec. V.21 (in *Orange Books,* Vol. VIII) recommends:

Frequency shift + and − 100 Hz.
Center frequency of channel 1, 1080 Hz.
Center frequency of channel 2, 1750 Hz.
In each case space (0) is the higher frequency.

Recommendation V.21 also provides for a disabling tone on echo suppressors, which is a very important consideration on long circuits. Recommendation V.22 (also in *Orange Books,* Vol. VIII) standardizes modulation rates for synchronous data transmission at 600 and 1200 bps. Recommendation V.23 stipulates 600/1200 bps modem standardized for use in the general switched telephone network for application to synchronous or asynchronous systems. Provision is made for an optional backward channel for error control (ARQ). For the forward channel, the following modulation rates and characteristic frequencies are presented:

	F_0	F_z	F_a
Mode 1 : up to 600 bps	1500 Hz	1300 Hz	1700 Hz
Mode 2 : up to 1200 bps	1700 Hz	1300 Hz	2100 Hz

The backward channel for error control is capable of modulation rates of up to 75 bps. Its mark and space frequencies are

F_z	F_a
390 Hz	450 Hz

Refer to Table 8.1 for the mark–space convention (F_z = mark or binary "1," F_a = space or binary "0," and F_0 = center frequency). The CCITT has tried to achieve a universality, recommending a modem that can be used nearly anywhere in the world "in the general switched telephone network." It considers worst-case conditions of amplitude–frequency response and EDD.

10.7.3 High Data-Rate Modems

Section 10.4 gave a limit on the data rate in a bandwidth of 3000 Hz. The Nyquist limit was 6000 bps for binary transmission. It was also noted that a practical limit for this bandwidth is about 3000 bps without automatic equalization. For the binary case, this would be equivalent to 1 bit per hertz.* For the quaternary case, a data rate of 6000 bps may be reached (3000 bauds $\simeq$ 6000 bps) or 2 bits per hertz. However, most telephone circuits do not have 3000 Hz of usable bandwidth available. See Table 8.2 for the modems and their required bandwidths and modulation schemes, which permit an improved data rate on a telephone channel.

10.8 Serial-to-Parallel Conversion for Transmission on Impaired Media

The transmission medium, in most cases the voice channel, often cannot support a high data rate, even with conditioning. The impairments may be due to poor amplitude–frequency response, EDD, or excessive impulse noise. One possible solution is to convert the high-speed serial bit stream at the dc level (e.g., demodulated) to a number of lower-speed parallel bit streams. A technique widely used is to divide a 2400-bit serial stream into 16 parallel streams, each carrying 150 bps. If each slower stream is di-bit coded (2 bits at a time; discussed in Section 10.5) and applied to a QPSK tone modulator, the modulation rate on each subchannel is reduced in this case to 75 bauds. The equivalent period for a di-bit interval is 1/75 s or 13 ms.

There are two obvious advantages to this technique. First, each subchannel has a comparatively small bandwidth and thus looks at a small and tolerable segment of the total delay across the channel. The EDD impairment is less on slower-speed channels. Second, there is less chance of a noise burst or hit of impulse noise smearing the subchannel signal beyond

*In this case 3000 bauds is equivalent to 3000 bps.

Table 8.2 High Data-rate Modems for a Telephone Channel

Data Rate (bps)	Modulation Rate (bauds)	Modulation	Bits per hertz	Bandwidth Required (Hz)
1. 2400 Synchronous	1200	Differential four-phase	2	1200
2. 4800 Synchronous	1600	Differential four-phase	3	1600
3. 3600 Synchronous	1200	Differential four-phase, two-level (combined PSK–AM)	3	1200
4. 2400 Synchronous	800	Differential eight-phase	3	800
5. 9600 Synchronous	4800	Differential two-phase, two-level	2	2400[a]

[a]Uses automatic equalizer.

recognition. If the duration of noise burst is less than half the pulse width, the data pulse can be regenerated and the pulse will not be in error. The longer the pulse width, the less chance of disturbance from impulse noise. In this case the symbol interval or pulse width has had an equivalent lengthening by a factor of 32.

10.9 Parallel-to-Serial Conversion for Improved Economy of Circuit Usage

Long, high-quality (conditioned) toll telephone circuits are costly to lease or are a costly investment. The user is often faced with a large number of slow-speed circuits (50 to 300 bps) that originate in one general geographic location, with a general destination to another common geographic location. If we assume these to be 75-bps circuits (100 wpm), which are commonly encountered in practice, then only 18 to 24 telegraph channels can be transmitted on a high-grade telephone channel by conventional voice-frequency telegraph techniques (see Section 10.7.1).

Circuit economy can be affected using a data–telegraph time-division multiplexer. A typical application of this type is illustrated in Figure 8.17, which shows one direction of transmission only. Here incoming, slow-speed VFCT channels are converted to equivalent dc bit streams. Up to 32 of these bit streams serve as input to a time-division multiplexer in the application illustrated in Figure 8.17. The output of the multiplexer is a 2400-bit synchronous serial bit stream. This output is fed to a conventional 2400-bit

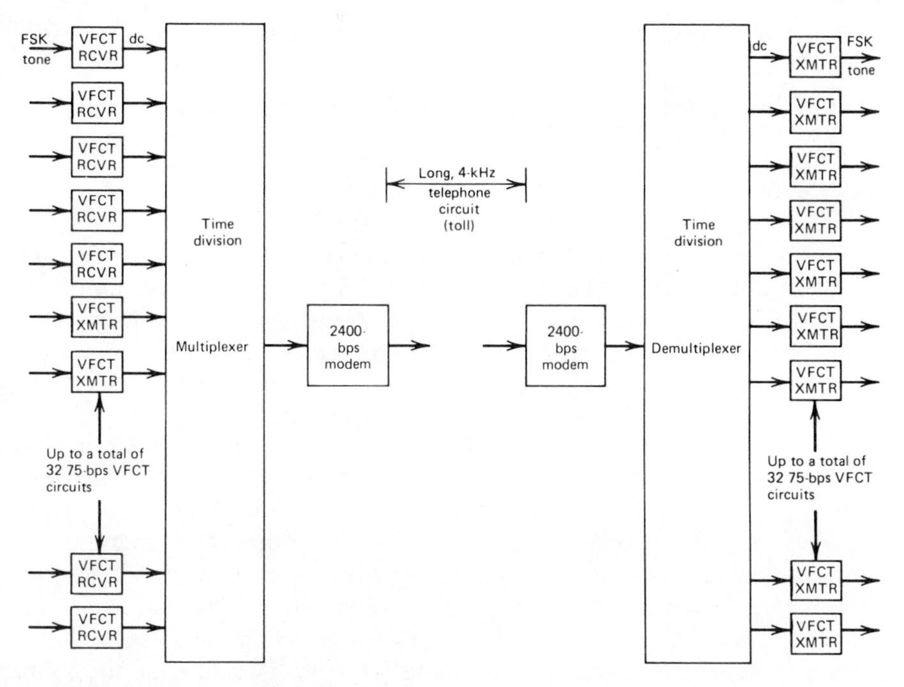

Figure 8.17 Typical application of parallel-to-serial conversion: VFCT, voice frequency carrier telegraph; rec., receiver (converter); xmt, transmitter (keyer); bps, bits per second; FSK, frequency shift keying.

modem. At the receiving end the 2400 bps serial stream is demodulated to dc and fed to the equivalent demultiplexer. The demultiplexer breaks the serial stream down back to the original 75-bps circuits.

Figure 8.17, which illustrates the concept of parallel-to-serial conversion, does not show clocking or other interconnecting circuitry. By use of a time-division multiplexer, the circuit utilization can be increased by a factor of nearly 2. Whereas only about 18 75-bps circuits can be transmitted on a good telephone channel by conventional VFCT means, an equivalent of up to 32 such circuits can be transmitted on the same channel by means of the multiplexer.

11 FACSIMILE

11.1 Background

Facsimile (fax) is a method of graphics communication. It permits the electrical transmission of printed or pictorial matter from one location to another with a reasonably faithful copy permanently recorded at the receiving end. Facsimile transmission over telephone lines has had widespread application since the 1920s. Before World War II its principal application was the transmission of weather maps and "wirephoto," which is the transmission of pictures for newspapers.

Today facsimile is showing new vigor, particularly in the office environment. Of course, it is still used for the transmission of weather maps and news photographs. In the commercial field facsimile is used for the transmission of way bills by trucking firms, signature verification by banks, and payment notices. In publishing it is used to expedite graphic communication, to dispatch news copy from satellite offices or bureaus to a newspaper's main newsroom, and to eliminate duplication of typesetting effort between separate printing facilities. It is also used in engineering and production for the transmission of engineering drawings. In law enforcement fax is used to transmit mugshots, and high-resolution fax is used for the transmission of fingerprints. "Electronic mail" is yet another application.

It may be said that data–telegraph methods of transmitting graphics are more rapid than fax. Whereas a standard printed page may take only 3 min to transmit by digital data techniques (transmitting 300 characters per second), that same page may take up to 6 min to transmit by facsimile. However, in the case of data transmission, we often forget that to the 3 min of transmit time we must add the operator keyboard time to compose the page from original copy. With fax, the original copy is only inserted into the machine. Further, fax is less error prone—considerably less, in fact. However, once we enter the domain of computer-to-computer or terminal-to-computer operations, such communication should be left to the techniques of data transmission explored earlier in this chapter.

11.2 Introduction to Facsimile Systems

A fax system consists of some method of converting graphic copy on paper to an electrical equivalent suitable for transmission on a telephone pair, transmission and connection of the pair–telephone circuit to the desired distant user, and the recording–printing of the copy by that user. Basically, we are dealing with analog technology. However, there is a marked trend toward digital techniques. Figure 8.18 is a simplified block diagram of a facsimile system.

11.3 Scanning

The purpose of scanning is to produce an electrical analog signal representing the graphic copy to be transmitted. Scanning can be carried out by (1) a spot of light scanning a fixed copy or (2) the copy moving across a fixed spot of light. In either method the light is reflected from the copy to a photoelectric cell that senses the tonal variations as a function of the mirror

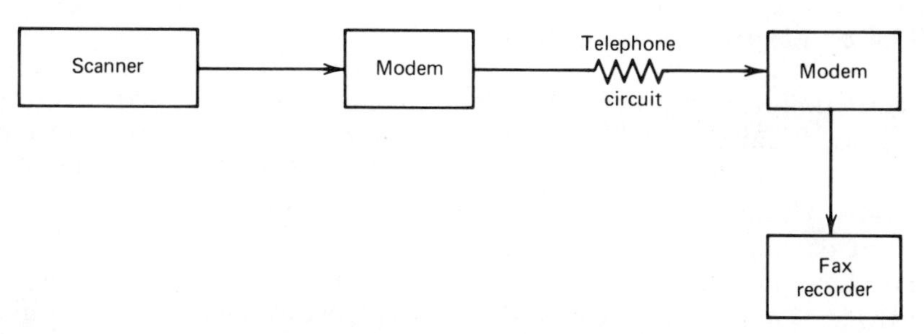

Figure 8.18 A facsimile (fax) system.

reflection from the copy. Thus a scanner is simply a photoelectric transducer.

There are two approaches to lighting the copy on the scanner, spot and flood. Both employ the technique of "bouncing" (reflecting) light off the graphic material to be transmitted. The spot type of scanning projects a tiny spot of light onto the surface of the printed copy. The reflection of the spot is then picked up directly by a photoelectric cell or transducer. With flood projection, the copy is illuminated with diffuse light in the area of scan. The reflected light is then optically projected through a very small aperture onto the cathode of a photoelectric transducer.

Scanning in modern facsimile systems is electromechanical, either with cylinder scanning or flat-bed scanning. In the former case the copy to be transmitted is wrapped around a cylinder. The cylinder is rotated such as to effect a continuous helical scan of the entire copy. Flat-bed scanners use a feed mechanism. Copy to be transmitted is fed into a slot where the mechanism takes over slowly advancing the copy through the machine. In another flat-bed method a flying spot is used where the copy remains at rest and the spot does all the movement. Optical transducers such as those used on facsimile scanner are based on the photoelectric cell or the photomultiplier tube. Most electronic scanners require a highly stabilized dc source.

11.4 Recording

11.4.1 Introduction

Recording is the reproduction from an electric signal of visual copy of graphic material. Like scanning at the transmit end of a fax circuit, recording at the receive end is nearly always electromechanical—drum or flatbed. There are four basic electromechanical processes of recording in use today: (1) electrolytic, (2) electrothermal, (3) electropercussive, and (4) electrostatic. There is one well-established recording process that is not electromechanical. It is photographic and is used in the facsimile reproduction of newspapers.

11.4.2 Electrolytic Recording

Electrolytic recording, the oldest and one of the most popular recording processes, requires a special type of paper that is saturated with an electrolyte. When an electric current is passed through the paper, it tends to discolor. The amount of discoloration or darkness is a function of the current passing through the paper.

For facsimile recorders, the electrolytic paper is passed between two electrodes. One is a fixed electrode, a backplate or platen on the machine; the other is the moving stylus. Horizontal lines of varying darkness appear on the paper as the stylus sweeps across the sheet. As each recorded horizontal line is displaced one line width per sweep of the stylus, a pattern begins to take shape. This pattern is a facsimile of the original pattern transmitted from the distant-end scanner.

This older, conventional stylus–backplate recording method is now giving away to what is called "helix-blade" recorders. The concept is basically the same, except that a special drum containing a helix at the rear and a stationary blade in front are used as the two electrodes. The drum-helix makes one complete revolution per scanning line. The rotating drum moving the helix carries out the same function as the moving stylus of the more conventional electrolytic recorders.

11.4.3 Electrothermal Recording

This process of facsimile recording is called "thermal" only because it appears to be so. It is similar to the electrolytic process in that the recording paper is interposed between two electrodes. The recorded pattern is made by arcing electric current through the paper. The arc gives the appearance of burning or a thermal process. A major characteristic of electrothermal recording is the high contrast achieved.

11.4.4 Electropercussive Recording

Electropercussive recording in facsimile is similar to the recording of audio on a record. An amplified facsimile signal is fed to an electromagnetic transducer that actuates a stylus in response to electrical signal variation. If a sheet of carbon paper is interposed between the stylus and a sheet of plain paper, a carbon impression is made on the paper by the vibrations of the stylus in accordance with the signal variation. The amount or intensity of darkness varies in proportion to the variations in strength of the picture signal. Electropercussive recording is also known by the terms *pigment transfer, impact,* or *impression recording.*

11.4.5 Electrostatic Recording

There are essentially two types of electrostatic recording, both of which are based on printing the image from a cathode ray tube (CRT): (1) transfer xerography and (2) a direct copy requiring special recording paper. It must be pointed out that the image tube need not necessarily be a CRT. Other image devices with the proper scanning mechanism can be used, such as the crater tube.

11.5 Technical Requirements for Facsimile Transmission

11.5.1 Phasing and Synchronization

Phasing and synchronization of the far-end fax receiver with the transmitting scanner permit the reassembly of picture elements in the same spatial order as when the picture was scanned by the transmitter. Phasing and synchronization must be distinguished. Phasing ensures that the receiving recorder stylus coincides in time and position on the copy at the start of transmission. Synchronization keeps the two this way throughout the transmission of a single graphic copy.

In many systems phasing is carried out by what may be termed a "stop–start" technique, where the receiving end is not really stopped, only retarded until a start-of-stroke signal is received from the transmitting scanner. This is the phasing signal, which usually is a pulse at the scan line rate of fixed amplitude and duration. A phasing sequence often lasts for several seconds before scanning on one end and printing on the other commence. Thus phasing may be called the "retarding of recording stroke until the start of recording stroke coincides with the start of scan stroke." Of course, once the recorder is retarded to permit coincidence, it must be allowed to speed up or "catch up" with the scanner.

Synchronization assures that the fax recorder remains in perfect step with the transmit scanner. There are three methods of synchronization: (1) common ac power-source frequency, (2) individual stabilized frequency power sources, and (3) transmission of sync signal during picture transmission. Methods 1 and 2 use 50 Hz (in Europe) or 60 Hz (in North America) synchronous motors for scanning and recording. Frequency stability of the ac power grid or the individual ac stabilized frequency power sources should be 1×10^{-5} per day as a minimum. Any degradation in stability from this figure will cause skew of the received picture.

"Slaving" the fax recorder to the distant-end transmitter essentially circumvents the stability problem inherent in the first two methods of synchronization. There are two ways of accomplishing sync. The first uses a sync tone, usually a multiple of 50 or 60 Hz, that is transmitted above or below the picture signal in the voice-frequency passband. Effective filtering is required to separate picture from sync signal. The second method is to transmit pulses in the amplitude domain where the sync pulses are below the white level [or the black level, depending on signal sense (polarity)].

Another method of pulse synchronization that can be continuous throughout picture transmission uses short sync pulses that provide a check on the speed of the recorder drive motor. The pulses are transmitted at the scan line rate and do not interfere with the picture signal by transmitting them during line flyback or at start-of-line.

11.5.2 Index of Cooperation

The index of cooperation is defined by the IEEE as the product of scan density measured in lines per inch (LPI) times the effective stroke length. This is a basic standard for facsimile transmission. If a scanner–transmitter and receiver–recorder have the same index of cooperation, they are considered to be compatible. This only means that the received copy is a faithful reproduction of the transmitted copy. It does not mean, however, that they are the same size.

To convert IEEE standards to the CCITT index of cooperation, multiply the IEEE index by 0.318. CCITT (Rec. T.1) recommends an index of cooperation of 352 or, alternatively, 264. These are equivalent to the IEEE indices of 1105 and 829, respectively. The World Meteorological Organization (WMO) specifies 576 and 288, which are equivalent to IEEE 1809 and 904, respectively. The EIA recommends IEEE 829, which is equivalent to CCITT 264.

11.5.3 Transmission Methods and Impairments

General. The output of a fax scanner contains electrical transitions representing reflectance changes of scanned copy. These transitions have a frequency component from subaudio, very nearly dc, up through the lower audio range. The transmission problem that the telecommunication engineer must face is to condition or convert this equivalent frequency spectrum so it can pass over a telephone circuit. We remember that the CCITT voice channel is encompassed in the band 300 to 3400 Hz. The critical fax frequencies are in the subaudio range, as low as 20 Hz.

Modulation. To overcome the frequency-response problem, simple carrier techniques have been adopted. With facsimile transmission the output of the scanner modulates an audio carrier using vestigial sideband modulation with an 1800-Hz carrier. When modulated by a fax signal, the carrier contains about 1300 Hz of information in the lower sideband, with the upper sideband vestiges extending to about 2300 Hz. With this approach the vital frequencies of the output of a facsimile scanner can be transmitted over a telephone voice channel.

Frequency-modulation techniques generally are more desirable than amplitude modulation (as described earlier) for transmission over a switched public telephone network or over a radio transmission media. The reason basically is that FM tends to be more impervious to noise. A common standard for FM transmission of fax is 1500 Hz for white and 2300 Hz for black.

Impairments. For good fax signal quality, a signal-to-noise ratio of 30 dB or better is required. This value is relative to maximum signal power level. Level should be maintained to at least 0.4 dB or better during transmission. Level variation has little effect on FM; this, then, is another good reason to use FM over most switched telephone circuits and on HF radio. Delay distortion in the audio range of interest should be maintained within ±300 μs. This requirement may be difficult to meet on some switched telephone circuits.

11.6 Reduction of Redundancy

The transmission of facsimile is extremely redundant, particularly the transmission of white. With the advent of electronic data processing (EDP) and the microprocessor, methods of reducing redundancy have been developed. To apply this technology, the first step is to convert the analog facsimile signal to digital and then to process the digital information. One such method is called *run-length encoding*. It is based on the binary technique of viewing all digital elements as either black or white. In this system periodic sampling is carried out on the content of a scan stroke, determining whether black or white exists during a particular instant. A code is then transmitted indicating the color (e.g., black or white) and the length of time the color lasts. This information can be contained in only two code words. Another system uses a form of PCM to do this.

Yet another on-line digital system is multilevel, with the following binary notation:

11—White
10—Light gray
01—Dark gray
00—Black

For instance, 11 could be represented by a level of +2 V, 10 by +1 V, 01 by − 1 V, and so forth.

REFERENCES

1. International Telephone and Telegraph Corporation, *Reference Data for Radio Engineers*, 6th ed., Howard W. Sams, Indianapolis, 1976.
2. *Transmission Systems for Communications*, 4th ed., Bell Telephone Laboratories, Holmdel, N.J., 1974.
3. W. R. Bennett and J. R. Davey, "Data Transmission," McGraw Hill, New York, 1965.
4. A. M. Rosie, *Information and Communication Theory*, Van Nostrand Reinhold, London, 1973.
5. E. R. Berlekamp, Ed., *Key Papers in the Development of Coding Theory*, IEEE Press, New York, 1974.

6. *Understanding Telegraph Distortion,* Stelma, Stamford, Conn., 1962.
7. *IEEE Standard Dictionary of Electrical and Electronic Terms,* IEEE Press, New York, 1977.
8. CCITT, *Orange Books,* Vol. III, Geneva, 1976, G.Recommendations.
9. CCITT, *Orange Books,* Vol. VII, R. Recommendations, Geneva, 1976. (telegraph).
10. CCITT, *Orange Books,* Vol. VIII V. Recommendations, Geneva, 1976. (data).
11. R. W. Lucky, J. Salz, and E. J. Weldon, *Principles of Data Communication,* McGraw-Hill, New York, 1968.
12. H. Nyquist, "Certain Topics in Telegraph Transmission Theory," *BSTJ*, 617–644 (April 1928).
13. C. E. Shannon, "A Mathematical Theory of Communication," *BSTJ*, 379–428 (July 1948); 623–656 (October 1948).
14. J. Martin, *System Analysis for Data Transmission,* Prentice-Hall, Englewood Cliffs, N.J., 1972.
15. W. P. Davenport, *Modern Data Communications,* Hayden, New York, 1971.
16. S. Goldman, *Information Theory,* Dover, New York, 1968.
17. *DCS Autodin Interface and Control Criteria,* DCAC 370-D-175-1, Defense Communication Agency, Washington D. C., 1965.
18. M. P. Ristenhall, "Alternative to Digital Communications," *Proc. IEEE,* **61** (6) (June 1973).
19. D. R. McGlynn, *Distributed Processing and Data Communications*, Wiley, New York, 1978.
20. C. L. Cuccia, "Subnanosecond Switching and Ultra-speed Communications," *Data Commun.* (November 1971).
21. D. R. Doll, "Controlling Data Transmission Errors," *Data Dynamics,* (July 1971).
22. E. N. Gilbert, *Information Theory after Eighteen Years,* Bell Telephone Monograph, Bell Telephone Laboratories, Holmdel, N.J., 1965.
23. EIA Standard RS-232C, Electronic Industries Association, August 1969, Washington, D.C.
24. *Analog Parameters Affecting Voiceband Data Transmission—Description of Parameters,* Bell Systems Technical Reference Publication No. 41008, American Telephone and Telegraph Corporation, New York, October 1971.
25. A. Kreithen, *Data Communications Consultant,* Data Tactics, St. Paul, Minn., 1976.
26. R. L. Freeman, *Telecommunication Transmission Handbook,* Wiley, New York, 1975.
27. C. Machover, "Display Terminals—Status and Standards," *Data Commun.,* (October 1973).
28. MIL-STD-188C with Notice 1, June 1976, U.S. Department of Defense, Washington, D.C.
29. D. R. Doll, *Data Communications: Facilities, Networks and Systems Design*, Wiley, New York, 1977.
30. W. R. Bennett, *Lecture Notes on Digital Communication Systems,* Michigan Univ. Press, East Lansing, July 1969.
31. A. Lender, "The Duobinary Technique for High Speed Data Transmission," *IEEE Transact. Commun. Electron.* (May 1963).
32. *General Information—Binary Synchronous Communications,* IBM report no. GA27-3004-1, December 1969.
33. J. R. Edwards, "The Choice of Data Transmission System for Efficient Use of Transmitter Power," *Radio Electron. Eng.,* **43** (10), (October 1973).
34. John E. McNamara, *Technical Aspects of Data Communication,* Digital Equipment Corp., Maynard, Mass., 1977.
35. Madhu S. Gupta, Ed., *Electrical Noise: Fundamentals and Sources,* IEEE Press, New York, 1977.

Chapter 9

DIGITAL TRANSMISSION AND SWITCHING SYSTEMS

1 DIGITAL VERSUS ANALOG TRANSMISSION

There are two notable advantages to digital transmission that make it extremely attractive to the telecommunication system engineer when compared to its analog counterpart. Dealing in generalities, we can say that:

1. Noise does not accumulate at repeaters and thus becomes a secondary consideration in system design, whereas in analog systems it is the primary consideration.
2. The digital format lends itself ideally to solid-state technology and in particular to integrated circuits.

The major portion of information to be transmitted in a common-carrier network is analog in nature, such as voice and video. Now convert these signals to a digital format, and we can take advantage of the two important features listed. Some readers justifiably will ask about the apparent data–telegraph disparity covered in the previous chapter. Is that not digital?

The apparent ambiguity stems from the input–output (I/O) devices. The telephone microphone generates an electrical-signal equivalent of the voice actuating the diaphram, and this is analog in nature. On the other hand, the data–telegraph keyboard, tape reader, or computer delivers digital "1"s and "0"s to the line. To transmit this information over the telephone network, the digital signal is converted to an analog signal compatible with that network's facilities. The modem carries out this function.

The objective now is to do the reverse to the analog voice (telephone) signal, that is, to convert it to a digital signal that may be transmitted electrically. There are two different modulation methods commonly used to do this: pulse-code modulation (PCM), which is widely used for transmission in common-carrier communications, and delta modulation, which is finding broad application in military communications. Digital switches are now being implemented to accommodate these modulation types. The following paragraphs emphasize PCM because of its applicability to common-carrier communication. Delta modulation is reviewed in Section 7 of the chapter.

2 BASIS OF PULSE-CODE MODULATION

Pulse-code modulation is a method of modulation in which a continuous analog wave is transmitted in an equivalent digital mode. The cornerstone of an explanation of the functioning of PCM is the sampling theorem, which states [Ref. 5, Section 21]:

> If a band-limited signal is sampled at regular intervals of time and at a rate equal to or higher than twice the highest significant signal frequency, then the sample contains all the information of the original signal. The original signal may then be reconstructed by use of a low-pass filter.

As an example of the sampling theorem, the nominal 4-kHz channel would be sampled at a rate of 8000 samples per second (i.e., 4000 × 2).

To develop a PCM signal from one or several analog signals, three processing steps are required: *sampling, quantization,* and *coding*. The result is a serial binary signal or bit stream, which may or may not be applied to the line without additional modulation steps. At this point a short review of Chapter 8 may be in order to clarify use of terminology such as mark, space, regeneration, and information bandwidth. One major advantage of digital transmission is that signals may be regenerated at intermediate points of links involved in transmission. The price for this advantage is the increased bandwidth required for PCM. Common systems in broad use require 16 times the bandwidth of their analog counterpart (e.g., a 4-kHz analog voice channel requires 16 × 4 or 64 kHz when transmitted by PCM). Regeneration of a digital signal is simplified and particularly effective when the transmitted line signal is binary, whether neutral, polar, or bipolar. An example of a bipolar bit stream is shown in Figure 9.1.

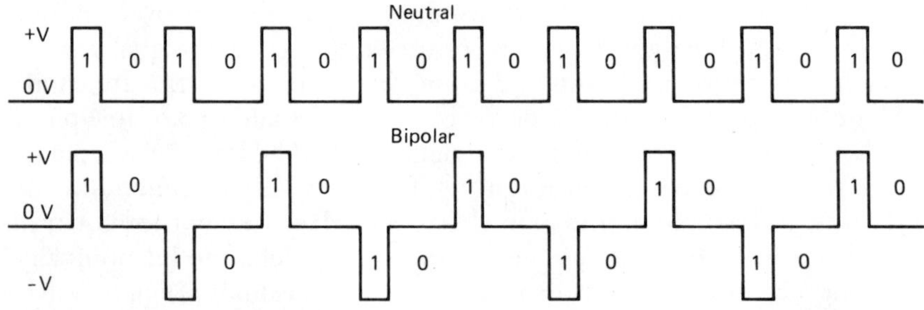

Figure 9.1 Neutral versus bipolar bit streams. The upper drawing illustrates alternate "1"s and "0"s transmitted in the neutral mode; the lower drawing illustrates the equivalent in a bipolar mode.

Binary transmission tolerates considerably higher noise levels (i.e., degraded signal-to-noise ratios), when compared to its analog counterpart (i.e., FDM, Chapter 5). This fact, in addition to the regeneration capability, is a great step forward in transmission engineering. The regeneration that takes place at each repeater by definition recreates a new digital signal; therefore, noise, as we know it, does not accumulate.

Error rate is another important factor in the design of PCM systems. If the error rate on a PCM system can be maintained end-to-end to one error in 10^5 bits, intelligibility will not be degraded. Even with an error rate of one bit in 10^3, intelligibility is fairly good. However, when errors exceed one in 10^2, intelligibility is lost. Another factor important in the design of PCM cable installations is crosstalk, which can degrade error performance. This is crosstalk spilling from one PCM system to another or in the same system from the send path to the receive path inside the same cable sheath.

3 DEVELOPMENT OF A PULSE-CODE-MODULATION SIGNAL

3.1 Sampling

Consider the sampling theorem given previously. If we now sample the standard CCITT voice channel, 300 to 3400 Hz (a bandwidth of 3100 Hz), at a rate of 8000 samples per second, we will have complied with the theorem and can expect to recover all the information in the original analog signal. Therefore, a sample is taken every 1/8000 s, or every 125 μs. These are key parameters for our future argument.

Another example may be a 15-kHz program channel. Here the lowest sampling rate would be 30,000 times per second. Samples would be taken at 1/30,000-s intervals, or at 33.3 μs.

3.2 The Pulse-Amplitude-Modulation Wave

With at least one exception (i.e., SPADE, Chapter 7, Section 7.7.3), practical PCM systems involve time division multiplexing. Sampling in these cases does not involve just one voice channel, but several. In practice, one system (see following paragraph) samples 24 voice channels in sequence, and another samples 32 channels. The result of the multiple sampling is a PAM (pulse amplitude modulation) wave. A simplified PAM wave is shown in Figure 9.2, in this case a single sinusoid. A simplified diagram of the processing involved to derive a multiplexed PAM wave is shown in Figure 9.3.

If the nominal 4-kHz voice channel must be sampled 8000 times per second, and a group of 24 such voice channels are to be sampled sequentially to interleave them forming a PAM multiplexed wave, this could be done by gating. The gate should be open for 5.2 μs (125/24) for each voice

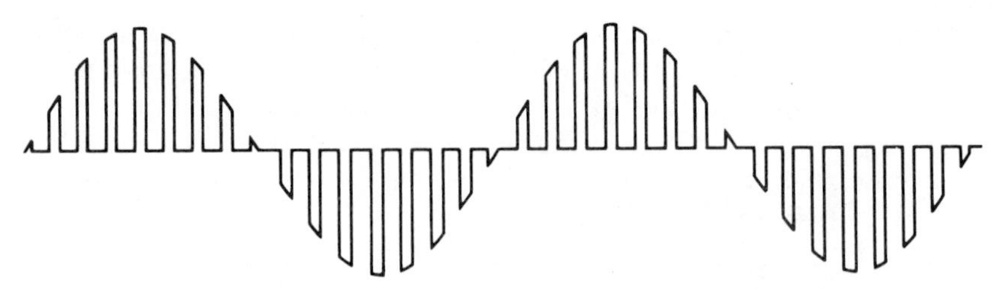

Figure 9.2 A PAM wave as a result of sampling a single sunusoid.

channel to be sampled successively from channels 1 through 24. This full sequence must be done in a 125-μs period ($1 \times 10^6/8000$). We call this 125-μs period a *frame*, and inside the frame all 24 channels are successively sampled once.

3.3 Quantization

It would appear that the next step in the process of forming a PCM serial bit stream would be to assign a binary code to each sample as it is presented to the coder.

Remember from Chapter 8 the discussion of code lengths, or what is more properly called coding "level." For instance, a binary code with four discrete elements (a four-level code) could code 2^4 separate and distinct meanings or 16 characters, not enough for the 26 letters in our alphabet; a five-level code would provide 2^5 or 32 characters or meanings. The ASCII is basically a seven-level code allowing 128 discrete meaning for each code combination ($2^7 = 128$). An eight-level code would yield 256 possibilities.

Another concept that must be kept in mind as the discussion leads into coding is that bandwidth is related to information rate (more exactly to modulation rate) or, for this discussion, to the number of bits per second transmitted. The goal is to keep some control over the amount of bandwidth necessary. It follows, then, that the coding length (number of levels) must be limited. As it stands, an infinite number of amplitude levels are being presented to the coder on the PAM highway. If the excursion of the PAM wave is 0 to +1 V, the reader should ask himself how many discrete values there are between 0 and 1. All values must be considered, even 0.0176487892 V.

The intensity range of voice signals over an analog telephone channel is on the order of 50 dB. The range −1 to 0 to 1 V of the PAM highway at the coder input may represent that 50-dB range. Further, it is obvious that the coder cannot provide a code of infinite length (e.g., an infinite number of coded levels) to satisfy every level in the 50-dB range (or a range from −1 to +1 V). The key is to assign discrete levels from −1 to +1 V (50-dB range). The assignment of discrete values to the PAM samples is called quantization. To cite an example, consider Figure 9.4.

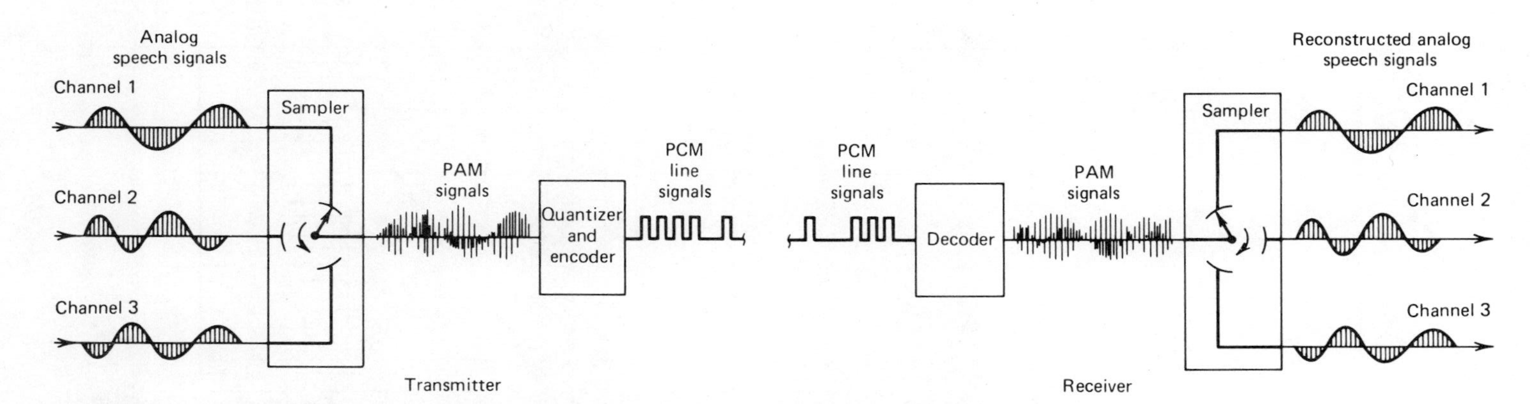

Figure 9.3 A simplified analogy of formation of a PAM wave. Courtesy of GTE Lenkurt Demodulator, San Carlos, Calif.

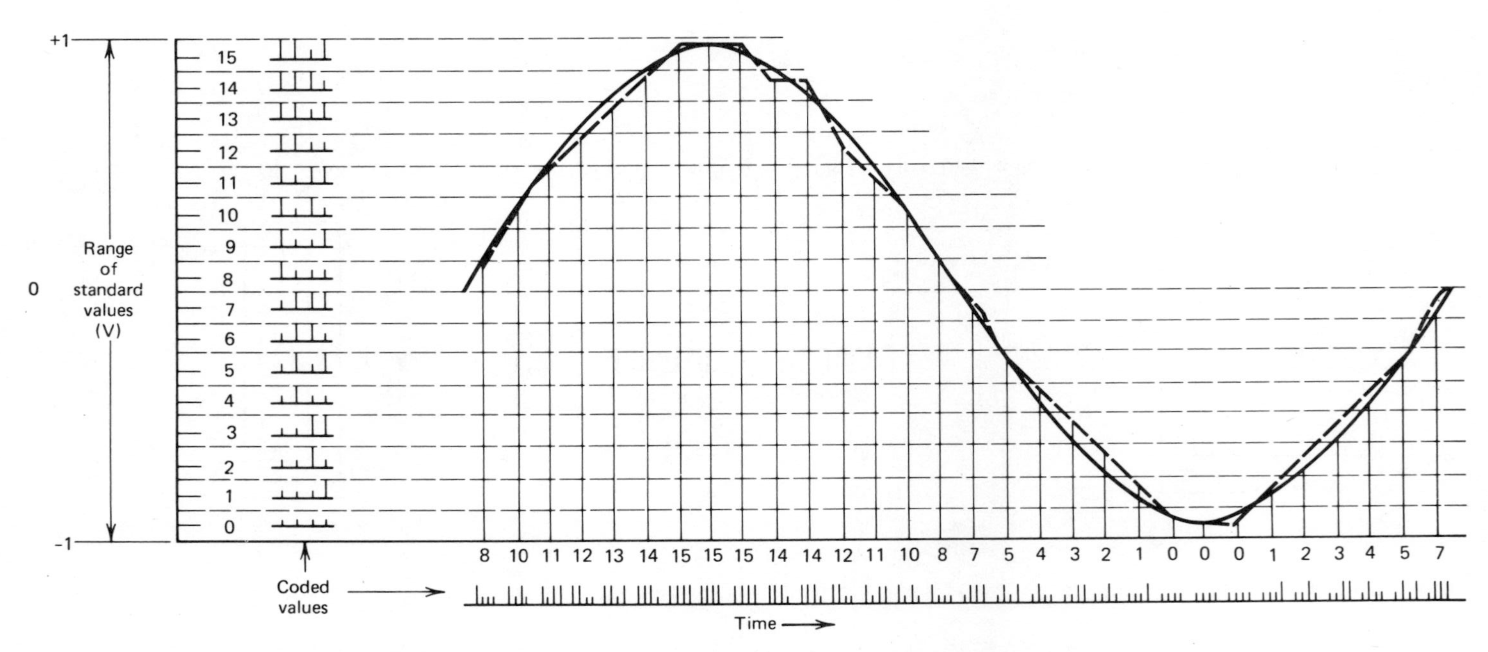

Figure 9.4 Quantization and resulting coding using 16 quantizing steps.

Sixteen quantum steps exist between −1 and +1 V and are coded as follows:

Step Number	Code	Step Number	Code
0	0000	8	1000
1	0001	9	1001
2	0010	10	1010
3	0011	11	1011
4	0100	12	1100
5	0101	13	1101
6	0110	14	1110
7	0111	15	1111

Examination of Figure 9.4 shows that step 12 is used twice. Neither time it is used is it the true value of the impinging sinusoid. It is a rounded-off value. These rounded-off values are shown with the dashed line in Figure 9.4, which follows the general outline of the sinusoid. The horizontal dashed lines show the point where the quantum changes to the next higher or next lower level if the sinusoid curve is above or below that value. Take step 14 in the curve, for example. The curve, dropping from its maximum, is given two values of 14 consecutively. For the first, the curve is above 14, and for the second, below. That error, in the case of "14," for instance, from the quantum value to the true value, is called *quantizing distortion*. This distortion is the major source of imperfection in PCM systems.

In Figure 9.4, maintaining the −1-0- + 1 V relationship, let us double the number of quantum steps from 16 to 32. What improvement would we achieve in quantization distortion? First determine the step increment in millivolts in each case. In the first case the total range of 2000 mV would be divided into 16 steps, or 125 mV/step. The second case would have 2000/32 or 62.5 mV/step. For the 16-step case, the worst quantizing error (distortion) would occur when an input to be quantized was at the half-step level, or in this case, 125/2 or 62.5 mV above or below the nearest quantizing step. For the 32-step case, the worst quantizing error (distortion) would again be at the half-step level, or 62.5/2 or 31.25 mV. Thus the improvement in decibels for doubling the number of quantizing steps is:

$$20 \log \frac{62.5}{31.25} = 20 \log 2 \text{ or } 6 \text{ dB (approximately)}$$

This is valid for linear quantization only (see Section 3.6). Thus increasing the number of quantizing steps for a fixed range of input values reduces quantizing distortion accordingly. Experiments have shown that if 2048 uniform quantizing steps are provided, sufficient voice signal quality is achieved.

For 2048 quantizing steps, a coder will be required to code the 2048 discrete meanings (steps). Reviewing Chapter 8, we find that a binary code with 2048 separate characters or meanings (one for each quantum step) requires an 11-element code or $2^n = 2048$; thus $n = 11$. With a sampling rate of 8000 per second for each voice channel, the binary information rate per voice channel will be 88,000 bps. Consider that equivalent bandwidth is a function of information rate; thus the desirability of reducing this figure is obvious.

3.4 Coding

Practical PCM systems use seven- and eight-level binary codes, or

$$2^7 = 128 \text{ quantum steps}$$
$$2^8 = 256 \text{ quantum steps}$$

Two methods are used to reduce the quantum steps to 128 or 256 without sacrificing fidelity. These are nonuniform quantizing steps and companding prior to quantizing, followed by uniform quantizing. Keep in mind that the primary concern of digital transmission using PCM techniques is to transmit speech, as distinct from the digital transmission covered in Chapter 8, which dealt with the transmission of data and message information. Unlike data transmission, in speech transmission there is a much greater likelihood of encountering signals of small amplitudes than those of large amplitudes.

A secondary, but equally important, aspect is that coded signals are designed to convey maximum information considering that all quantum steps (meanings or characters) will have an equally probable occurrence (i.e., the signal-level amplitude is assumed to follow a uniform probability distribution between 0 and ± the maximum voltage of the channel). To circumvent the problem of nonequiprobability of signal level for voice signals, specifically, that lower-level signals are more probable than are higher-level signals, larger quantum steps are used for the larger-amplitude portion of the signal, and finer steps are used for the signals with low amplitudes. The two methods of reducing the total number of quantum steps can now be more precisely labeled:

- Nonuniform quantizing performed in the coding process.
- Companding (compression) before the signals enter the coder, which now performs uniform quantizing on the resulting signal before coding. At the receive end, expansion is carried out after decoding.

An example of nonuniform quantizing could be derived from Figure 9.4 by changing the step assignment. For instance, 20 steps may be assigned between 0.0 and +0.1 V (another 20 between 0.0 and −0.1, etc.), 15 between 0.1 and 0.2 V, 10 between 0.2 and 0.35 V, eight between 0.35 and 0.5 V, seven between 0.5 and 0.75 V, and four between 0.75 and 1.0 V.

Most practical PCM systems use companding to give finer granularity (more steps) to the smaller amplitude signals. This is instantaneous companding, as compared to the syllabic companding used in analog carrier telephony. Compression imparts more gain to lower amplitude signals. The compression and later expansion functions are logarithmic and follow one of two laws, the A law or the "mu" (μ) law. The curve for the A law may be plotted from the formula

$$Y = \frac{AX}{(1+\log A)} \qquad 0 \leqslant v \leqslant \frac{V}{A}$$

$$Y = \frac{1+\log (AX)}{(1+\log A)} \qquad \frac{V}{A} \leqslant v \leqslant V$$

where $A = 87.6$. The curve for the mu law may be plotted from the formula:

$$Y = \frac{\log (1 + \mu x)}{\log (1 + \mu)}$$

where $\mu = 100$ for the original North American T1 system and 255 for later North American (D2) systems and the CCITT 24-channel system (CCITT Rec. G.733). In these formulas:

$$X = \frac{v}{V}$$

$$Y = \frac{i}{B}$$

where v is the instantaneous input voltage, V is the maximum input voltage for which peak limitation is absent, i is the number of the quantization step starting from the center of the range, and B is the number of quantization steps on each side of the center of the range (CCITT Rec. G.711).

A common expression used in dealing with the "quality" of a PCM signal is the signal-to-distortion ratio (expressed in decibels). Parameters A and μ determine the range over which the signal-to-distortion ratio is comparatively constant. This is the dynamic range. Using a μ of 100 can provide a dynamic range of 40 dB of relative linearity in the signal-to-distortion ratio.

In actual PCM systems the companding circuitry does not provide an exact replica of the logarithmic curves shown. The circuitry produces approximate equivalents using a segmented curve, and each segment is linear. The more segments the curve has, the more it approaches the true logarithmic curve desired. Such a segmented curve is shown in Figure 9.5. If the μ law were implemented using a seven (height)-segment linear approximate equivalent, it would appear as shown in Figure 9.5. Thus on coding, the first three coded digits would indicate the segment number (e.g., $2^3 = 8$). Of the seven-digit code, the remaining four digits would

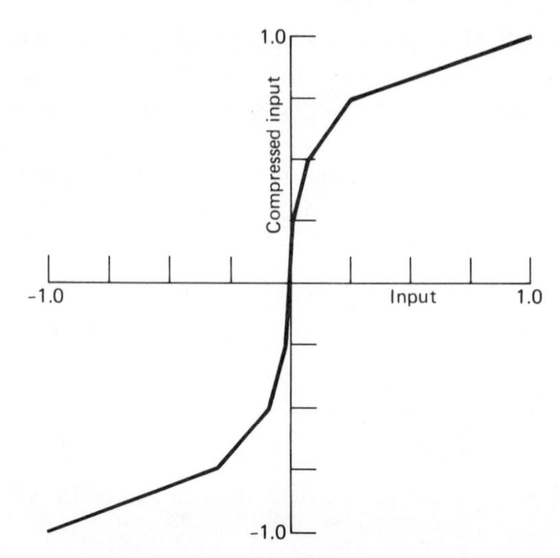

Figure 9.5 Seven-segment linear approximate of the logarithmic curve for μ law (μ = 100) [6]. Copy © 1970 Bell Telephone Laboratories.

divide each segment in 16 equal parts to further identify the exact quantum step (e.g., 2^4 = 16). For small signals, the companding improvement is approximately

A law: 24 dB
μ law: 30 dB

using a seven-level code. These values derive from the equation of companding improvement or

$$G_{\mathrm{dB}} = 20 \log \frac{\text{Uniform (linear) scale}}{\text{Companded scale}}$$

Coding in PCM systems utilizes straightforward binary codes. Examples of such coding are shown in Figure 9.6, which is expanded in Figure 9.7, and in Figure 9.8, which is expanded in Figure 9.9.

The coding process is closely related to quantizing. In practical systems, whether the A-law or the μ-law is used, quantizing employs segmented equivalents of the companding curve (Figures 9.6 and 9.8), as discussed earlier. Such segmenting is a handy aid to coding. Consider the European 30 + 2 PCM system, which uses a 13-segment approximation of the A-law curve (Figure 9.6). The first code element indicates whether the quantum step is in the negative or positive half of the curve. For example, if the first code element were a 1, it would indicate a positive value (e.g., the quantum step is located above the origin). The following three-code elements (bits) identify the segment, as there are seven segments above and seven segments below the origin (horizontal axis).

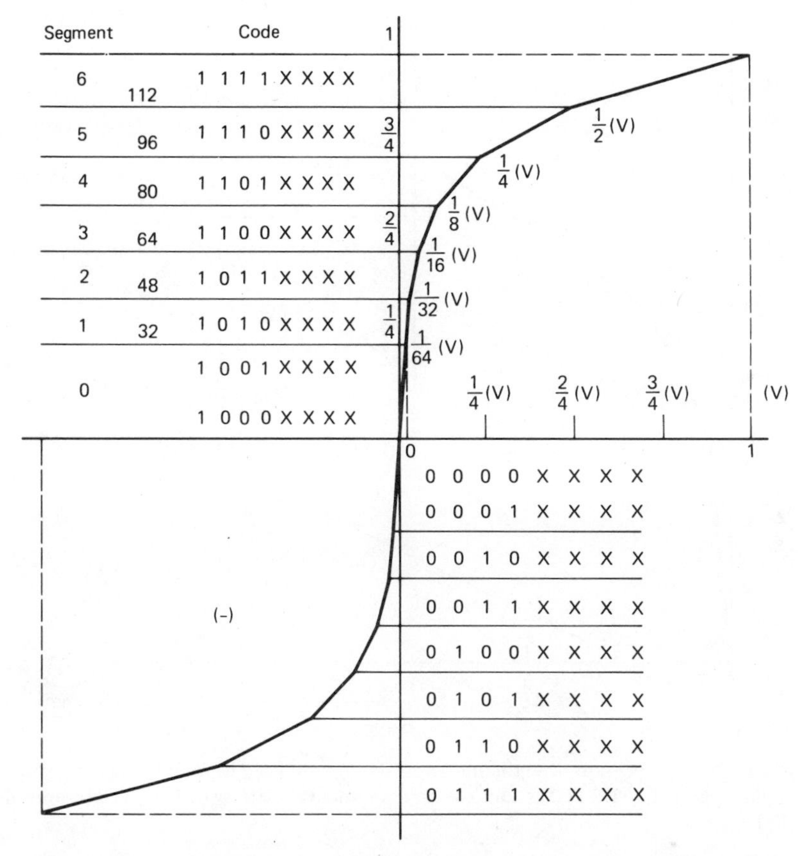

Figure 9.6 Quantization and coding used in the CEPT 30 + 2 PCM system.

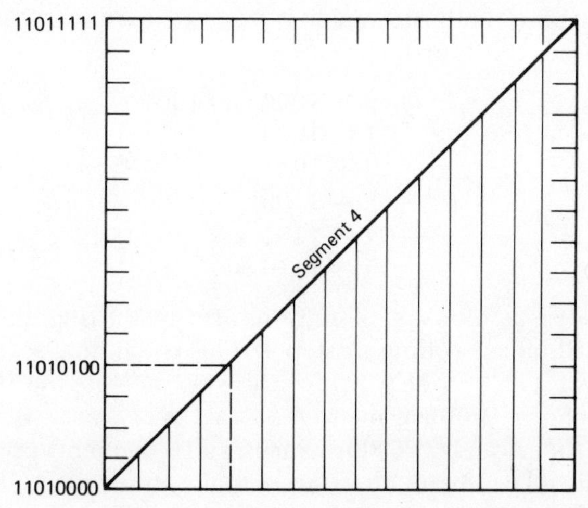

Figure 9.7 The CEPT 30 + 2 PCM system, coding of segment 4 (positive).

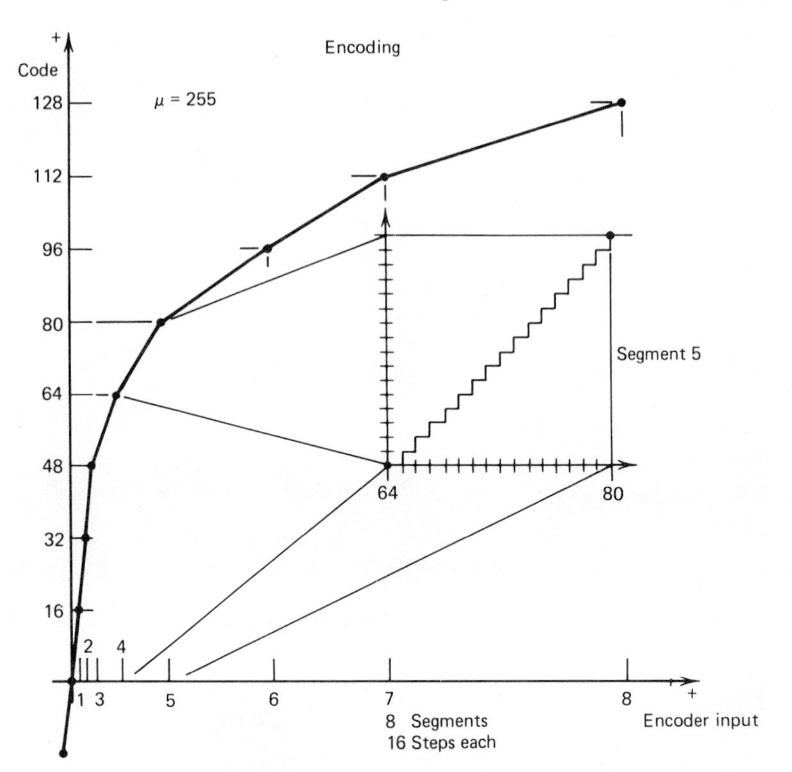

Figure 9.8 Positive portion of segmented approximation of μ law quantizing curve used in North American (ATT) D2 PCM channelizing equipment. Courtesy of ITT Telecommunications, Raleigh, N. C.

The first four elements of the fourth + segment are 1101. The first "1" indicates it is above the horizontal axis (e.g., it is positive). The next three elements indicate the fourth step or

0 — 1000 and 1001
1 — 1010
2 — 1011
3 — 1100
⟶ 4 — 1101
5 — 1110 etc.

Figure 9.7 shows a "blowup" of the uniform quantizing and subsequent straightforward binary coding of step 4. This is the final segment coding, the last 4 bits of a PCM code word for this system. Note the 16 steps in the segment, which are uniform in size.

The North American D2 PCM system uses a 15-segment approximation of the logarithmic μ law. Again, there are actually 16 segments. The segments cutting the origin are colinear and counted as one. The quantization in the

Code Level		Digit Number							
		1	2	3	4	5	6	7	8
255	(Peak positive level)	1	0	0	0	0	0	0	0
239		1	0	0	1	0	0	0	0
223		1	0	1	0	0	0	0	0
207		1	0	1	1	0	0	0	0
191		1	1	0	0	0	0	0	0
175		1	1	0	1	0	0	0	0
159		1	1	1	0	0	0	0	0
143		1	1	1	1	0	0	0	0
127	(Center levels)	1	1	1	1	1	1	1	1
126	(Nominal zero)	0	1	1	1	1	1	1	1
111		0	1	1	1	0	0	0	0
95		0	1	1	0	0	0	0	0
79		0	1	0	1	0	0	0	0
63		0	1	0	0	0	0	0	0
47		0	0	1	1	0	0	0	0
31		0	0	1	0	0	0	0	0
15		0	0	0	1	0	0	0	0
2		0	0	0	0	0	0	1	1
1		0	0	0	0	0	0	1	0
0	(Peak negative level)	0	0	0	0	0	0	1*	0

*One digit is added to ensure that the timing content of the transmitted pattern is maintained.

Figure 9.9 Eight-level coding of North American (ATT) D2 PCM system. Note that there are actually only 255 quantizing steps because steps "0" and "1" use the same bit sequence, thus avoiding a code sequence with no transitions (i.e., "0"s only).

D2 system is shown in Figure 9.8 for the positive portion of the curve. Segment 5, representing quantizing steps 64 through 80, is shown blown up in Figure 9.8. Figure 9.9 shows the D2 coding. As can be seen in this figure, again the first code element, whether a "1" or whether a "0," indicates whether the quantum step is above or below the horizontal axis. The next three elements identify the segment, and the last four elements (bits) identify the actual quantum level inside that segment. Of course, we see that the D2 is a basic 24-channel system using eight-level coding with μ-law quantization characteristic where $\mu = 255$.

3.5 The Concept of Frame

As shown in Figure 9.3, PCM multiplexing is carried out in the sampling process, sampling sources sequentially. These sources may be the nominal 4-kHz voice channels or other information sources, possibly data or video.

The final result of the sampling and subsequent quantization and coding is a series of pulses, a serial bit stream ("1"s and "0"s) that requires some indication or identification of the beginning of a scanning sequence. This identification is necessary at the far-end receiver so it will know exactly when each sampling sequence starts and ends; it times the receiver. Such identification is called *framing*, and a full sequence or cycle of samples is called a *frame* in PCM terminology.

Consider the framing structure of several practical PCM systems: the ATT D1 System is a 24-channel PCM system using a seven-level code (e.g., $2^7 = 128$ quantizing steps). To each 7 bits representing a coded quantum step, 1 bit is added for signaling. To the full sequence, 1 bit is added, and this is called a *framing bit*. Thus a D1 frame consists of

$$(7+1) \times 24 + 1 = 193 \text{ bits}$$

making up a full sequence or frame. By definition 8000 frames are transmitted so the bit rate is

$$193 \times 8000 = 1{,}544{,}000 \text{ bps}$$

The CEPT* 30 + 2 system is a 32-channel system where 30 channels transmit speech derived from incoming telephone trunks and the remaining two channels transmit signaling and synchronization information. Each channel is allotted a time slot (TS), and we can tabulate TS 0 through 31 as follows:

TS	Type of Information
0	Synchronizing (framing)
1–15	Speech
16	Signaling
17–31	Speech

In TS 0 a synchronizing code or word is transmitted every second frame, occupying digits 2 through 8 as follows:

0011011

In those frames without the synchronizing word, the second bit of TS 0 is frozen at a 1 so that in these frames the synchronizing word cannot be imitated. The remaining bits of time slot 0 can be used for the transmission of supervisory information signals (see Chapter 4).

The North American (ATT) D2 system is a 96-voice channel system made up of four groups of 24 channels each. A multiplexer is required to

*This is the Conference Européene des Postes et Télécommunications.

bring these four groups into a serial bit stream system. The 24-channel basic building block of the D2 system has the following characteristics:

255 quantizing steps, μ-law companding, 15-segment approximation, with $\mu = 255$ using an 8-element code

The makeup of the frame is similar to that of the D1 system, or

$$8 \times 24 + 1 : 193 \text{ bits per frame}$$

The frame structure is shown in Figure 9.10. Note that signaling is provided by "robbing" bit 8 from every channel in every sixth frame. For all other frames, all bits are used to transmit information coding.

Framing and basic timing should be distinguished. "Framing" ensures that the PCM receiver is aligned regarding the beginning (and end) of a sequence or frame; "timing" refers to the synchronization of the receiver clock, specifically, that it is in step with its companion (far-end) transmit clock. Timing at the receiver is corrected via the incoming mark-to-space

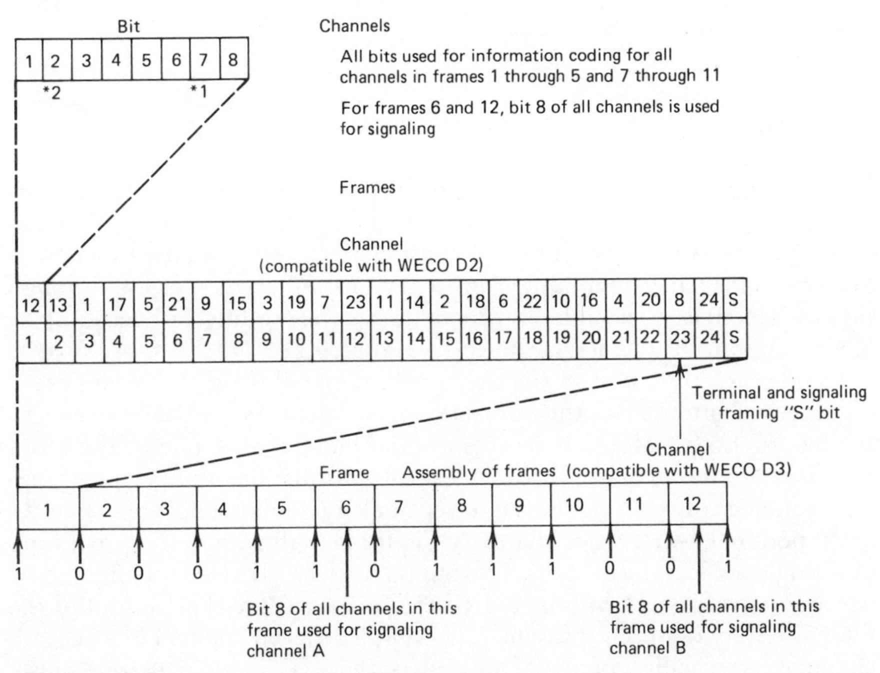

Figure 9.10 Frame structure of North American (ATT) D2 PCM system for channel bank. Note bit "robbing" technique used on each sixth frame to provide signaling information. Courtesy of ITT Telecommunications, Raleigh, N. C. [*Notes:* (1) If bits 1 to 6 and 8 are 0, then bit 7 is transmitted as 1; (2) bit 2 is transmitted as 0 on all channels for transmission of end-to-end alarm; (3) composite pattern 000110111001, etc.]

(and space-to-mark) transitions. It is important, then, that long periods of no transitions do not occur. This point is discussed later in reference to line codes and digit inversion.

3.6 Quantizing Distortion

Quantizing distortion has been defined as the difference between the signal waveform as presented to the PCM multiplex (codec*) and its equivalent quantized value. For a linear codec with n binary digits per sample, the ratio of the full-load sine wave power to quantizing distortion power (S/D) is [6]

$$\frac{S}{D} = 6n + 1.8 \text{ dB}$$

where n is the number of bits per PCM word, the word expressing the sample. For instance, the ATT D1 system uses a 7-bit word to express a sample (level), and the 30 + 2 and D2 systems use essentially 8 bits. If we had a 7-bit word and uniform quantizing, S/D would be 43.8 dB. Practical S/D values range in the order of 33 to 38 dB, depending largely on the talker levels (using 8-bit words).

4 PULSE-CODE-MODULATION SYSTEM OPERATION

Pulse-code modulation (PCM) is four-wire. Voice-channel inputs and outputs to and from a PCM multiplex are on a four-wire basis. The term "codec" is used to describe a unit of equipment carrying out the function of PCM multiplex and demultiplex and stands for *co*der-*dec*oder even though the equipment carries out more functions than just coding and decoding. A block diagram of a codec is shown in Figure 9.11.

A codec accepts 24 or 30 voice channels, depending on the system used; digitizes and multiplexes the information; and delivers a serial bit stream to the line of 1.544† Mbps. It accepts a serial bit stream at one or the other modulation rate, demultiplexes the digital information, and performs digital to analog conversion. Output to the analog telephone network is the 24 or 30 nominal 4-kHz voice channels. Figure 9.11 illustrates the processing of a single analog voice channel through a codec. The voice channel to be transmitted is passed through a 3.4-kHz low-pass filter. The output of the filter is fed to a sampling circuit. The sample of each channel of a set of n channels (n usually equals 24 or 30) is released in turn to the pulse-amplitude-modulation (PAM) highway. The release of samples is under

*Codec is a term used in PCM meaning COder–DECoder and is analogous to the modem described in Chapter 8.

†This is the rate for the ATT D2 24-channel bank.

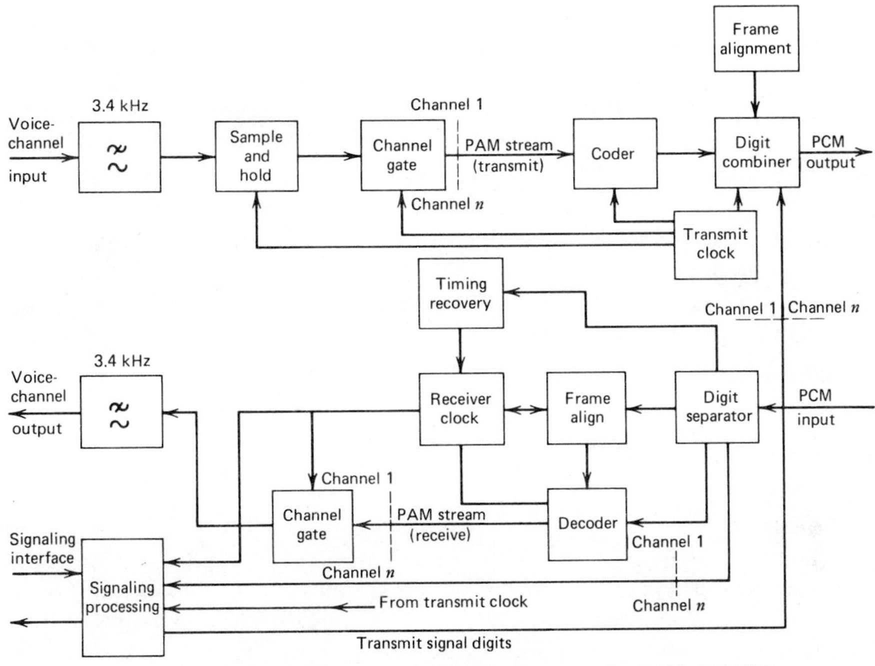

Figure 9.11 Simplified functional block diagram of a PCM CODEC.

control of a channel gating pulse derived from the transmit clock. The input to the coder is the PAM highway. The coder accepts a sample of each channel in sequence and then generates the appropriate signal character corresponding to each sample presented. The coder output is the basic PCM signal that is fed to the digit combiner where framing-alignment signals are inserted in the appropriate time slots, as well as the necessary supervisory signaling digits corresponding to each channel (European approach), and are placed on a common signaling highway that makes up one equivalent channel of the multiplex serial bit stream transmitted to the line. In North American practice supervisory signaling is carried out somewhat differently by "bit robbing" such as bit 8 in frame 6 and bit 8 in frame 12. Thus each equivalent voice channel carries its own signaling (see Figure 9.10).

On the receive side the codec accepts the serial PCM bit stream, inputting the digit separator where the signal is regenerated and split delivering the PCM signal to four locations to carry out the following processing functions: (1) timing recovery, (2) decoding, (3) frame alignment, and (4) signaling (supervisory). Timing recovery keeps the receive clock in synchronism with the far-end transmit clock. The receive clock provides the necessary gating pulses for the receive side of the PCM codec. The frame-alignment circuit senses the presence of the frame-alignment signal at the correct time interval, thus providing the receive terminal with frame alignment. The decoder, under control of the receive clock, decodes the code character signals corresponding to each channel. The output of the decoder is the reconstituted pulses making up a PAM highway. The channel gate accepts the PAM highway, gating the *n*-channel PAM highway in sequence under control of the receive clock. The output of the channel gate is fed in turn to each channel filter, thus enabling the reconstituted analog voice signal to reach to the appropriate voice path. Gating pulses extract signaling information in the signaling processor and apply this information to each of the reconstituted voice channels with the supervisory signaling interface as required by the analog telephone system in question.

5 PRACTICAL APPLICATION

5.1 General

Pulse-code modulation has found widest application in expanding interoffice trunks (junctions) that have reached or will reach exhaust* in the near future. An interoffice trunk is one pair of a circuit group that connects two switching points (exchanges). Figure 9.12 sketches the interoffice trunk concept. Depending on the particular application, at some point where distance *d* is exceeded it will be more economical to install PCM on existing VF cable plant than to rip up streets and add more VF cable pairs. For the planning engineer, the distance *d* where PCM becomes an economic alternative is called the "prove-in" distance. The distance *d* may vary from 8 to 16 km (5 to 10 mi), depending on the location and other circumstances. For distances less than *d*, additional VF cable pairs should be used for expanding plant.

The general rule for measuring expansion capacity of a given VF cable is as follows:

- For ATT D1/D2 channelizing equipment, two VF pairs will carry 24 PCM channels.
- For the CEPT 30 + 2 system as configured by ITT, two VF pairs plus a phantom pair will carry 30 PCM speech channels.

All pairs in a VF cable may not necessarily be usable for PCM transmission, partly because there is a possibility of excessive crosstalk between PCM carrying pairs. The effect of high crosstalk levels is to introduce digital errors in the PCM bit stream. Error rate may be related on a statistical basis

*"Exhaust" is an outside-plant term meaning that the useful pairs of a cable have been used up (assigned) from a planning point of view.

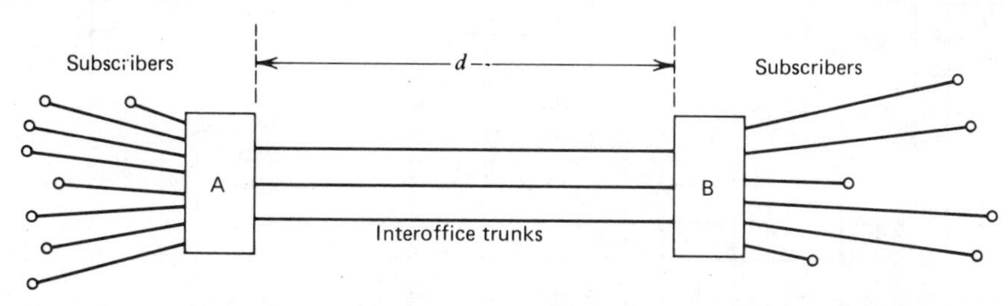

Figure 9.12 Simplified application diagram of PCM as applied to interoffice (interexchange) plant; *A* and *B* are switching centers.

to crosstalk, which, in turn, is dependent on the characteristics of the cable and the number of PCM carrying pairs.

One method for reducing crosstalk and thereby increasing VF pair usage is to turn to two-cable working, rather than have the "go" and "return" PCM cable pairs in the same cable. Another factor that can limit cable pair usage is the incompatibility of FDM and PCM carrier systems in the same cable. On the cable pairs that will be used for PCM, the following should be taken into consideration:

- All load coils must be removed.
- Build-out networks and bridged taps must also be removed.
- No crosses, grounds, splits, high-resistance splices, nor moisture are permitted.

The frequency response of the pair should be measured out to 1 MHz and considered as far out as 2.5 MHz. Insulation should be checked with a megger. A pulse reflection test using a radar test set is also recommended. Such a test will indicate opens, shorts, and high-impedance mismatches. A resistance test and balance test using a Wheatstone bridge may also be in order. Some special PCM test sets are available such as the GTE Lenkurt 91100 PCM cable test set using pseudorandom PCM test signals and the conventional digital test eye pattern.

5.2 Practical System Block Diagram

A block diagram showing the elemental blocks of a PCM transmission link used to expand installed VF cable capacity is shown in Figure 9.13. Most telephone administrations (companies) distinguish between the terminal area of a PCM system and the repeatered line. The term "span" comes into play here. A span line is composed of a number of repeater sections permanently connected in tandem at repeater apparatus cases mounted in manholes or on pole lines along the span. A "span" is defined as the group

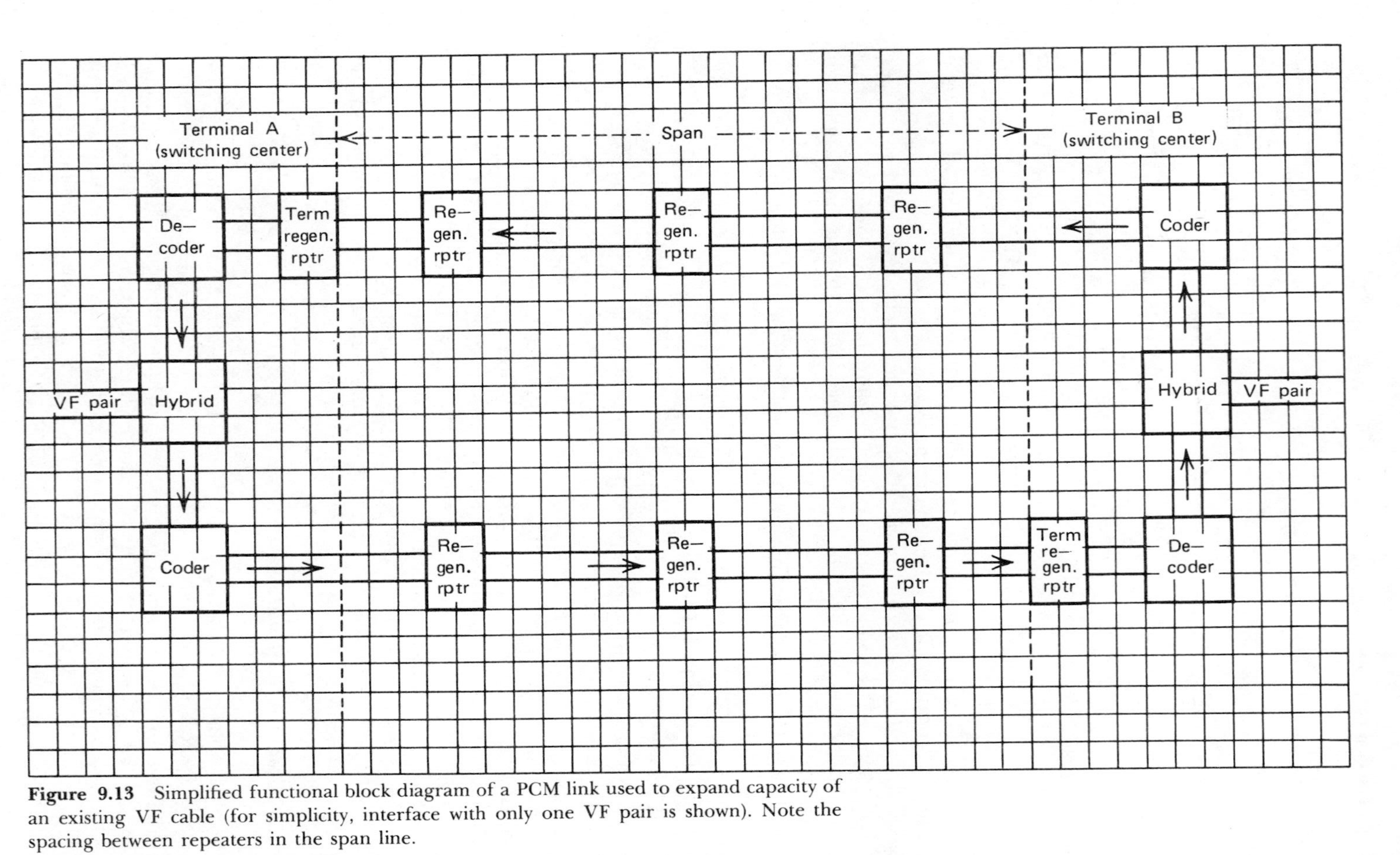

Figure 9.13 Simplified functional block diagram of a PCM link used to expand capacity of an existing VF cable (for simplicity, interface with only one VF pair is shown). Note the spacing between repeaters in the span line.

of span lines that extend between two exchange (switching center) repeater points.

A typical span is shown in Figure 9.13. The spacing between regenerative repeaters is important. Section 5.1 mentioned the necessity of removing load coils from those trunk (junction) cable pairs that were to be used for PCM transmission. It is at these load points that the PCM regenerative repeater should be installed. On a VF line with H-type loading (see Chapter 2, Section 2.6), spacing between load points is normally about 6000 ft (1830 m). It will be remembered from Chapter 2 that the first load coil out from the exchange on a trunk pair is at half-distance or 3000 ft (915 m). This is provident, for a regenerative repeater also must be installed at this point. Such spacing is shown in Figure 9.13 (1 space = 1000 ft). The purpose of installing a repeater at this location is to increase the pulse level before entering the environment of an exchange area where the levels of impulse noise may be quite high. High levels of impulse noise introduced into the system may cause significant increases in digital error rate of the incoming PCM bit streams, particularly when the bit stream is of a comparatively low level. Generally, the amplitude of a PCM pulse output of a regenerative repeater is on the order of 3 V. Likewise, 3 V is the voltage on the PCM line cross connect field at the exchange (terminal area).

A guideline used by Bell Telephone Manufacturing Company (Belgium) is that the maximum distance separating regenerative repeaters is that corresponding to a cable-pair attenuation of 36 dB at 1024 kHz at the maximum expected temperature. This frequency is equivalent to the half-bit rate for the CEPT systems (e.g., 2048 kb/s). Actually, repeater design permits operation on lines with attenuations anywhere from 4 to 36 dB, allowing considerable leeway in placing repeater points. Table 9.1 gives some other practical repeater-spacing parameters for the CEPT-ITT-BTM 30 + 2 system. The maximum distance is limited by the maximum number of repeaters, which in this case is a function of power feeding and supervisory considerations. For instance, the fault-location (i.e., troubleshooting) system can handle up to a maximum of 18 tandem repeaters for the BTM (ITT) configuration.

Table 9.1 Line Parameters for ITT/BTM PCM Configuration

Pair Diameter (mm)	Loop Attenuation at 1 MHz (dB/km)	Loop Resistance (Ω/km)	Voltage Drop (V/km)	Maximum Distance[a] (km)	Total Repeaters	Maximum Distance System (km)
0.9	12	60	1.5	3	18	54
0.6	16	100	2.6	2.25	16	36

[a]Between adjacent repeaters.

Power for the BTM system is fed through a constant-current feeding arrangement over a phantom pair serving both the "go" and related "return" repeaters, providing up to 150 V dc at the power feed point. The voltage drop per regenerative repeater is 5.1 V; thus for a "go" and "return" repeater configuration the drop is 10.2 V. For example, let us determine the maximum number of regenerative repeaters in tandem that may be fed from one power feed point by this system, using 0.8-mm-diameter pairs with a 3-V voltage drop in an 1830-m spacing between adjacent repeaters:

$$\frac{150}{(10.2 + 3)} = 11$$

Assuming power fed from both ends and an 1800-m "dead" section in the middle, the maximum distance between power feed points is approximately

$$(2 \times 11 + 1)\ 1.8\ \text{km} = 41.4\ \text{km}$$

Fault tracing for the North American (ATT) T1 system is carried out by means of monitoring the framing signal, the 193rd bit (Section 3.5). The framing signal (amplified) normally holds a relay closed when the system is operative. With loss of framing signal, the relay opens actuating alarms, and thus a faulty system is identified, isolated, and dropped from "traffic."

To locate a defective regenerator on the BTM (Belgium)-CEPT system, traffic is removed from the system and a special pattern generator is connected to the line. The pattern generator transmits a digital pattern with the same bit rate as does the 30 + 2 PCM signal, but the test pattern can be varied to contain selected low-frequency spectral elements. Each regenerator on the repeatered line is equipped with a special audio filter, each with a distinctive passband. Up to 18 different filters may be provided in a system. The filter is bridged across the output of the regenerator, sampling the output pattern. The output of the filter is amplified and transformer-coupled to a fault-transmission pair, which is normally common to all PCM systems on the route, span, or section. To determine which regenerator is faulty, the special test pattern is tuned over the spectrum of interest. As the pattern is tuned through the frequency of the distinct filter of each operative repeater, a return signal will derive from the fault-transmission pair at a minimum specified level. Defective repeaters will be identified by absence of return signal or a return level under specification. The distinctive spectral content of the return signal is indicative of the regenerator undergoing test.

5.3 The Line Code

Pulse-code-modulation signals are transmitted to the cable and are in the bipolar mode, as shown in Figure 9.1. The marks, or "1"s, have only a 50% duty cycle. There are several advantages to this mode of transmission:

- No dc return is required; thus transformer coupling can be used on the line.
- The power spectrum of the transmitted signal is centered at a frequency equivalent to half the bit rate.

It will be noted in bipolar transmission that the "0"s are coded as absence of pulses and "1"s are alternately coded as positive and negative pulses, with the alternation taking place at every occurrence of a "1." This mode of transmission is also called *alternate mark inversion* (AMI).

One drawback to straightforward AMI transmission is that when a long string of "0"s is transmitted (e.g., no transitions), a timing problem may arise because repeaters and decoders have no way of extracting timing without transitions. The problem can be alleviated by forbidding long strings of "0"s. Codes have been developed that are bipolar but with N zeros substitution; they are called "BNZS" codes. For instance, a B6ZS code substitutes a particular signal for a string of six "0"s.

Another such code is the HDB3 code (high-density binary 3), where the "3" indicates substitution for binary formations with more than three consecutive "0"s. With HDB3, the second and third zeros of the string are transmitted unchanged. The fourth "0" is transmitted to the line with the same polarity as the previous mark sent, which is a "violation" of the AMI concept. The first "0" may or may not be modified to a "1" to assure that the successive violations are of opposite polarity.

5.4 Signal-to-Gaussian-Noise Ratio on Pulse-Code-Modulation Repeater Lines

As we mentioned earlier, noise accumulation on PCM systems is not an important consideration. However, this does not mean that Gaussian noise (nor crosstalk or impulse noise) is unimportant. Indeed, it may affect error performance expressed as error rate (see Chapter 8). Errors are cumulative, as is the error rate. A decision in error, whether "1" or "0," made anywhere in the digital system is not recoverable. Thus such an incorrect decision made by one regenerative repeater adds to the existing error rate on the line, and errors taking place in subsequent repeaters further down the line add in a cumulative manner, thus tending to deteriorate the received signal.

In a purely binary transmission system, if a 20-dB signal-to-noise ratio is maintained, the system operates nearly error free. In this respect, consider Table 9.2.

As discussed in Section 5.3, PCM, in practice, is transmitted on-line with alternate mark inversion. The marks have a 50% duty cycle, permitting energy concentration at a frequency of half the transmitted bit rate. Thus it is advisable to add 1 or 2 dB to the values shown in Table 9.2 to achieve a desired error rate in a practical system.

Table 9.2 Error Rate of a Binary Transmission System Versus Signal : rms Noise Ratio

Error Rate	S/N(dB)	Error Rate	S/N(dB)
10^{-2}	13.5	10^{-7}	20.3
10^{-3}	16.0	10^{-8}	21.0
10^{-4}	17.5	10^{-9}	21.6
10^{-5}	18.7	10^{-10}	22.0
10^{-6}	19.6	10^{-11}	22.2

6 HIGHER-ORDER PCM MULTIPLEX SYSTEMS

Using the 24-channel D2 channel bank as a basic building block, higher-order PCM multiplex systems are being developed in North America. For instance, four D2 channel banks are multiplexed by a M1-2 multiplexer, placing 6.312 Mbps on a single wire pair (T2 digital line). Figure 9.14 is a simplified block diagram of the first step in the development of a higher-order PCM multiplex configuration in North America.

The North American PCM multiplex hierarchy is shown in the following list and diagrammatically in Figure 9.15.

DS1 (Deriving from D1 or D2 channel banks)	1.544 Mbps
DS2 (Output of multiplexer M1–2)	6.312 Mbps
DS3 (Output of multiplexer M2–3)	44.736 Mbps
DS4 (Output of multiplexer M3–4)	274.176 Mbps

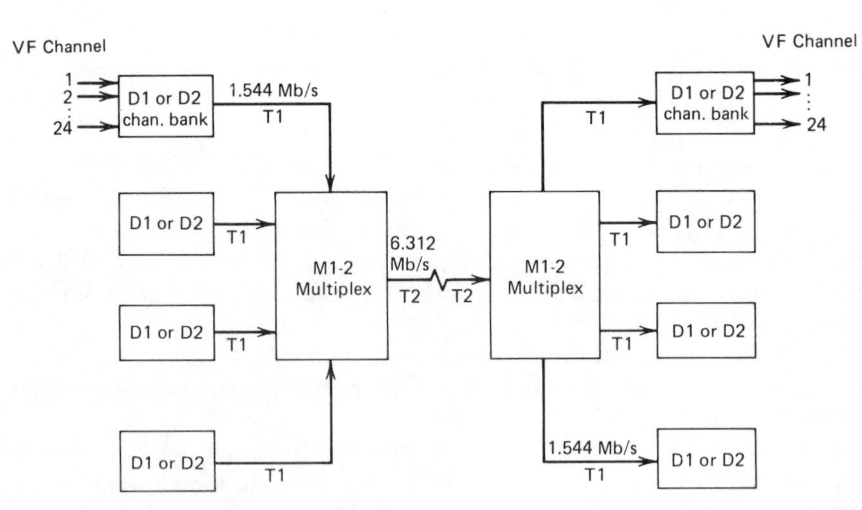

Figure 9.14 Development of the 96-channel T2 (ATT) system by multiplexing the 24-channel D1 or D2 channel bank outputs.

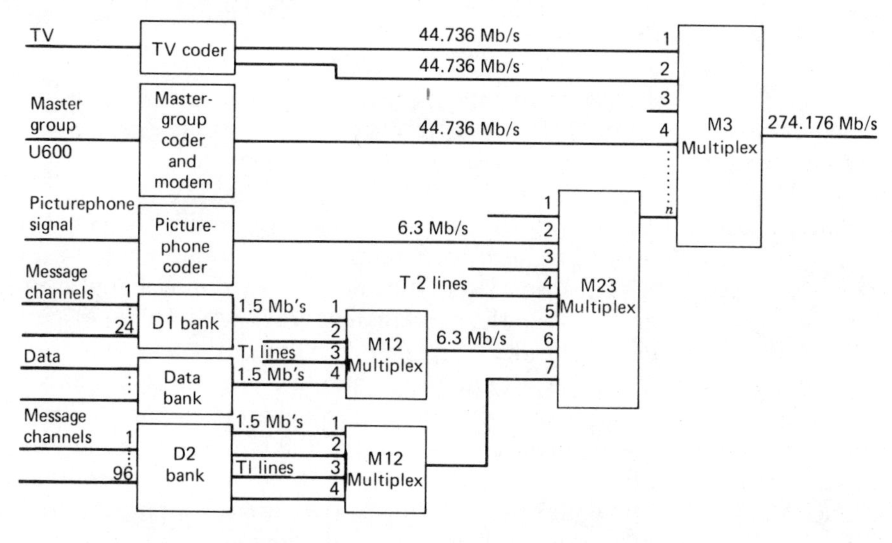

Figure 9.15 North American PCM hierarchy.

As we have seen previously, there are basically two types of PCM system now in use in the world. The North American type is based on 24 voice channels. Japan follows a similar system inasmuch as its basic line rate is 1.544 Mbps based on 24 voice channels. The other system is European, based on 32 channels (30 channels of voice plus a signaling channel and a synchronization channel). The differences between these three systems for higher-level multiplex are shown in Table 9.3.

Table 9.3 Higher-Level PCM Multiplex Comparison

	Level				
System Type	1	2	3	4	5
North American T/D type	1	2	3	4	
Number of voice channels	24	96	672	4032	
Line bit rate (Mbps)	1.544	6.312	44.736	274.176	
Japan					
Number of voice channels	24	96	480	1440.0	5760.0
Line bit rate (Mbps)	1.544	6.312	32.064	97.728	400.352
Europe					
Number of voice channels	30	120	480	1920.0	7680.0
Line bit rate (Mbps)	2.048	8.448	34.368	139.264	560.0

7 DELTA MODULATION

Delta modulation is another method of transmitting an audio signal in a digital format. It is quite different from PCM in that coding is carried out before multiplexing and the code is far more elemental, actually coding at only 1 bit at a time.

The delta modulation code is a one-element code and differential in nature. Of course, we mean here that comparison is always made to the prior condition. A "1" is transmitted to the line if the incoming signal at the sampling instant is greater than the immediately previous sampling instant. It is "0" if of a smaller amplitude. Here the derivative of the analog input is transmitted rather than the instantaneous amplitude as in PCM. This is achieved by integrating the digitally encoded signal and comparing it with the analog input to decide which of the two has the larger amplitude. The polarity of the next binary digit placed on line is either plus or minus, to reduce the amplitude of the two waveforms [i.e., analog input and integrated digital output (previous digit)].

We thus see the delta encoder basically as a feedback circuit as shown in Figure 9.16.

Let's see how this feedback concept is applied to the delta encoder. Figure 9.17 illustrates the application. The switch is the double NAND gate and flip-flop. The comparator is the amplifier in Figure 9.16, and the feedback network is the integrating network.

The basic delta decoder consists of a current-source, integrating network, amplifier, and low-pass filter. Figure 9.18 illustrates a simplified delta decorder.

We have seen that the digital output signal of the delta coder is indicative of the slope of the analog input signal (its derivative is the slope)—a "1" for positive slope and "0" for a negative slope. But the "1" and "0" give no idea of instantaneous or even semiinstantaneous steepness of slope. This leads

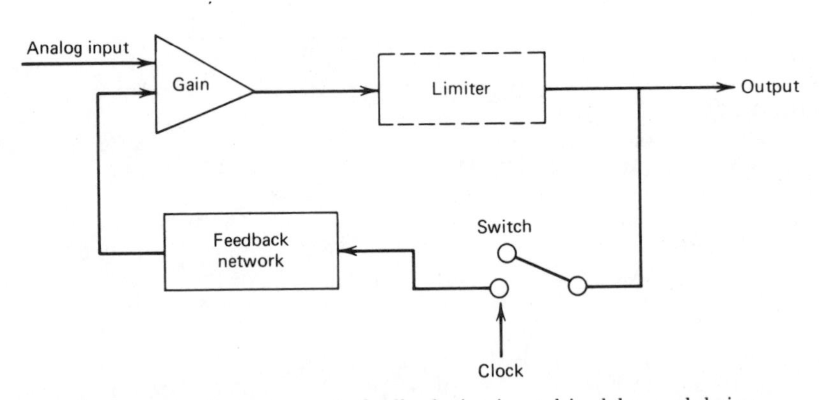

Figure 9.16 Basic electronic feedback circuit used in delta modulation.

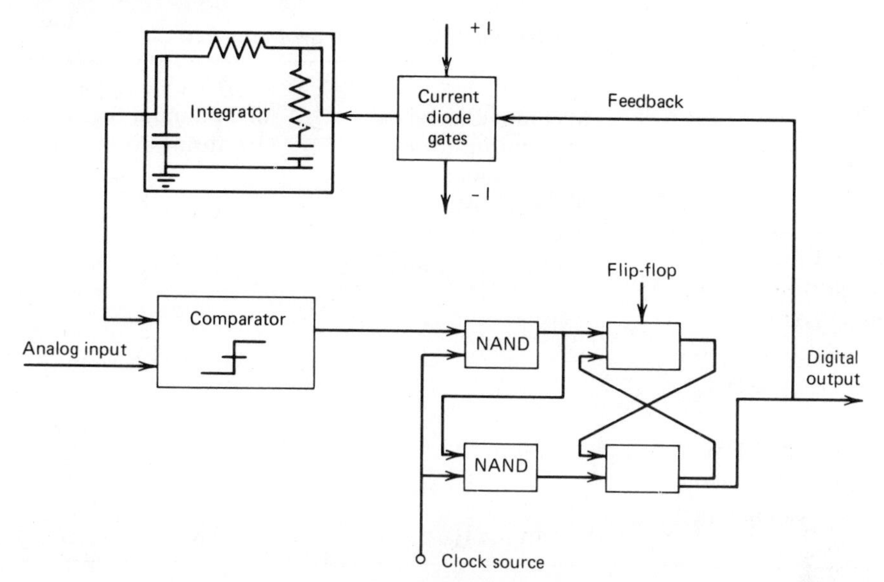

Figure 9.17 Delta encoder.

to the basic weakness found in the development of delta modulation systems, namely, poor dynamic range or poor dynamic response given a satisfactory signal-to-quantizing noise ratio. For delta circuits, this limit is given as 26 dB. A number numerically greater than 26 dB is satisfactory, and a number numerically less than 26 dB is unsatisfactory. The reader is cautioned not to numerically equate quantizing noise in PCM to quantizing noise in delta modulation (DM), although the concept is the same.

One method used to improve the dynamic range of a delta-modulation system is by using two integrator circuits (we showed just one Figure 9.17) in the codec. This is called *double integration*. Companding provides further improvement. Table 9.4 compares several 56-kbps digital systems with a 3-kHz bandwidth-input analog signal regarding dynamic range.

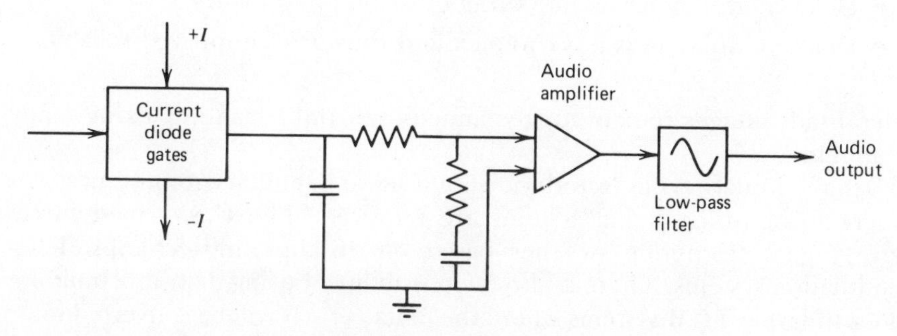

Figure 9.18 Delta decoder.

Table 9.4 Dynamic Range—Digital Modulation Systems Compared

System	Maximum Signal-to-Quantizing Noise Ratio (dB)	Dynamic Range for Minimum Signal-to-Quantizing Noise Ratio of 26 dB (dB)
Basic DM, Single Integration	34	8
Basic DM, Double Integration	44	18
Companded DM, Single Integration	34	15
Companded DM, Double Integration	44	31
7-Digit Companded PCM	30 dB UVR[a]	At *S/N* of 31 dB

[a]Useful volume range.

As seen from Table 9.4, companded delta modulation has equal properties to PCM. One reason for the good coding nature of companded delta is that voice signals are predictable, unlike band-limited random signals. These can be based on knowledge of the speech spectrum or on the autocorrelation function. Thus delta coders can be designed on the principle of prediction.

The advantages of delta modulation over PCM are as follows:

- Multiplexing is carried out by simple digital multiplexers, whereas PCM interleaves analog samples.
- Delta modulation is essentially more economical for small numbers of channels.
- Delta modulation has few varieties of building blocks.
- Delta modulation is less complex and thus gives improved reliability.

The disadvantages remain for dynamic range and for multiplexing many channels.

Delta modulation has found wide application in military communications where low-bit-rate digital systems are required. The U.S. armed forces have digital telephone switches based on 16-kbps and 32-kbps delta-modulation systems. There is also the possibility of using delta as a building block in larger PCM systems where the delta format will be converted to or from PCM at points of concentration or at local switches on the subscriber side.

8 LONG-DISTANCE DIGITAL TRANSMISSION

8.1 General

Binary digital transmission, which is capable of regeneration that essentially eliminates the accumulation of noise as a signal traverses its transmission media, would appear to be the choice for long-distance (toll) transmission systems for backbone, long-haul routes. However, this has not been the case. The disadvantages of digital systems, both PCM and delta, must be considered as well. Most important is the competition with FDM systems, the L5 system, for instance (see Chapter 5). The ATT L5 provides 10,800-VF channel capacity over a pair of coaxial cables. The required bandwidth for this capacity is 60 MHz. To transmit the same number of channels by PCM would require up to 16 times the bandwidth. Delta modulation would require up to eight times as much bandwidth.

This discussion all boils down to the fact that where bandwidth is at a premium, FDM is more cost-effective, even with the consequent noise accumulation. It is also important to note that the telephone company (administration) world is still analog. Once more tandem digital switches are cut over, there will be the incentive to install more long-haul digital transmission systems. Waveguide, satellite millimeter systems, and optical-fiber techniques will offer the necessary bandwidth.

8.2 Jitter

There is one other important limitation of present-day technology on using PCM or delta modulation (DM) as a vehicle for long-haul transmission. This is jitter, more particularly timing jitter. A general definition of jitter is "the movement of zero crossings of a signal (digital or analog) from their expected time of occurrence." In Chapter 8 it was called "unwanted phase modulation or incidental FM." Such jitter or phase jitter affected the decision process of the zero crossing in a digital data modem. Much of this sort of jitter can be traced to the intervening FDM equipment between one end of a data circuit and the other.

Pulse-code modulation has no intervening FDM equipment, and jitter in PCM systems takes on different characteristics. However, the effect is essentially the same—uncertainty in a decision circuit as to when a zero crossing (transition) took place, or the shifting of a zero crossing from its proper location. In PCM it is more proper to refer to jitter as "timing jitter." The primary source of timing jitter is the regenerative repeater. In the repeatered line, jitter may be systematic or nonsystematic. Systematic jitter may be caused by offset pulses (i.e., where the pulse peak does not coincide with regenerator timing peaks or where transitions are offset), intersymbol interference (dependent on specific pulse patterns), and local clock threshold offset. Nonsystematic jitter may be traced to timing variations from repeater to repeater and to crosstalk.

In long chains of regenerative repeaters, systematic jitter is predominant and cumulative, increasing in rms value as $N^{1/2}$, where N is the number of repeaters in the chain. Jitter is also proportional to a repeater's timing filter bandwidth. Increasing the Q of these filters tends to reduce jitter of the regenerated signal, but it also increases error rate caused by sampling the incoming signal at nonoptimum times.

The principal effect of jitter on the resulting analog signal after decoding is to distort the signal. The analog signal derives from a PAM pulse train, which is then passed through a low-pass filter. Jitter displaces the PAM pulses from their proper location, showing up as undesired pulse-position modulation (PPM).

Because jitter varies with the number of repeaters in tandem, it is one of the major restricting parameters of long-haul, high-bit-rate PCM systems. Jitter can be reduced in future systems by using an elastic store at each regenerative repeater (which is costly) and high-Q phase locked loops.

9 DIGITAL TRANSMISSION BY RADIO LINK

9.1 General

Digital transmission by radio link is becoming increasingly important in both civilian and military communication. Civilian communication organizations such as telephone companies (administrations) and specialized common carriers have opted for PCM rather than DM, so our discussion from here onward stresses PCM.

The transmission of PCM by radio link is a viable alternative to PCM VF cable pair transmission under the following circumstances:

- Where physical or natural obstructions make cable laying impractical.
- For relatively long metropolitan trunk groups under 600 voice channels where cable laying is very costly.
- As an alternative routing of a cable system.
- Between PCM local switches to avoid requirements of A/D–D/A (analog-to-digital–digital-to-analog) conversion, thus eliminating FDM as an economically viable possibility.

Consider a situation where a large number of trunk (junction) routes are presently equipped with PCM. An FDM–FM radio link is contemplated, and the system engineer is faced with one or several of the preceding circumstances. The use of FM radio links with FDM multiplex will prove expensive. The existing PCM will have to be brought to VF (demultiplexed–demodulated) to interface with the new FDM equipment. Use of PCM eliminates the additional multiplex equipment cost. Further, PCM channelizing equipment, if we accept groups of 24 or 30 channels at a time, is less expensive on a per channel basis than is FDM equipment.

9.2 Modulation, RF Bandwidth, and Performance

As we have discussed previously, PCM requires large amounts of information bandwidth. Electromagnetic spectrum (i.e., radio spectrum) is at a premium, particularly where there is the greatest demand for radio facilities, namely, in industrially built-up areas. Consider a 672-voice-channel PCM radio system. Such a system transmits at 44.736 Mbps (Table 9.3). At 1 bit per hertz of RF bandwidth, about 45 MHz would then be required. On an FDM–FM conventional radiolink, about 6 MHz would be required. The system design engineer must resort to modulation methods that are bandwidth conservative or must resort to FDM–FM.

There are three basic methods of modulating a radiowave: by amplitude, by frequency, or by phase. With PCM we are dealing with binary conditions. The simplest approach would be to two-state modulate a carrier such that one of the states represents the "1" and the other the "0" of the PCM bit stream. Again, we arrive at a bandwidth of 1 bit per hertz. This is described in the literature as one symbol per hertz. For AM, this would represent a discrete level (amplitude) for the 1 and another for the 0; or for PSK, say, 0° phase for the 1 and 180° for the 0; or for FSK a discrete frequency for the 1 and another for the 0.

Suppose the carrier can take on four discrete states, coding the bit stream 2 bits at a time. A symbol represents a transition in state. In the case of phase 0° is assigned the symbol 00, 90° the symbol 01, 180° the symbol 10, and 270° the symbol 11. By this method we transmit 2 bits of information per symbol, or, if you will, change of state. Essentially, we have cut the required bandwidth in half. This is done at the expense of signal-to-noise ratio. To achieve the same error rate, a 6-dB signal-to-noise-ratio improvement is required for four-state system over the two-state system. Table 9.5 reviews modulation possibilities with the resulting bandwidth required.

Table 9.5 Digital Modulation Techniques versus Spectral Efficiency

Type of Modulation	Number of Logic Levels	Number of Bits per Symbol
Amplitude modulation	2	1
Frequency-shift keying	2	1
Two-phase shift keying	2	1
Four-phase shift keying	4	2
Eight-phase shift keying	8	3
Sixteen-phase shift keying	16	4
Quadrature partial response signaling AM	16	4

Of course, the major advantage of digital modulation of radio systems is the ability to periodically regenerate the waveform at each repeater site. Another advantage is that it is little affected by traffic loading, and any mix of voice and data traffic has no effect on system performance. Loading is a major problem on conventional FDM systems.

10 PULSE-CODE-MODULATION (PCM) SWITCHING

10.1 Introduction

In Chapter 3 we dealt with space-division analog switching, in which a metallic (conductive) path is set up between calling and called subscribers. "Space division" refers to the fact that speech paths are physically separated (in space). Time division switching permits a common metallic path to be used by numerous calls separated one from the other in the time domain. In time-division switching the speech or other information to be switched is digital in nature, either PCM or delta modulation. Samples of each call are assigned time slots, as described previously. Pulse-code-modulation switching involves the distribution of these slots in sequence to the desired destination port. The incoming and outgoing ports of the switch are connected by digital "highways." Here the speech path is a time slot.

The conceptual difference between space-division switching and time division switching is shown in Figure 9.19. Figure 9.19A is the familiar switching matrix of Chapter 3, and Figure 9.19B shows the time-division switching concept: memory and gates.

For a simplified description, consider a switch with four 24-channel PCM highways (A, B, C, D) at the input ports of a time-division switch and four PCM highways as output ports (W, X, Y, Z). A subscriber connected to channel 3 of the B highway is to be connected to channel 12 on highway X. Initial signaling information would program the call such that all time slots of channel 3 on the B highway would be stored, released, and distributed to time slot channel 12 on the X highway.

10.2 Essential Functions of a Pulse-Code-Modulation Switch

A PCM switch must be designed to carry out the following five functions:

1. To identify the PCM system (incoming highway) and the channel in that system that has requested service for a connection through the switch.
2. To identify the outgoing routing required for a connection and allocate the channel in the appropriate PCM system (outgoing highway).
3. Where necessary, to rearrange by time switching the incoming channel time slot so that it corresponds in time with the allocated outgoing channel time slot.

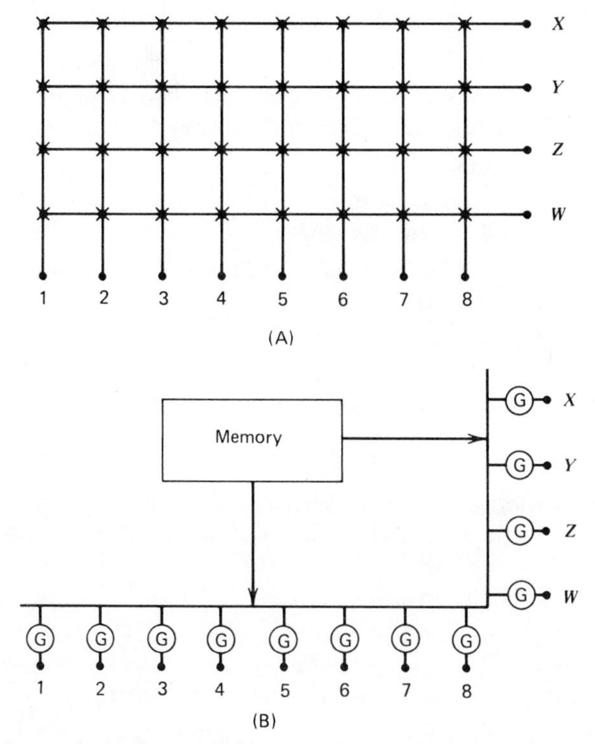

Figure 9.19 Comparison of space-division and time-division switching: (A) space division (32 cross points); (B) time division [12 gates (G)].

4. To allocate a path through the switching matrix between the two channel time slots that are to be interconnected and to check that the allocated path is valid.
5. To make the required connection between the two channel time slots.

10.3 Advantages of Digital Switching

There are both economic and technical advantages to digital switching; in this context we refer to PCM switching (of course, most of the arguments hold for delta switching as well). The economic advantages of time-division switching (PCM) include:

- There are fewer equivalent cross points for a given number of lines and trunks than in space division.
- It (a PCM switch) is of considerably smaller size.
- It has more common circuitry (i.e., common modules).
- It has full availability (i.e., essentially nonblocking), in general leading to better traffic performance for less expense.
- It stimulates the installation of PCM transmission systems.

The technical advantages include:

- It is regenerative (i.e., the switch does not distort the signal. In fact, the output signal is "cleaner" than the input).
- It is noise resistant.
- A digital exchange is lossless (i.e., there is no insertion loss as a result of a switch inserted in the network).
- The message format (binary) is compatible with digital computers. And also with signaling.
- It exploits the continuing cost erosion of logic and memory.

Two technical problems that may be listed as disadvantages are:

- The switch may deteriorate the error performance of the system.
- Switch and network synchronization and the elimination of wander and jitter are problems that have not yet been optimally solved.

It should also be borne in mind that a PCM switch is inherently an SPC switch offering all the benefits of SPC switching set out in Chapter 3.

10.4 Approaches to Pulse-Code-Modulation Switching

10.4.1 Introduction

In a PCM switch information is transferred by opening semiconductor gates (i.e., causing diodes or other semiconductors to conduct current). The gates are controlled by a processor opening them at the exact and proper time interval to correspond to the group of bits of a particular input channel such that these bits are transferred to the required output channel. However, these advanced digital switches still require some conducting medium to carry the digital information. Usually more than one highway must be provided with one hard-wire medium for each highway. The highways, depending on switch design, form a switch matrix with semiconductor cross points; thus, in effect, we are dealing with a form of space division. There are two basic ways to integrate space division into a time-division PCM switch. These are commonly called "time–space–time" and "space–time–space."

Considering the necessity for cross points on PCM highways for the distribution of time slots, the more highways there are, the more gated crosspoints are required. An example would be eight 24-channel input ports to be distributed in eight 24-channel output ports. An 8 × 8 matrix would be required. The number of cross points can be materially reduced by supermultiplexing. In this simplified example we can multiplex four PCM groups into one highway. Instead of eight ports at the input and output, the matrix can be reduced to 2 × 2. There is a trade-off then between supermultiplexing forming "superhighways" and reducing cross points or accepting the larger number of cross points and increased individual control. In the following discussion all switch inputs and outputs are considered in PCM format.

10.4.2 Time–Space–Time Switching

With the time–space–time configuration, a memory is required at each input port. Memories should be large enough to accommodate all customers simultaneously. A processor with memory controls the store and release to the matrix as well as closing the proper gates for the correct time interval. The actual distribution is carried out in the matrix. At the outputs of the matrix buffer memories controlled by the processors reconstitute, by storage and release, each outgoing digital highway. Figure 9.20 illustrates a basic time–space–time PCM switch.

10.4.3 Space–Time–Space Switching

In this type of PCM switching the time memory is located between an incoming and an outgoing space switch where only sufficient storage is required to handle traffic peaks. Some sort of memory buffering is still required at each input port to synchronize the bit-stream timing to that of the digital exchange. Usually this buffering requires at least one frame of memory. Figure 9.21 details the concept of space–time–space PCM switching.

Suppose, for the sake of argument, that each incoming highway con-

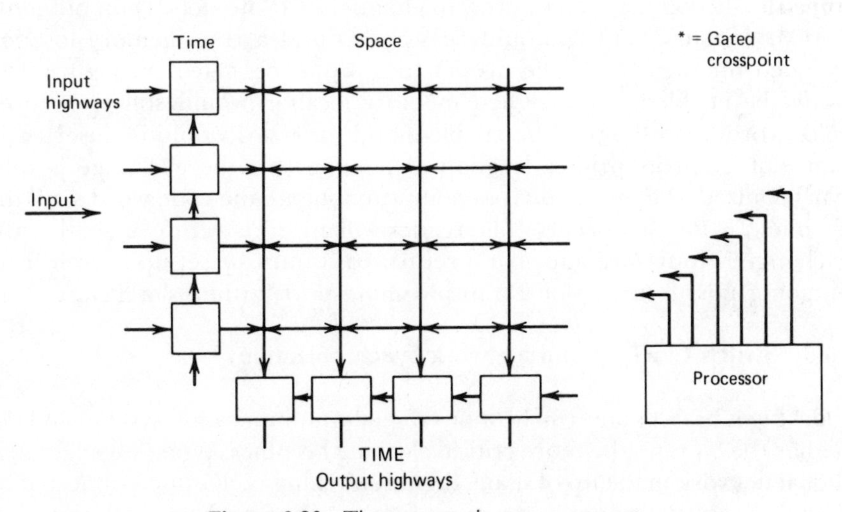

Figure 9.20 Time–space–time arrangement.

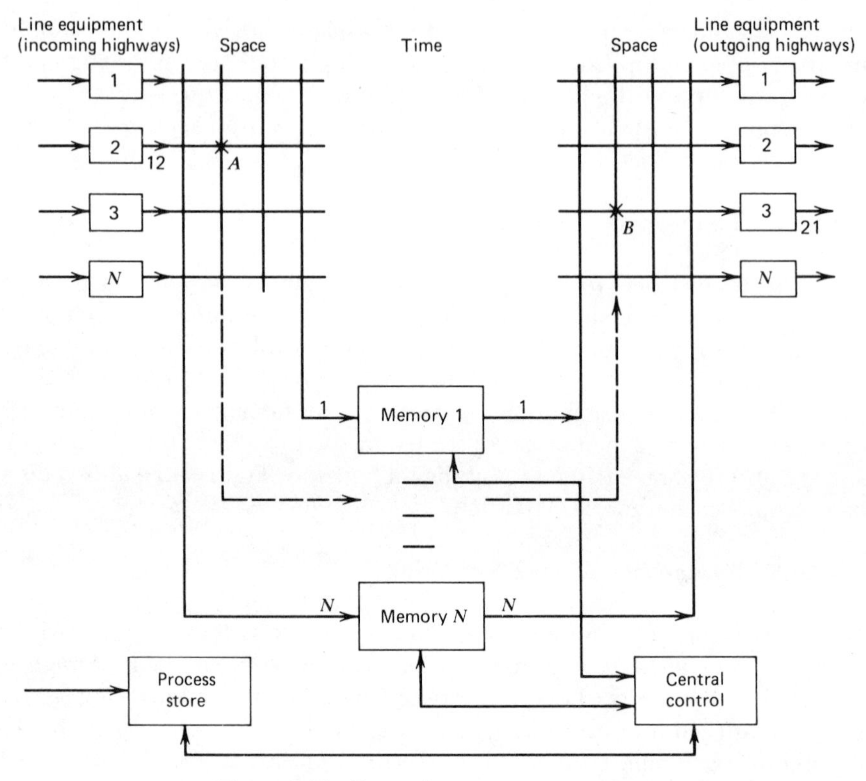

Figure 9.21 Space–time–space arrangement.

tained 30 time-multiplexed voice channels. To connect voice channel 12 (time slot 12) on incoming port 2 to channel 21 (time slot 21) on outgoing port 3, memory bank 2 would be scanned and a free memory location assigned for the call. At the proper time sequence, gate *A* in Figure 9.21 would be enabled, the assigned memory location would spill the stored voice word, gate *B* would be enabled, and the word would be inserted in time slot 21 in its proper location. Assuming that the exchange is fully synchronized, the assigned cross point cannot gate the code word until the channel 21 time slot occurs. The reader will appreciate, of course, that the exchange is four-wire and that a return path must be set up as well from outgoing port 3 (time slot 21) to incoming port 2 (time slot 12).

10.5 Switch Clocking and Network Synchronization

Clocking is basic to any synchronous digital communication system, and the higher the bit rate, the more critical clocking becomes. When we examine a digital network made up of many digital switching nodes interconnected by digital transmission links, we must face another problem, network synchronization. A "clock" is defined as a frequency source coupled to a divider or counter. It provides a time base to control the timing in a digital exchange (or may be used to control an entire digital network). The most desirable situation would be to have all network clocks exactly synchronized, each clock identical with identical stabilities and resettabilities. Such a situation is unworkable because of delay. A signal sent from point X to point Y suffers delay.

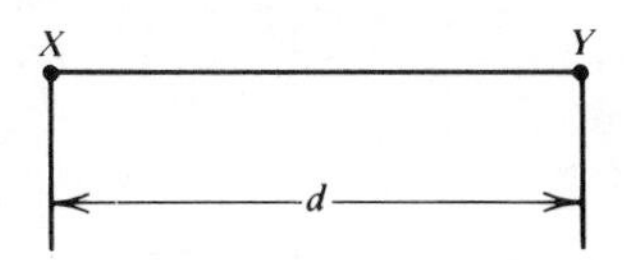

Such delay is a function of the distance and the medium–media traversed. Velocity of propagation varies not only with the type of medium of transmission, but also with temperature. The velocity of propagation can vary between 10,000 mi/s to nearly 186,000 mi/s (16,000 km/s to nearly 300,000 km/s). If the velocity of propagation were 180,000 mi/s over a certain transmission medium and the distance d were 18 mi from one digital exchange to another (transmission route miles), then the signal delay would be 18/180,000, or 0.1 ms (100 μs). At 2 Mbps (2×10^6 bps), 100 μs would represent 200 bits.

If a digital exchange is operating with 100-Mbps PCM highways, timing must be considerably better than 50% of the period of 1 bit or $0.5[1/(100 \times 10^6)]$ s or 0.005 μsec or 5×10^{-9} clock stability per day. With such a clock rate, we would lose 1 bit per day. Such a loss is called "slip." Keep in mind that this thinking deals only with an isolated PCM switch. Slip, in reference to network synchronization, will have to be better defined.

Table 9.6 lists several off-the-shelf frequency standards that may be used to slave a master clock, showing the adjustment interval required to maintain an accuracy (stability) of 1 part in 10^{10}.

A network operated on the basis of individual high-stability clocks is called a *plesiochronous network.* In some areas of the world the term *quasisynchronous* is used. Another form of network timing uses a master clock or clocks and is referred to as a *synchronous network.*

Table 9.6 Comparison of Frequency Standards

Type	Drift per Month	Adjustment Interval
Cesium atomic	3×10^{-12} (no drift)	>1000 years
Rubidium	2×10^{-11}	5 months
Good oven quartz crystal	1.5×10^{-9}	2 days
Medium oven quartz crystal	1.5×10^{-8}	5 h
Quartz crystal (no oven)	5×10^{-7}	12 min

The plesiochronous network uses the "slip" to maintain synchronization. If, between digital exchanges A and B, we had 8-bit buffer on 2-Mbps trunks, what sort of daily slip could be expected given two different clocks at exchange *A*? The transmission rate again is 2×10^6 bps.

Situation	Clock Stability	Bits Accumulated in 24 h	Slips per Day (8 bits at a Time)
1	1×10^{-7}	17,280	2160
2	1×10^{-12}	0.17	0.02

We calculate bits accumulated by $2 \times 10^6 \times 60 \times 60 \times 24/10^7$ for situation 1 and divide by 10^{12} for situation 2.

In a plesiochronous network, slip rates should be specified. Reference 1 states that a slip rate of 300 per hour can be tolerated in a speech network. Such a slip rate would be intolerable if the network were to carry data, particularly without error correction. The use of atomic clocks at digital switching nodes would reduce the slip rate to 1 in 3 months.

Variations in velocity of propagation (delay) are compensated by elastic buffers. Such buffers provide a very small amount of delay. This delay is controlled in such a way that the buffer output is constant.

In synchronous digital networks a method of frequency–phase control is used, and hence these networks are without slip. Two available approaches to the clocking design of digital networks are despotic control and mutual control. With despotic control, one master clock controls all other clocks (the despot, if you will). Mutual control is based on an interdependence of clocks. The master–slave method is perhaps the most straightforward method of maintaining synchronism. One digital exchange is denominated "master." Synchronism is maintained using phase-lock techniques. Slave terminals use elastic buffers where information is contained on phase differences from the master. This information is used to control a phase-lock loop. Security is an important factor to consider. Loss of the single master could cause a full loss of synchronism.

Another method is the hierarchical master–slave method, which overcomes much of the security problem. For clocking control, exchanges are assigned levels in a hierarchy. A single digital exchange clock is assigned the role of master clock. Loss of the master causes fall-back to an exchange clock at a secondary level and so forth. Still another feasible method is the use of an external clock with a separate transmission media to disseminate time [29] via radio or coaxial cable. Clocking riding on the public television network is another practical possibility. Another method is mutual synchronization, where a switching node clock averages inputs from all other clocks, and we can imagine such a network to operate on an "averaged" time of all clocks. The cooperating clocks must also take into account phase delay between exchanges. Again, such delay can be compensated in elastic buffers.

11 DIGITAL NETWORKS—AN INTRODUCTION

11.1 Evolution of a Pulse-Code-Modulation Network

Because of the many advantages of digital communications, it would be ideal if the present analog telephone network could be scrapped and replaced by a PCM network. This assumes that the several technical problems discussed in Sections 8 and 10 have been overcome. For instance, the jitter and timing problems are well on their way to being solved. Transmission media with much wider information bandwidths are nearly within reach. Nevertheless, we cannot scrap the multi-billion-dollar existing telephone plant overnight, for a number of cogent reasons:

- The investment capital for such a very large expenditure is not available.
- The digital equipment cannot be produced in such a short time span to replace its analog counterpart.
- The demand for such a drastic change does not exist.

We can safely assume that the telephone network will be fully digital after some 20 to 50 years. It will become so by the slow phase-out of its analog counterpart. The phase-in is indeed being done on a slow, piecemeal basis. The rationale for phase-in now and in the future will be basically economic.

First, let us consider switching. For local switches, we could say that there are two basic functions carried out, line selection and group (trunk group) selection. A tandem or transit exchange carries out the latter function only; it selects trunk groups. Thus in the switching environment PCM is very suitable for tandem exchanges. It lends itself very well to carry out the group selection function. A PCM tandem switch is more economic than its analog counterpart if all we had to compare were the switches themselves. Where this rationale falls down is in the neglect of the cost of the interface equipment. This is the expense of analog-to-digital and digital-to-analog conversion of the incoming and outgoing trunk groups. Signaling conversion may also add considerably to the cost.

Now consider one point where the world of switching and transmission meet. At an analog transit exchange,* racks of FDM equipment are required to multiplex and demultiplex trunk groups down to the voice channel for switching on both incoming and outgoing circuits. In a fully

*We differentiate between a tandem exchange and a transit exchange in that a tandem exchange switches trunk traffic in the local area; a transit exchange serves the same purpose in the long-distance plant.

PCM network at a purely transit exchange, no multiplex equipment would be required at all. Remember that the ports of a PCM exchange accept multiplexed PCM groups. Of course, some digital multiplex equipment may be required if the switch ports require a higher-order multiplex than what is provided by incoming and outgoing trunk groups (or vice-versa). The virtual elimination of multiplex at a transit exchange makes the PCM approach very attractive. The same thinking will hold for a tandem exchange.

In a conventional local analog exchange, that important concentration–expansion is carried out by the switch. With a digital exchange this function is transferred to the transmission area and is done by the interface equipment, namely, by the codecs and higher-order multiplexers. As we mention in Subsection 11.2, this interface can be extended into the subscriber plant by PCM subscriber carrier or remote PCM concentrators. The loading of the subscriber plant PCM is much lower than on the trunk plant, say, from 0.6 erlangs for each trunk circuit to less than 0.1 erlang for the subscriber application. Thus on a per erlang basis the cost of PCM in the subscriber plant appears to be greater but is recouped because:

- Considerably less pair subscriber cable is required.
- Subscriber loop attenuation is drastically reduced; that attenuation is only the loss of the loop from subset to nearby codec.
- The digital-switch–subscriber interface is eliminated.

The integration of PCM and the slow phase-out of analog transmission and switching will be carried out in a number of rational steps. In the local area one scheme may well be the following:

1. Install PCM trunks between local exchanges. This is being done at an accelerated pace today.
2. Cutover PCM tandem exchanges served by PCM trunks.
3. Install local exchanges with PCM group selectors and space-division line selectors. Interface equipment would be required in the exchange, but trunks would be PCM.
4. Move the interface equipment at the local exchanges out into the subscriber plant where the codec will, in a sense, act as a concentrator.
5. Implement all digital switching. All trunk signaling is common channel.
6. Provide PCM subscriber to subscriber service over the entire network. All subscriber subsets utilize single-line codecs.

The economy of step 6 may be difficult to justify in the near term. Of course, with direct digital service to household and office, other, more sophisticated services now may be offered more readily, such as high-speed data, digital facsimile, and more effective speech privacy. Yet the justification of the universal provision of this service remains a matter of conjecture.

11.2 Some Digital Network Concepts

Exchanges in an all digital network will be much smaller than their analog counterparts. One of the major limiting factors regarding the size of a conventional exchange is the mainframe where subscriber lines enter. Take subscriber lines 100 at a time. Where 100 pairs entered a mainframe on an analog exchange, this same number now enters multiplexed on eight pairs or less. A 10,000-line exchange may be reduced in size by a factor of from 12 to 20 over conventional voice-pair exchanges. As a consequence, switching centers can become much larger, serving 100 to 500,000 subscribers. The principal limitation on exchange size in this case would be security. A loss of a switching plant through fire, explosion, or sabotage affects more customers. If such a loss occurs, the time to effect replacement service is considerably extended than if the exchange were smaller.

Larger exchanges tend toward networks of lesser complexity. Trunking connections in a given local area in full mesh for four exchanges is far less than if 24 exchanges served the same area. Eventually, 24 and 30 channel PCM trunks will be replaced by second- and third-level PCM multiplex highways. The desirability of higher-level multiplex switching ports has been demonstrated in Section 9. Coaxial cable tubes will replace wire-pair trunks to accommodate the high-level bit rates. In the future, optical fiber links promise digital highways of even greater capacity.

Concentrator size is another aspect. In the case of analog line concentrators, typically 100 subscribers are served. There is no valid reason why statistical PCM concentrators cannot be used serving 2000 or more subscribers. Because each concentrator is under the control of a mother digital exchange, we are in fact then looking at distributed switching.

A basic premise in local-area design (discussed in Chapter 2) is how limitations on subscriber loop length tend to constrict an exchange's serving area. One method is to minimize the number of specially conditioned loops to serve subscribers comparatively far from an exchange. Using the concentrator approach, exchange serving areas can be greatly extended. In most cases loops will be short to local concentrators. The distance of concentrators to their homing exchanges can be 50 mi (80 km) or more using repeatered PCM lines. As we know, these lines could well be voice pairs being used to serve 24- or 30-channel systems.

National networks present another arena. There are problems of accumulated phase jitter on multirepeatered terrestrial links. Radio bandwidth is another consideration. Pulse-code-modulation trunks on direct toll routes via satellite, particularly satellites operating in the millimetric region on a TDMA basis, will allow such operation in the near term.

Table 9.7 Benefits of Integrated Switching and Transmission

Economic
Size savings can defer new construction
The PCM carrier proves in at almost zero distance
Fewer four-wire term sets required (in toll network)
Fewer trunks required because PCM switches are nonblocking
Shorter installation time for switches
Maintenance costs are a fraction of those of analog switches
No load balancing required
Simplified dial administration
Technical
No switching impulse noise
Essentially lossless switching and transmission; thus reference equivalent improved
Improved echo grade of service because far better balance and return loss and fewer two-wire to four-wire conversions
Nonblocking PCM switches increase call completions
On data service, no modems required
Many new services available to subscriber

11.3 Advantages of Integrated Switching and Transmission

In a digital (PCM) network, the two traditional disciplines of telecommunications, transmission and switching, have lost their identity. In essence, they have become one or "integrated." For instance, the timing and framing functions that are so important in digital transmission are equally as important to the PCM switch. Actually, in a PCM network, master timing has been established at a switch, as we have shown. Switching architecture is based on the frame. This integration has many far-reaching advantages when we compare it to its analog counterpart. Some of these advantages are summarized in Table 9.7.

REFERENCES

1. *Digital Telephony,* L. M. Ericcson Telephone Company, Stockholm, 1977.
2. Marvin Hobbs, *Modern Communication Switching Systems,* TAB Books, Blue Ridge Summit, PA 1974.
3. G. H. Bennett, *PCM and Digital Transmission,* Marconi Instruments, St. Albans, Herts (UK), July 1976.
4. Roger L. Freeman, *Telecommunication Transmission Handbook,* Wiley, New York, 1975.
5. International Telephone and Telegraph Corporation, *Reference Data for Radio Engineers,* 6th ed., Howard W. Sams, Indianapolis, 1976.

6. *Transmission Systems for Communications,* 4th ed., Bell Telephone Laboratories, Holmdel, N.J., 1973.
7. K. W. Catermole, *Principles of Pulse Code Modulation,* Illiffe, London 1969.
8. W. C. Sain, "Pulse Code Modulation Systems in North America," *Electr. Commun.,* **48** (1, 2) (1973).
9. *PCM—System Application—30 + 2TS,* BTM/ITT, Antwerp, Belgium.
10. *Technical Manual—Operations and Maintenance Manual for T324 PCM Cable Carrier Systems,* ITT Telecommunications, Raleigh, N. C., April 1973.
11. J. V. Marten and E. Brading, "30-Channel Pulse Code Modulation System," *Electr. Commun.,* **48** (1, 2) (1973).
12. R. B. Moore, "T2 Digital Line System," *Proceedings IEEE International Conference on Communications,* June 1973, IEEE, Seattle, Washington.
13. J. R. Davis, "T2 Repeater and Equalization," *Proceedings IEEE International Conference on Communications,* June 1973, IEEE, Seattle, Washington.
14. T. H. Flowers, *Introduction to Exchange Systems,* Wiley, London, 1976.
15. CCITT, *Orange Books,* Vol. III, Geneva, 1976, in particular the 700 series recommendations.
16. Borje Andersson, *Planning of Local Networks with Digital Exchanges,* L. M. Ericcson Telephone Company, Stockholm, 1977.
17. W. H. Smith, "PCM Microwave Links," *Telecommunications* (April 1973).
18. A. E. Pinet, "Telecommunication Integrated Network," *IEEE Transact. Commun.* (August 1973).
19. F. S. Boxal, "Digital Transmission Via Microwave Radio (Parts I and II)," *Telecommunications* (April and May 1972).
20. E. Cookson and C. Volkland, "Taking the Mystery out of Phase Jitter Measurement," *Telephony* (September 25, 1972).
21. M. E. Collier and G. Williams, *Transmission Performance of Evolving Analogue–Digital Switched Networks,* ITT/STL, London, 1977 (ITT System Confidential).
22. *Bell System Technical Journal* "No. 4 ESS," entire issue (September 1977).
23. GTE Lenkurt Demodulator, *PCM Update,* Lenkurt Electric Company, San Carlos, Calif., January–February 1975, Parts 1 and 2.
24. GTE Lenkurt Demodulator, *Increasing PCM Span-line Capacity,* Lenkurt Electric Company, San Carlos, Calif., May–June 1976.
25. GTE Lenkurt Demodulator, *Switching and PCM System Interfaces,* Lenkurt Electric Company, San Carlos, Calif., July–August 1976.
26. H. R. Schindler, "Delta Modulation," *IEEE Spectrum* (October 1970).
27. GTE Lenkurt Demodulator, *Digital Radio,* Lenkurt Electric Company, San Carlos, Calif., November–December 1977.
28. GTE Lenkurt Demodulator, *PCM over Microwave,* Lenkurt Electric Company, San Carlos, Calif., September–November 1974, Parts 1, 2, and 3.
29. Roger L. Freeman, "The Science of Time and its Inverse," *Telecommun. J.* (February 1977).
30. K. W. Catermole, "The Impact of Pulse Code Modulation on the Telecommunication Network," *Radio Electron. Eng.* (January 1969).
31. J. C. McDonald and J. R. Baichtal, "Digital Toll Switching Brings Big-time Service to Small Towns," *Telephony* (July 26, 1976).
32. M. Mulcay, "Should There be a Light Route Digital Radio in Your Future?" *Telephony* (August 29, 1977).

Chapter 10

DATA NETWORKS AND THEIR OPERATION

1 INTRODUCTION

Data communications is the fastest growing segment in telecommunications. Its orientation is different from that of telephony. An outstanding difference is the human interface. Whereas conventional telephony transmits from the mouth in the form of speech and is received by the ear, data transmits through the hands and receives through the eyes. In fact, some data operations have no direct human interface at all, or nearly none. We humans do, indeed, feel its effects. An example of the latter is automated control systems.

A major stimulus to advance data communications is the increasing use of electronic data processing, which is commonly referred to as EDP; thus it is computer oriented, although in some applications the computer may be nothing more than a microprocessor.

Unlike telephony, where the conventional telephone implies the spoken word, data communications handles the written word. It not only deals with "words" but other symbols such as numbers, graphics, punctuation, or just bit sequences that have no direct meaning to us but that act as a specific stimulus to a "machine" (or network of machines) to bring about a desired reaction.

The following are 10 illustrative examples where the communication of alphanumerics and graphics is required and exemplifies some everyday problems solved by EDP and data communications networks.

1. Multipoint inventory control.
2. Airline and train reservations.
3. Banking transactions.
4. Truck–train–ship cargo control.
5. Air traffic control.
6. Defense: air defense, ship defense, ASW, remote radar.
7. Research and development resource sharing.

8. Remote text composition (newspapers and magazines).
9. Police: anticrime EDP.
10. Weather forecasting: computerized automatic collection of weather data.

Each application requires a distinct approach. A principal consideration is whether the data processing itself will be centralized or distributed processing. This and many other items must be quantified or qualified before proceeding in network design. The material in this chapter uses Chapter 8 as a base. The reader is advised to review Chapter 8 before continuing in the succeeding discussion.

2 INITIAL DESIGN CONSIDERATIONS

2.1 General

Data-communication networks can vary from an elemental two-terminal system shown in Figure 10.1 to a multiterminal, distributed processing network such as the US ARPA network shown in Figure 10.2. Between these extremes there is a large variation of data networks regarding size, configuration, and capability. A network configuration depends on user requirements. The following considerations enter into network design:

- Type of network organization.
- Tariffs and tariff structures.
- Service reliability.
- Type of communication services (switched lines, leased lines, private lines, or combinations thereof).
- Line routings.
- Types of terminal equipment used at remote sites.
- Protocols (location and types of communication-control procedures).
- Error-control procedures.

There are two possibilities for organizing a data network, centralized and distributed. "Centralized" implies one main processing location, with all

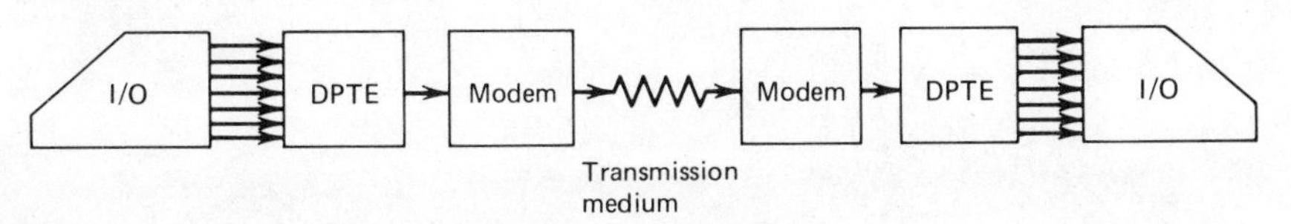

Figure 10.1 Elemental point-to-point data-communication system (two terminals) I/O = input–output device; DPTE = data-processing terminal equipment).

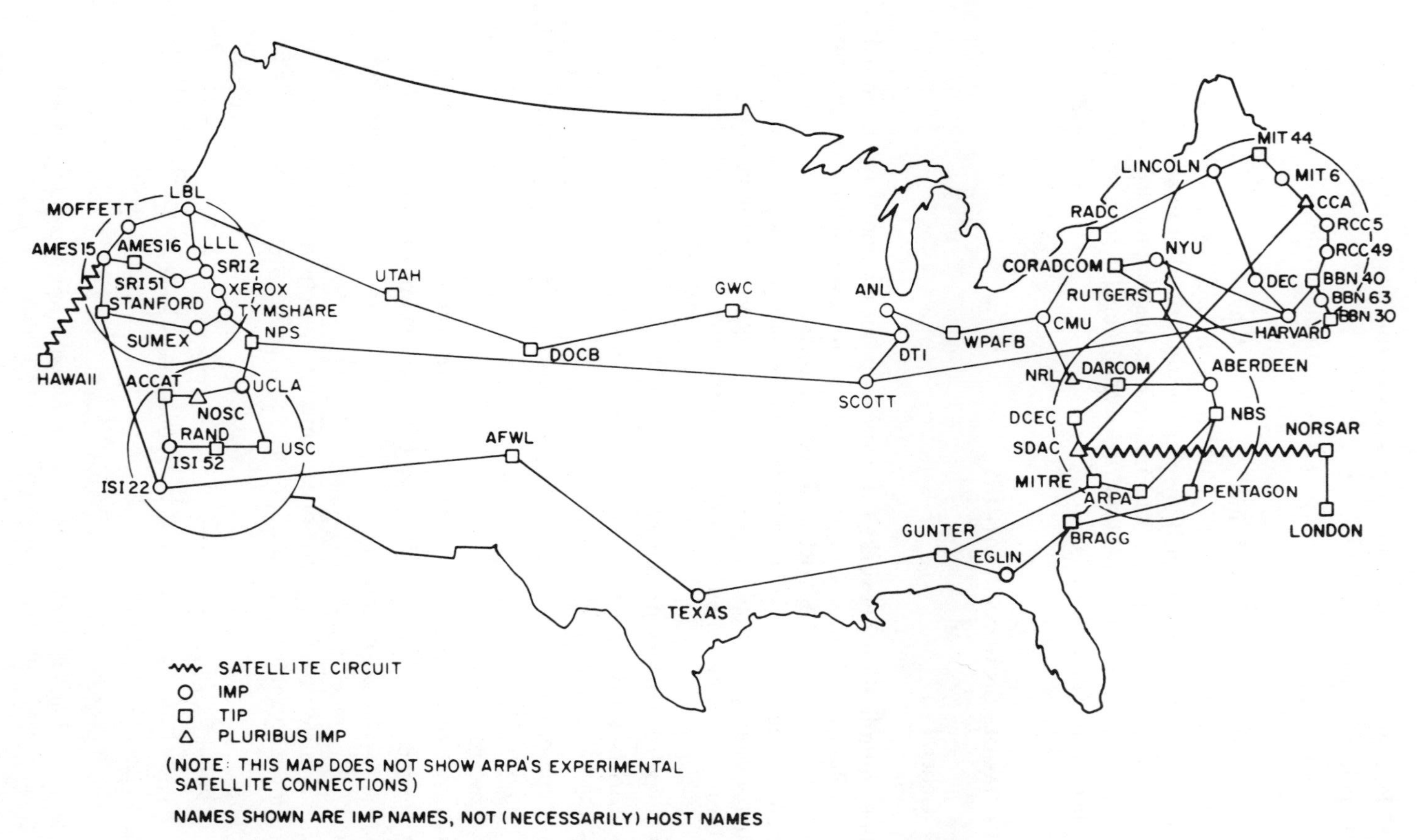

Figure 10.2 The ARPANet geographic map, March 1979. Courtesy Bolt, Beranek and Newman, Inc.

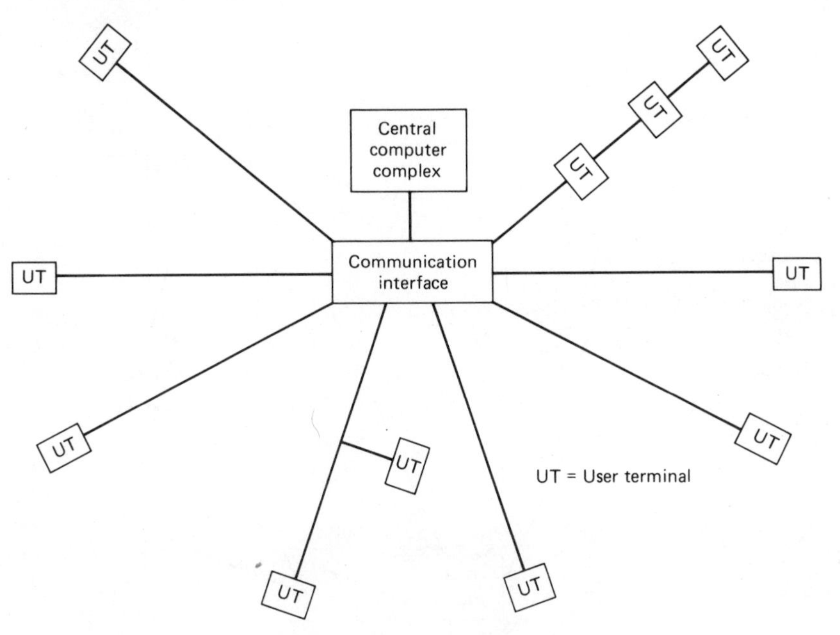

Figure 10.3 A centralized data network.

traffic between remote terminals and the single CPU (central processing unit). A centralized network is shown in Figure 10.3. In a distributed network major data processing capabilities are located in more than one location. Each of the several or multiple processing centers is often controlled by a different operating system. A distributed network is shown in Figure 10.4. There are arguments both in favor of and against each approach. With distributed processing, communication costs may be markedly lower, yet processing costs may rise. This rise is not only due to the increase in processing equipment, but in EDP administration as well. The corporate chief of EDP may opt for the centralized approach. The telecommunication manager may show major savings in communication costs for the distributed approach, as we see later on.

With the advent of microprocessors and minicomputers, a compromise between centralized and distributed approaches has been made available to the network designer. "Intelligent" terminals at remote locations can carry out some of the basic communication processing and control functions, such as code conversion, error control, and display control. Also consider possibilities of more advanced local processing functions that are well within reach for remote terminals such as extensive local preprocessing and transaction preparation functions. Both will result in less usage of the communication media.

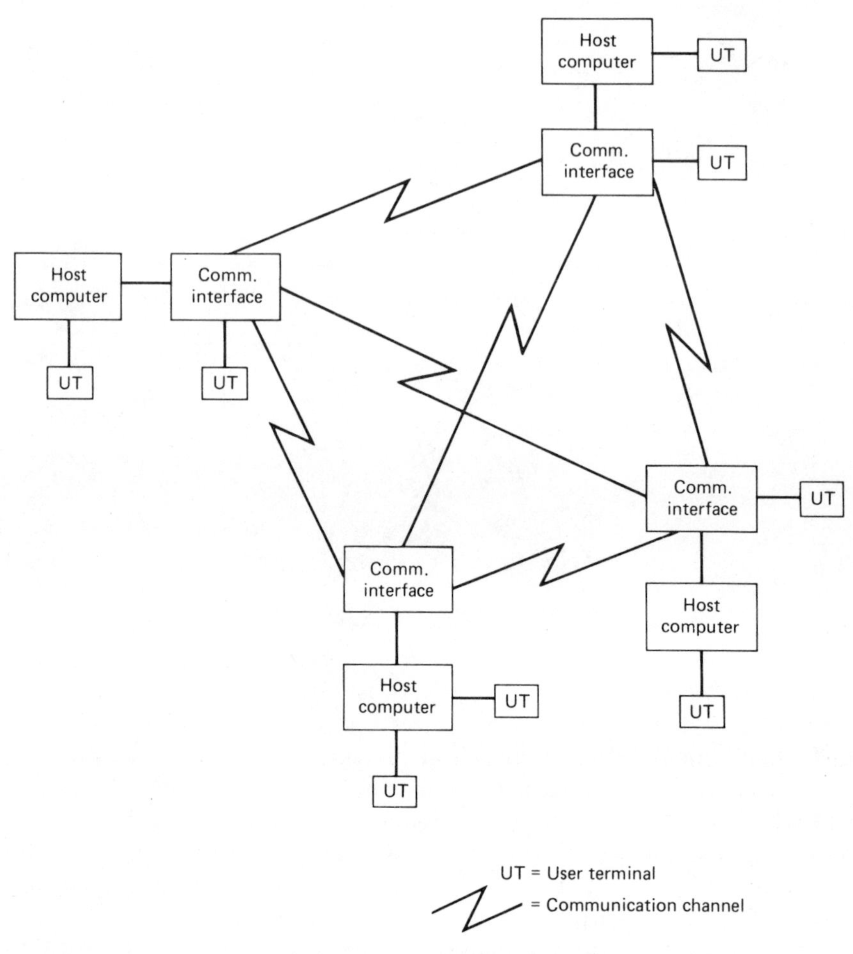

Figure 10.4 A distributed data network. (Note mesh connection.)

2.2 User Requirements

A data network is built to fulfill a need. The technical specifications of the network must be based on the users' business application requirements. Prior to initial network design, we must know and quantify well the following:

1. Number and locations of processing sites.
2. Number and locations of remote terminals.
3. Types of transactions to be processed.
4. Traffic intensities for each type of transaction by type of terminal.
5. Urgency of information to be transmitted (timeliness).

6. Patterns of traffic flow.
7. Acceptable error rates.
8. Required availability of the system.

First let us consider one of the simpler approaches to carrying out EDP requirements for a particular corporation. Assume that these requirements are accomplished by one centralized processing center. Processing will be done by the batch, or by what is better known as "batch processing."* Cards will be punched at remote locations—a remote location may be down the hall or hundreds of miles away. The card decks are delivered by vehicle or by mail. The processing may be completed at the computer center within hours, but it may sometimes take days. The results, the output, are returned by the same means. Time or timeliness is no object. Payrolls, certain inventory operations, and sales analyses may fit into this category. It is simple, economic, and easy to administer. In other applications timeliness is very important. The user may need faster response such as with airline or hotel reservations. If results of processing are required in something less than "hours," wire communications may well have to be considered. There is also the old adage, "time is money." Within the "urgency" requirements, the total system cost should be optimized: communication costs versus processing costs.

The next expedient is the telephone. Many telephone administrations permit either acoustic coupling or direct coupling to the switched telephone network. Thus the user dials the telephone number of the computer center, connects a modem to the line, and dumps a serial data stream from 75 to 2400 bps and even up to 4800 bps to the processing center. Some minutes (or hours) later, the results may be returned to the remote user by the same means. The weaknesses to this approach are:

- There is no immediate assurance that the error rate is inside the user's specifications (because it is half-duplex).
- There is a lack of high-speed duplex service, thus preventing interaction, or at least immediate interaction unless a second line is dialed up to the center.
- The bit-transfer rate is limited.
- Telephone-setup time may be a crucial factor.
- The queue of incoming batch jobs may require a larger buffer storage at the processing center.
- A fairly large amount of human intervention is required.

*Batch processing is a method of processing data in which transactions are collected over a period of time and prepared for input to the computer for processing as a single input. Contrast this with on-line processing, in which transactions are dealt with as they arise.

- There is a high risk of communication dropout in the midst of a transaction.

Figure 10.5 illustrates a typical "telephone connection."

The next level of increased "communication cost" is a leased-line system where, indeed, we configure our own network. The term "overlay network" may be used to describe a data-communication system that utilizes leased lines. With a leased line the telephone company or administration provides full-period connection of a telephone circuit. There is no intervention of a switch. The connecting patches are installed at switch mainframes and are permanent. The data transmitted on a leased telephone line still suffer the limitations or transmission constraints expected on such a medium. For instance, the highest practical bit rate is 9600 bps (see Chapter 8).

2.2.1 Communication Needs—Capacities

A computer is an extremely high speed device and is also very costly. To derive optimum benefit from a central computer or distributed processing system, it would be uneconomic to permit processing capability to remain dormant or be underemployed. Mainframe computers can accept inputs from 200 kbits to many Mbits per second. On new, well-engineered systems with remote terminal preprocessing, the equivalent input–output (I/O) rate can be increased even more. Limitations on conditioned leased telephone lines permit no more than about 10 kbps and special broadband circuits up to 50 kbps.

Computer transaction time varies widely from less than a millisecond to many minutes; thus user requirements must be carefully specified. How much can we improve on communication costs by the use of "intelligent" terminals at remote sites? An "intelligent" terminal is a device that has considerable computing capability to execute local control, display, or program functions without requiring participation or intervention by a larger

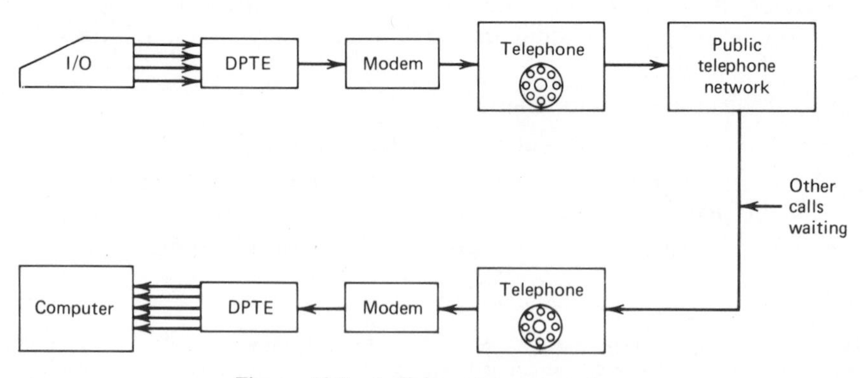

Figure 10.5 A dial-up data connection.

computer. It can format, word process, and carry out certain preprocessing routines.

Another consideration is the trade-off between capacity and urgency. For instance, high-urgency data traffic may be transmitted via a dial-up connection. Low urgency data may be handled in batch and be physically transported to the central processing location. Telex, of course, is another expedient. International telex circuits operate at 50 bps. Suppose a voice-grade circuit capable of a 4800-bps data-transmission is to be leased. Can we transmit enough data traffic to pay for the line lease? Several factors enter the picture, such as the "value" of the data, its urgency, alternative means available, and, of course, the "quantity" of data and the error rate that can be tolerated. When a 4800-bps line is leased, not just one is leased, but more often two, to provide full duplex operation, thus permitting immediate interaction between the remote terminal and the central-processing unit (CPU) or between different CPUs. This may be vital to the processing operation such as in airline-reservation systems. However, two dial up lines may also serve. Leased lines are often conditioned for data supporting a higher data rate with a certain error performance. If ARQ operation (Chapter 8, Section 5.5) is necessary, full duplex facilities are desirable but not absolutely necessary. The user must describe his needs regarding capacity such as message length, bits per day, error rate, access time, traffic flows, response time, interaction requirements, and type of communication—whether conversational, interactive, or batch.

Table 10.1 gives a relative cost to transmit a megabit of graphic information in the United States. The distance factor is 1400 statute miles (2200 km). Table 10.1 assumes leased equipment. Usage is 8 hs per working day. A cost index of 100 is set for standard 50-baud telex service.

2.2.2 Geographic Dispersal of Activities

The user of a data network, the entity with a data-processing requirement, must carefully quantify locations or points requiring EDP. Compare the needs of the U.S. Department of Defense Supply System versus a small electronic firm with two plants in Eastern Massachusetts. The National Police (Guardia Civil) of Spain present an entirely different problem compared to Madrid's traffic police force. If we determine that a data communication network is required, where will the entry points be of that network? Of course, the types of transaction and their number per time period must also be identified at each point.

3 POSSIBILITIES WITH REMOTE TERMINALS

The expression "data terminal" is widely used and can mean different things to different people. Doll [5] broadly categorizes the terminals tra-

Table 10.1 Cost to Transmit 1 Mbit of Graphic Information 1400 mi

Medium[a]	Relative Cost to Index	Comments
Telegram	1650	Daytime rate 30 bits/word, 100-word messages
Night letter	280	Overnight delivery 30 bits/word
Telex	100	50 bits/s
DDD (103A)	11	Direct-distance dial, 300 bps, daytime (dial-up)
Autodin	4	Full use during working hours (U.S. military network), 2400 bps
DDD (202)	1.75	Data sets (dial-up), 2000 bps
Letter (airmail)	1.15	30 bits/word, 250 words/page
Western Union Broadband	1.00	Full duplex, 2400 bps
WATS	0.75	Special ATT telephone service, 200 bps, 8-h working day
Leased line (201)	0.28	Full duplex, 2000 bps commercial
Leased Line (303)	0.11	Full duplex, 50 kbps, commercial
Mail DEC tape	0.10	Airmail, 2.5-Mbit tape
Mail IBM tape	0.017	Airmail, 100-Mbit tape

[a]Numbers in parentheses identify Bell System modems used in the indicated service.

ditionally used in data-communication applications as those devices employed to convert source data into machine-processible bit groupings by using a two-step procedure involving (1) a human being who converts the data from a source document into a key stroke (on a keyboard) and (2) a machine (the terminal) that converts the keystroke into the appropriate group of binary digits for internal representation and subsequent transmission to a computer or data base. Doll further states, "Since the early days of teleprocessing, applications for terminals have broadened materially to include many situations where there is no requirement for the direct involvement of a human being for data entry." Some examples are optical character readers (OCR), facsimile readers, and remote batch processing. A remote data terminal station (such as that shown in Figure 10.1) consists, as a maximum, of three distinguishable elements:

1. An input–output device (I/O).
2. Data-processing terminal equipment (DPTE).
3. A modem or other device that conditions the data signal to make it compatible with the transmission media and/or the converse.

The data terminal is the basic entry port to a data network. It can be a source or a sink and in most cases serves both purposes simultaneously. A data terminal is essentially a computer peripheral. The difference depends on the length of the transmission line. A peripheral communicates with its host computer by bit-parallel transmission on multipair cable (see Chapter 8, Section 8). However, over longer distances the parallel transmission mode is converted to serial for transmission on a two-wire electrical connection. When the computer peripheral becomes remote and accepts and transmits a serial bit stream, it is referred to as a "terminal."

3.1 Types of Terminal

There are seven general types of terminals in use: (1) teleprinter, (2) alphanumeric cathode ray tube (CRT), (3) graphic CRT, (4) remote batch, (5) data preparation, (6) point-of-sale (POS), and (7) industrial data collection. The teleprinter was discussed in Chapter 8. It can serve as input, as output, or as both. A teleprinter may be made up of the following configurations:

- Receive only (RO) providing printed copy.
- Keyboard send–receive (KSR) providing a keyboard to send and a hard-copy printer to receive.
- Automatic send receive (ASR) with keyboard, paper tape reader, paper tape perforator, and a hard-copy printer.

Teleprinters are generally asynchronous (start–stop), operating at a rate of up to 100 words per minute. However, some operate at 200 words per minute or more. Alphanumeric CRTs are functionally analogous to teleprinters. Instead of a hard-copy printer, there is a CRT capable of displaying alphanumeric characters. Some alphanumeric terminals also offer a limited number of graphic symbols. These are usually the graphics of the ASCII, EBCDIC, or some of the BCD code variations. Cursors are provided on most displays to show the position of the next character with the cursor positioning control. Features of editing and checking ease keyboard data entry. A graphic CRT provides direct display and input of pictorial data representations. Such displays can be more meaningful to the human operator than tabulated data. Input control of the presentation is via a "joystick." Terminals with editing (partial erasure of presentation) require fast logic and large memories making the units more complex. Remote batch terminals provide such I/O facilities as card reader (and card punch), perforated tape reader, and magnetic tape units to provide batch computer processing input from a remote terminal. Line operation is a bit serial from 1200 to 9600 bps. A common term for this type of terminal is "remote job entry" (RJE).

"Data preparation" usually infers traditional key punching, that is, "IBM card" preparation facilities. The cards are collected in batches and transmitted in sequence via a card reader. Data preparation also extends from keyboard to tape or disk systems where the operator works from source documents. Such procedures are often done off-line with several operations going on simultaneously. Cassettes and cartridges are coming more into use as a storage–transfer medium.

Point of sale (POS) is a terminal system designed for retail-store applications. With these devices the clerk at the store's checkout counter is also acting as the equivalent of a keypunch operator. Each transaction is either directly entered into a computer or placed in off-line storage for later input. Inventories are automated, as are recorders from the warehouse, as well as cash transactions, cash flow, and, in fact, an entire retail accounting system. A POS terminal resembles an enhanced cash register. Other input media include badge readers, credit cards, and encoded merchandise tags. Many POS terminals have internal processing to calculate taxes, perform multiplication, and verify check digits. As pointed out, a POS terminal may be on-line or off-line.

Industrial data-collection terminals are used in industrial applications such as in the storeroom, receiving department, tool stations, assembly lines, and outgoing quality control (QC). Data input is obtained either optically (by optical character reader or bar code reader), magnetically, or by card punch and/or keyboard.

3.2 Programmable Terminals

Programmable terminals perform certain communication interface functions and are increasing alleviating the CPU of certain preprocessing functions. The programming capability is provided by a built-in microprocessor, which can help the user acquire, edit, sort, update, file, calculate, and manipulate source data off-line. The programmable terminal typically has a memory capacity from 2*k* to 8*k* bytes. We define a "byte" as 8 bits. (Remember from Chapter 8 that the more common line codes for data transmission have seven information bits to represent a character plus 1 parity bit.) The terminal may also have ROM (read-only memory) defining up to some 50 macroinstructions. A macroinstruction is a group of microinstructions that perform logical, arithmetic, and I/O functions such as "read one record." Realize that once we deal with programmable terminals, we are dealing with a form of distributed processing. They are versatile but also have their weaknesses. For example, the added complexity gives rise to additional maintenance. It also requires programming capabilities at remote locations with people available who can work at machine-language levels. Some terminals provide ROM capability where only factory-provided programs are available to the user.

Cluster terminals are now in common use, where a cluster accesses the

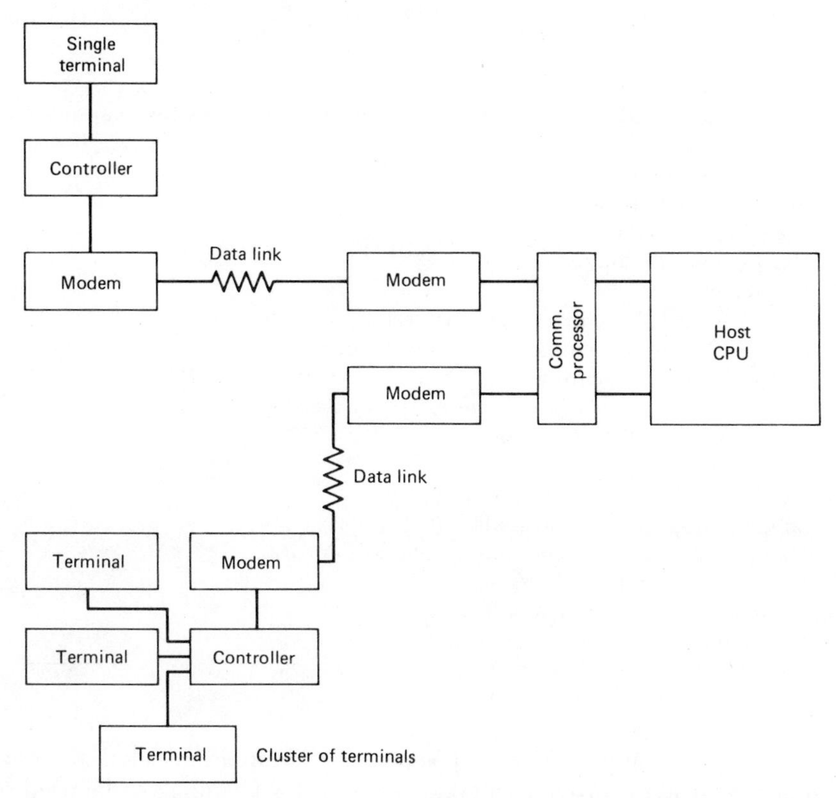

Figure 10.6 A stand-alone or single terminal and a terminal cluster accessing a host computer.

communication link via a common data-link controller. The cluster concept is illustrated in Figure 10.6.

Consider that if more local intelligence is provided at the remote terminal, it will depend less on interaction intelligence with the distant CPU. Editing on a CRT display is one important function where local intelligence can provide a major advantage. If the entire message can be composed locally without resorting to the intelligence from the distant CPU, two advantages accrue: (1) less communication link usage and (2) less CPU usage. These advantages must be traded off against the cost of the intelligence in hardware, software, and EDP administration at the remote terminal.

The transaction terminals are used in applications such as on-line POS; banking, especially credit-card banking; and hotel, airline, and train reservations. These require a rapid reaction from the CPU (host computer). Most applications are for short inquiry responses. A typical response time should not be longer than 2 or 3 s. Such short response times require a leased-line network because a dial-up connection takes 5 to 25 s or more

Table 10.2 Major Terminal Requirements

1. Buffered or unbuffered
2. Human factors (the human interface of the I/O), facilities for man–machine functions
3. Send only, receive only, send–receive, and simultaneous send–receive
4. Operating speeds; selectable speeds
5. Code compatibility
6. Hard copy, soft copy
7. Portability
8. Leased line, dial-up, or special service lines
9. Code and code formatting, code conversion
10. Local message formatting
11. Communication-network operational requirements
12. Error control (FEC or ARQ)
13. Security features
14. Possibility of peripherals, clustering, off-line
15. Simplex, duplex lines; possibility of dial-up backup
16. Maintenance routines, checkout local and network
17. User programmability
18. Local storage and storage access
19. Unattended operation

between inquiry and response if separate dial-up is initiated for each transaction. Doll [5] provides a good checklist for terminals to be used on data networks. Table 10.2 is derived from this checklist.

4 SOME SIMPLIFIED NETWORK CONFIGURATIONS

4.1 General

In Section 1 we distinguished between centralized and distributed data networks and stressed processing rather than structure. In this section configurations or structures are examined.

4.2 Centralized Networks

The centralized network is reminiscent of the star network for telephony, or simply various remote terminals connect to a central computer, as suggested in the following diagram. There are four basic characteristics of a centralized network:

- Its computing and switching facilities are centrally located at one site.
- It has a treelike appearance, although in some cases it is a ring or loop.

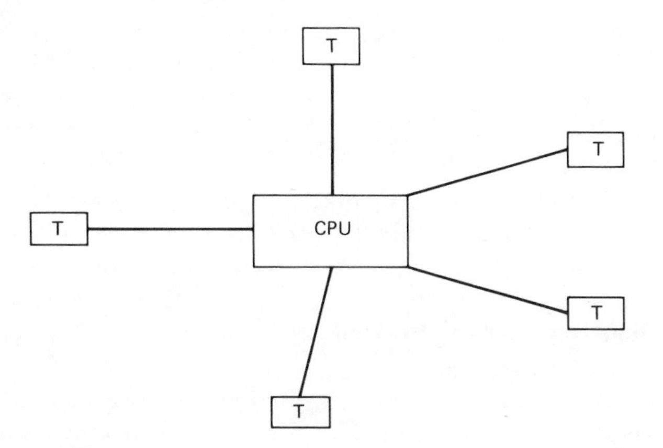

- There is only one unique communication path between the terminal and the CPU (dial-up telephone lines may be used for backup).
- It is a terminal-oriented system. Traffic flow is between the terminals and the CPU.

4.3 Ring Configuration

The ring network is one in which data input-output points are connected by a ring structure. A simplified ring network is shown in the following diagram. Of course, there are many ways of configuring the ring. Traffic

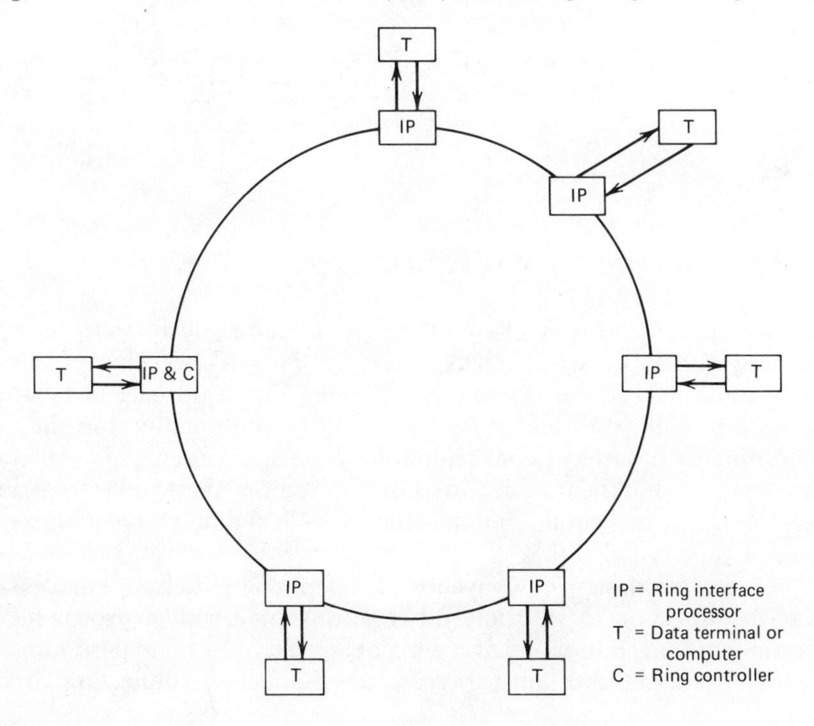

could be unidirectional, say, clockwise. It could be bidirectional, thereby adding considerably to reliability. The processors could be simple or complex. In the simple configuration only one pair of stations could communicate at a time. In the complex configuration using a TDM (time-division multiplex) bit stream in both directions, all stations could intercommunicate quasisimultaneously. A ring-switched network is more suitable for the interconnection of local terminals and computers than for the long-distance network.

4.4 Multipoint versus Point-to-Point

A multipoint (or multidrop) system is one in which two or more terminals share a dedicated or leased line. The multipoint concept is shown in the following diagram. The format of this multipoint connection is like the

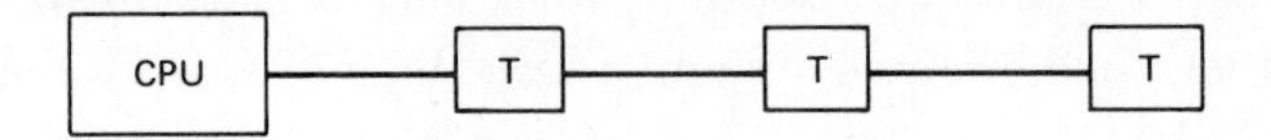

rural party line in telephony. A point-to-point connection is any connection between a source–sink pair. The point-to-point concept is shown in the following diagram.

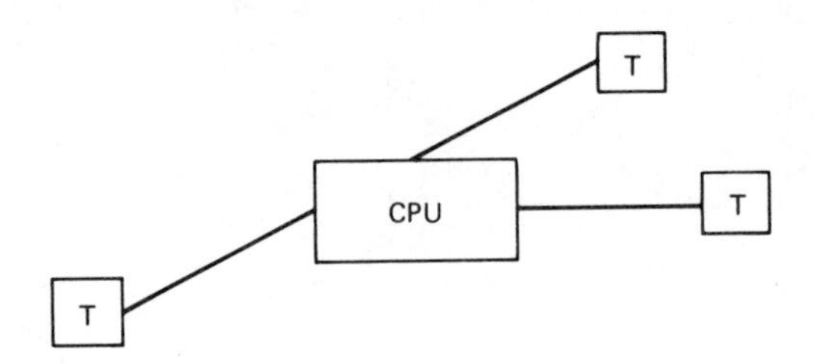

5 TWO APPROACHES TO DATA SWITCHING

The two approaches to data switching are (1) store and forward message switching and (2) packet switching. Store and forward switching usually involves some delay, whereas packet switching does approach a real-time connection. Fully real time switching would be uneconomical in the data environment. A conventional telephone connection occurs in real time. Some may say that the time required for call setup makes it quasi-real time. Nevertheless, a real circuit in either the space or the time domain is set up between source and sink.

The network design of conventional telephone switching enables any subscriber to connect with any other at any time with a probability of effecting a connection stated as a percentage. To reduce the percentage or chance of blockage, we must increase the number of connecting circuits and augment the switching capability. Basically, store and forward message switching stores data (or telex) messages at switching nodes and forwards that traffic to the next node or addressee(s) when circuits become available. There is no direct connection (in space or time) from source to sink as there is in present telephony. With this type of network, messages are sent in their entirety along a predetermined path from source to sink with delays at switching nodes from seconds to hours. These delays are often due to blockage, and the message waits its turn to traverse the blocked link. In such a system connecting links are highly efficient. Storage media at switching nodes must be considerable. It may be on-line, with mass storage devices, or off-line, with magnetic or paper tape. Store-and-forward message switching has the following disadvantages when compared to packet switching:

- Expensive switch costs.
- Often long message delays.
- Less efficiency in utilization of network (not link) resources.
- Less flexibility in adjusting to traffic conditions.

In packet switching a data message is broken down into parts called "packets." These packets could be called "short little messages," each with a "header".* The packet can be likened somewhat to the frame of the PCM system described in Chapter 9. But because they may be transmitted on diverse routes, they may not arrive at the far-end switch sequentially. A major advantage is that these packets can be stored in the main switch memory core because of their shortness. This can substantially reduce switch cost and delivery delays. Packet switching can really show a comparatively high efficiency when multiple routing possibilities exist between switching nodes. Adaptive routing becomes possible where a path between nodes to pass a packet has not been selected in advance.† Path selection is a dynamic function of real-time network conditions. Packets advancing through the network can bypass trunks and nodes where congestion or failure exists.

Each switch in a packet-switched network is a special computer, usually a minicomputer. Packets are forwarded over the best routes, by the switching node, which also supplies error control and notifies the originator of packet receipt. The message originator (the source) divides a data message into packets and places the appropriate header information on each packet. Each packet is handled as a short message and makes its own way through

*"Header" or message (packet) heading identifies the message or packet and its sequence, originator, and destination(s). It may also contain priority, message-length information, sync timing, and so on.

†A multiplicity of paths from source to sink is assumed and is basic to an efficient packet-switched network.

the data network. At the final switching node, packets are reassembled and forwarded to the sink. Because the packets of one data message can travel over different routings one from another, some later packets can arrive at the sink node before earlier ones. It is the responsibility of the switch associated with the sink to reassemble the packets in their proper order as configured at the source. Packet switching approaches the intent of real-time switching. Of course, there is no *real* connection from source to sink. The connection is referred to as a "virtual" connection. Section 9 of this chapter describes a typical packet-switched network.

6 CIRCUIT OPTIMIZATION

Data networks, whether packet switched or message switched, have source–sinks with varying requirements regarding "quantity" of data and its urgency. If a source–sink has access to a node in a network, we would expect the connection to be leased, a so-called voice-grade line. Such a line can support 2400 bps, 4800 bps, or even 9600 bps. It would be uneconomical to underutilize such an expensive line. Suppose that urgency were not a consideration in a particular application, and that the only requirements were to send and receive several thousand bits of data per day. At 75 bps the *maximum* amount of effective bits that can be sent would be as follows:

10 sec	750 bits
1 min	4500 bits
10 min	45,000 bits
1 h	270,000 bits

Another common data rate is 110 bps (see Chapter 8, asynchronous ASCII with a 2-bit stop element). With this data rate the maximum amount of effective data that can be sent is as follows:

1 s	110 bits
10 s	1100 bits
1 min	6600 bits
10 min	66,000 bits
1 h	396,000 bits

Here the reader should beware. An asynchronous ASCII character, as stated, has 11 bits as shown below. Three bits per character are really overhead bits, needed for asynchronous operation of the circuit, but they are unneeded by the far-end sink. The parity bit can be put into the same category; thus we could send 11 bits to transmit 7 bits of source information.

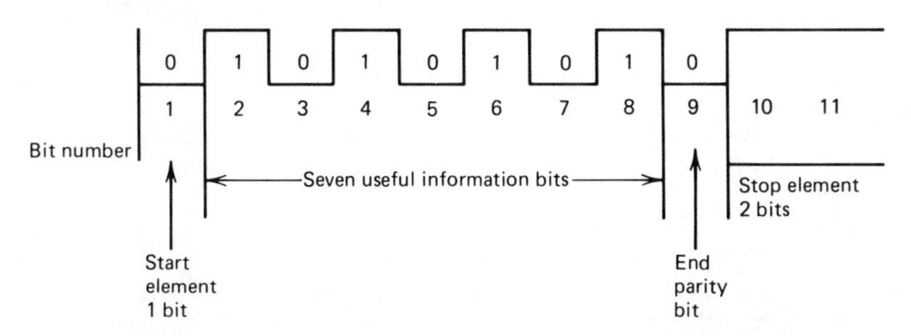

Now consider a data message or packet (block). A typical frame of SDLC (IBM) (refer subsection 8.4.3) for a data message is shown in the following diagram. In this case the text, containing the useful information for the

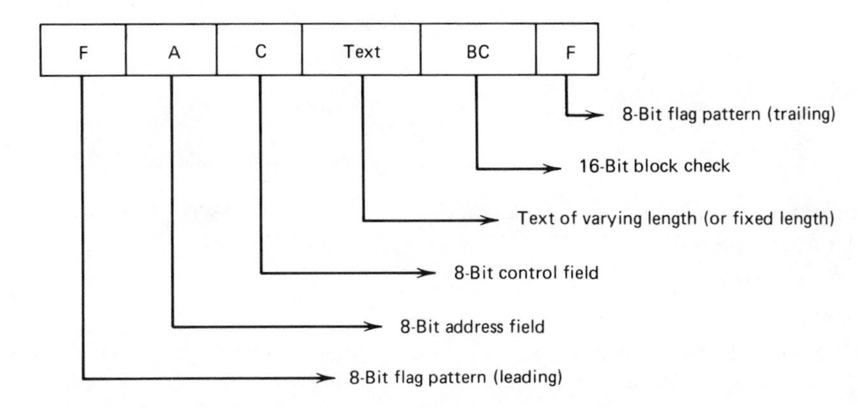

sink user, must amortize the 48 overhead bits for the example shown. Of course, the longer is the text, the better we can amortize the overhead bits. But there is a limit to this length, as we see later. This problem is partially covered in the next section when we discuss net transfer rate of bits. The direct concern here is to describe several methods of optimizing data-communication links. This can be done by frequency division and time division, assuming that a full-period voice-grade line is not required and can, therefore, be shared by various users.

If a corporation had its computer center in city *A* and a number of facilities in city *B*, there might be several ways of configuring data links *B* to *A* (and *A* to *B*) (see Figure 10.7). If no facility at city *B* required more than 75-bps service, the design in Figure 10.7 would serve adequately. If the tariff for a leased line varied with distance, as it almost always does, the longer is the distance from *A* to *B*, the more economic this alternative becomes. The operation of voice-frequency carrier telegraph (VFCT) is described in Chapter 8, Section 10.7.1.

There are other configurations available as well for remote users requiring higher-data-rate service. Figure 10.8, which is taken from CCITT Rec. R.20, shows five possible ways of configuring voice-frequency carrier telegraph. Still another approach to this same problem is the use of TDM in

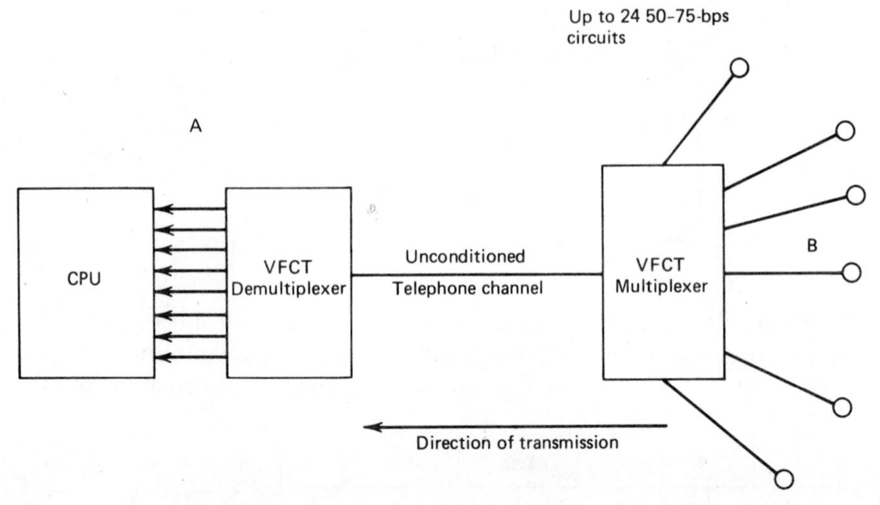

Figure 10.7 Frequency sharing of a voice-grade circuit.

conjunction with high-data-rate modems. This is discussed in Section 10.9 of Chapter 8.

7 EFFECTIVE DATA TRANSFER

The old adage, "All that glitters is not gold," applies to effective data transfer. In a large data network, suppose a link were established from source to sink operating at 2400 bps. How efficient is the circuit? This subject was introduced in the previous section. At first glance we would say that in 1 h we could deliver 2400 × 60 × 60 or 8.64 Mbits. However, how many bits in a time period are really useful to the CPU at the sink? The American National Standards Institute recommends the use of the term "transfer rate of information bits" (TRIB) to quantify the net data transfer rate.

$$\text{TRIB} = \frac{\text{Number of information bits accepted by the sink}}{\text{Total time required to get those bits accepted}}$$

This formula has been used by Doll [5] with block transmission assumed. The formula is rewritten as:

$$\text{TRIB} = \frac{K_1\,(M - C)}{N_t\,(M/R) + \Delta T}$$

where K_1 = information bits per character
M = message block length in characters

Mean frequency (Hz)	420	540	660	780	900	1020	1140	1260	1380	1500	1620	1740	1860	1980	2100	2220	2340	2460	2580	2700	2820	2940	3060	3180
Channel No.	001 101	002 102	003 103	004 104	005 105	006 106	007 107	008 108	009 109	010 110	011 111	012 112	013 113	014 114	015 115	016 116	017 117	018 118	019 119	020 120	021 121	022 122	023 123	024 124

In accordance with
Recommendation R.31 } 50 bauds/
Recommendation R.35 } 120 Hz

Mean frequency (Hz)	480	720	960	1200	1440	1680	1920	2160	2400	2640	2880	3120
Channel No.	201	202	203	204	205	206	207	208	209	210	211	212

Recommendation R.37
50 bauds } 240 Hz
100 bauds }

Mean frequency (Hz)	600	1080	1560	2040	2520	3000
Channel No.	401	402	403	404	405	406

Recommendation R.38 A
200 bauds/480 Hz

Mean frequency (Hz)	540	900	1260	1620	1980	2340	2700	3060
Channel No.	301	302	303	304	305	306	307	308

Recommendation R.38 B
200 bauds/360 Hz

Mean frequency (Hz)	420	540	660	780	900	1020	1140	1260	1560	2040	2340	2460	2640	2880	3120
Channel No.	101	102	103	104	105	106	107	108	403	404	117	118	210	211	212

One example of the application of Recommendation R.36

2 channels-200 bauds/480 Hz
3 channels-100 bauds/240 Hz
10 channels-50 bauds/120 Hz

Figure 10.8 Five possible configurations of VFCT equipment (taken from CCITT Rec. R.70). Courtesy of the International Telecommunication Union–CCITT.

R = line transmission rate in characters per second
C = average number of noninformation characters per block
N_t = average number of transmissions required to get block accepted at the sink
ΔT = time between blocks in seconds

If P is the probability of having to retransmit a block, then N_t can be expressed as

$$N_t = \frac{1}{1 - P}$$

Then

$$\text{TRIB} = \frac{K_1\,(M - C)\,(1 - P)}{(M/R) + \Delta T}$$

There is no direct reference to error rate in the formula; it is implied in the term $(1 - P)$. The reader will appreciate, after some reflection, that there is an optimum block length (given the channel error performance) that will optimize data transfer, at least as far as M is concerned, all other terms remaining constant.

As we see in the next section and as shown in the previous section, the term C, the noninformation bits, can reduce the TRIB considerably. Other delays are also contributory to a poorer overall TRIB such as:

- Dial-up time (if the switched telephone network is used).
- Satellite channels and their inherent propagation delay.
- Modem synchronization delays.
- Type of ARQ, whether stop and wait or continuous running on full duplex.

A data link can be so overburdened with inefficiencies that, for example, a 2400-bps data link may only afford 100 bps of TRIB source to sink.

8 DATA NETWORK OPERATION

8.1 Introduction

When more than one remote data terminal is required to input data to a CPU, the connecting data link(s) may be point-to-point or (and) multipoint. Of course, if there were only two or three terminals, operation could simply be on a "first-come, first-served" basis. This is known as *contention,* where terminals effectively compete for access to a line or a CPU. On multipoint lines the CPU may organize a queue in its communication-control pro-

Table 10.3 Data Network Interface Characteristics

1. Synchronous or asynchronous
   If asynchronous, stop element characteristic
   If synchronous, sync time and pattern for all users
2. Transmission speed or data rate in bps and speed variation limits (i.e., 4800 bps ± 0.1%)
3. Error-rate criteria, with or without correction
4. Error detection and correction: parity type, horizontal, vertical; CRC. special codes; ARQ and type
5. EIA-232C, EIA-449, or V-24, including sense convention
6. Modulation (i.e., FSK: modulation plan interface: BPSK or QPSK: shift, center frequency, sense, etc.)

gram. The problem is most apparent under multipoint conditions (see Section 4.4) when contention is used because there is not only competition for a slot (time) on the CPU but also competition for the communication facility connecting the remote terminal to the CPU.

With large, heavy-usage networks, some kind of procedure and organization must be followed to operate the network and achieve optimum results. Before setting up the rules, generally called "protocol," we would assume a network structure. In this structure certain interface characteristics must be layed down; if not, the "rules" must allow for the variations in interface. One such aspect of the interface is the line code, where some terminals communicate in ASCII, others in EBCDIC, or in a variation of BCD. Are all terminals complying with the convention of sense (Chapter 8, Section 3)? Other interface characteristics are listed in Table 10.3.

Of course the protocol is another important interface characteristic for data network operation. Before proceeding with actual network operations, let us list the items making up a protocol. In other words, what are the topics that must be covered in the rules? These topics are reviewed in Table 10.4. We have introduced the problem of data network operation and have briefly reviewed the simplest network operational technique, namely, contention. Section 8.2 introduces the reader to another fairly simple network control and access technique, polling. Section 8.3 covers advanced operational and control procedures implied under the broad heading of "protocol."

8.2 Polling—An Operational Routine

Polling is an operational procedure mainly used on multipoint lines. The ring or loop network may also use polling. Of course, on a multipoint line at any one instant communication can be established with only one remote station. The order of use of the common line or who gets to use it and when is established in a polling procedure. There are two types of polling now in

Table 10.4 Protocol Topics

1. Framing; frame makeup; or block makeup, message or packet
2. Error control (note that this is included in Table 10.3 as well)
3. Sequence control—the numbering of messages (or blocks) to eliminate duplication, and to maintain a record for proper identification of messages, especially in ARQ systems or for message servicing
4. Transparency—of the communication links, link control equipment, multiplexers concentrators, modems, and so on to any bit pattern the user wishes to transmit, even though these patterns resemble control characters or "prohibited" bit sequences such as long series of "1"s or "0"s
5. Line control—determination, in the case of a half-duplex or multipoint line, of which station is going to transmit and which station(s) is (are) going to receive
6. Idle patterns to maintain network sync (see Table 10.3)
7. Time-out control—which procedures to follow if message (block or packet) flow ceases entirely
8. Startup control—getting a network into operation initially or after some period of remaining idle for one reason or another
9. Sign-off control—under normal conditions, the process of ending a communication or transaction before starting the next transaction or message exchange

use, roll-call polling and hub polling. With roll-call polling, a control station, probably a CPU controller, queries each remote station on the multipoint line in a prearranged order of sequence. On receipt of a query directed to it, each station replies with a "no traffic" signal. If traffic is on hand to send, the polling stops and the message is transmitted. Polling is resumed after message receipt. It will be appreciated that message traffic can pass either way, from terminal *x* to controller (CPU) or vice versa. The polling sequence can be varied as well. If we know that station *B* has twice the traffic as station *A,* then station *B* can be polled twice as often as *A*. Likewise, there can be dynamic changes to the polling list, such as varying the list repetition of stations with the time of day. Hub polling is a method of handing over the polling inquiry. Station *A* is polled and then sends its traffic to the host, and hands over the polling inquiry to station *B,* and then from *B* to *C* and so forth. In this case a strict, invariable sequence is followed. However, there is considerably less overhead involved in call and response.

A procedure known as "selection" is used with polling. The more common type is called "select hold," where the master or control station follows the roll call. Each station is double queried. First it is asked about its ability to receive, and if the answer is affirmative, the station is asked to pass traffic (or traffic is passed to it). The second method is called "fast select," where traffic operations start without any prior checkout. The second method achieves more operating time on the circuit but may require more rigorous error control.

8.3 Advanced Data Network Access and Control Procedures

The term "communications protocol" defines a set of procedures by which communication is accomplished within standard constraints. As shown in Table 10.4 protocol deals with control functions. We must distinguish between data-link control and user-device control. The "data link" is defined as the configuration of equipment enabling end terminals in two different stations to communicate directly. The data link includes the paired DPTEs, modems, or other signal converters and the interconnecting communication facilities. The user device, of course, is a CPU, business machine, or some other peripheral. Pictorially, we can illustrate the difference between data-link control and user device control by the following diagram. The current literature distinguishes between "circuit connection"

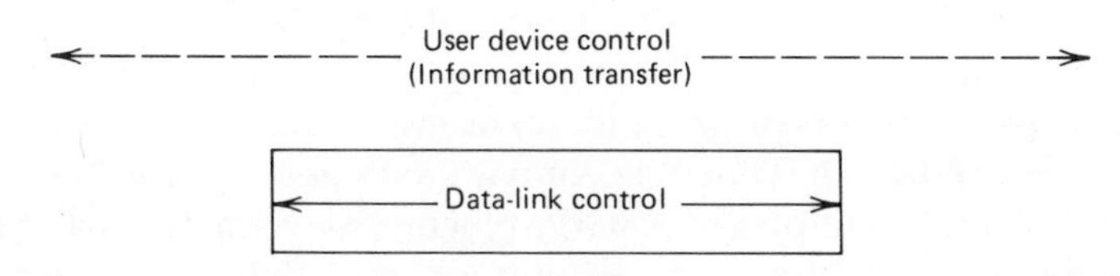

and "link connection." "Circuit connection" is simply the establishment of an electrical path between two points (or multipoints) that want to communicate. We know from our previous discussion that the connection may be metallic (i.e., wire or cable), and/or radio and in either the frequency or the time domain (i.e., a time slot in a frame). The mere establishment of an "electrical" connection does not necessarily mean that data communication can take place. Link establishment is a group of procedures that prepare the source to send data and the sink in a posture to receive that data.

In conventional telephony there are three distinct phases to a telephone call: (1) call setup, (2) information transfer, where subscribers at each end of the connection carry on their conversation, and (3) call termination. The data-link control is analogous to call setup, supervision, and call termination. The user control is analogous to the information-transfer portion. We might expect the following sequence in telephony (C = calling subscriber; A = answering subscriber):

1. C: "Hello."
2. A: "Hello."
3. C: "Good morning. May I speak to Paul Jones?"
4. A: "Just a moment, I'll see if he is in; who may I say is calling?"
5. C: "John Doe."
6. A: "Just a moment, please."
7. A: "Good morning, John."
8. C: "Good morning, Paul. If you can spare a minute, perhaps we can settle the matter on. . . ."

9. . . . .
10. . . .
11. C: "Then that's settled, and I'll confirm by letter."
12. A: "Thank you, John; I appreciate the call, and we will be talking again soon."
13. C: "Goodbye."
14. A: "So long for now" (conversation terminated) (Both parties hang up; connection terminated)

Note that steps 1 through 8 initiate conversation; calling and called parties are identified. On data networks this is called "hand shaking." Step 8 is start of text, and step 11 is end of text. Step 12 may be considered the ACK, and steps 13 through 15 are analogous to EOM, or end of message.

Before going into data link operation, the reader should distinguish three types of character in data-oriented codes such as the ASCII. There are (1) printed characters or graphics (2) user control bit sequences such as space, carriage return, and line feed, and (3) data-link-control sequences or characters. Several examples of ASCII link-control characters are shown in the following list (see Chapter 8, Section 4.3, and Ref. 11 for a complete listing).

- ACK (Acknowledgment). A control character sent by a sink to sources as an affirmative response.
- NAK (Negative acknowledgment). A control character sent by a sink to a source to indicate that the received message contains an error.
- SOH (Start of heading). A control character to indicate the beginning of a sequence of information dealing with routing and allied information.
- STX (Start of text). This control character also means end of heading. It separates heading information from text information.
- ETX (End of text). Terminates the sequence of information started by STX.
- EOT (End of transmission). A control character indicating termination of transmission that may have had one or more headings and texts.

The link is established by initial communication between send and receive modems. The reader should understand here that we are discussing full-duplex operation. The sending modem or calling station transmits "request to send," and the receiving modem replies "clear to send." Of course, if no reply is forthcoming on a "request to send," all operation ceases since the receiving station, for one reason or another, is in no condition to receive traffic. On most dedicated (leased-line) data circuits the "request to send" message is the ENQ (inquiry) sequence and "proceed to send," the ACK sequence. A NAK sequence from the sink would indicate that the receiving station was not ready for traffic. Once the ENQ was sent and the sink replied with the ACK sequence, message transfer between the two stations can commence.

Suppose a metallic (or equivalent) connection has been made between a data source and sink and that data link has been established. Let us now discuss message communication, where we deal with yet a third level of control requirements. Pictorially, we can represent the present location of the system by the heavy lines in the following diagram. The message-

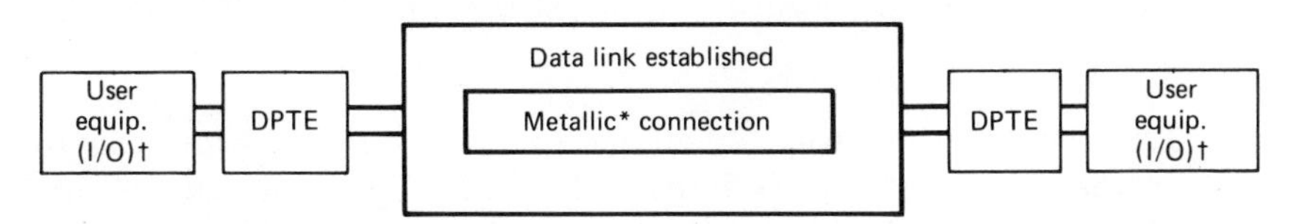

transfer phase may now commence. Messages will be transferred by one or more blocks. Assume synchronous transmission to be a duplex circuit. There are two levels of synchronization; the first is purely a transmission problem between modems, which must be bit synchronized. To maintain bit synchronization, a certain minimum number of mark-to-space and space-to-mark transitions must be continuously received. The second level of synchronization is between DPTEs, or what might be called "format synchronization" or "framing." The makeup of the frame or block and its rules of use are the basis of this level of synchronization. Figure 10.9 shows a typical data block (or frame) and its barest essentials consisting of "header," text, and "trailer." Once the blocks have been sent, each is acknowledged by the sink with an ACK if error free. Then the link is taken down or terminated, by an EOT sequence. We have assumed ARQ operation. There are two operational types of ARQ: (1) stop–wait and (2) continuous. Stop–wait ARQ is the simplest, where an ACK (or NAK) is sent after each received block. If an ACK is received from the sink, the source proceeds with the next block; if not, that is, if a NAK signal is received at the source, the previous block is repeated. The full block in the sending process must be held in buffer storage until an ACK is received from the distant end. A NAK signal from the distant end stops the uniform flow of data traffic from the source and the "bad" block is repeated, possibly several times, until the ACK is received. This can be costly for the stop–wait decode time—costly in overhead bits and channel idle time.

The second method of ARQ permits almost continuous data flow but requires more storage and processing. It is called "continuous ARQ" in that when the distant end or sink is receiving $N + 1$ block, it is processing the N block already received. There are two methods of handling continuous

*This is a metallic (or equivalent) connection.
†This may be remote terminal or CPU.

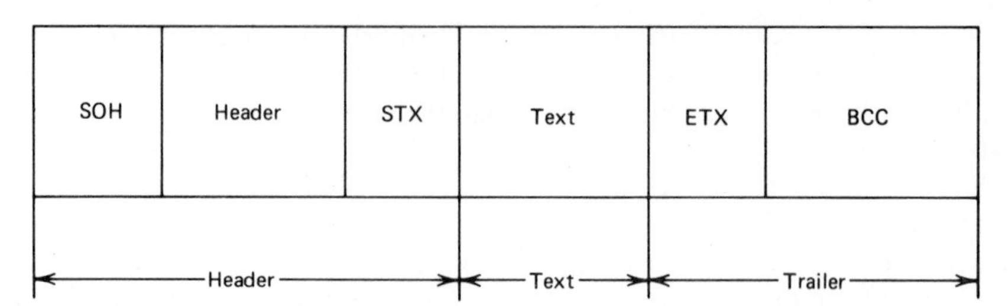

Figure 10.9 A data-message block and its components (BCC = block check character; see Chapter 8, Section 5.5).

ARQ. The first, called "go back N," allows continuous transmission from the source until a NAK is received from the sink. Then the source stops sending its normal sequence and begins repetition of the last two or three (or even four) previous blocks. The number of blocks that must be repeated is a function of the propagation time and processing time at the sink. For instance, the source could already be sending block $N + 2$ when it is notified that block N was in error. In this case blocks N, $N + 1$, and $N + 2$ will be repeated. Thus we can see that the source must have sufficient buffer storage for three blocks in this particular case. The impact on a circuit's TRIB is evidenced by the requirement to retransmit blocks $N + 1$ and $N + 2$, which may well have been received error free in the first place. Another method of continuous ARQ is a form of selective retransmission and involves repeating only the block in error. This requires buffering and additional processing at the sink or receiving end for the proper reassembly of blocks to maintain block sequence so that retransmitted blocks are inserted in their proper locations. Thus some form of block identification is required for this type of ARQ.

8.4 Protocol

A convenient method of classifying different types of protocol is by the message framing techniques used. There are character-oriented, byte-oriented, and bit-oriented types of protocol. Typical character-oriented protocol is IBM's BISYNC, or binary synchronous protocol. Character-oriented protocol uses special characters to indicate such events as SOH for start of heading, STX for start of text, and ETX for end of text. Byte oriented protocol also uses character-sequence delimiters in the header similar to character-oriented protocol and include a "count," which indicates the number of data characters in the message.

Bit-oriented protocol utilizes a special flag character to delineate which bits constitute messages. Such a flag may be the bit sequence 01111110, and no series of six "1"s in a row are permitted, except for the transmission of the flag character. The flag character tells the receiving station at a given instant whether it is receiving text or BCC. For instance, as we see further on, the 16 sequential bits prior to receipt of flag is the BCC and all bits before the BCC and the previous flag were message bits.

As we have seen in prior paragraphs, every data link protocol must provide the operational rules and procedures of a data network. These rules are based on the topics covered in Table 10.4. The following paragraphs review some of these operational rules and other features of some of the more commonly known protocols.

8.4.1 The BISYNC Protocol

The BISYNC (BSC) protocol, developed by IBM, is designed for half-duplex communication only. It is applicable to point-to-point and multipoint network configurations. On a particular network only one of the following three data-transmission codes is permitted by the BSC protocol: (1) EBCDIC, (2) ASCII, and (3) 6 bit transcode. A system designed around BSC must select the most appropriate of the three listed codes. Error detection employed in BSC is CRC or LRC/VRC. The CRC-12 is used for the 6-bit transcode; CRC-16 is used on the 8-bit transmission codes (EBCDIC and ASCII). The LRC/VRC is also used with the ASCII code. As we mentioned earlier, BISYNC is a character-oriented protocol. Table 10.5 is a listing of character-control sequences used in BISYNC. For the three transmission codes admitted in BISYNC, five to seven of the character sequences shown in Table 10.5 must be modified to fit the code properties. For instance, WACK is DLE in EBCDIC and ASCII and is "DLE W" in the 6-bit transcode.

Table 10.5 BISYNC Character-control Sequences

Character Sequence	Meaning
SYN	Synchronous idle
SOH	Start of heading
STX	Start of text
ITB	End of intermediate transmission block
ETB	End of transmission block
EOT	End of transmission
ENQ	Enquiry
ACK0/ACK1	Alternating affirmative acknowledgments
WACK	Wait-before-transmit positive acknowledgment
NAK	Negative acknowledgment
DLE	Data-link escape
RVI	Reverse interrupt
TTD	Temporary text delay
DLE EOT	Disconnect sequence for switched line.

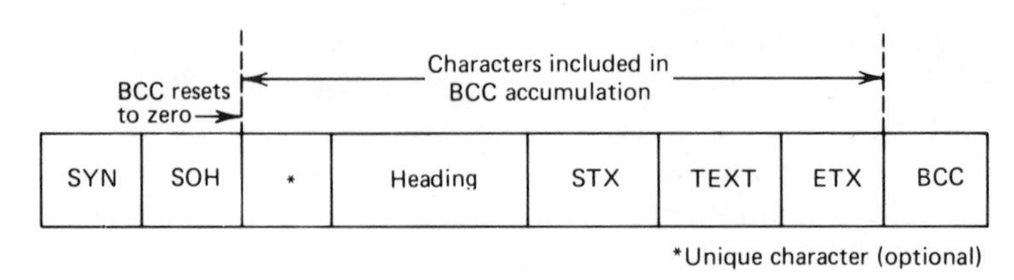

Figure 10.10 Basic BSC frame.

A typical (nontransparent mode) BSC frame is shown in Figure 10.10. The BSC has numerous limitations for the more advanced data network. For instance, it is designed for use only with "stop–wait ARQ." Transparent text operations are available but are cumbersome and inefficient when compared to more modern protocols. Because specific control character sequences are used for data-link control at some times, at other times for error-control recovery, and still at other times for the control of end-user I/O devices, ambiguities in implementation of telecommunication software routines for control are liable to occur and, in fact, do occur.

8.4.2 Digital Data-communication Message Protocol

Digital data-communication message protocol (DDCMP) is a data network protocol developed by the Digital Equipment Corporation (DEC). It is a byte-oriented protocol and can be used on synchronous or on asynchronous circuits. It can be used on half-duplex or full-duplex circuits, on point-to-point or multipoint circuits (serial or parallel). A DDCMP data message consists of two parts: a header containing control information and the text followed by a 16-bit BCC. Figure 10.11 illustrates the message format of DDCMP. The DDCMP protocol has three classes of messages: SOH, ENQ, and DLE, meaning "data," "control," and "maintenance," respectively. The class of message is indicated by the sequence SOH, and so on, and this sequence appears in the block in Figure 10.11 immediately after the synchronization sequences shown as "SYN." After the class, in Figure 10.11, is the count, which gives the number of characters that follow for "data" or "maintenance" class messages. The count gives a counting of the text characters. For control messages in lieu of the "count" per se, the

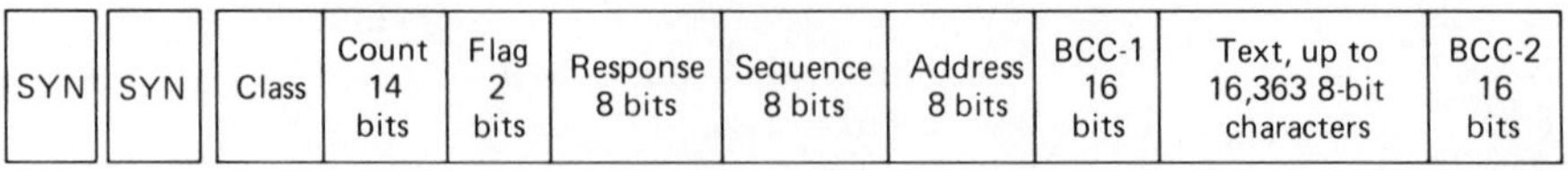

SYN	SYN	Class	Count 14 bits	Flag 2 bits	Response 8 bits	Sequence 8 bits	Address 8 bits	BCC-1 16 bits	Text, up to 16,363 8-bit characters	BCC-2 16 bits

Figure 10.11 Basic DDCMP message format (frame format). (*Note:* BCC1 and BCC2 are referred to in the DEC literature as CRC1 and CRC2. Both BCCs use CRC-16 error detection.)

14 bits take on another meaning. In this case the first 8 of the 14 bits designate the type of control message. The remaining 6 bits are filler bits, usually all zeros. However, in the case of NAK (negative acknowledgment) messages, these last six bits give the reason for the NAK. For instance:

000001	BCC header error
000010	BCC data error
001000	Receiver overrun
010001	Header format error

The first bit of the 2-bit flag informs the sink that the message will be followed by sync characters. These are used, of course, to maintain synchronization and avoid the filling of receiver buffers until the next block or frame comes on line. The second bit is the select flag, which indicates that the last message in a sequence of messages is the final message from a particular transmitting station. Such a flag is useful on multipoint nets or half-duplex circuits where transmitters need to be turned on or off. The response field indicates the message number of the last message correctly received. This 8-bit field is used in the data class of message and for ACK and NAK control messages. The sequence field contains information on message sequence numbers. It is useful for certain types of control message and in the data class of message. In the case of message control, it asks the distant end whether it has received all messages correctly up through a certain sequence number. On data messages it contains the message sequence number of the message as assigned by the transmitting end. In multipoint operation the address field is used to identify the station to which the message is directed. For point-to-point operation, by placing a "1" in the first bit position, the receiving end ignores all other information in that field.

With DDCMP, data are transmitted in blocks or data messages that are sequentially numbered. If the receiving end sees no errors, it accepts the message as correct. It sends no immediate ACK message on correct receipt and will continue receiving a string of up to 255 messages. However, if the receiving end finds a message in error, a NAK message is sent to the transmitting end, indicating the sequence number of the last "good" message (or block) received. Errors may be found in the header by the header CRC or "out of sequence" or in the data text CRC (BCC). The text of a DDCMP message may be of varying length, with a maximum length of 16,363 8-bit characters.

8.4.3 Synchronous Data-Link Control Protocol

Synchronous data-link control (SDLC) is a bit-oriented protocol that establishes data-link control procedures for IBM's System Network Architecture

Beginning flag 01111110 (8 bits)	Address 8 bits	Control (8 bits) includes P/F	Information text—any number of bits	Frame check (BCC) 16 bits	Ending flag 0111110 (8 bits)

Figure 10.12 Basic SDLC frame (P/F = poll/final).

(SNA). Four essential architectural features are basic to SDLC:

- The use of a common grammar
- Increased reliance on a data-link facility for error detection and recovery.
- Two-level hierarchy made up of primary and secondary stations.
- Each data-transmission block, called "frame," has a specific format.

The SDLC frame is shown in Figure 10.12. The flags in the SDLC frame serve as reference points delineating the frame, positioning the address and control fields, and initiating the error check. The flags tell the receiving station the sequence of transmission. An ending flag serves as the beginning flag of a second frame if two frames are transmitted one directly following the other. It also indicates that the previous 16-bit group was the frame check. Thus the sequence 01111110 must be unique so that the flag can be recognized at any time. The system architecture is such that this particular sequence is prevented from appearing anywhere else other than the flag position. To assure this, SDLC procedures require that a binary "0" be inserted by the transmitter after any succession of five continuous binary "1"s. The "0" is removed at the receiver. Inserted and removed "0"s are not included in the error check.

The address field designates the particular station or stations to which the frame is directed. The control field is 8 bits in length and can have three basic formats for the makeup of the 8 bits, as shown in the following diagram. The P/F (poll-final) is common to all three formats. The P/F bit is

Format					
• Information format	Receive count	P/F	Send count		0
• Supervisory format	Receive count	P/F	Code	0	1
• Nonsequenced format	Code	P/F	Code	1	1

the send–receive control. For instance, suppose a "1" is sent by a primary station in the P/F position. This forces the addressed secondary station to respond. This then is called a "poll." On the other hand, a "1" sent by a secondary station indicates that the frame containing the "1" in the P/F position is the final frame (i.e., no more traffic). A sending station with a "0" in the P/F position indicates that more frames follow. Again a "1" in the P/F position is indicative of the last frame and response requested. As depicted in the control field diagrams, the P/F bit is bit 4 in the control field.

In the information (transfer) format, bit positions 1, 2, 3, 5, 6, and 7 provide information on frame sequence numbering. This is vital to ARQ "accounting" for "go back *N*" (*N* frames) operation. For a station that transmits sequenced frames counts and numbers each frame, this count is known as N_s. A station receiving sequenced frames counts each error-free sequenced frame that it receives. This count is called N_r (note the use here of "s" for sending and "r" for receiving). The N_r count advances one when a frame is checked and found to be error-free.

- N_s Information is placed in bit positions 5, 6, and 7.
- N_r Information is placed in bit positions 1, 2, and 3.

Of course, the P/F bit occupies bit position 4 in the control field. The counting capacity for N_r or $\mathrm{N_s}$ is 8, using the digits zero through seven. Up to seven frames may be sent before the receiver reports its N_r count to the transmitter. All unconfirmed frames must be stored by the transmitting end because it may be necessary to repeat some or all of them. The reported N_r count is the sequence number of the next frame that the receiver expects to receive. Thus if, at a checkpoint, it is not the same number as the transmitter's next sequence number, some of the frames already sent must be repeated. In fact, the N_r number received by the transmitter in the "acknowledment" sequence gives the transmitting end the information necessary for frame repetition when errors occur. For instance, if seven frames were sent and the N_r indicates only the first five received (e.g., the last two frames or at least frame 6 in error), the transmitter will repeat the last two frames.

The supervisory format is used in conjunction with the information format to initiate and control information transfer of the information format. Supervisory frames are used to acknowledge transmissions, prohibit transmissions and request retransmissions. This information is contained in bit positions 5 and 6 of the supervisory frame as follows:

00	RR = Receive ready (acknowledgement)
10	REJ = Reject (negative acknowledgement)
01	RNR = Receive not ready (wait)
11	Reserved (a reserved sequence)

Responses use the same format in bit positions 5 and 6. It should be noted that N_s is used both to send information (information frames) and data-link control (supervisory frames).

The nonsequenced format is used for initializing stations and for setting operating modes. It is nonsequenced and thus it does not use N_s and N_r. Such a format is identified by the appearance of a "1" in bit positions 7 and 8 (as shown). A supervisory frame is identified by having a "0" in bit position 7 and a "1" in bit position 8. There are five bits available in the nonsequenced (NS) format for encoding commands and responses. Table 10.6 outlines the commands and responses for the three types of control frames. The information field can be of any length but must be in multiples of 8 bits. It is the information field that contains the data message that is to be transferred from one station to another in a data network. The frame-check sequence (FCS) or block check contains 16 bits employing a form of cyclic redundancy check (CRC-16). The transmitting side sends the 16 digits of the FCS to the receiver after performing the cyclic redundancy check on a specific frame. The receiver performs a similar computation on the frame received. If the results are not the same as those for the transmitted value, the frame in question is discarded as erroneous.

Synchronous data-link control has a two-level hierarchy made up of a primary station and one or more secondary stations. The primary station has system control. The secondary stations may be connected to the primary on a point-to-point basis, multipoint, ring or loop, or combinations thereof. An example of combinations is shown in Figure 10.13, where a secondary station serves as a primary station as well. With the SDLC protocol, a primary station can carry out two time-out functions that are used to control operations. These are "idle detect" and "nonproductive receive." "Time-out" is a term indicating that an expected action on a circuit has not happened, usually because something has gone wrong. Circuit recovery or retransmission is brought about when a time-out occurs.

"Idle-detect" is a condition indicating that events on a circuit are not occurring in their proper frame of time. In this case the primary station detects a nonresponse condition from a secondary station when a response would have been received. The time period in which a response should be received should account for a number of factors, including the propagation time to and from the secondary station, the processing time at the secondary station, and the operational characteristics (through-put rate) of the modems involved. Once a reasonable time period has been exceeded (and this time period should be set for the particular circuits and systems involved), the idle-detect function operates, initiating the necessary action. This may be retransmission or circuit recovery. The second time-out function, "nonproductive receive," is initiated when the received signal from a secondary station is highly distorted, thus causing an excessive error rate.

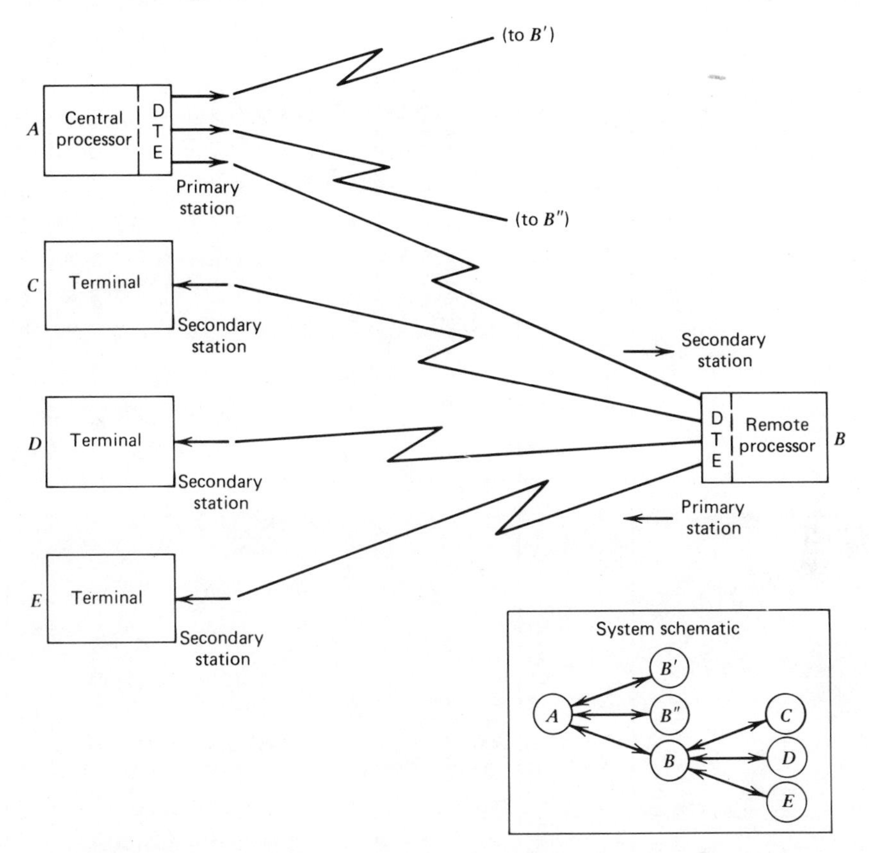

Figure 10.13 Example of a complex system. Note dual role of Station *B*.

When this occurs, actions similar to those described for idle-detect are taken.

An "abort" is a premature termination of a data line in an "active state." In SDLC an abort signal consists of the transmission of eight binary "1"s by the transmitting station, whether primary or secondary. An abort returns the circuit to the "idle state." Only a primary station can reactivate the link.

A major advantage of SDLC over older types of protocol, such as BISYNC, is that it may be used on a full-duplex basis. In fact, a primary station can transmit to secondary station *A* while receiving from secondary station *B*. Because SDLC is a bit-oriented protocol, different devices and/or different code formats can be mixed on the same multipoint or loop communication line. Also, because of the positional conventions used in its frame structure, transparency can be accommodated more easily. Synchronous data-link control also accommodates both contention and polling-type procedures to control station activity on full duplex links. The fact that

Table 10.6 Summary of Command or Response Control Fields

Format	Sent Last ↓	Binary Configuration	Sent First ↓	Acronym	Command	Response	I-Field Prohibited	Resets Nr and Ns	Confirms frames through Nr-1	Defining Characteristics
Nonsequenced (NS)	000	P/F	0011	NSI[a]	X	X				Command or response that requires nonsequenced information
	000	F	0111	RQI[e]		X	X			Initialization needed; expect SIM
	0C0	P	0111	SIM[f]	X		X	X		Set initialization mode; using system prescribed procedures
	100	P	0011	SNRM[b]	X		X	X		Set normal response mode; transmit on command
	000	F	1111	ROL[g]		X	X			This station is off-line.
	010	P	0011	DISC[c]	X		X			Do not transmit or receive information
	011	F	0011	NSA[d]		X	X			Acknowledge NS commands
	100	F	0111	CMDR[h]		X				Nonvalid command received; must receive SNRM, DISC, or SIM
	101	P/F	1111	XID	X	X				System identification in I field
	001	0/1	0011	NSP	X		X			Response optional if no *P*-bit
	111	P/F	0011	TEST[i]	X	X				Check pattern in I field
Supervisory (S)	N_r	P/F	0001	RR	X	X	X		X	Ready to receive
	N_r	P/F	0101	RNR	X	X	X		X	Not ready to receive
	N_r	P/F	1001	REJ	X	X	X		X	Transmit or retransmit, starting with frame N_r
Information (I)	N_r	P/F	N_s 0	I	X	X			X	Sequenced I-frame

[a]*Nonsequenced information:* As a command, an NSI frame is the vehicle for nonsequenced information. An NSI frame is also a vehicle for nonsequenced information sent to the primary station; NSI is not acknowledged.
[b]*Set normal response mode:* This command subordinates the receiving secondary station to the transmitting primary station. No unsolicited transmissions are allowed from a secondary station that is in normal response mode. The expected response is NSA. The primary and secondary station N_r and N_s counts are reset to 0. The secondary station remains in normal response mode until it receives a DISC or SIM command.
[c]*Disconnect:* This command terminates other modes and places the receiving secondary station effectively off-line. The expected response is NSA. (A switched data-link station then disconnects or goes "on hook".) A disconnected secondary understation cannot receive or transmit information frames; it remains disconnected until it receives an SNRM or SIM command.
[d]*Nonsequenced acknowledgment:* This is the affirmative response to an SNRM, DISC, or SIM command. Further transmissions are at the option of the primary station.
[e]*Request for initialization:* An RQI frame is transmitted by a secondary station to notify the primary station of the need for an SIM command. Any command other than SIM causes repetition of RQI by the secondary station.
[f]*Set initialization mode:* This command initiates system-specified procedures at the receiving secondary station, for the purpose of initializing link-level functions. NSA is the expected response. The primary and secondary station N_r and N_s counts are reset to 0.
[g]*Request on-line:* This response is transmitted by a secondary station, to indicate that it is disconnected. A secondary station in NDM transmits this response if the received command is not implemented or not valid.
[h]*CMDR (Command Reject):* This response is transmitted by a secondary station in NRM, when it receives a nonvalid command. A received command may be nonvalid because (1) it is not implemented at the receiving station (this category includes unassigned commands), (2) the I field is too long to fit into the receiving station buffers (this use is optional), (3) the command received does not allow the I field that was also received, and (4) the N_r received from the primary station is incongruous with the N_s sent to it. The secondary station cannot release itself from the CMDR condition, nor does it act on the command that caused the condition. It repeats CMDR whenever it responds, except to an acceptable mode-setting command: SNRM, DISC, or SIM. The secondary station sends an I field containing status information as part of the CMDR response frame. This I field provides the secondary station status data that the primary station needs to select appropriate recovery action.
[i]*Test:* As a command, a TEST frame may be sent to a secondary station in any mode to solicit a TEST response. If an I field is included in the command, it is returned in the response (unless the I field cannot be stored in the secondary station buffer).
Note: NRM = normal response mode; NDM = normal disconnect mode.
Source: Reference 4, courtesy of IBM Corp.

SDLC is limited to sending only seven blocks or frames at a time without acknowledgment may be a drawback, particularly when communications are CPU–CPU; DDCMP, on the other hand, permits as many as 255 unacknowledged blocks, each with up to 16,000 characters.

8.4.4 The CCITT X.25 Protocol

The X.25 protocol is covered under CCITT Rec. X.25. It has been widely accepted in Europe and by the ISO (International Standards Organization). The recommendation states, "The establishment in various countries of public data networks providing packet-switched data transmission services, creates a need to produce standards to facilitate international interworking." This statement establishes the goal of the protocol, namely, standardized international packet switching of data. The protocol has notable similarity with SDLC regarding frame format. A typical X.25 frame is shown in Figure 10.14. For instance, the flags and the address field are the same as SDLC. However, in the case of the latter, two types of addresses are delineated: A and B, as follows:

Address	Bit Position							
	1	2	3	4	5	6	7	8
A	1	1	0	0	0	0	0	0
B	1	0	0	0	0	0	0	0

Address A includes frames containing commands transferred from the DCE to the DTE and frames containing responses transferred from the DTE to the DCE. Address B includes frames containing commands transferred from the DTE to the DCE and frames containing responses transferred from the DCE to the DTE. The control field contains a command or response and sequence numbers where applicable. There are three types of control-field format, which are shown in Table 10.7. The three formats are briefly described as follows:

Information transfer format (I) (see SDLC, Section 8.4.3).

Supervisory format (S). The S format is used to perform link supervisory control functions such as acknowledge information frames, request transmission of information frames, and request a temporary suspension of transmission of information frames.

Flag 01111110	Address	Control	Data text (optional)	Frame check sequence (16 bits)	Flag 01111110

Figure 10.14 The CCITT X.25 frame format.

Table 10.7 The CCITT X.25 Control Field Formats

Control field bits	1	2	3	4	5	6	7	8
I Frame			N(S)[a]		P/F[e]		N(R)[b]	
S Frame	0		S[c]		P/F		N(R)[b]	
U Frame	1		M[d]		P/F		M	

[a]Transmitter send sequence count.
[b]Transmitter receiver sequence count.
[c]Supervisory function bits.
[d]Modified function bits.
[e]Poll bit when issued by primary station; final bit when issued by secondary station.

Unnumbered format (U). The U format is used to provide additional link-control functions such as SARM (set asynchronous response mode), and DISC (disconnect), and responses such as UA (unnumbered acknowledge) and CMDR (command reject).

The CCITT X.25 recommends that the maximum data field length of packets be 128 octets (8-bit bytes). Additionally, some telecommunication administrations may opt for other maximum lengths. It is recommended that these be taken from the following powers of two: 16, 32, 64, 256, 512, and 1024 octets and exceptionally 255 octets, according to the CCITT recommendation. In data packet networks, CCITT X.25 protocol uses its information frame as the vehicle for a packet. The packet itself is carried in the data (information) field as shown in Figure 10.14A, which repeats Figure 10.14, expanding the information (data) field to show the packet

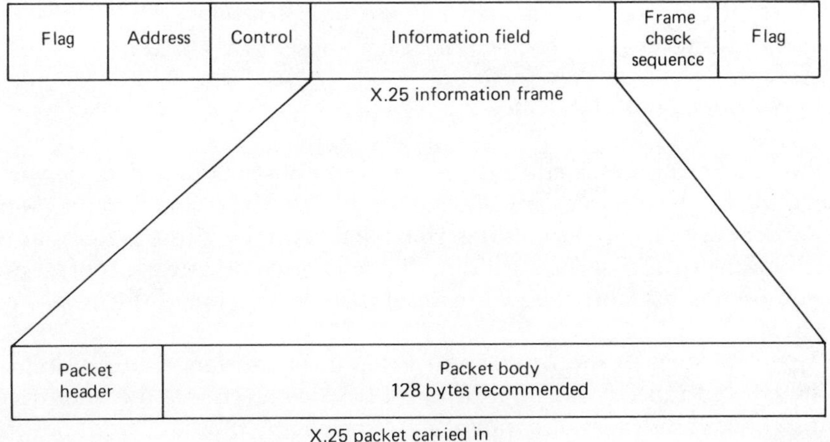

Figure 10.14A A data packet is carried in an information frame. (Ref. CCITT Rec. X.25).

header and body. Unlike SDLC, CCITT X.25 permits two different modes of operation: primary–primary and primary–secondary. In primary–primary operation each of two stations connected by a data link can act as primary station. Keep in mind from SDLC that primary station initiates command and control functions and a secondary station executes command and control functions. The frame-check sequence uses a CRC polynomial that is different from SDLC. The generating polynomial in this case is $X^{16} + X^{12} + X^5 + 1$.

8.4.5 Other Types of Protocol

High-level data-link control (HDLC) is a protocol being formulated by ISO (International Standards Organization). The frame structure portion of the standard has been approved by the organization body and published as Standard IS-3309.

Advanced data-communications control protocol (ADCCP) is a protocol developed by the American National Standards Institute (ANSI). The reader should refer to ANSI Standard X3.28 (1976) titled "Procedures for the Use of the Communication Control Characters of ASCII in Specified Data Communication Links." The frame structure of ADCCP data-link control protocol is the same as that of SDLC.

Burroughs data-link control (BDLC) is a bit-oriented protocol especially designed for use with Burroughs data equipment. Its frame format is very similar to that of SDLC. However, the address, which is normally 8 bits long, may be extended in 8-bit increments to accommodate additional secondary stations. The control field is also expandable, in this case to 16 bits, of which the sequence number field is expandable to 7 bits, thereby increasing the potential number of unacknowledged frames from 7 to 127.

The BOLD program is a bit-oriented protocol of the National Cash Register Corporation. The protocol is a subset of ADCCP.

The CDCCP program is a bit-oriented protocol developed by the Control Data Corporation. This protocol is also a subset of ADCCP.

8.5 Network Control Architecture

Network control architecture defines the components of a data network, including hardware, software, and communication links and how they work and interwork. The protocols described above carry out the actual control mechanism. In this section we describe and discuss network control and operation not just from the viewpoint of a single link, but of the network as a whole.

Again we turn to the protocol in general. For instance, in a terminal-oriented, centralized system, a protocol has two levels: (1) the line procedure that administers the physical transmission medium and possibly detects and corrects errors and (2) the procedure that manages the information flow between a terminal (or a concentrator) and host computer. In a packet-switched data network a protocol is called up to handle special duties, to establish useful communication from terminal to computer or from computer to computer. The problem is further complicated by the fact that in large networks various communication functions are performed among a diversity of terminals and computer operating systems. A protocol can then be characterized in five levels, although every computer network need not necessarily have all of them. A brief description of each level (or layer) follows:

1. Level of DTE/DCE interface characteristics. It defines the electrical characteristics and control procedures between the data terminal equipment and data-communication equipment, including character alignment, interface procedures, and timing of events; selection, call progress, and line identification; failure detection and isolation (CCITT Recs. X.25 and X.21).
2. Line-control procedures. We saw how BSC, SDLC, DDCMP, and X.25 carried out this task (the frame).
3. Control procedures between a pair of communication processors, such as error detection and correction, flow, and routing control. CCITT Recommendation X.25 describes level 3 as "description of the packet level DTE/DCE interface for virtual call and permanent virtual circuit facilities" (the packet).
4. Control between a communication processor and a host computer or between a communication processor and a data terminal. This level provides the operating rules that permit the host computer or data terminal to send messages to other specified host computers or data terminals and to be informed of the disposition of those messages. In particular, this level constrains the host computer or terminal to make optimum use of the available communication facilities without denying the availability to others.
5. Control procedures between a pair of host computers, often called "user procedures."

This last level is not covered in the protocols discussed above earlier in this section.

The reason for this "layering" is to provide a clear-cut separation of functional responsibility in the distributed modules that control the network. Such layering permits the full separation of functions associated with data-link control, with the modules performing the path control and the managing of logical channels. With this clear-cut separation, a certain transparency is obtained where changes carried out in one layer do not affect other layers (i.e., there is no "ripple" effect.). A typical architecture utilizing this concept is the IBM SNA (system network architecture) which is based on the SDLC protocol described in Section 8.4.3. It has a layered structure of only three levels. The layers listed as follows that are applicable to SNA are from user side to transmission side.

- Application layer.
- Function management layer.
- Transmission subsystem layer.

These layers are shown in Figure 10.15, as well as communication possibilities among layers and between layers. As shown in Figure 10.15, the layer closest to the user is the application layer, which includes the hardware and software functions for implementing application programs. The function management layer manages the transfer of information between the layers, defined by discrete devices distributed throughout the network. The transmission subsystem layer describes the generalized routing and transfer of information between system nodal points. The transmission subsystem has a three-level logical structure:

- Data-link control.
- Path control.
- Transmission control.

The IBM SNA stresses the separation of functions into well-defined, logical layers and the distribution of functions across different points of intelligence (nodes). The source and sink in SNA are interfaced with the network by means of "network-addressable units" (NAU). These units are the logical network ports that user programs refer to symbolically and the network refers to by "network address." An NAU may be likened to a subscriber port on a local telephone exchange. That port is uniquely identified with a telephone number. If we consult the telephone directory, that number is referenced to a specific name. A "network address" can be reached from any other "network address" in the network. The connection is made by specification of a "path." The transmission subsystem selects and controls a physical path between two NAUs. The integrity of transmission is assured along each data link by the data-link control mechanism.

A formally bound pairing called a "session" must be established between two NAUs before their end users can communicate. The communication system does this when an end user invokes a connection protocol for initiating a session. This basic SNA structure is shown in Figure 10.16.

9 AN EXAMPLE OF A PACKET NETWORK—THE ARPANet

9.1 General

The ARPANet is the best example available of an advanced packet-switching data network for resource sharing. The abbreviation ARPA

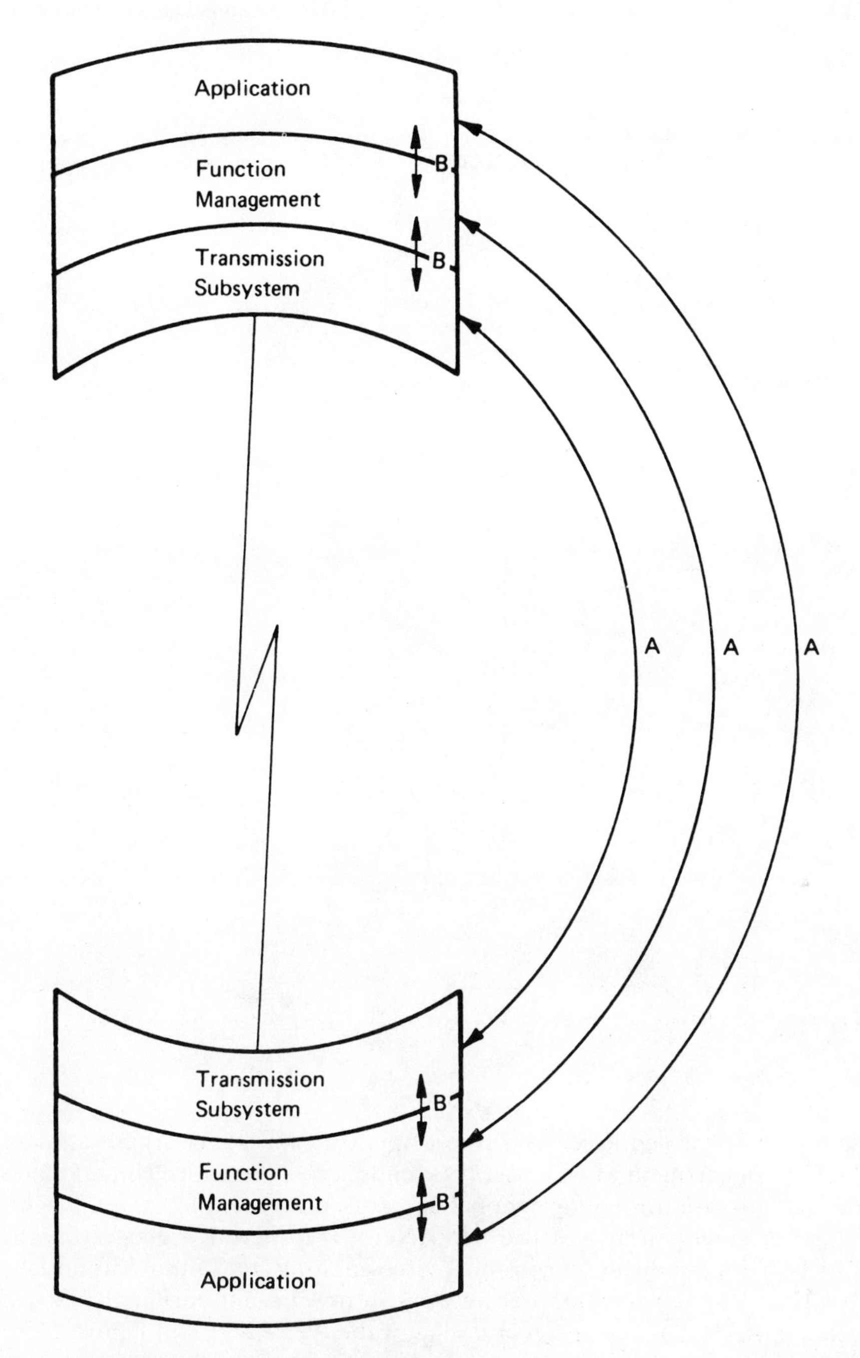

Figure 10.15 The three levels of the SNA architecture. Courtesy of IBM Corporation, Armonk, New York.

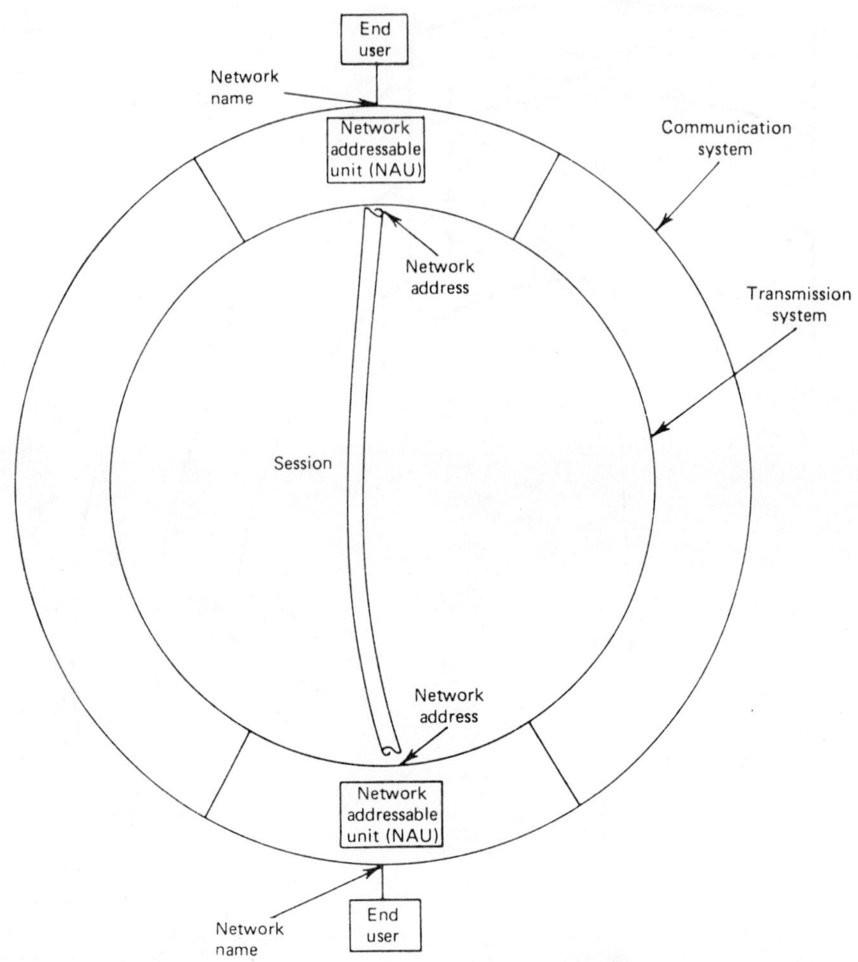

Figure 10.16 Basic SNA structure. Courtesy of IBM Corporation, Armonk, New York.

stands for Advanced Research Project Agency, a subsidiary organization of the U.S. Department of Defense. It extends across the entire United States and has provision for international access (see Figure 10.2).

The original objective of the ARPANet was to permit a geographically distributed set of different computers to communicate. Thus it was initially strictly a computer-to-computer network. It now permits terminal access as well. Figure 10.2 is a geographical map of the ARPANet, and Figure 10.17 is a logical map of the network. The interconnecting communication system is based on very fast response (interactive) message switching. This communication network is called the communication "subnet," or just subnet for short.

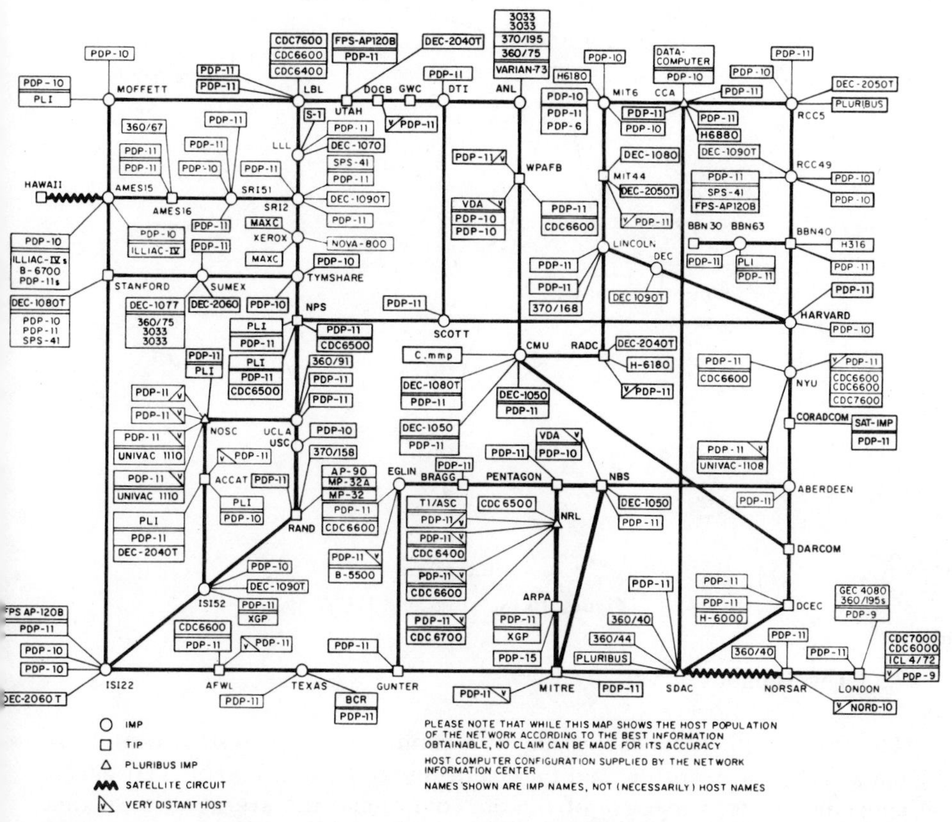

Figure 10.17 Logical map of the ARPANet (March 1979). Courtesy of Bolt, Beranek and Newman, Inc.

9.2 The Concept of Host and Interface Message Processors

The ARPANet connects dispersed computers of various manufacture and varying design. The subnet providing that connection is a form of store and forward system and must deal with such problems as routing, buffering, synchronization, error control, reliability, and other related issues. To isolate the computers from these problems of communication and to isolate the subnet from the problems of computer interface, small identical processors are placed at each node to serve a (host) corresponding computer or computers. These processors and their connecting transmission media then form the network. In this arrangment ARPA defines the research computer center as a *host.* The small processors isolating the host from the subnet are called IMPs, or Interface Message Processors. A simplified diagram of hosts and IMPs is shown in Figure 10.18.

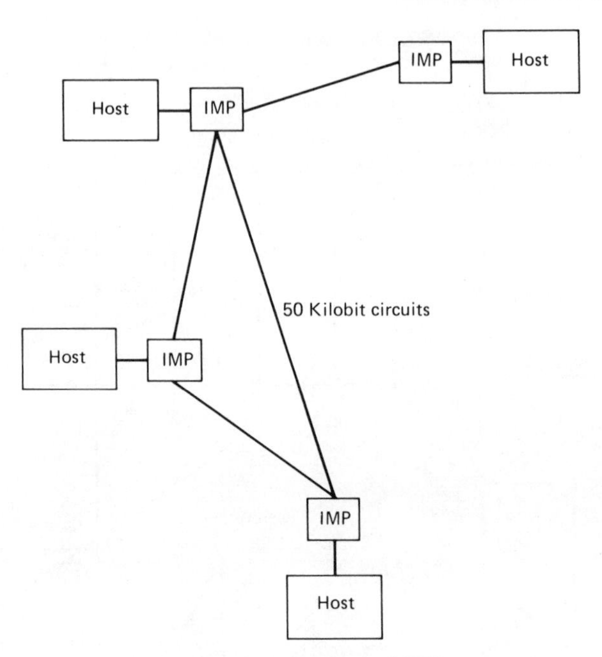

Figure 10.18 Hosts and IMPs.

9.3 The Subnet

The subnet functions exclusively as a communications system whose primary task is to transfer data bits from source to sink reliably. To permit convenient interactive use of remote computers, the average transit time through the subnet is kept to about 0.2 s. The subnet is completely autonomous. It functions as a store-and-forward system; thus the IMP cannot be dependent on its local host (as circut switching is to the telephone subscriber). As a consequence, host-to-host communications is a separate, isolated issue constrained only by the limitations of the serving subnet.

9.3.1 Subnet Operation

The ARPANet has been designed to accommodate a wide range of data text lengths, from a few bits to text arbitrarily long, such as a very long file. Because of buffering limitations at nodes, maximum message length is limited to 8192 bits. Messages can be of any length up to this maximum. Of course, a data text to be transmitted can be made up of one or many such "messages." The message is what a host presents to its corresponding IMP. The IMP, then, breaks the message up into data packets of no more than 1024 bits each.

The subnet transmits messages over unidirectional virtual paths between hosts known as "links." In this context a link is a conceptual path, not a physical path as found in typical circuit switched networks. Only one message at a time is accepted by the subnet on a given link. Other messages entering the link are blocked until the source IMP has learned of the arrival at the proper sink of the message being serviced. Once this occurs (e.g., the link becomes unblocked), a special service message is set to the source host. This message is called an "RFNM" (ready for next message). While waiting for an RFNM on one link, a host may send its next message on yet another link. The design of the ARPANet allows for up to 63 separate outgoing links to serve a specific host site. Messages inserted into the subnet for transmission have a "leader." This is similar to the header on a telex message or as mentioned earlier in the chapter. In the "leader" the originating host specifies the destination host and a link number in the first 32 bits of the message. The IMP takes the message and provides route selection, delivery, and notification of receipt when effected. Such utilization of links and corresponding RFNMs allow the delivery of sequences of messages from terminal IMP to destination host in proper order. This is because the subnet permits only one message at a time on a specific link. Figure 10.19 shows the basic aspects of message handling from source to sink. We can see that messages are broken up into packets; each packet is structured by the originating IMP with a header that contains routing information, packet number, check digits, and so on. Packets are individually routed from IMP-to-IMP through the network toward final destination. Each intervening IMP reframes the packet with initial and terminal framing characters and check digits for the use of the next IMP along the route. As the packet moves through the network, each IMP stores the packet until a "positive acknowledge" is received from the succeeding IMP, meaning that the packet was received complete without error. Once an IMP sends positive acknowledge to the preceding IMP, it retains responsibility for the packet until it, in turn, receives a positive acknowledge from the next IMP it was routed to in the subnet. An IMP can refuse a packet. It does this by not responding with a positive acknowledge. It would not respond for several reasons such as buffer full, busy, or receipt of packet in error.

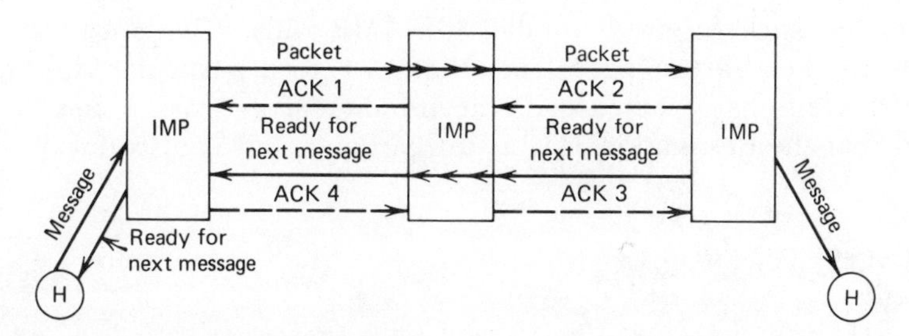

Figure 10.19 Basic ARPANet massage handling.

Exactly what is absence of positive acknowledge? A return positive acknowledge must be sent in a reasonable time interval or we can expect that it will not be sent. That reasonable time interval is usually 100 ms.

At the destination (sink), IMP packets could arrive out of order. It is the responsibility of the destination IMP to reassemble packets, placing them in proper order, to strip off headers of each packet, fabricate and attach a leader, and deliver to the destination host the complete message as a single unit with source host and link identified.

9.3.2 Routing

Routing is based on an algorithm implemented at the IMP. A packet is directed to destination along a path where the total transit time is shortest. This is the basis of the routing algorithm, namely, shortest transit time. The IMPs are independent. Paths are not determined in advance. Each IMP individually decides on which of its output lines to transmit a packet for onward delivery. The selection is made by means of a fast look-up table, which is a routing table covering all possible destinations and their appropriate next leg from that particular IMP. The table is kept current, updated every 0.5 s. Table entries reflect line trouble, IMP failure, congestion and connectivity. Updating is carried out essentially based on the following:

> An IMP transmits about twice a second to all its contiguous neighbors its minimum delay estimates for each destination in the network. This is the delay it expects each packet to encounter in reaching every possible destination over each of its output lines. The estimated delay is based on recent performance of the connecting communication circuits and queue lengths. For a selected destination the table states the sum of delays from the IMP to neighbor and neighbor's delay to destination where the delay is shortest.

Here we can see that an IMP need not know the topology of the network. The system is self-adaptive to congestion, route failure, and so forth. Faulty or disconnected IMPs are detected by the transmission of additional information to neighbors. Time to destination is measured by the number of IMPs required to be traversed from the IMP in question to destination. To the smallest such received number, the IMP adds 1. This number then represents the shortest path from that IMP to a particular destination. When this number ever exceeds the number of network nodes, it is assumed that the destination IMP is unreachable and is disconnected.

REFERENCES

1. Daniel R. McGlynn, *Distributed Processing and Data Communications,* Wiley, New York, 1978.
2. Paul E. Green, Jr., and Robert W. Lucky, Eds., *Computer Communications, IEEE Press,* New York, 1974.

3. John E. McNamara, *Technical Aspects of Data Communication,* Digital Equipment Corporation, Maynard, Mass., 1978.
4. *IBM Synchronous Data Link Control,* IBM publication No. GA27-3093-1, 1974.
5. Dixon R. Doll, *Data Communications Facilities, Networks and Systems Design,* Wiley, New York, 1978.
6. Ralph Glasgal, *Advanced Techniques in Data Communications,* Artech House, Dedham, Mass., 1976.
7. Robert P. Blanc and Ira W. Cotton, Eds., *Computer Networking,* IEEE Press, New York, 1976.
8. CCITT Recommendations, Geneva, 1972, amended Geneva 1976, X Recommendations, in particular X.20, X.21, X.21 bis, X.24, X.25, and all of Volume VIII.2.
9. *Synchronous Network Architecture, General Information,* IBM publication No. GA27-3102-0, 1974.
10. *General Information—Binary Synchronous Communications,* IBM publication No. GA27-3004-2, 1970.
11. American National Standards Institute (ANSI), document No. X3.28, 1976.
12. *Proc. IEEE* (Special Issue on Computer Communications) (November 1973).
13. Roger L. Freeman, *Telecommunication Transmission Handbook,* Wiley, New York, 1975.
14. Elwyn R. Berlekamp, Ed., *Key Papers in the Development of Coding Theory,* IEEE Press, New York, 1974.
15. J. C. R. Licklider and Albert Vezza, "Applications of Information Networks," *Proc. IEEE,* **66** (11) (November 1978).
16. Louis Pouzin and Hubert Zimmermann, "A Tutorial on Protocols," *Proc. IEEE,* **66** (11) (November 1978).
17. Robert Sproull and Dan Cohen, "High Level Protocols," *Proc. IEEE,* **66** (11) (November 1978).
18. Vinton Cerf and Peter Kirstein, "Issues in Packet-Network Interconnection," *Proc. IEEE,* **66** (11) (November 1978).

Chapter 11

TELECOMMUNICATION PLANNING

1 GENERAL

The "planning function" is vital to a telecommunication operating company or administration. In the last 20 years or so it has been found equally as vital in the industrial and government environment. However, the two areas must be distinguished and treated separately.

An operating company is often a monopoly faced with a demand for service yet with limited resources to satisfy the demand. Ideally, because it is a monopoly, it is overseen by a watchdog commission such as the Federal Communications Commission (FCC) in the United States. The industrial–government entity requires an optimal communication facility (with communications taking on an ever widening meaning). Such facilities must again be under the constraint of limited resources. The industrial–government telecommunication manager has the prime responsibility of making communications serve the organization with minimum cost and maximum results. The principal portion of the chapter deals with the operating telephone company or administration and its planning function. The latter part of the chapter briefly discusses the industrial government problem.

2 GROSS PLANNING ON A NATIONAL SCALE

2.1 Economics

Telecommunication planning is largely economic. For this discussion, we consider this "economics" in really three levels: (1) macroeconomics, (2) midrange economy, and (3) microeconomics. Macroeconomics (our definition) deals with the amount of wealth a country spends in the telecommunication sector. Of course, the stage of development of a particular country and whether the overall economy is controlled, free-running, or somewhere in between will be strongly contributing factors. It is very important for evolving nations to achieve a balance among sectors for efficient development. Telecommunications should be given its proper place with the other sectors such as transport, industry, agriculture, and social programs. We define "midrange economy" as the method used by an operating company (or administration) to raise capital. Distinguish between capital and tariff revenue. Oversimplifying, we could say that for a privately owned telephone company in a free economy that capital investment (e.g., new plant) will be financed by internally generated funds* and from the sale of securities. The dividends and interest to be paid on those securities will derive from the tariff revenue.

There is the other extreme, where the telecommunication operating administration is owned by the government. In this case (oversimplifying again) capital for new plant may be taken from tax coffers, probably a legislated budget for the year. Then tariff revenues (all or a portion thereof) are returned to the national treasury. Many administrations or "telephone" companies are in a "gray" area. Spain is an example. Its telephone company (CTNE) is government controlled but issues stock and is operated as any other industry with an annual report to the stockholders.

We define "microeconomics" as plant extension. It deals with costs and return on investment for the expansion or upgrading of a telephone plant. It is project oriented or area oriented. An example may be the installation and cutover of several new exchanges with revamped serving areas to meet demand in a certain metropolitan area several years hence. Such projects and improvements or plant extension follow economic and technical plans. Economic plans permit the extending of capital and will show payback to management. Technical plans assure compatibility and coordinated upgrading and modernization. The "microeconomics" issue deals with the problem of getting the best for the least expenditure. Therefore, the telecommunication planner, even at this low stage, is just as involved with finances as he is with engineering. In the case of commercial common carrier systems (telephone administrations) the telecommunication planner will probably resort to expressing the cost of projects in terms of present value (worth) of annual charges, usually expressed as PVAC (PWAC), depending on which side of the ocean he is.

3 BASIC PLANNING

We would expect upper management of a telephone company to make certain policy decisions that will affect the planning group. For instance, one policy decision might state that 99% of telephones may be reached by direct subscriber dialing within 5 years, or that the entire toll network will be conditioned to support 4800 bps of data transmission with an error rate

*In most cases these funds are retained earnings.

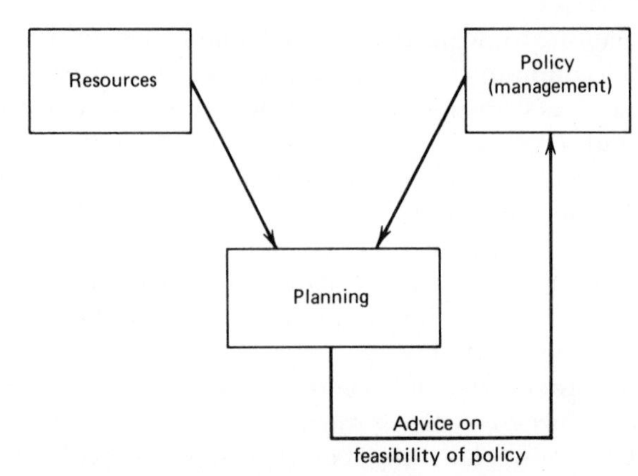

Figure 11.1 How "planning" interworks with management and budgetary limitations inside a telecommunication company (administration).

better than 10^{-5}. The financial impact of such plans must be verified by the planning group for feasibility before they are made policy. The successful planning group interworks with management and other departments in the preparation of plans and on plant extension projects. This interrelationship is shown in Figure 11.1.

The basis of all planning is economic. The greater part of technical planning consists of selecting from various possible schemes to achieve a level of development and extension, those schemes that are the most economic. However, economy is not the only consideration. Planning includes overall organization, work-force expansion, and training; controlling the flow of work; estimating the impact on revenue of tariff changes; advising on the improvement of service quality (Chapter 1, Section 13); and assisting and advising other departments on the estimation of overall profitability, cash-flow requirements, and future growth. In some cases the financial resources may not be available to carry out certain planning objectives, thus requiring their deferral or cancellation. In other cases the money may be on hand and the planning group may be asked how best it may be spent to encourage development and/or improve quality of service. Engineering planning, therefore, is interactive and part of a complex cycle of operations whose primary task is to keep the business running in sound financial condition.

Engineering planning starts with (1) subscriber and traffic forecasts and (2) technology forecasts. With the present network as a basis, fundamental plans are developed. Actual plant extension and modernization projects result from annual capital expenditure programs and short-term plans. Inputs to the formation of these programs and action plans result from guidelines and objectives set down in fundamental plans. These plans state

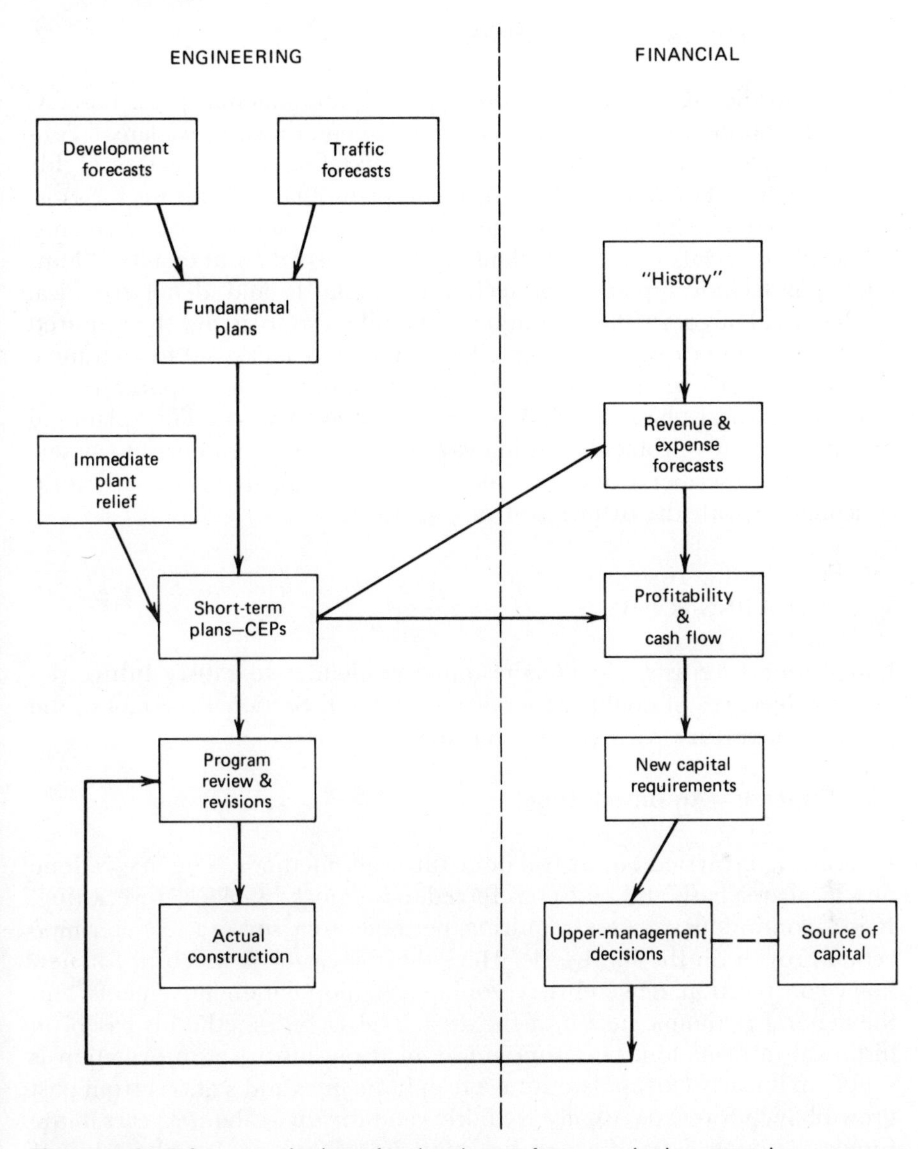

Figure 11.2 Telecommunication planning in a telecommunication operating company (administration)—a plan from inception to execution with actual construction (CEP = capital expenditure program).

general policy and milestones and they also reflect the pressures requiring short term plant relief. Before these plans and projects are put into action, a financial review is carried out to determine impact on revenue and to allot resources. These basic planning functions are illustrated in the flow-chart in Figure 11.2.

4 TYPES OF FUNDAMENTAL PLANNING

We like to distinguish between two types of fundamental plan, namely, fundamental development plans and fundamental technical plans. "Fundamental plans" are plans that serve as a foundation for detailed or highly specific plans. "Fundamental development plans" state the means for satisfying expected demand for new lines and services or service improvements. These plans specify *quantity* of plant (as distinguished from quality). "Fundamental technical plans" set technical standards and detail technical guidelines. They state the techniques to be followed to assure the required flexibility of a network and compatibility among its parts and to guarantee that service performance meets the desired standard. The important attribute of technical plans is that they specify *quality* of plant. The "planning group" faces two problems simultaneously: (1) satisfying demand for service (new subscribers) and (2) upgrading and improving the network, concomitant with the offering of new services.

5 THE STARTING POINTS

Long-range forecasts provide the initial guidelines to satisfy future demand. Objectives of quality of service (Chapter 1, Section 13) establish the point of departure for network improvement.

5.1 Forecasts—An Introduction

Forecasting is carried out by two quite different methods. The first is done on a local area basis and is often referred to as "block-by-block" forecasting. It is a continuous process requiring periodic area surveys to determine actual growth on the local scale. Here the forecaster is watchful for new home construction, new industry, zoning laws, population movements, and the general economic health of the area. The second method is based on historical information. One simple way of using historical information is simply to linearly extrapolate future telephone lines and stations from past growth. Such forecasts usually are fairly valid for up to about 3 years in the future. Another way of using historical information is on the basis of "analogy." With "analogy," telephone growth forecast is based on other quantities whose growth is more predictable, such as the number of automobiles per capita, the business index, and the use of electrical power. Another quite acceptable method, when working on a national basis, is to select another country with similar characteristics but that is 5 or 10 years ahead in telephone development. Own country trend in growth will probably fairly well match that country. Another favorite index is the gross domestic product (GDP), which is used in the CCITT *Handbook—Economic Studies in Telecommunications* [4]. Once cyclic and seasonal factors have been removed from past statistics, an underlying trend can often be recognized, and growth can be expressed mathematically by one of the five equations listed as follows:

Linear	$Y = A + Bt$
Parabolic	$Y = A + BC + Ct^2$
Exponential	$Y = Ae^{Bt}$
Gompertz function	$\log Y = A - Br^t$
Logistic function	$Y = \dfrac{1}{A + Br^t}$

where A, B, and C are constants, the value of r should be taken as something less than unity. t is time and Y is the value to be forecast (e.g., subscriber density). Linear growth is the extrapolation we mentioned previously. Line growths are rarely linear and approximate a constant percentage per year (e.g., exponential).

Application of the parabolic function to express growth trends is attractive because it is a compromise between the linear and the exponential functions. It adjusts better to past data because it has three constants: A, B, and C. Further, the parabolic curve can be fitted to approximate exponential growth initially and less rapid growth deeper in the forecast period. However, in the final analysis the parabolic curve will not prove any more accurate than the exponential if the same historical data are used to construct both functions. The Gompertz and logistic functions are applied when saturation is expected. When there is some doubt regarding the accuracy of the theoretical functions that we have reviewed, it is often helpful to resort to finding the best regression line (i.e., the line that gives the least mean square error when calculated in terms of the growth equation selected). Then the telecommunication planner selects the growth equation with the greatest correlation coefficient. Of course, we are trying to correlate historical data and future growth with current and future telephone stations and lines, which is our final interest.

When there are waiting lists for telephone service, accurate forecasting is made even more difficult. The existence of waiting lists is the rule, not the exception, even in many advanced industrialized nations. Nevertheless, where there is a waiting list, immediate pressure is taken off the back of the forecaster for accurate, short-term forecasts. This is because upper management will set a policy regarding short-term relief of the waiting list. It might state that it is the policy of the administration to install 100,000 new lines each year. Waiting lists do not reflect true demand. Short-term demand (2 or 3 years) equals waiting lists, plus hidden demand, minus abandonments. Hidden demand may reach 40% of the waiting list and abandonments, 20%.

Traffic forecasting (see Chapter 2, Sections 1, 11, and 12) is yet another important aspect of basic planning, although the actual traffic forecasts are usually made by specialists in traffic departments. In the case of local exchanges, future calling rate and holding times per line remains (see Table 2.7) fairly constant over the years. Distinguish, though, among business, residence, and rural lines. With care, future traffic can be forecast quite accurately from past history in the local area on a per line basis. Again, as discussed in Chapters 1 and 2, the importance of keeping good traffic data is patently evident. Leaving aside cyclic calling (holidays) and emergencies (flood, blackouts, hurricanes, etc.), two occurrences that can vary the traffic forecasts are (1) service improvements and (2) tariff changes to stimulate usage. One example of the former that was quite striking was the traffic jump when earth-station service was instituted in Chile and Argentina. International traffic jumped four to six times prior values. On the othe hand, tariff changes do not usually affect the busy hour. Such changes are usually to stimulate usage of the telephone system during periods of low usage such as on nights, weekends, and holidays. Also, local traffic forecasts sometimes err if care is not taken with cutovers of new exchanges and the impact on traffic forecast accuracy. There is a particular liability on the side of error when there is a mix of new service and transferred service from existing exchanges needing relief.

Long-distance or toll-traffic forecasting also extrapolates past history (traffic data) to determine future calling rate and erlangs of traffic. Long-distance service is also sensitive to tariffs. One rule of thumb valid for a large number of countries outside of North America [2] is that 4 to 6% of the local calling rate will give some measure of the toll calling rate. Or we could also say that the average subscriber pays about the same for local as for long-distance service (half of the typical bill for each). Of course, this figure will vary from country to country, depending on toll tariff structure and particularly on the size of the local area and tariff area increments. Toll service usage shows constant growth in more developed nations and, in particular, North America. The calling rate for toll calls in the United States runs from 10 to 15% or more of the local rate on a typical bill. If calling rate were counted on bulk service situations such as the wide-area telephone service (WATS) in the United States, the equivalent toll figures would be even larger. In all these latter cases more than half the telephone bill is for toll service. (*Note:* Compare this with accumulated capital investment share shown in Chapter 2, Section 1, where we show over 65% in local area.)

5.2 Quality of Service

Standards and objectives of quality of service (described in Chapter 1, Section 13) serve as the principal starting point guideline for the formation of fundamental technical plans. Consider the several service quality factors listed here:

1. Transmission quality (level, crosstalk, echo, etc.).
2. Dial-tone delay, and postdial delay.
3. Grade of service (lost calls).
4. Fault incidence and service deficiency.
5. Adaptation of the system to the subscriber.
6. Billing errors (method of billing and its administration).

When a quality-of-service requirement is set, one of two possible approaches can be taken:

1. Design for a maximum permitted impairment in the most unfavorable case.
2. Design for a certain range of impairments occurring as a result of chance combination of elements, such as a majority of subscribers' opinions being favorable.

The latter is often referred to as "statistical design." Both approaches have weaknesses. The first may require unnecessarily high performance to satisfy those rare unfavorable cases. Some subscribers may be very unhappy with the second approach, which will be considerably more influenced by *variability* of plant performance. Such variability is particularly felt in the areas of signaling and transmission and should be taken into account in the planning of those technical areas. There are three possible causes of variability: (1) manufacturing standardization and quality control, (2) multiple usage of elements (e.g., links switched in tandem), and (3) day-to-day variability (e.g., temperature, loading, and other variations). The third item is also strongly influenced by maintenance and its intervals, standards, and quality. Technical planning in North America essentially follows the lines of statistical design. In Europe, those countries following European practice and with CCITT, maximum impairment design is used, although often modified where, for instance, a certain standard will be met in perhaps 95% of the situations.

6 FUNDAMENTAL TECHNICAL PLANS

There are at least six fundamental technical plans that should be prepared and periodically updated by a telephone company or administration:

1. Numbering (Chapter 3, Section 16).
2. Routing (Chapters 1, 2, and 6).
3. Transmission (Chapters 5 and 9).
4. Switching (Chapters 3 and 9).

5. Charging (Chapter 3) (sometimes coupled with switching)—rates and tariffs.
6. Signaling (Chapter 4).

6.1 Numbering

"Numbering" provides for the assignment of telephone numbers over a period of growth for the plant. That period should be 40 years or more, with a review of the Numbering Plan every 10 years to check its validity and conformity to updated forecasts. The plan should meet three major constraints:

- The numbering should be easily understood by the subscriber.
- It should be compatible with existing and planned switching equipment.
- It must be fully interworkable with international numbering schemes.

In developing a numbering plan, the planner will be limited by:

- Existing numbering practices.
- Switching equipment installed and in use.
- Pertinent CCITT recommendations.
- Services offered such as PABX; in-dialing; abbreviated dialing; and special services such as fire, police, and "information," direct dial, and others that tend to block numbers.
- Economics and economic trade-offs.

6.2 Routing

Chapters 1, 2, and 6 discussed routing techniques and network design. A routing plan is closely related to the switching plan. For most planning purposes, a network is designed on 20-year development figures without reference to existing equipment. One rule of thumb [1] states that a system will generally grow to four times its size in 20 years. To comply with a 20-year objective of network design to the meeting of certain routing philosophies, intermediate network designs may be advisable for 10-year and even 5-year periods.

A routing plan should include as a minimum:

1. Description of hierarchy.
2. Definition of full direct, high usage, and overflow routes and the criteria for choosing between them.
3. Specifications of grades of service on trunk and local routes.

4. Principal route layouts geographically and rules for survivability (route diversity) to lessen consequences of breakdowns or disaster.
5. Guidelines for the selection of transmission media for major and minor routes.

6.3 Switching and Charging Plans

A "switching plan" is closely related to a "routing plan." One essential element of a switching plan is to define the number of links for various connections and the dependencies of one class of exchange on another. The plan must also specify the facilities required at each class of exchange. Likewise it sets out rules for combination exchanges. One example of a combination is the direct connection of local subscribers to higher level exchanges, where a local exchange may serve with a toll exchange, in the same building and may even use some or most of the same equipment. At least three headings are required in a switching plan: (1) national long-distance switches, (2) local urban switches, and (3) rural switches. The amortization of a switch is 20 to 25 years (although some switches see more than a 40-year lifetime). A switching plan is often written for 20 years. Many administrations cover the charging criteria and methods in their switching plan.

6.4 The Transmission Plan

The most essential requirement of a transmission plan for a telephone company or administration is that it enable all subscribers to talk to each other satisfactorily. It should reflect or improve on CCITT requirements for international calling. We would expect that for a telephone administration, the transmission plan deals with various aspects of telephone transmission impairments. It will state standards and objectives for volume, noise, crosstalk, bandwidth, and amplitude distortion. We find that the most important transmission factor for a telephone system is volume (e.g., the receive level at the subscriber instrument). The internationally accepted unit of perceived volume is the "reference equivalent" (see Chapter 2, Section 2.3). Loss reduces subscriber receive level. Inherently, telephone systems require loss to avoid singing and reduce echo on long circuits. Economy dictates the type of subscriber loop now in use. They are lossy. A transmission plan assigns losses across the network. As we discussed in Chapter 6, key to reducing loss in the overall network is the balance return loss achievable in two-wire- to four-wire conversions (e.g., hybrids or "term" sets). The concept of virtual switching point used by the CCITT is vital to the understanding of a transmission plan. The plan should explain this and the basic signal levels to be adopted in four-wire switches. Existing and future design standards of subscriber loops should also be included in the transmission plan.

Modern telephone networks are designed primarily for speech transmission. But more and more these networks are being given the task of carrying other information such as telegraph, data, and facsimile (see Chapter 8). Thus the transmission plan must also cover these aspects. Loading of carrier systems, envelope delay distortion (group delay), impulse noise, signal-to-noise ratio, the characteristics of conditioned circuits, and so on must be clearly expressed in the plan. Absolute delay has become important and limits the use of earth-to-satellite transmission systems for data transmission and certain types of telephone signaling.

6.5 Signaling

Signaling is a particularly ticklish matter for the planner. A well-thought-out and properly implemented fundamental signaling plan can pay off in spades to the administration. Signaling is dependent on switching, and vice versa. Remember that the planner has inherited a switching plant that is from 0 to 30 years or more of age. With the variation of plant, he has inherited a number, perhaps five or more, signaling systems. These require various kinds of applique units (black boxes) for compatibility. Thus the fundamental signaling plan should fix at the outset standard signaling systems, for both the long-distance (toll) and local areas. A policy should be expressed in the plan for milestones to convert to the standard. Likewise, a standard approach should be adopted for signal conversions. One example may be in the local area where we would find exchanges arranged for loop signaling. When a pulse-code-modulation (PCM) trunk system is installed, a choice is available to substitute loop signaling by E and M signaling (see Chapter 4), or of converting E and M to loop signaling with appliqué units and remaining with a standard incoming relay set at the switch.

We remember from Chapter 4 that a telephone call involves two types of signaling, line signaling for call supervision and interregister signaling. Both require special study for the plan. That plan, among other items, must consider special facilities for operators, antispoofing devices, and malicious call interception. The transmission aspects of signaling are also a major consideration of the plan such as levels, signal-to-noise ratio, talk-down, and the use of out-of-band signaling.

A signaling plan should have a 10-to-15 year duration. With the phase-in of digital transmission and switching, the shorter period may be more practical. As a minimum, a signaling plan should include:

- A standard interregister signaling system for adoption across the entire network and a milestone plan for total phase-in.
- A standard line-signaling system and a plan for total phase-in.
- A standard local-area signaling criterion and its interface with the toll network (both line and interregister).

It is most important that the signaling plan be coordinated with the charging and tariff plans, and the numbering plans, the transmission plans, the switching plans, and the routing plans.

7 NOTES ON OUTSIDE PLANT PLANNING AND DESIGN

7.1 Importance of the Outside Plant

Although about 40% of the total telephone company (administration) investment is in outside plant, it is often relegated to the position of passing mention or just neglected entirely in the literature. It is often forgotten that all telephone conversations originate and terminate in subscriber instruments that are connected to local exchanges by wire or cable. Thus it is important that the outside plant be just as well designed as the switching and transmission subsystems if service is to be satisfactory. Likewise, as we saw in Chapter 2, the close coordination of outside plant planning with exchange placement cannot be overemphasized. For this discussion, *outside plant* is defined as all telephone plant between exchange buildings and between exchange buildings and subscriber premises. It also includes the subscriber subset. Outside plant is primarily concerned with cable such as service connections or drop wires (to subscriber equipment), subscriber cables, interexchange trunk (junction) cables, and long-distance cables. It also includes terminal cabinets, pole lines, conduit systems, duct work, and rural distribution systems.

7.2 Subscriber Plant

Figure 11.3 illustrates the principal elements of a subscriber distribution network. In essence it is the outside plant cable system connecting the subscriber to his local exchange. This alone represents from 12 to 16% of the total telephone plant investment (see Chapter 2, Section 1). Working our way out from the exchange, there is the main cable that connects to a branch splice or cabinet (for cross connects). Feeder cables connect outward from the main cable from the cross-connection cabinets to the distribution cables. Drop wires connect from the various distribution points and the distribution cable. It is the drop wire that serves the subscriber.

The selection of cable route, the cable sizing and assignment, requires a refined engineering effort to minimize cost and optimize service. The outside plant engineer uses the term "cable fill" when he refers to the number of pairs in use in a specific cable. The percentage of fill is the number of pairs in use divided by the total number of pairs available in the cable. The period of fill is the time to fill the cable to capacity. For small growth rates, the period of fill is 12 to 15 years. Many feeder cables have a period of fill of 7 to 8 years, and for main cables, which have an annual

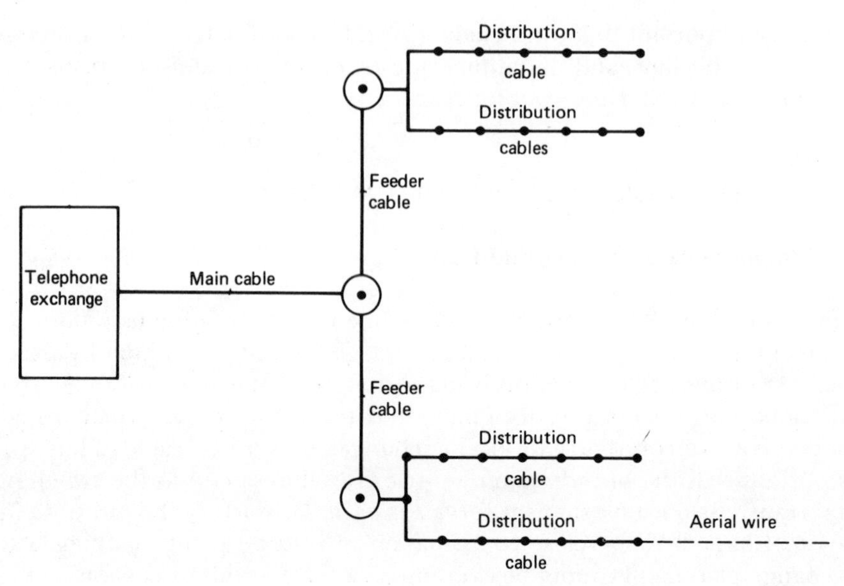

Figure 11.3 Principal elements of a subscriber distribution network: • distribution point; ⊙ cross-connect box or branch splice.

growth rate exceeding 300 pairs per year, an optimum period of fill may be 5 years. Total fill of a cable beyond 85% is found to be uneconomic. The remaining 15% are saved for spares or are pairs found to be unserviceable.

Subscriber loop design from the transmission viewpoint was discussed in Chapter 2, where we found that the principal limiting factor to exchange area size was economic, depending more on the maximum permissible length of subscriber loops without conditioning to achieve longest possible extension. A typical local exchange serving area could be defined as one with a population of 10,000 inhabitants at the end of the forecast period. Assume that the area meets transmission and signaling requirements in all cases. Thus the subscriber plant design would involve the selection (of):

- Distribution design—the choices regarding the use of pairs or star-quaded conductors, rigid and flexible layouts, flexibility joints or cabinets, multiple tees and bridged taps, dedicated conductors, and the amount of dedication.
- Construction design—the selection of construction criteria (i.e., aerial, ducted or buried, choice of materials for sheaths, conductors, insulation, and block or drop wires).

7.2.1 Distribution Design

There are two basic approaches in the design of subscriber loop network, *rigid* and *flexible*. Rigid networks are those where all pairs are continuous from the exchange MDF (main distribution frame) out to the distribution point (see Figure 11.3). No provision is made for intermediate distribution points. The flexible network uses intermediate cross-connection points that interconnect feeder and distribution cables. As the term implies, such a network is more flexible, but more expensive. However, in the long run the flexible network may turn out more economic because feeder cables can be operated at a higher fill than distribution cable. On a rigid network the overall fill is usually lower. Telephone companies usually use some of both design approaches or what might be called a "compromise." The rigid approach is used to connect subscribers comparatively close to the exchange out to a cable route distance of 600 to 3000 ft (200 to 1000 m). Beyond these distances, the outer portions of the network use cabinets and the flexible approach. The use of multiple tees and bridge taps is deprecated. The multiple tee has application where extensive multiparty service exists, but such service offerings have become less and less in most parts of the world. Bridge taps offer another form of flexibility but place additional transmission impairment in loops when used in the order of 0.8 dB/km. The needed flexibility can be better provided by using cross connects with the flexible approach.

In many areas telephone subscribers are highly mobile, moving in, moving out, changing requirements, and so forth. This has resulted in the expenditure of large amounts by telephone companies for rearrangements. To reduce this cost, administrations have attempted various schemes to "dedicate" plant. That is to dedicate MDF space, cable pairs and station equipment to each premise even before its occupancy. Full dedication has not proved economical, especially for optimum use of main and feeder cable pairs. Dedication does prove out in many cases in flexible areas where plant is dedicated in distribution pairs with the use of cross-connect cabinets at the junction of distribution and feeder cables. This allows expensive main and feeder cable to fill more to exhaust before new cable must be put into service. In other words, dedicated pairs that are unused are not sitting idle in this portion of the network.

Flexible systems require careful planning and reasonably accurate 20-year forecasts to be successful. Forecast demand should be broken down into units of demand consisting of 10 to 15 acres (4 to 6 ha), except where population density is very low. These "units" represent cabinet districts at the end of the forecast period. Ideally, a cabinet installed at the beginning of a forecast period would serve throughout the entire period. Cabinet unit districts and their size and shape should be chosen for minimum rearrangement of feeder and distribution cables as new cabinets are added during the period to satisfy growth requirements.

Once cabinet districts have been established, the outside plant planner then must concern himself with routing main and feeder cables to the cabinet locations and the sizing of each cable. Estimates of the approximate year of installation of each cable is part of the long-range plan. The

Table 11.1 Distribution Points and Their Application

Terminal type	Application	Distribution-cable type	Service-wire type
Main distribution frame	Large office buildings, commercial enterprises	Duct	Internal cable
Cross-connection cabinet	Apartment buildings, medium-sized office buildings	Duct	Internal cable
Box with unsealed or sealed chamber on pole or building	Residential or light business	All types	Drop or block wire
Box with unsealed chamber flush with ground surface	Medium residential	Buried	Buried
Pedestal	Medium residential	Buried	Buried
Ready-access messenger support	Light residential	Aerial	Aerial
Encapsulated in cable splice	All residential	Buried	Buried

Source: Ref. 14.

short-range plan(s) define(s) specific expansion years. Main cables normally have a 5-year planning interval and feeder cables, a 10-year interval. Distribution cables are added to care for year-to-year growth and sized to provide for the estimated useful life of the cable in its district under the working conditions found.

The type and size of distribution points vary with the application or concentration of subscribers. Table 11.1 gives the basic types of distribution point now available and their application. Large buildings with a fairly high telephone population will use some form of distribution frame similar to the main distribution frame used at the exchange. For any type of distribution point, service or drop wires should not exceed about 150 feet (50 m) in length. If that length is exceeded, cable extension should be considered. Distribution points external to buildings should accommodate a maximum of 25 subscribers.

7.2.2 Notes on Outside Plant Construction

There is underground and buried construction. Underground construction refers to cables in conduit or duct works. Conduit or duct works offer the cable added protection. Also, on routes over which a succession of cables are required, say, at 5- to 10-year intervals, rather than excavating streets with possible resulting cable damage, advanced provision of spare ducts is advisable. Such spare ducts in place permit the placing of new cables without movement or damage to cable already in place. Conduit systems, rather than direct buried cable, are almost always indicated in metropolitan areas where the cable lays are near other services, such as power, sewage, water, and steam.

Buried construction, that is, the direct laying of cable underground, is cheaper than conduit cable. Buried cable is used where there is less chance of damage or where there is little possibility that future additional cables will be layed. Buried cable is recommended, therefore, in areas of lower population density such as residential neighborhoods where there is no pole line. Cable depth is about 18 to 20 in. (60 cm) below the surface in residential areas, whether in ducts or directly buried. Main route cable with a high concentration of service is placed at a greater depth, from 3 to 5 ft (100 to 130 cm).

Modern cable laying is done by either "trenching" or "plowing." Trenching is where a trench is dug, and the cable is placed in the trench, which is subsequently back-filled. "Plowing" refers to the insertion of a cable into the ground through a hollow plow blade or by pulling it into the ground just behind the plow blade. Plowing is more economic than trenching for depths down to 25 to 30 in. (ca. 75 cm). For depths greater than this, larger tractors are required, and the resulting cost is often greater than that for trenching. Of course, plowing is difficult when there are many underground obstacles, particularly rock ledges. Aerial construction may be indicated as more economic where there is an existing pole line and where such aerial plant can be maintained for at least 20 years. These conditions are now found almost exclusively in rural areas.

8 INSTITUTIONAL, CORPORATE, AND INDUSTRIAL COMMUNICATIONS

Telecommunications planning and management today in the institutional, corporate, and industrial areas is a far cry from what it was 30 or 20, or even 10 years ago. Once it consisted of seeing to the provision of a PBX large enough to satisfy office needs, to make periodic changes in extensions, and to pay the telephone and telegraph bills.

It is a far different story today. Now the corporate communication manager is involved (or should be involved) with over a dozen distinct activities:

1. Private automatic branch exchange (PABX; probably EPABX, electronic PABX) and its dimensioning, its multiple service offerings, its cost, and its interface with the local telephone company or administration.
2. To qualify and quantify the local and long-distance telephone ser-

vice required. The planner must optimize that service with the services offered by the telephone company (or administration), such as taking North America as an example, WATS, foreign exchange lines, local access trunks, local tie lines, and special billing arrangements.

3. Telex service and other special telegraph services.
4. Own data services without telephone-company intervention. This probably would be data from local terminals to CPU on a loop basis (see Chapters 8 and 10).
5. Data services with distant terminals, CPUs, or both. He will make decisions on centralized or distributed processing in conjunction with the corporate EDP manager. A compromise between the two may be arrived at. Then consideration would be given to smart terminals, network design and protocols, data administration, choice of dial-up or leased line, type of line, and possibilities of concentration or TDM or both, to achieve the lowest net EDP-communication cost with the most efficient service. (See Chapter 10).
6. Facsimile service and its possible integration with the office environment. Use of specialized common carrier and trade-offs with item 5 must also be considered.
7. Page-boy communications.
8. Office security systems, CCTV, handie-talkie, alarms, and so on.
9. Cryptographic systems and computer security.
10. Transport dispatching and other mobile applications.
11. Private corporate, industrial and institutional networks such as own VHF–UHF and microwave broadband, multichannel systems, and in some cases cable systems. Compare savings with cost of use of common carrier and its security, grade of service, and so on.
12. Savings on long-distance service with specialized common carriers, some of which supply switching as well. These could be dedicated corporate circuits, semidedicated, or equivalent common usage circuits.
13. Specialized common carrier without local telephone company access. Satellite Business Systems is an example.

The list is not exhaustive, especially considering how the entire office procedures, including accounting, are being melded into an overall "communication" system (see Section 9). The list also has been made considering a minimally regulated society, especially in the telecommunication sector. For instance, in most countries in the world today even the idea of a specialized common carrier is unthought of. All forms of communication in such places are government monopolies. Changes may be forced on the more "controlled" societies when the competitiveness of the more "open" societies is demonstrated, perhaps due to no small part on more efficient communications.

8.1 Typical EPABX Considerations

A major responsibility of the corporate, industrial, and institutional telecommunication planner–manager is the selection and later maintenance of a PABX system (or systems). In almost every case he will select an electronic exchange. Many of these exchanges, probably the greater majority, use principles of TDM switching. There are many advantages to this type of switching, as we discussed in Chaptter 9—small-scale, low-power consumption, economy in so much common circuitry, and suitability for microprocessor control with all the attributes available of SPC (Chapter 3). A typical small EPABX is ITT's TD-100, which provides service for up to 100 lines and 24 trunks. It is a time-division switch using pulse amplitude modulation (PAM). It fits into a cabinet 58 in. high, 39-½ in. wide, and 24 in. deep. The only other equipment is the attendant console. But one of the most important items for the planner's consideration is the features offered by an EPABX such as the TD-100, as small as it is. These features include:

1. Station-to-station calling, three-digit dialing.
2. Station-to-trunk calling:
   a. Direct outward dialling (DOD), foreign exchange (FX), and WATS trunks can also be provided. Second dial tone is used.
   b. Through dialing. Restricted lines may place calls through attendant or attendant gives direct dial connection on request.
   c. WATS and CCSA (common control switching arrangement) using single or two-digit access.
3. Trunk-to-station calling:
   a. Incoming trunk calls via attendant.
   b. Direct inward dialing (DID).
4. Tie trunks. These are one-way or two-way circuits for interconnecting two PABX systems.
5. Tandem switching. This is another tie trunk arrangement that permits tie trunk to tie trunk connections and tie trunk to central office (exchange) or special-service trunk connections. Such calls may be completed with or without attendant intervention depending on strapping of the EPABX.
6. Special power failure arrangements.
7. Station hunting. With this option a call is routed to an idle EPABX station in a prearranged group when the called station is busy.
8. Camp-on busy. When the attendant extends a call to a busy station, the trunk will automatically camp on the busy line. The busy station

and the party to which it is connected will hear a distinct tone indicating a camp-on condition; the calling trunk party will hear a ring back tone. Connection between the trunk calling party and the station is made when the called station goes off hook.

9. Station restriction. This is a class-of-service item denying access of certain stations (extensions) to some of the special features, such as direct access to WATS or FX lines, or other features such as camp-on or call transfer.
10. Toll restriction. Several variations are available; for instance, access may be permitted by all stations to local area dial tone and some stations may be prohibited from making automatically dialed toll calls.
11. Key system features. Key telephones serving up to 20 circuits, in the case of the TD-100.
12. Off-premise extensions. Some EPABX loops can be extended from the 1000-Ω limit to a 2000-Ω limit.
13. Night answering service.
14. Line lockout. The TD-100 employs 100% line lockout to prevent control equipment from being held when it is no longer required. Busy tone or reorder tone is applied to the line to indicate lockout. (In European parlance, lockout is referred to as "time out.")
15. Timed recall. Puts attendant on line when party does not answer call on incoming trunk for some predetermined time, often 30 s.
16. Consultation hold. The act of putting a calling party on hold while the called party drops off the line to carry out another task, for a consultation, a document, or whatever.
17. Call transfer, automatic by called party or by attendant.
18. Add-on conference.
19. Intercept. Vacant level intercept and vacant number intercept.
20. Music on hold.
21. "Tel-touch" dialing ("touch-tone" dialing).
22. Paging service ("page-boy" service).
23. Call-forwarding (follow-me) and call forward on no answer.
24. Busy line call back and busy trunk call back.
25. Priority override.
26. Automatic identification outward dialing (AIOD). Records calling number and called number and time for billing or for supervisors to check on spurious usage of company telephones by employees.
27. Dial call hold. The placing of a second call after another call on the same line is placed on hold.

9 THE INFORMATION MANAGER—INTEGRATED OFFICE COMMUNICATIONS

The modern office is the office of the future. Advances in office communications are occurring so rapidly that what is on the design boards this morning is fact this afternoon and becomes obsolescent tomorrow. The text that follows will be past history; yet we would be remiss to the reader if we did not outline the dynamic trends in office information systems. Lest we forget: the commodity the communicator really deals with is information.

Mention was made in passing in Chapter 10 of the "pull," or competition between the EDP manager and the telecommunications manager. Enter now the office manager, and the three become one. The office of yesterday rendered the electric typewriter obsolete or relegated it to typing assistants. Document preparation was carried out on a word processor. This device consisted of a keyboard input, a VDU display, storage (memory), and a microprocessor. Printout of the final document was made at a local or centralized printing station. Printout was final copy. Editing, proofreading, and corrections were made on the VDU with the keyboard. If the document was a working paper, the document was kept in memory, either magnetic tape, bubble, or disk.

The CWP (communicating word processor) entered the field, as did the "electronic mailbox." All line and staff members of the organization were assigned an electronic mailbox. This was provided in memory compartments of a large central storage. This storage was accessed by individual keyboard/VDU units, which were the I/Os (see Figure 11.4). An individual wishing to access his mailbox would enter his or her identication number, password, and delimiter. The delimiter delimits both in time and content so that the recipient is not inundated with a long flow of documents on his VDU that are not pertinent to his/her immediate needs. One possibility is that once the recipient of a document accesses a specific communication from his mailbox, the matter of informing the originator of receipt becomes elementary.

The archival or filing problem, not only personal but corporate, is han-

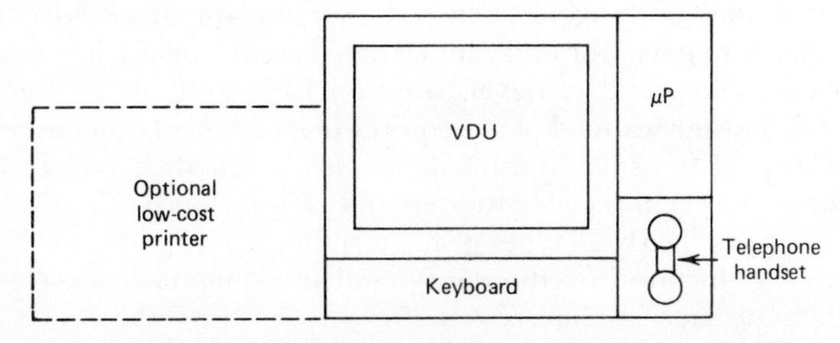

Figure 11.4 Office subscriber terminal, replacing or augmenting EPABX extensions.

dled in a similar manner. A great deal of space is saved, and access is made more rapid and efficient. Labor is also another big savings. Document searches may not only be made, say, horizontally by chronology and identification number, but vertically, as well by the use of key words. A user may request a search for, say, all correspondence and other documents on file dealing with "optical fiber attenuation" from September onward. Not only do word processors communicate (CWP) in the local office domain, they also can communicate at a distance. For example, a memo addressed to 100 recipients on both sides of the ocean would be an ideal application of CWP if each of the offices involved were compatibly CWP equipped. Computer message service (CMS) provides an allied service. In fact, many computer nets in the past, such as the ARPANet (See Chapter 10), show a large portion of traffic to be message service rather than true terminal–computer or computer–computer data transfer. If we allow a CWP network to be mini- or microprocessor controlled with error detection–correction and message or packet switching, it would now converge into the realm of CMS. If a CMS I/O were given the capability of text editing and storage plus microprocessor-controlled high-quality printing, this technology would converge on CWP.

Facsimile was, is, and continues to be yet another contender in the office environment (see Chapter 8, Section 11). It has certain advantages and disadvantages over CWP and CMS. First off, it is simpler, at least operationally. The encoding of information on a page by a raster scan permits the sending of whatever is on that page without operator intervention, which is required with telex and TWX (and to a degree with data messages). Furthermore, it can be used to send pictures, maps, and other drawings. Another attribute of facsimile is that the sender retains control over the format of the output, including type font, layout, and, to a degree, print quality. Facsimile is also comparatively insensitive to transmission errors because of its inherent redundancy in the raster scanning of the graphic material to be transmitted. It also has been made compatible for transmission over the telephone network so that every telephone extension is a potential candidate location for a fax terminal (I/O device). Facsimile does have a number of drawbacks for integrated office use, however. For instance, it does nothing to reduce the amount of office "paper." It is difficult, as is, to parse for entry into a data base or for use in constructing an index for filing and retrieval based on EDP, and, as we discussed in Chapter 8, a raster scanned page requires from 20 to 40 times as many bits as a similar page transmitted using data–telex-encoded characters. None of these constraints is impossible to overcome. For instance the ITT ComPak network accepts nearly all types of fax terminal inputs, makes in the order of a 4-to-1 reduction in redundancy, and transmits fax messages as any other data message in packets.

Of the three basic system approaches toward office communications (i.e., CMS, CWP, and fax), CMS (computer message system) comes out ahead.

Probably its greatest drawback is to have it accepted by office users or potential users. Such a system, with a costly data-base management software for retrieving messages, can be of greatest utility to the sophisticated user. But for the naive user, such sophistication may not be accepted nor used, and if used, used improperly. This could well dilute the impact of CMS. Computer message service (CMS) must also take on the attribute of sending a message to a person rather than to a terminal. Once this is accomplished, the rest of the technology that is cost effective is with us now, namely, a universal low-cost data network such as the ARPANet (see Chapter 10) and TYMNET. The idea is to be able to log into a computer from any terminal to get one's mail (communications). Of course, we refer to here ready access to one's virtual mailbox inside the computer memory. The ATT ACS (advanced communication system) seems to be leaning in this direction.

Figure 11.5 is a functional block diagram of the electronic office based on the subscriber terminal shown in Figure 11.4. The heart of the system is the office central computer. Instead of filing cabinets containing ream upon ream of paper storing all sorts of information, this office uses computer-associated multimegabit memories for storage. The system provides a terminal at all functioning office locations where once there was just a PABX extension. Of course, with such a system, most communications, particu-

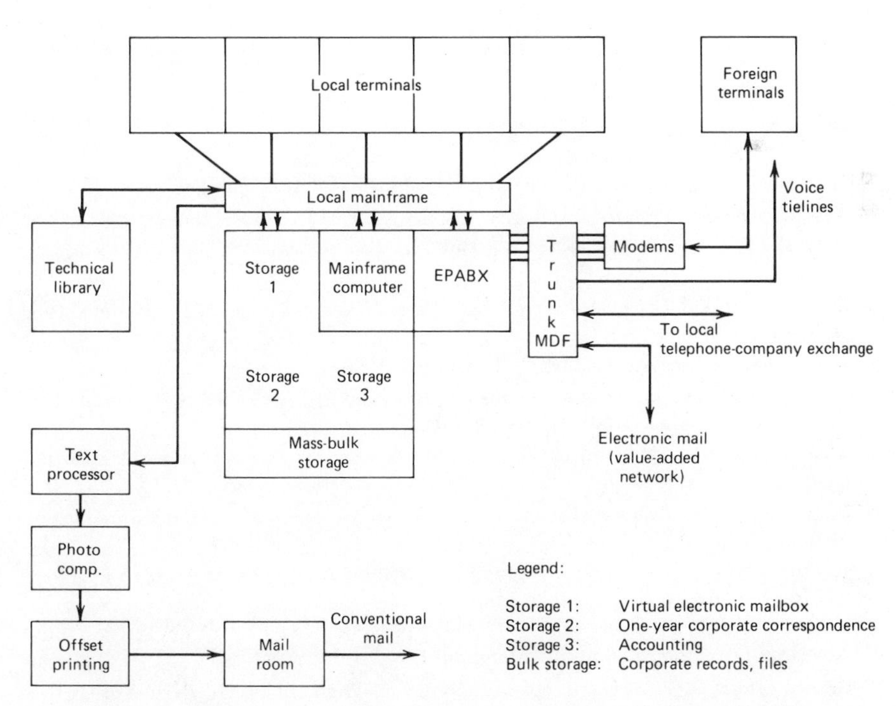

Figure 11.5 A functional block diagram of an integrated electronic office.

larly intraoffice, are never put on paper. A memorandum from, say, the sales manager to a vice-president would be composed by keyboard and displayed on the VDU. Once the capacity of the display has been reached, the material is entered into local memory. The completed memo is redisplayed for editing, correction, and review. It is then locally transmitted to the manager for his approval and signature. Once this occurs, the electronic memorandum is forwarded to the virtual mailboxes of the vice president and other recipients. A status light appears at each of their terminals, indicating that there is "mail" in their "box." To receive such mail, one only needs to sign into the computer with his password (for security), mailbox address, identifier, and delimiter. The memo now appears on the vice president's screen. He then has the option of storing it in his own local memory or placing it in central corporate files simply by entering a short selective sequence on the keyboard and executing the operation.

As one might imagine, security becomes of prime importance. Unless careful, well-thought-out measures are taken, such a system is open to all kinds of industrial espionage, spoofing, and vandalism. This is causing a whole new group of systems to evolve for industrial security of computer operations. Federal standards are coming on line for cryptographic systems. Others are being developed for passwords and other "coded signatures." Systems for authentication are also in operation, and more sophisticated ones are on the planning boards.

REFERENCES

1. *Telecommunication Planning,* ITT Laboratories (Spain), Madrid, January 1974.
2. *National Telephone Networks for the Automatic Service,* CCITT, ITU, Geneva, 1964.
3. R. Chapius, "Common Carrier Telecommunications in the World Economy," *Telecommun. J.* (October 1972).
4. *Economic Studies at the National Level in the Field of Telecommunications,* ITU, Geneva, 1968 (Amended 1972).
5. *Local Telephone Networks,* CCITT, ITU, Geneva, 1968.
6. U.S. Federal Communications Commission, *Rules and Regulations,* Parts 31 and 32, U.S. Government Printing Office, Washington, D. C.
7. *General Information* (bulletin) (ITT), TD-100 Electronic Private Automatic Branch Exchange, Section OA-000-001, Issue 3, ITT, Des Plaines, Ill., March 1977.
8. *PCM—IST Regional Studies, Definition of Transition Scenario,* ITT Laboratories (Spain), Madrid, 1974.
9. International Telephone and Telegraph Corporation, *Reference Data for Radio Engineers,* 6th ed., Howard W. Sams, Indianapolis, 1976.
10. Raul Schkolnick, *The Economics of Telephone Networks—The Case of Costa Rica,* Inter-American Development Bank, Washington, D. C., 1976.
11. *Generic Central Office Switching System Specification—Draft*, ITTWHQ New York, 1971. (unpublished).

12. T. A. Morgan, *Telecommunications Economics,* MacDonald, London, 1958.

13. O. Smidt, *Engineering Economics,* Telephony Publishing Corporation, Chicago, 1970.

14. Outside Plant, Telecommunication Planning Series, ITT Laboratories (Spain), Madrid, 1972.

15. Robert J. Potter, "Electronic Mail", *Science,* **195** (1978).

16. F. W. Miller, "Electronic Mail Comes of Age," *InfoSystems* **11** (1977).

17. Marvin A. Sirbu, Jr., *Innovation Strategies in the Electronic Mail Marketplace,* MIT Industrial Liaison Program, Cambridge, Mass., 1978.

18. "Information Processing and the Office of Tomorrow," *Fortune* (spec. suppl.) (October 1977).

19. V. E. Guilian, "The Office of Tomorrow Can be Here Today," *Time* (spec. suppl.) (November 14, 1977).

20. *Application to the FCC,* ITT Domestic Transmission Systems, December 1975, New York.

21. Daniel M. Costigan, *Electronic Delivery of Documents and Graphics,* Van Nostrand Reinhold, New York, 1978.

22. Robert F. Gellerman, *Subscriber Financing of Telecommunication Investments: IntelCom 1979,* Horizon House, Dedham, Mass, 1979.

INDEX